EARTHQUAKE RESEARCH AND ANALYSIS – NEW FRONTIERS IN SEISMOLOGY

Edited by **Sebastiano D'Amico**

INTECHOPEN.COM

Earthquake Research and Analysis – New Frontiers in Seismology
Edited by Sebastiano D'Amico

Published by InTech

Janeza Trdine 9, 51000 Rijeka, Croatia

Edition 2014

Publishing Process Manager Igor Babic

Technical Editor Teodora Smiljanic

Cover Designer InTech Design Team

Additional hard copies can be obtained from orders@intechopen.com

Earthquake Research and Analysis – New Frontiers in Seismology,
Edited by Sebastiano D'Amico
p. cm.
ISBN 978-953-307-840-3

INTECH

open science | open minds

Contents

Preface

The assessment of seismic hazard is probably the most important contribution of seismology to society. The prediction of the earthquake ground motion has always been of primary interest to seismologists and structural engineers. Large earthquakes that have occurred in densely populated areas of the world in recent years (eg Izmit, Turkey, August 17, 1999; Duzce, Turkey, November 12, 1999; Chi-Chi, Taiwan, September 20, 1999; Bhuj, India, January 26, 2001; Sumatra, Indonesia, December 26, 2004; Wenchuan, China, May 12, 2008; L'Aquila, Italy, April 6, 2009; Haiti, January 2010; Turkey 2011) highlight the dramatic inadequacy of a massive portion of the buildings erected in and around the epicentral areas. It has been observed that many houses were unable to withstand the ground shaking. Building earthquake-resistant structures and retrofitting old buildings on a national scale can be extremely expensive and can represent an economic challenge even for developed Western countries. Planning and design should be based on available national hazard maps which, in turn, must be produced after a careful calibration of ground motion predictive relationships for the region, estimation of seismic site effects as well as studies for the characterization of seismicity, seismogenic sources etc. Updating existing hazard maps represents one of the highest priorities for seismologists who contribute by refining the ground motion scaling relations and reducing the related uncertainties. The chapters in this book are devoted to various aspects of earthquake research and analysis, from theoretical advances to practical applications.

Chapters one to four are dedicated to ground motion studies, spanning the seismic source characterization to estimation of ground motion parameters.

Chapters in the site characterization section tackle the significance of the local seismic response. This topic is of increasing importance in earthquake seismology and in the seismic microzonation since regional geology can have a large effect on the characteristics of ground motion. The site response of the ground motion may vary in different locations of the city according to the local geology. Some chapters are dedicated to seismic hazard and early warning systems.

The final chapter presents a study in investigating land mass movements. Such studies could be integrated with the seismic hazard estimation and microzonation since ground deformations, along with severe shaking, could also be responsible for the devastation.

I would like to express my special thanks to Mr. Igor Babic and Ms. Ivana Lorkovic. Last but not least, I would like to thank the whole staff of InTech Open Access Publishing, especially Mr. Igor Babic, for their professional assistance and technical support during the entire publishing process that has led to the realization of this book.

Sebastiano D'Amico
Research Officer III
Physics Department
University of Malta
Malta

Part 1

Ground Motion Studies

1

Strong Ground Motion Estimation

Daniel R.H. O'Connell[1], Jon. P. Ake[2], Fabian Bonilla[3], Pengcheng Liu[4], Roland LaForge[1] and Dean Ostenaa[1]

[1]Fugro Consultants, Inc.
[2]Nuclear Regulatory Commission
[3]IFSTTAR
[4]Bureau of Reclamation
[1,2,4]USA
[3]France

1. Introduction

At the time of its founding, only a few months after the great 1906 **M** 7.7 San Francisco Earthquake, the Seismological Society of America noted in their timeless statement of purpose "that earthquakes are dangerous chiefly because we do not take adequate precautions against their effects, whereas it is possible to insure ourselves against damage by proper studies of their geographic distribution, historical sequence, activities, and effects on buildings." Seismic source characterization, strong ground motion recordings of past earthquakes, and physical understanding of the radiation and propagation of seismic waves from earthquakes provide the basis to estimate strong ground motions to support engineering analyses and design to reduce risks to life, property, and economic health associated with earthquakes.

When a building is subjected to ground shaking from an earthquake, elastic waves travel through the structure and the building begins to vibrate at various frequencies characteristic of the stiffness and shape of the building. Earthquakes generate ground motions over a wide range of frequencies, from static displacements to tens of cycles per second [Hertz (Hz)]. Most structures have resonant vibration frequencies in the 0.1 Hz to 10 Hz range. A structure is most sensitive to ground motions with frequencies near its natural resonant frequency. Damage to a building thus depends on its properties and the character of the earthquake ground motions, such as peak acceleration and velocity, duration, frequency content, kinetic energy, phasing, and spatial coherence. Strong ground motion estimation must provide estimates of all these ground motion parameters as well as realistic ground motion time histories needed for nonlinear dynamic analysis of structures to engineer earthquake-resistant buildings and critical structures, such as dams, bridges, and lifelines.

Strong ground motion estimation is a relatively new science. Virtually every **M** > 6 earthquake in the past 35 years that provided new strong ground motion recordings produced a paradigm shift in strong motion seismology. The 1979 **M** 6.9 Imperial Valley, California, earthquake showed that rupture velocities could exceed shear-wave velocities over a significant portion of a fault, and produced a peak vertical acceleration > 1.5 g (Spudich and Cranswick, 1984; Archuleta; 1984). The 1983 **M** 6.5 Coalinga, California, earthquake revealed a new class of seismic sources, blind thrust faults (Stein and Ekström,

1992). The 1985 **M** 6.9 Nahanni earthquake produced horizontal accelerations of 1.2 g and a peak vertical acceleration > 2 g (Weichert et al., 1986). The 1989 **M** 7.0 Loma Prieta, California, earthquake occurred on an unidentified steeply-dipping fault adjacent to the San Andreas fault, with reverse-slip on half of the fault (Hanks and Krawinkler, 1991), and produced significant damage > 100 km away related to critical reflections of shear-waves off the Moho (Somerville and Yoshimura, 1990; Catchings and Kohler, 1996). The 1992 **M** 7.0 Petrolia, California, earthquake produced peak horizontal accelerations > 1.4 g (Oglesby and Archuleta, 1997). The 1992 **M** 7.4 Landers, California, earthquake demonstrated that multi-segment fault rupture could occur on fault segments with substantially different orientations that are separated by several km (Li et al., 1994). The 1994 **M** 6.7 Northridge, California, earthquake produced a then world-record peak horizontal velocity (> 1.8 m/s) associated with rupture directivity (O'Connell, 1999a), widespread nonlinear soil responses (Field et al., 1997; Cultera et al., 1999), and resulted in substantial revision of existing ground motion-attenuation relationships (Abrahamson and Shedlock, 1997). The 1995 **M** 6.9 Hyogo-ken Nanbu (Kobe) earthquake revealed that basin-edge generated waves can strongly amplify strong ground motions (Kawase, 1996; Pitarka et al., 1998) and provided ground motion recordings demonstrating time-dependent nonlinear soil responses that amplified and extended the durations of strong ground motions (Archuleta et al., 2000). The 1999 **M** > 7.5 Izmit, Turkey, earthquakes produced asymmetric rupture velocities, including rupture velocities ~40% faster than shear-wave velocities, which may be associated with a strong velocity contrast across the faults (Bouchon et al., 2001). The 1999 **M** 7.6 Chi-Chi, Taiwan, earthquake produced a world-record peak velocity > 3 m/s with unusually low peak accelerations (Shin et al., 2000). The 2001 **M** 7.7 Bhuj India demonstrated that **M** > 7.5 blind thrust earthquakes can occur in intraplate regions. The **M** 6.9 2008 Iwate-Miyagi, Japan, earthquake produced a current world-record peak vector acceleration > 4 g, with a vertical acceleration > 3.8 g (Aoi et al., 2008). The 2011 M 9.1 Tohoku, Japan, earthquake had a world-record peak slip on the order of 60 m (Shao et al., 2011) and produced a world-record peak horizontal acceleration of 2.7 g at > 60 km from the fault (NIED, 2011).

This progressive sequence of ground motion surprises suggests that the current state of knowledge in strong motion seismology is probably not adequate to make unequivocal strong ground motion predictions. However, with these caveats in mind, strong ground motion estimation provides substantial value by reducing risks associated with earthquakes and engineered structures. We present the current state of earthquake ground motion estimation. We start with seismic source characterization, because this is the most important and challenging part of the problem. To better understand the challenges of developing ground motion prediction equations (GMPE) using strong motion data, we present the physical factors that influence strong ground shaking. New calculations are presented to illustrate potential pitfalls and identify key issues relevant to ground motion estimation and future ground motion research and applications. Particular attention is devoted to probabilistic implications of all aspects of ground motion estimation.

2. Seismic source characterization

The strongest ground shaking generally occurs close to an earthquake fault rupture because geometric spreading reduces ground shaking amplitudes as distance from the fault increases. Robust ground motion estimation at a specific site or over a broad region is predicated on the availability of detailed geological and geophysical information about locations, geometries, and rupture characteristics of earthquake faults. These characteristics are not random, but are dictated by the physical properties of the upper crust including rock types, pre-existing faults and fractures, and strain rates and orientations. Because such information is often not readily available or complete, the resultant uncertainties of source characterization can be the dominant contributions to uncertainty in ground motion estimation. Lettis et al. (1997) showed that intraplate blind thrust earthquakes with moment magnitudes up to 7 have occurred in intraplate regions where often there was no previously known direct surface evidence to suggest the existence of the buried faults. This observation has been repeatedly confirmed, even in plate boundary settings, by numerous large earthquakes of the past 30 years including several which have provided rich sets of ground motion data from faults for which neither the locations, geometries, or other seismic source characterization properties were known prior to the earthquake. Regional seismicity and geodetic measurements may provide some indication of the likely rate of earthquake occurrence in a region, but generally do not demonstrate where that deformation localizes fault displacement. Thus, an integral and necessary step in reducing ground motion estimation uncertainties in most regions remains the identification and characterization of earthquake source faults at a sufficiently detailed scale to fully exploit the full range of ground motion modelling capabilities. In the absence of detailed source characterizations, ground motion uncertainties remain large, with the likely consequence of overestimation of hazard at most locations, and potentially severe underestimation of hazard in those few locations where a future earthquake ultimately reveals the source characteristics of a nearby, currently unknown fault. The latter case is amply demonstrated by the effects of the 1983 **M** 6.5 Coalinga, 1986 M 6.0 Whittier Narrows, 1989 **M** 6.6 Sierra Madre, 1989 **M** 7.0 Loma Prieta, 1992 **M** 7.4 Landers, 1994 **M** 6.7 Northridge, 1999 **M** 7.6 Chi-Chi Taiwan, 2001 **M** 7.7 Bhuj, India, 2010 **M** 7.0 Canterbury, New Zealand, and 2011 **M** 6.1 Christchurch, New Zealand, earthquakes. The devastating 2011 M 9.1 Tohoku, Japan, earthquake and tsunami were the result of unusually large fault displacement over a relatively small fault area (Shao et al., 2011), a source characteristic that was not forseen, but profoundly influenced strong ground shaking (NIED, 2011) and tsunami responses (SIAM News, 2011). All these earthquakes occurred in regions where the source faults were either unknown or major source characteristics were not recognized prior to the occurrence of these earthquakes.

3. Physical basis for ground motion prediction

In this section we present the physical factors that influence ground shaking in response to earthquakes. A discrete representation is used to emphasize the discrete building blocks or factors that interact to produce strong ground motions. For simplicity, we start with linear stress-strain. Nonlinear stress-strain is most commonly observed in soils and evaluated in terms of site response. This is the approach we use here; nonlinear site response is discussed in Section 4. The ground motions produced at any site by an earthquake are the result of seismic radiation associated with the dynamic faulting process and the manner in which seismic energy propagates from positions on the fault to a site of interest. We assume that fault rupture initiates at some point on the fault (the hypocenter) and proceeds outward along the fault surface. Using the representation theorem (Spudich and Archuleta, 1987), ground velocity, $\dot{u}_k(t)$, depends on the convolution of the time evolution of the slip-time functions, $s_{ij}(t)$, and the Green's functions, $g_{kij}(t)$, the impulse responses between the fault and the site (Figure 3.1) as,

$$\dot{u}_k(t) = \sum_{ij}^{nm} \dot{s}_{ij}(t) \otimes g_{kij}(t) \tag{1}$$

where k is the component of ground motion, ij are the indices of the discrete fault elements, n is the number of fault elements in the strike direction and m is the number of elements in dip direction (Figure 3.1). We use the notation $F(\omega)$ to indicate the modulus of the Fourier

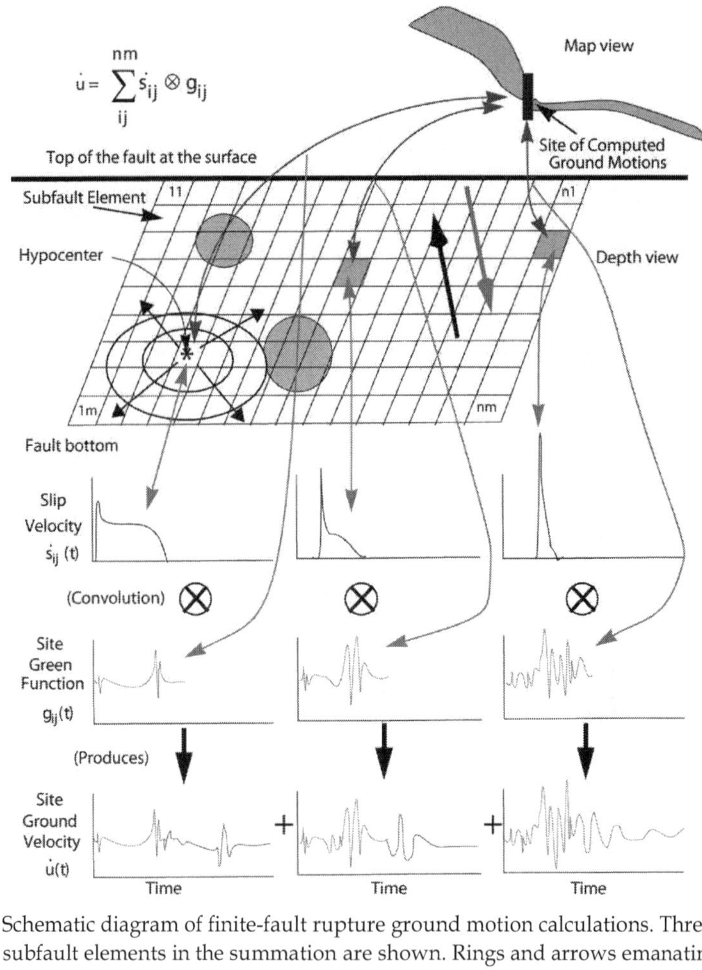

Fig. 3.1. Schematic diagram of finite-fault rupture ground motion calculations. Three discrete subfault elements in the summation are shown. Rings and arrows emanating from the hypocenter represent the time evolution of the rupture. The Green functions actually consist of eight components of ground motion and three components of site ground velocities. Large arrows denote fault slip orientation, which is shown as predominantly reverse slip with a small component of right-lateral strike slip. Hatched circles schematically represent regions of high stress drop.

transform of $f(t)$. It is instructive to take the Fourier transform of (1) and pursue a discussion similar to Hutchings and Wu (1990) and Hutchings (1994) using,

$$\dot{U}_k(\omega) = \sum_{ij}^{nm} \dot{S}_{ij}(\omega)\, e^{-i\omega\tau_{ij}(\omega)} \cdot G_{kij}(\omega)\, e^{-i\omega\phi_{kij}(\omega)} \tag{2}$$

where at each element ij, $\dot{S}_{ij}(\omega)$ is the source slip-velocity amplitude spectrum, $\tau_{ij}(\omega)$ is the source phase spectrum, $G_{kij}(\omega)$ is the Green's function amplitude spectrum, and $\phi_{kij}(\omega)$ is the Green's function phase spectrum. The maximum peak ground motions are produced by a combination of factors that produce constant or linear phase variations with frequency over a large frequency band. While the relations in (1) and (2) are useful for synthesizing ground motions, they don't provide particularly intuitive physical insights into the factors that contribute to produce specific ground motion characteristics, particularly large peak accelerations, velocities, and displacements. We introduce isochrones as a fundamental forensic tool for understanding the genesis of ground motions. Isochrones are then used to provide simple geometric illustrations of how directivity varies between dipping dip-slip and vertical strike-slip faults.

3.1 Isochrones analysis of rupture directivity

Bernard and Madariaga (1984) and Spudich and Frazer (1984; 1987) developed the isochrone integration method to compute near-source ground motions for finite-fault rupture models. Isochrones are all the positions on a fault that contribute seismic energy that arrives at a specific receiver at the same time. By plotting isochrones projected on a fault, times of large amplitudes in a ground motion time history can be associated with specific regions and characteristics of fault rupture and healing.

A simple and reasonable way to employ the isochrone method for sites located near faults is to assume that all significant seismic radiation from the fault consists of first shear-wave arrivals. A further simplification is to use a simple trapezoidal slip-velocity pulse. Let f(t) be the slip function, For simplicity we assume where t_r is rupture time, and t_h is healing time. Then, all seismic radiation from a fault can be described with rupture and healing isochrones. Ground velocities (v) and accelerations (a) produced by rupture or healing of each point on a fault can be calculated from (Spudich and Frazer, 1984; Zeng et al., 1991; Smedes and Archuleta, 2008)

$$v(x,t) = \ddot{f}(t) \otimes \int_{y(t,x)} (s \cdot G)\, c\, dl \tag{3}$$

$$a(x,t) = \dddot{f}(t) \otimes \int_{y(t,x)} \left[c^2 \left(\frac{ds}{dq} \cdot G \right) + c^2 \left(\frac{dG}{dq} \cdot s \right) + \frac{dc}{dq} \cdot (s \cdot G) + \kappa s G c^2 \right] dl \tag{4}$$

where $\boldsymbol{c}$ is isochrone velocity, $\boldsymbol{s}$ is slip velocity (either rupture or healing), $\boldsymbol{G}$ is a ray theory Green function, $\boldsymbol{x}$ are position vectors, $y(t,x)$ are isochrones, κ is the curvature of the isochrone, dl denotes isochrone line integral integration increment, and dq denotes a spatial derivative.

Since isochrones are central to understanding ground motions, we provide explicit expressions for rupture and healing isochrones to illustrate how source and propagation factors can combine to affect ground motions. The arrival times of rupture at a specific receiver are

$$T_r(x) = t_\beta(x,\xi) + t_r(\xi) \tag{5}$$

where x is the receiver position, ξ are all fault positions, t_β are shear-wave propagation times between the receiver and all fault positions, and t_r are rupture times at all fault positions. The arrival times of healing at a specific receiver are

$$T_h(x) = T_r(x,\xi) + R(\xi) \tag{6}$$

where R are the rise times (the durations of slip) at all fault positions.

Archuleta (1984) showed that variations in rupture velocity had pronounced effects on calculated ground motions, whereas variations in rise times and slip-rate amplitudes cause small or predictable changes on calculated ground motions. The effect of changing slip-velocity amplitudes on ground motions is strongly governed by the geometrical attenuation (1/r for far-field terms). Any change in the slip-velocity amplitudes affects most the ground motions for sites closest to the region on the fault where large slip-velocities occurred (Spudich and Archuleta, 1987). This is not the case with rupture velocity or rise time; these quantities influence ground motions at all sites. However, as Anderson and Richards (1975) showed, it takes a 300% change in rise time to compensate for a 17% change in rupture time. Spudich and Oppenheimer (1986) show why this is so. Spatial variability of rupture velocity causes the integrand in (3) to become quite rough, thereby adding considerable high-frequency energy to ground motions. The roughness of the integrand in (3) is caused by variations of isochrone velocity c, where

$$c = \left|\underline{\nabla}_S \boldsymbol{T_r}\right|^{-1} \tag{7}$$

where $\boldsymbol{T_r}$ are the isochrones from (5) and $\underline{\nabla}_S$ is the surface gradient operator. Variations of $\boldsymbol{T_r}$ on the fault surface associated with supershear rupture velocities, or regions on the fault where rupture jumps discontinuously can cause large or singular values of c, called critical points by Farra et al. (1986). Spudich and Frazer (1984) showed that the reciprocal of c, isochrone slowness is equivalent to the seismic directivity function in the two-dimensional case. Thus, by definition, critical points produce rupture directivity, and as is shown with simulations later, need not be associated strictly with forward rupture directivity, but can occur for any site located normal to a portion of a fault plane where rupture velocities are supershear.

It is useful to interpret (3) and (4) in the context of the discrete point-source summations in (1) and (2). When isochrone velocities become large on a substantial area of a fault it simply means that all the seismic energy from that portion of the fault arrives at nearly the same time at the receiver; the summation of a finite, but large number of band-limited Green's functions means that peak velocities remain finite, but potentially large. Large isochrone velocities or small isochrone slownesses over significant region of a fault are diagnostic of ground motion amplification associated with rupture directivity; the focusing of a significant fraction of the seismic energy radiated from a fault at a particular site in a short time interval. In this way isochrones are a powerful tool to dissect ground motions in relation to various characteristics of fault rupture. Times of large ground motion amplitudes can be directly associated with the regions of the fault that have corresponding large isochrone velocities or unusually large slip velocities. From (5) and (6) it is clear that both fault rupture variations, and shear-wave propagation time variations, combine to determine isochrones and isochrone velocities.

3.1.1 The fundamental difference between strike-slip and dip-slip directivity

Boore and Joyner (1989) and Joyner (1991) discussed directivity using a simple line source model. A similar approach is used here to illustrate how directivity differs between vertical strike-slip faults and dipping dip-slip faults. To focus on source effects, we consider unilateral, one-dimensional ruptures in a homogenous half-space (Figure 3.2). The influence of the free surface on amplitudes is ignored. The rupture velocity is set equal to the shear-wave velocity to minimize time delays and to maximize rupture directivity. To eliminate geometric spreading, stress drops increase linearly with distance from the site in a manner that produces uniform ground motion velocity contribution to the surface site for all points on the faults. Healing is ignored; only the rupture pulse is considered. Thrust dip-slip faulting is used to produce coincident rake and rupture directions. Seismic radiation is simplified to triangular slip-velocity pulses with widths of one second.

For the strike-slip fault, the fault orientation and rupture directional are coincident. But, as fault rupture approaches the site, takeoff angles increase, so the radiation pattern reduces amplitudes, and total propagation distances (rupture length plus propagation distance) increase to disperse shear-wave arrivals in time (Figures 3.2a and 3.2b). The surface site located along the projection of the thrust fault to the surface receives all seismic energy from the fault at the same time, and c is infinity because the fault orientation, rupture, and shear-wave propagation directions are all coincident for the entire length of the fault (Figures 3.2c and 2d). Consequently, although the strike-slip fault is 50% longer than the thrust fault, the thrust fault produces a peak amplitude 58% larger than the strike-slip fault. The thrust fault site receives maximum amplitudes over the entire radiated frequency band. High-frequency amplitudes are reduced for the strike-slip site relative to the thrust fault site because shear-waves along the strike-slip fault become increasingly delayed as rupture approaches the site, producing a broadened ground motion velocity pulse. The geometric interaction between dip-slip faults and propagation paths to surface sites located above those faults produces a kinematic recipe for maximizing both isochrone velocities and radiation patterns for surface sites that is unique to dip-slip faults. In contrast, Schmedes and Archuleta (2008) use kinematic rupture simulations and isochrone analyses to show why directivity becomes bounded during strike-slip fault along long faults. Schmedes and Archuleta (2008) consider the case of subshear rupture velocities and use critical point analyses with (3) and (4) to show that for long strike-slip ruptures there is a saturation effect for peak velocities and accelerations at sites close to the fault located at increasing distances along strike relative to the epicenter, consistent with empirical observations (Cua, 2004; Abrahamson and Silva, 2008; Boore and Atkinson, 2008; Campbell and Bozorgnia, 2008; Chiou and Youngs, 2008).

Dynamic fault rupture processes during dip-slip rupture complicate dip-slip directivity by switching the region of maximum fault-normal horizontal motion from the hangingwall to the footwall as fault dips increase from 50° to 60° (O'Connell et al., 2007).

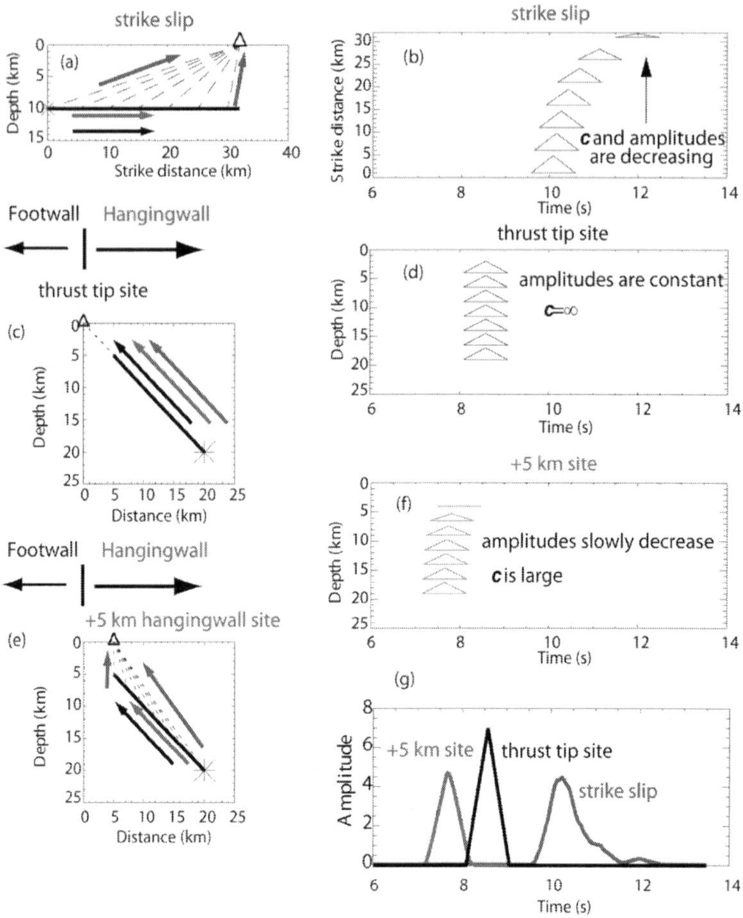

Fig. 3.2. Schematics of line source orientations for strike-slip (a) and thrust faults (c) and (e) relative to ground motion sites (triangles). Black arrows show the orientation of the faults, red arrows show fault rupture directions, and blue arrows show shear-wave propagation directions (dashed lines) to the sites. Discrete velocity contributions for seven evenly-spaced positions along the fault are shown to the right of each rupture model (b, d, f) as triangles with amplitudes (heights) scaled by the radiation pattern. The output ground motions for each fault rupture are shown in (g). Isochrone velocity, *c*, is infinity in (d), is large, but finite, in (f), and decreases as the fault nears the ground motion site in (b).

Typically, seismic velocities increase with depth, which changes positions of maximum rupture directivity compared to Figure 3.2. For dip-slip faults, the region of maximum directivity is moved away from the projection of the fault to the surface, toward the hanging wall. This bias is dependent on velocity gradients, and the dip and depth of the fault. For strike-slip faults, a refracting velocity geometry can increase directivity by reducing takeoff angle deviations relative to the rupture direction for depth intervals that depend on the velocity structure and position of the surface site (Smedes and Archuleta, 2008).

When the two-dimensional nature of finite-fault rupture is considered, rupture directivity is not as strong as suggested by this one-dimensional analysis (Bernard et al., 1996), but the distinct amplitude and frequency differences between ground motions produced by strike-slip and dip-slip faulting directivity remain. Full two-dimensional analyses are presented in a subsequent section. A more complete discussion of source and propagation factors influencing ground motions is presented next to provide a foundation for discussion of amplification associated with rupture directivity. The approach here is to discuss ground motions separately in terms of source and propagation factors and then to discuss how source and propagation factors can jointly interact to strongly influence ground motion behavior.

3.2 Seismic source amplitude and phase factors

Table 1 list factors influencing source amplitudes, $\dot{S}_{ij}(\omega)$. Table 2 lists factors influencing source phase, $\tau_{ij}(\omega)$. The flat portion of an amplitude spectrum is composed of the frequencies less than a corner frequency, ω_c, which is defined as the intersection of low- and high-frequency asymptotes following Brune (1970). The stress drop, $\Delta\sigma$, defined as the difference between an initial stress, σ_0, minus the dynamic frictional stress, σ_f, is the stress available to drive fault slip (Aki, 1983). Rise time, R, is the duration of slip at any particular point on the fault. Rise times are heterogeneous over a fault rupture surface. Because the radiation pattern for seismic phases such as body waves and surface waves are imposed by specification of rake (slip direction) at the source and are a function of focal mechanism, radiation pattern is included in the source discussion.

Regressions between moment and fault area (Wells and Coppersmith, 1994; Somerville et al., 1999; Leonard, 2010) show that uncertainties in moment magnitude and fault area are sufficient to produce moment uncertainties of 50% or more for any particular fault area. Consequently, the absolute scaling of synthesized ground motions for any faulting scenario have about factor of two uncertainties related to seismic moment (equivalently, average stress drop) uncertainties. Thus, moment-fault area uncertainties introduce a significant source of uncertainty in ground motion estimation.

Andrews (1981) and Frankel (1991) showed that correlated-random variations of stress drop over fault surfaces that produce self-similar spatial distributions of fault slip are required to explain observed ground motion frequency amplitude responses. Somerville et al. (1999) showed that a self-similar slip model can explain inferred slip distributions for many large earthquakes and they derive relations between many fault rupture parameters and seismic moment. Their results provide support for specifying fault rupture models using a stochastic spatially varying stress drop where stress drop amplitude decays as the inverse of wavenumber to produce self-similar slip distributions. They assume that mean stress drop is independent of seismic moment. Based on their analysis and assumptions, Somerville et al. (1999) provide recipes for specifying fault rupture parameters such as slip, rise times, and asperity dimensions as a function of moment. Mai and Beroza (2000) showed that $5.3 < M < 8.1$ magnitude range dip-slip earthquakes follow self-similar scaling as suggest by Somerville et al. (1999). However, for strike-slip earthquakes, as moment increases in this magnitude range, they showed that seismic moments scale as the cube of fault length, but fault width saturates. Thus, for large strike slip earthquakes average slip increases with fault rupture length, stress drop increases with magnitude, and self-similar slip scaling does not hold. The large stress drops observed for the M 7.7 1999 Chi-Chi, Taiwan thrust-faulting earthquake (Oglesby et al., 2000) suggest that self-similar slip scaling relations may also breakdown at larger moments for dip-slip events.

Factor	Influence
Moment rate, $\dot{S}_{ij}{}^{M_0}(\omega)$	Moment rate scales peak velocities and accelerations. Moment determines the average slip for a fixed fault area and known shear moduli.
Stress drop, $\dot{S}_{ij}{}^{\Delta\sigma}(\omega)$	Since $\dot{S}_{ij}(\omega) \propto \Delta\sigma$, Δσ scales peak slip velocities. Spatial variations of stress drop introduce frequency dependent amplitude variations.
Rupture velocity, $\dot{S}_{ij}{}^{V_r}(\omega)$	High rupture velocities increase directivity. Rupture velocities interact with stress drops and rise times to modify the amplitude spectrum. Supershear rupture velocities can increase directivity far from the fault (Andrews, 2010).
Healing velocity, $\dot{S}_{ij}{}^{V_h}(\omega)$	High healing velocities increase amplification associated with directivity. Healing velocities interact with stress drop and rise time variations to modify the amplitude spectrum, although to a smaller degree than rupture velocities, since rupture slip veloc ities are typically several times larger than healing slip velocities.
Rake and radiation pattern, $\dot{S}_{ij}{}^{A}(\omega)$	Rake and spatial and temporal rake variations scale amplitudes as a function of azimuth and take-off angle. Rake spatial and temporal variations over a fault increase the spatial complexity of radiation pattern amplitude variations and produce frequency-dependent amplitude variability.
Rise time, $\dot{S}_{ij}{}^{R}(\omega)$	Since $\omega_c \propto \frac{1}{R}$, spatially variable rise times produce a frequency dependence of the amplitude spectrum.
Crack diffraction, $\dot{S}_{ij}{}^{C}(\omega)$	Diffraction at the crack tip introduces a frequency dependent amplitude to the radiation pattern (Madariaga, 1977; Boatwright, 1982; Fukuyama and Madariaga, 1995).
Dynamics, $\dot{S}_{ij}{}^{D}(\omega)$	Fault rupture in heterogeneous velocity structure can produce anisotropic slip velocities relative to rupture direction (Harris and Day, 1997) and slip velocities and directivity are a function of rake and dip for dip-slip faults (Oglesby et al., 1998; 2000; O'Connell et al., 2007). Frictional heating, fault zone fluids, and melting may also influence radiated energy (Kanamori and Brodsky, 2001; Andrews, 200X).
Hypocentral depth, $\dot{S}_{ij}{}^{H_z}(\omega)$	For crustal earthquakes, deeper hypocenters tend to increase peak ground motions by increasing isochrone velocities (O'Connell, 1999c. Smedes and Archuleta, 200X).

Table 1. Seismic Source Amplitude Factors ($\dot{S}_{ij}(\omega)$)

Factor	Influence
Rupture velocity, $\tau_{ij}^{V_r}(\omega)$	High rupture velocities increase directivity. Rupture velocities interact with stress drops and rise times to modify the amplitude spectrum. Supershear rupture velocities can increase directivity far from the fault (Andrews, 2010).
Healing velocity, $\tau_{ij}^{V_h}(\omega)$	High healing velocities increase amplification associated with directivity. Healing velocities interact with stress drop and rise time variations to modify the amplitude spectrum, although to a smaller degree than rupture velocities, since rupture slip veloc ities are typically several times larger than healing slip velocities.
Rake, $\tau_{ij}^{A}(\omega)$	Rake and spatial and temporal rake variations scale amplitudes as a function of azimuth and take-off angle. Rake spatial and temporal variations over a fault increase the spatial complexity of radiation pattern amplitude variations and produce frequency-dependent amplitude variability.
Rise time, $\tau_{ij}^{R}(\omega)$	Since $\omega_c \propto \frac{1}{R}$, spatially variable rise times produce a frequency dependence of the amplitude spectrum.
Crack diffraction, $\tau_{ij}^{C}(\omega)$	Diffraction at the crack tip introduces a frequency dependent amplitude to the radiation pattern (Madariaga, 1977; Boatwright, 1982; Fukuyama and Madariaga, 1995).
Dynamics, $\tau_{ij}^{D}(\omega)$	The same dynamic processes identified in Table 1 produce corresponding phase variability.

Table 2. Seismic Source Phase Factors ($\tau_{ij}(\omega)$)

Oglesby et al. (1998; 2000) showed that stress drop behaviors are fundamentally different between dipping reverse and normal faults. These results suggest that stress drop may be focal mechanism and magnitude dependent. There are still significant uncertainties as to the appropriate specifications of fault rupture parameters to simulate strong ground motions, particularly for larger magnitude earthquakes. O'Connell et al. (2007) used dynamic rupture simulations to show that homogeneous and weakly heterogeneous half-spaces with faults dipping ≲50°, maximum fault-normal peak velocities occurred on the hanging wall. However, for fault dips ≳50°, maximum fault-normal peak velocities occurred on the footwall. Their results indicate that simple amplitude parameterizations based on the hanging wall and/or footwall and the fault normal and/or fault parallel currently used in ground motion prediction relations may not be appropriate for some faults with dips > 50°. Thus, the details of appropriate spatial specification of stress drops and/or slip velocities as a function of focal mechanism, magnitude, and fault dip are yet to be fully resolved.

Day (1982) showed that intersonic rupture velocities ($\beta < V^r < \alpha$) can occur during earthquakes, particularly in regions of high prestress (asperities), and that peak slip velocity is strongly coupled to rupture velocity for non-uniform prestresses. While average rupture velocities typically remain subshear, high-stress asperities can produce local regions of supershear rupture combined with high slip velocities. Supershear rupture velocities have been observed or inferred to have occurred during several earthquakes, including the M 6.9

1979 Imperial Valley strike-slip earthquake (Olson and Apsel, 1982; Spudich and Cranswick, 1984; Archuleta, 1984), the M 6.9 1980 Irpinia normal-faulting earthquake (Belardinelli et al., 1999), the M 7.0 1992 Petrolia thrust-faulting earthquake (Oglesby and Archuleta, 1997), the M 7.3 Landers strike-slip earthquake (Olsen et al., 1997; Bouchon et al., 1998; Hernandez et al., 1999) the M 6.7 1994 Northridge thrust-faulting earthquake (O'Connell, 1999b), and the 1999 M 7.5 Izmit and M 7.3 Duzce Turkey strike-slip earthquakes (Bouchon et al., 2001). Bouchon et al. (2010) find that surface trace of the portions of strike-slip faults with inferred supershear rupture velocities are remarkably linear, continuous and narrow, that segmentation features along these segments are small or absent, and the deformation is highly localized. O'Connell (1999b) postulates that subshear rupture on the faster footwall in the deeper portion of the Northridge fault relative to the hangingwall produced supershear rupture in relation to hangingwall velocities and contributed to the large peak velocities observed on the hangingwall.

Harris and Day (1997) showed that rupture velocities and slip-velocity functions are significantly modified when a fault is bounded on one side by a low-velocity zone. The low-velocity zone can produce asymmetry of rupture velocity and slip velocity. This type of velocity heterogeneity produces an asymmetry in seismic radiation pattern and abrupt and/or systematic spatial variations in rupture velocity. These differences are most significant in regions subject to rupture directivity, and may lead to substantially different peak ground motions occurring at either end of a strike slip fault (Bouchon et al., 2001). Thus, the position of a site relative to the fast and slow sides of a fault and rupture direction may be significant in terms of the dynamic stress drops and rupture velocities that are attainable in the direction of the site. Observations and numerical modeling show that the details of stress distribution on the fault can produce complex rupture velocity distributions and even discontinuous rupture, factors not typically accounted for in kinematic rupture models used to predict ground motions (e.g. Somerville et al., 1991; Schneider et al., 1993; Hutchings, 1994; Tumarkin et al., 1994; Zeng et al., 1994; Beresnev and Atkinson, 1997; O'Connell, 1999c). Even if only smooth variations of subshear rupture velocities are considered ($0.6{*}\beta < Vr < 1.0{*}\beta$), rupture velocity variability introduces ground motion estimation uncertainties of at least a factor of two (Beresnev and Atkinson, 1997), and larger uncertainties for sites subject to directivity.

Rupture direction may change due to strength or stress heterogeneities on a fault. Beroza and Spudich (1988) inferred that rupture was delayed and then progressed back toward the hypocenter during the M 6.2 1984 Morgan Hill earthquake. Oglesby and Archuleta (1997) inferred that arcuate rupture of an asperity may have produced accelerations > 1.40 g at Cape Mendocino during the M 7.0 1992 Petrolia earthquake. These results are compatible with numerical simulations of fault rupture on a heterogeneous fault plane. Das and Aki (1977) modeled rupture for a fault plane with high-strength barriers and found that rupture could occur discontinuously beyond strong regions, which may subsequently rupture or remain unbroken. Day (1982) found that rupture was very complex for the case of non-uniform prestress and that rupture jumped beyond some points on the fault, leaving unbroken areas behind the rupture which subsequently ruptured. In the case of slip resistant asperity, Das and Kostrov (1983) found that when rupture began at the edge of the asperity, it proceeded first around the perimeter and then failed inward in a "double pincer movement". Thus, even the details of rupture propagation direction are not truly specified once a hypocenter position is selected.

Guatteri and Spudich (1998) showed that time-dependent dynamic rake rotations on a fault become more likely as stress states approach low stresses on a fault when combined with heterogeneous distributions of stress and nearly complete stress drops. Pitarka et al. (2000) found that eliminating radiation pattern coherence between 1 Hz and 3 Hz reproduced observed ground motions for the 1995 M 6.9 Hyogo-ken Nanbu (Kobe) earthquake. Spudich et al. (1998) used fault striations to infer that the Nojima fault slipped at low stress levels with substantial rake rotations occurring during the 1995 Hyogo-ken Nanbu earthquake. This dynamic rake rotation can reduce radiation-pattern coherence at increasing frequencies by increasingly randomizing rake directions for decreasing time intervals near the initiation of slip at each point on a fault, for increasingly complex initial stress distributions on faults. Vidale (1989) showed that the standard double-couple radiation pattern is observable to 6 Hz based on analysis of the mainshock and an aftershock from the Whittier Narrows, California, thrust-faulting earthquake sequence. In contrast, Liu and Helmberger (1985) found that a double-couple radiation pattern was only discernible for frequencies extending to 1 Hz based on analysis the 1979 Imperial Valley earthquake and an aftershock. Bent and Helmberger (1989) estimate a $\Delta\sigma$ of 75 MPa for the 1987 Whittier Narrows M 6.1 thrust faulting earthquake, but allow for a $\Delta\sigma$ as low as 15.5 MPa. The case of high initial, nearly homogeneous stresses that minimize rake rotations may produce high-frequency radiation pattern coherence as observed by Vidale (1989). These results suggest that there may be a correlation between the maximum frequency of radiation pattern coherence, initial stress state on a fault, focal mechanism, and stress drop.

3.3 Seismic wave propagation amplitude and phase factors

Table 3 lists factors influencing propagation amplitudes, $G_{kij}(\omega)$. Table 4 lists factors influencing propagation phase, ϕ_{ij}. Large-scale basin structure can substantially amplify and extend durations of strong ground motions (Frankel and Vidale, 1992; Frankel, 1993; Olsen and Archuleta, 1996; Wald and Graves; 1998; Frankel and Stephenson, 2000; Koketsu and Kikuchi, 2000; Frankel et al., 2001). Basin-edge waves can substantially amplify strong ground motions in basins (Liu and Heaton, 1984; Frankel et al., 1991; Phillips et al., 1993; Spudich and Iida, 1993; Kawase, 1996; Pitarka et al., 1998, Frankel et al., 2001). This is a particular concern for fault-bounded basins where rupture directivity can constructively interact with basin-edge waves to produce extended zones of extreme ground motions (Kawase, 1996; Pitarka et al., 1998), a topic revisited later in the paper. Even smaller scale basin or lens structures on the order of several kilometers in diameter can produce substantial amplification of strong ground motions (Alex and Olsen, 1998; Graves et al., 1998; Davis et al., 2000). Basin-edge waves can be composed of both body and surface waves (Spudich and Iida, 1993; Meremonte et al., 1996; Frankel et al., 2001) which provides a rich wavefield for constructive interference phenomena over a broad frequency range.

Critical reflections off the Moho can produce amplification at distances > ~75-100 km (Somerville and Yoshimura, 1990; Catchings and Kohler, 1996). The depth to the Moho, hypocentral depth, direction of rupture (updip versus downdip), and focal mechanism determine the amplification and distance range that Moho reflections may be important. For instance, Catchings and Mooney (1992) showed that Moho reflections amplify ground motions in the > 100 km distance range in the vicinity of the New Madrid seismic zone in the central United States.

Factor	Influence
Geometric spreading, $G_{kij}{}^{r}(\omega)$	Amplitudes decrease with distance at $1/r$, $1/r^2$, and $1/r^4$ for body waves and $1/\sqrt{r}$ for surface waves. The $1/r$ term has the strongest influence on high-frequency ground motions. The $1/\sqrt{r}$ term can be significant for locally generated surface waves.
Large-scale velocity structure, $G_{kij}{}^{V_{3D}}(\omega)$	Horizontal and vertical velocity gradients and velocity discontinuities can increase or decrease amplitudes and durations. Low-velocity basins can amplify and extend ground motion durations. Abrupt changes in lateral velocity structure can induce basin-edge-waves in the lower velocity material that amplify ground motions.
Near-surface resonant responses, $G_{kij}{}^{L}(\omega)$	Low-velocity materials near the surface amplify ground motions for frequencies > $\beta/(4{*}h)$, where h is the thickness of near-surface low velocity materials. Coupled interface modes can amplify and extend durations of ground motions.
Nonlinear soil responses, $G_{kij}{}^{N}(\ddot{u},\omega)$, (equivalent linear) $G_{kij}{}^{N}(\ddot{u},t)$ (fully nonlinear)	Depending on the dynamic soil properties and pore pressure responses, nonlinear soil responses can decrease intermediate- and high-frequency amplitudes, amplify low- and high-frequency amplitudes, and extend or reduce duration of large amplitudes. The equivlanet linear approximation is $G_{kij}{}^{N}(\ddot{u},\omega)$. The fully nonlinear form, $G_{kij}{}^{N}(\ddot{u},t)$, can incorporate any time-dependent behavior such as pore-pressure responses.
Frequency independent attenuation, $G_{kij}{}^{Q}(\omega)$	Linear hysteretic behavior that reduces amplitudes of the form $e^{\frac{-\pi\cdot f\cdot r}{\beta\cdot Q}}$.
High-frequency atten uation, $G_{kij}{}^{\kappa}(\omega)$	Strong attenuation of high-frequencies in the shallow crust of the form $e^{-\pi\cdot\kappa(r)\cdot f}$.
Scattering, $G_{kij}{}^{S}(\omega)$	Scattering tends to reduce amplitudes on average, but introduces high amplitude caustics and low-amplitude shadow zones and produces nearly log-normal distributions of amplitudes (O'Connell, 1999a).
Anisotropy, $G_{kij}{}^{A}(\omega)$	Complicates shear-wave amplitudes and modifies radiation pattern amplitudes and can introduce frequency-dependent amplification based on direction of polarization.
Topography, $G_{kij}{}^{T}(\omega)$	Can produce amplification near topographic highs and introduces an additional sources of scattering.

Table 3. Seismic Wave Propagation Amplitude Factors ($G_{kij}(\omega)$)

Numerous studies have demonstrated that the seismic velocities in the upper 30 to 60 m can greatly influence the amplitudes of earthquake grounds motions at the surface (e.g. Borcherdt et al., 1979; Joyner et al., 1981; Seed et al., 1988). Williams et al. (1999) showed that significant resonances can occur for impedance boundaries as shallow as 7-m depth. Boore and Joyner (1997) compared the amplification of generic rock sites with very hard rock sites for 30 m depth averaged velocities. They defined very hard rocks sites as sites that have shear-wave velocities at the surface > 2.7 km/s and generic rock sites as sites where shear-wave velocities at the surface are ~0.6 km/s and increase to > 1 km/s at 30 m depth. Boore and Joyner (1997) found that amplifications on generic rock sites can be in excess of 3.5 at high frequencies, in contrast to the amplifications of less than 1.2 on very hard rock sites. Considering the combined effect of attenuation and amplification, amplification for generic rocks sites peaks between 2 and 5 Hz at a maximum value less than 1.8 (Boore and Joyner, 1997).

Factor	Influence
Geometric spreading, $\phi_{kij}{}^{r}(\omega)$	Introduces frequency dependent propagation delays.
Large-scale velocity structure, $\phi_{kij}{}^{V_{3D}}(\omega)$	Horizontal and vertical velocity and density gradients and velocity and density discontinuities produce frequency dependent phase shifts.
Near-surface resonant responses, $\phi_{kij}{}^{L}(\omega)$	Interactions of shear-wave arrivals of varying angles of incidence and directions produce frequency dependent phase shifts.
Nonlinear soil responses, $\phi_{kij}{}^{N}(\ddot{u},\omega)$ (equivalent linear), $\phi_{kij}{}^{N}(\ddot{u},t)$ (fully nonlinear)	Depending on the dynamic soil properties and pore pressure responses, nonlinear responses can increase or reduce phase dispersion. In the case of coupled pore-pressure with dilatant materials can collapse phase producing intermittent amplification caus tics.
Frequency indepen dent attenuation, $\phi_{kij}{}^{Q}(\omega)$	Linear hysteretic behavior produces frequency-dependent velocity dispersion that produces frequency dependent phase variations.
Scattering, $\phi_{kij}{}^{S}(\omega)$	The scattering strength and scattering characteristics determine propagation distances required to randomize the phase of shear waves as a function of frequency.
Anisotropy, $\phi_{kij}{}^{A}(\omega)$	Complicates shear-wave polarizations and modifies radiation pattern polarizations.
Topography, $\phi_{kij}{}^{T}(\omega)$	Complicates phase as a function of topographic length scale and near-surface velocities.

Table 4. Seismic Wave Propagation Phase Factors ($\phi_{kij}(\omega)$)

A common site-response estimation method is to use the horizontal-to-vertical (*H/V*) spectral ratio method with shear waves (Lermo and Chavez-Garcia, 1993) to test for site resonances. The *H/V* method is similar to the receiver-function method of Langston (1979). Several investigations have shown the *H/V* approach provides robust estimates of resonant frequencies (e.g., Field and Jacob, 1995; Castro et al., 1997; Tsubio et al., 2001) although absolute amplification factors are less well resolved (Castro et al., 1997; Bonilla et al., 1997). One-dimensional site-response approaches may fail to quantify site amplification in cases when upper-crustal three-dimensional velocity structure is complex. In southern California, Field (2000) found that the basin effect had a stronger influence on peak acceleration than detailed geology used to classify site responses. Hartzell et al. (2000) found that site amplification characteristics at some sites in the Seattle region cannot be explained using 1D or 2D velocity models, but that 3D velocity structure must be considered to fully explain local site responses. Chavez-Garcia et al. (1999) showed that laterally propagating basin-generated surface waves can not be differentiated from 1D site effects using frequency domain techniques such as *H/V* ratios or reference site ratios. The ability to conduct site-specific ground motion investigations is predicated on the existence of geological, geophysical, and geotechnical engineering data to realistically characterize earthquake sources, crustal velocity structure, local site structure and conditions, and to estimate the resultant seismic responses at a site. Lack of information about 3D variations in local and crustal velocity structure are serious impediments to ground motion estimation.

It is now recognized that correlated-random 3D velocity heterogeneity is an intrinsic property of Earth's crust (see Sato and Fehler, 1998 for a discussion). Correlated-random means that random velocity fluctuations are dependent on surrounding velocities with the dependence being inversely proportional to distance. Weak (standard deviation, ε, of ~5%), random fractal crustal velocity variations are required to explain observed short-period ($T < 1$ s) body-wave travel time variations, coda amplitudes, and coda durations for ground motions recorded over length scales of tens of kilometers to tens of meters (Frankel and Clayton, 1986), most well-log data (Sato and Fehler, 1998), the frequency dependence of shear-wave attenuation (Sato and Fehler, 1998), and envelope broadening of shear waves with distance (Sato and Fehler, 1998). As a natural consequence of energy conservation, the excitation of coda waves in the crust means that direct waves (particularly direct shear waves that dominate peak ground motions) that propagate along the minimum travel-time path from the source to the receiver lose energy with increasing propagation distance as a result of the dispersion of energy in time and space.

Following Frankel and Clayton (1986) fractal, self-similar velocity fluctuations are described with an autocorrelation function, *P*, of the form,

$$P(k_r) \approx \frac{a^n}{(1+k_r a)^n} \tag{8}$$

where a is the correlation distance, k_r is radial wavenumber, n=2 in 2D, and n=3 in 3D. When n=4 an exponential power law results (Sato and Fehler, 1998). Smoothness increasing with distance as a increases in (8) and overall smoothness is proportional to n in (8). This is a more realistic model of spatial geologic material variations than completely uncorrelated, spatially independent, random velocity variations. "Correlated-random" is shortened here to "random" for brevity. Let λ denote wavelength. Forward scattering dominates when $\lambda << a$ (Sato and Fehler, 1998). The situation is complicated in self-similar fractal media when considering a broad frequency range relevant to strong motion seismology (0.1 to 10 Hz) because λ spans the range $\lambda >> a$ to $\lambda << a$ and both forward and backscattering become important, particularly as n decreases in (8). Thus, it is difficult to develop simple rigorous expressions to quantify amplitude and phase terms associated with wave propagation through the heterogeneous crust (see Sato and Fehler, 1998). O'Connell (1999a) showed that direct shear-wave scattering produced by *P-SV*-wave coupling associated with vertical velocity gradients typical of southern California, combined with 3D velocity variations with n=2 and a standard deviation of velocity variations of 5% in (8), reduce high-frequency peak ground motions for sediment sites close to earthquake faults. O'Connell (1999a) showed that crustal scattering could substantially influence the amplification of near-fault ground motions in areas subjected to significant directivity. Scattering also determines the propagation distances required to randomize phase as discussed later in this paper.

Dynamic reduction of soil moduli and increases in damping with increasing shear strain can substantially modify ground motion amplitudes as a function of frequency (Ishihara, 1996). While there has been evidence of nonlinear soil response in surface strong motion recordings (Field et al., 1997; Cultera et al., 1999), interpretation of these surface records solely in terms of soil nonlinearity is intrinsically non-unique (O'Connell, 1999a). In contrast, downhole strong motion arrays have provided definitive evidence of soil nonlinearity consistent with laboratory testing of soils (Chang et al., 1991; Wen et al., 1995, Ghayamghamain and Kawakami, 1996; Satoh et al, 1995, 1997, 2001).

Idriss and Seed (1968a, b) introduced the "equivalent linear method" to calculate nonlinear soil response, which is an iterative method based on the assumption that the response of soil can be approximated by the response of a linear model whose properties are selected in relation to the average strain that occurs at each depth interval in the model during excitation. Joyner and Chen (1975) used a direct nonlinear stress-strain relationship method to demonstrate that the equivalent linear method may significantly underestimate short-period motions for thick soil columns and large input motions. Archuleta et al. (2000) and Bonilla (2000) demonstrated that dynamic pore-pressure responses can substantially modify nonlinear soil response and actually amplify and extend the durations of strong ground motions for some soil conditions. When a site is situated on soil it is critical to determine whether soil response will decrease or increase ground amplitudes and durations, and to compare the expected frequency dependence of the seismic soil responses with the resonant frequencies of the engineered structure(s). When soils are not saturated, the equivalent linear method is usually adequate with consideration of the caveats of Joyner and Chen (1975). When soils are saturated and interbedding sands and/or gravels between clay layers is prevalent, a fully nonlinear evaluation of the site that accounts for dynamic pore pressure responses may be necessary (Archuleta et al., 2000).

Lomnitz et al. (1999) showed that for the condition $0.91\beta_1 < \alpha_0$, where β_1 is the shear-wave velocity of low-velocity material beneath saturated soils, and α_0 is the acoustic (compressional-wave) velocity in the near-surface material, a coupled mode between Rayleigh waves propagating along the interface and compressional waves in the near surface material propagates with phase velocity α_0. This mode can propagate over large distances with little attenuation. Lomnitz et al. (1999) note that this set of velocity conditions provides a "recipe" for severe earthquake damage on soft ground when combined with a large contrast in Poisson's ratio between the two layers, and when the resonant frequencies of the mode and engineering structures coincide. Linear 2D viscoelastic finite-difference calculations demonstrate the existence of this wave mode at small strains, but nonlinear 2D finite-difference calculations indicate that long-distance propagation of this mode is strongly attenuated (O'Connell et al., 2010).

Anisotropy complicates polarizations of shear waves. Coutant (1996) showed that shallow (< 200 m) shear-wave anisotropy strongly influences surface polarizations of shear waves for frequencies < 30 Hz. Chapman and Shearer (1989) show that quasi-shear (qS) wave polarizations typically twist along ray paths through gradient regions in anisotropic media, causing frequency-dependent coupling between the qS waves. They show that this coupling is much stronger than the analogous coupling between P and SV waves in isotropic gradients because of the small difference between the qS-wave velocities. Chapman and

Shearer (1989) show that in some cases, far-field excitation of both quasi-shear wave and shear-wave splitting will result from an incident wave composed of only one of the quasi-shear waves. The potential for stronger coupling of quasi-shear waves suggests that the influence of anisotropy on shear-wave polarizations and peak ground motion may be significant in some cases. While the influence of anisotropy on strong ground motions is unknown, it is prudent to avoid suggesting that only a limited class of shear-wave polarizations are likely for a particular site based on isotropic ground motion simulations of ground motion observations at other sites.

Velocity anisotropy in the crust can substantially distort the radiation pattern of body waves with shear-wave polarization angles diverging from those in an isotropic medium by as much as 90 degrees or more near directions where group velocities of quasi-*SH* and *SV* waves deviate from corresponding phase velocities (Kawasaki and Tanimoto, 1981). Thus, anisotropy has the potential to influence radiation pattern coherence as well as ground motion polarization. A common approach is to assume the double-couple radiation pattern disappears over a transition frequency band extending from 1 Hz to 3 Hz (Pitarka et al., 2000) or up to 10 Hz (Zeng and Anderson, 2000). The choice of frequency cutoff for the radiation pattern significantly influences estimates of peak response in regions prone to directivity for frequencies close to and greater than the cutoff frequency. This is a very important parameter for stiff (high-frequency) structures such as buildings that tend to have natural frequencies in the 0.5 to 5 Hz frequency band (see discussion in Frankel, 1999).

Topography can substantially influence peak ground motions (Boore, 1972; 1973). Schultz (1994) showed that an amplification factor of 2 can be easily achieved near the flanks of hills relative to the flatter portions of a basin and that substantial amplification and deamplification of shear-wave energy in the 1 to 10 Hz frequency range can occur over short distances. Bouchon et al. (1996) showed that shear-wave amplifications of 50% to 100% can occur in the 1.5 Hz to 20 Hz frequency band near the tops of hills, consistent with observations from the 1994 Northridge earthquake (Spudich et al., 1996). Topography may also contribute to amplification in adjacent basins as well as the contributing to differential ground motions with dilatational strains on the order of 0.003 (Hutchings and Jarpe, 1996). Topography has a significant influence on longer-period amplification and groundshaking durations. Ma et al. (2007) showed that topography of the San Gabriel Mountains scatters the surface waves generated by the rupture on the San Andreas fault, leading to less-efficient excitation of basin-edge generated waves and natural resonances within the Los Angeles Basin and reducing peak ground velocity in portions of the basin by up to 50% for frequencies 0.5 Hz or less.

These discussions of source and propagation influences on amplitudes and phase are necessarily abbreviated and are not complete, but do provide an indication of the challenges of ground motion estimation, and developing relatively simple, but sufficient ground motion prediction equations based on empirical strong ground motion data. Systematically evaluating all the source and wave propagation factors influencing site-specific ground motions is a daunting task, particularly since it's unlikely that one can know all the relevant source and propagation factors. Often, insufficient information exists to quantitatively evaluate many ground motion factors. Thus, it is useful to develop a susceptibility checklist for ground motion estimation at a particular site. The list would indicate available information for each factor on a scale ranging from ignorance to strong quantitative information and indicate how this state of information could influence ground motions at the site. The result of such a checklist would be a susceptibility rating for potential biases and errors for peak motion and duration estimates of site-specific ground motions.

4. Nonlinear site response

4.1 Introduction

The near surface geological site conditions in the upper tens of meters are one of the dominant factors in controlling the amplitude and variation of strong ground motion, and the damage patterns that result from large earthquakes. It has long been known that soft sediments amplify the earthquake ground motion. Superficial deposits, especially alluvium type, are responsible for a remarkable modification of the seismic waves. The amplification of the seismic ground motion basically originates from the strong contrast between the rock and soil physical properties (e.g. Kramer, 1996). At small deformations, the soil response is linear: strain and stress are related linearly by the rigidity modulus independently of the strain level (Hooke's law). Mainly because most of the first strong motion observations seemed to be consistent with linear elasticity, seismologists generally accept a linear model of ground motion response to seismic excitation even at the strong motion level. However, according to laboratory studies (e.g. Seed and Idriss, 1969), Hooke's law breaks down at larger strains and the nonlinear relation between strain and stress may significantly affect the strong ground motion at soil sites near the source of large earthquakes.

Since laboratory conditions are not the same as those in the field, several authors have tried to find field data to understand nonlinear soil behavior. In order to isolate the local site effects, the transfer function of seismic waves in soil layers has to be estimated by calculating the spectral ratio between the motion at the surface and the underlying soil layers. Variation of these spectral ratios between strong and weak motion has actively been searched in order to detect nonlinearity. For example, Darragh and Shakal (1991) observed an amplification reduction at the Treasure Island soft soil site in San Francisco. Beresnev and Wen (1996) also reported a decrease of amplification factors for the array data in the Lotung valley (Taiwan). Such a decrease has also been observed at different Japanese sites including the Port Island site (e.g. Satoh et al., 1997, Aguirre and Irikura, 1997). On the other hand, Darragh and Shakal (1991) also reported a quasi-linear behavior for a stiff soil site in the whole range from 0.006 g to 0.43g. According to these results there is a need to precise the thresholds corresponding to the onset of nonlinearity and the maximum strong motions amplification factors according to the nature and thickness of soil deposits (Field et al., 1998).

Nevertheless, the use of surface ground motion alone does not help to directly calculate the transfer function and these variations. Rock outcrop motion is then usually used to estimate the motion at the bedrock and to calculate sediments amplification for both weak and strong motion (e.g. Celebi et al., 1987; Singh et al., 1988; Darragh et al., 1991; Field et al., 1997; Beresnev, 2002). The accuracy of this approximation strongly depends on near surface rock weathering or topography complexity (Steidl et al., 1996). Moreover, the estimate of site response can be biased by any systematic difference for the path effects between stations located on soil and rock. One additional complication is also due to finite source effects such as directivity. In case of large earthquakes, waves arriving from different locations may interfere causing source effects to vary with site location (Oglesby and Archuleta, 1997). Since these finite source effects strongly depend on the source size, they could mimic the observations cited as evidence for soil nonlinearity. Finally, O'Connell (1999) and Hartzell et al. (2005) show that in the near-fault region of M > 6 earthquakes linear wave propagation in weakly heterogeneous, random three dimensional crustal velocity can mimic observed, apparently, nonlinear sediment response in regions with large vertical velocity gradients that persist from near the surface to several km depth, making it difficult to separate soil nonlinear responses from other larger-scale linear wave propagation effects solely using surface ground motion recordings.

Because of these difficulties, the most effective means for quantifying the modification in ground motion induced by soil sediments is to record the motion directly in boreholes that penetrate these layers. Using records from vertical arrays it is possible to separate the site from source and path effects and therefore clearly identify the nonlinear behavior and changes of the soil physical properties during the shaking (e.g. Zeghal and Elgamal, 1994; Aguirre and Irikura, 1997; Satoh et al., 2001; Assimaki et al., 2007; Assimaki et al., 2010; Bonilla et al. 2011).

4.2 Nonlinear soil behavior

For years, it has been established in geotechnical engineering that soils behave nonlinearly. This fact comes from numerous experiments with cyclic loading of soil samples. The stress-strain curve has a hysteretic behavior, which produces a reduction of shear modulus as well as an increasing in damping factor.

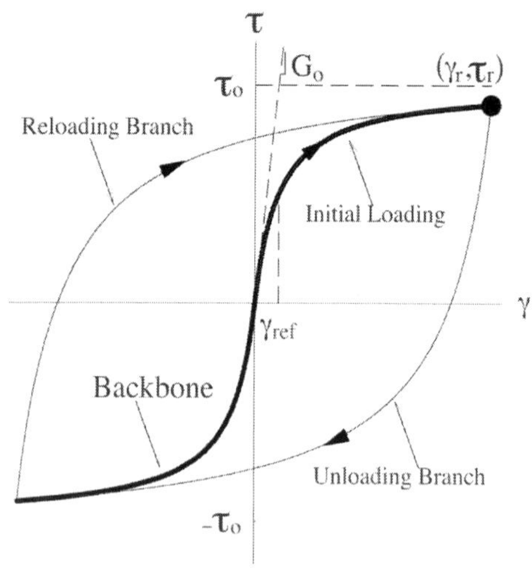

Fig. 4.1. Hyperbolic model of the stress-strain space for a soil under cyclic loading. Initial loading curve has a hyperbolic form, and the loading and unloading phases of the hysteresis path are formed following Masing's criterion.

Figure 4.1 shows a typical stress-strain curve with a loading phase and consequent hysteretic behavior for the later loading process. There have been several attempts to describe mathematically the shape of this curve, and among those models the hyperbolic is one of the easiest to use because of its mathematical formulation as well as for the number of parameters necessary to describe it (Ishihara, 1996; Kramer, 1996; Beresnev and Wen, 1996)

$$\tau = \frac{G_0\,\gamma}{1 + \frac{G_0}{\tau_0}\,|\gamma|}$$

where G_0 is the undisturbed shear modulus, and τ_0 is the maximum stress that the material can support in the initial state. G_0 is also known as G_{max} because it has the highest value of shear modulus at low strains.

In order to have the hysteretic behavior, the model follows the so-called Masing's rule, which in its basic form translates the origin and expands the horizontal and vertical axis by a factor of 2. Thus,

$$\frac{\tau - \tau_r}{2} = \tau\left(\frac{\gamma - \gamma_r}{2}\right)$$

where (γ_r, τ_r) is the reversal point for unloading and reloading curves.

This behavior produces two changes in the elastic parameters of the soil. First, the larger the maximum strain, the lower the secant shear modulus G obtained as the slope of the line between the origin and the reversal point of the hysteresis loop. Second, hysteresis shows a loss of energy in each cycle, and as it was mentioned above, the energy is proportional to the area of the loop. Thus, the larger the maximum strain, the larger the damping factor.

How can the changes in the elastic parameters be detected when looking at transfer functions? We know that the resonance frequencies f_m are proportional to $\beta\,(2m+1)/4H$ (the fundamental frequency corresponds to $m = 0$). Where β is the shear velocity and H is the soil thickness. Thus, if the shear modulus G is reduced then the resonance frequencies are also reduced because $G = \rho\,\beta^2$, where ρ is the material density. In other words, in the presence of nonlinearity the transfer function shifts the resonance frequencies toward lower frequencies. In addition, increased dissipation reduces soil amplification.

Figure 4.2 shows an example of nonlinear soil behavior at station TTRH02 (Vs_{30} = 340 m/s), KiK-net station that recorded the M_{JMA} 7.3 October 2000 Tottori in Japan. The orange shaded region represents the 95% borehole transfer function computed using events having a PGA less than 10 cm/s^2. Conversely, the solid line is the borehole transfer function obtained using the data from the Tottori mainshock. One can clearly see the difference between these two estimates of the transfer function, namely a broadband deamplification and a shift of resonance frequencies to lower values. The fact that the linear estimate is computed at the 95% confidence limits means that we are confident that this site underwent nonlinear site responses at a 95% probability level.

However, nonlinear effects can also directly be seen on acceleration time histories. Figure 4.3 shows acceleration records, surface and downhole, of the 1995 Kobe earthquake at Port Island (left) and the 1993 Kushiro-Oki earthquake at Kushiro Port (right). Both sites have shear wave velocity profiles relatively close each other, except in the first 30 meters depth. Yet, their response is completely different. Port Island is a man-made site composed of loose sands that liquefied during the Kobe event (Aguirre and Irikura, 1997). Practically there is no energy after the S-wave train in the record at the surface. Conversely, Kushiro Port is composed of dense sands and shows, in the accelerometer located at ground level, large acceleration spikes that are even higher than their counterpart at depth. Iai et al., (1995), Archuleta (1998), and Bonilla et al., (2005) showed that the appearance of large acceleration peak values riding a low frequency carrier are an indicator of soil nonlinearity known as cyclic mobility. Laboratory studies show that the physical mechanism that produces such

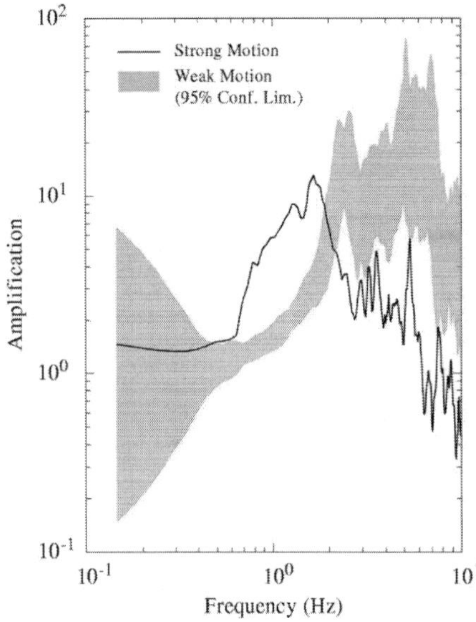

Fig. 4.2. Borehole transfer functions computed at KiK-net station TTRH02 in Japan. The orange shaded area represents the 95% confident limits of the transfer function using weak-motion events (PGA < $10cm/s^2$). The solid line is the transfer function computed using the October 2000 Tottori mainshock data.

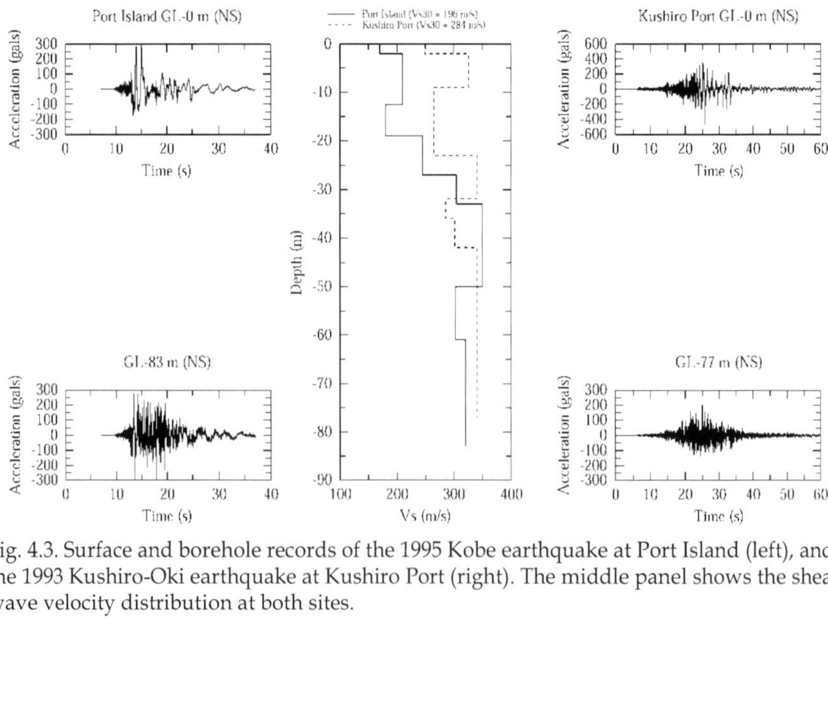

Fig. 4.3. Surface and borehole records of the 1995 Kobe earthquake at Port Island (left), and the 1993 Kushiro-Oki earthquake at Kushiro Port (right). The middle panel shows the shear wave velocity distribution at both sites.

phenomenon is the dilatant nature of cohesionless soils, which introduces the partial recovery of the shear strength under cyclic loads. This recovery translates into the ability to produce large deformations followed by large and spiky shear stresses. The spikes observed in the acceleration records are directly related to these periods of dilatancy and generation of pore pressure.

These examples indicate that nonlinear soil phenomena are complex. We cannot see the effects of nonlinear soil behavior on the transfer function only, but also on the acceleration time histories. This involves solving the wave equation by integrating nonlinear soil rheologies in the time domain, the subject treated in the next section.

4.3 The strain space multishear mechanism model

The multishear mechanism model (Towhata and Ishihara, 1985) is a plane strain formulation to simulate pore pressure generation in sands under cyclic loading and undrained conditions. Iai et al. (1990a, 1990b) modified the model to account for the cyclic mobility and dilatancy of sands. This method has the following strong points:

- It is relatively easy to implement. It has few parameters that can be obtained from simple laboratory tests that include pore pressure generation.
- This model represents the effect of rotation of principal stresses during cyclic behavior of anisotropically consolidated sands.
- Since the theory is a plane strain condition, it can be used to study problems in two dimensions, e.g. embankments, quay walls, among others.

In two dimensional cartesian coordinates and using vectorial notation, the effective stress σ' and strain ϵ tensors can be written as

$$\{\sigma'\}^T = \{\sigma'_x \; \sigma'_y \; \tau_{xy}\}$$
$$\{\epsilon\}^T = \{\epsilon_x \; \epsilon_y \; \gamma_{xy}\}$$

where the superscript T represents the vector transpose operation; σ'_x, σ'_y , ϵ_x, and ϵ_y represent the effective normal stresses and strains in the horizontal and vertical directions; τ_{xy} and γ_{xy} are the shear stress and shear strain, respectively.

The multiple mechanism model relates the stress and strain through the following incremental equation (Iai et al., 1990a, 1990b),

$$\{d\sigma'\} = [D](\{d\epsilon\} - \{d\epsilon_p\})$$

where the curly brackets represent the vector notation; ϵ_p is the volumetric strain produced by the pore pressure, and D is the tangential stiffness matrix given by

$$[D] = \mathrm{K} n^{(0)} n^{(0)T} + \sum_{i=1}^{I} R^{(i)} n^{(i)} n^{(i)T}$$

The first term is the volumetric mechanism represented by the bulk modulus K. The second part is the shear mechanism represented by the tangential shear modulus $R^{(i)}$ idealized as a collection of I springs (Figure 4.4). Each spring follows the hyperbolic stress-strain model (Konder and Zelasko, 1963) during the loading and unloading hysteresis process. The shear mechanism may also be considered as a combination of pure shear and shear by differential compression.

In addition,

$$
\begin{aligned}
n^{(0)T} &= \{1\ 1\ 0\} \\
n^{(i)T} &= \{\cos\theta_i\ \ -\cos\theta_i\ \ \sin\theta_i\} \\
\theta_i &= (i-1)\frac{\pi}{I}
\end{aligned}
$$

where $\Delta\theta = \pi/I$ is the angle between each spring as shown in Figure 4.4.
Towhata and Ishihara (1985) found, using laboratory data, that the pore pressure excess is correlated with the cumulative shear work produced during cyclic loading. Iai et al. (1990a, 1990b) developed a mathematical model that needs five parameters, called hereafter dilatancy parameters, to take into account this correlation. These parameters represent the initial and final phases of dilatancy, p_1 and p_2; overall dilatancy w_1; threshold limit and ultimate limit of dilatancy, c_1 and S_1. These parameters are obtained by fitting laboratory data, from either undrained stress controlled cyclic shear tests or from cyclic stress ratio curves. Details of this constitutive model can be found in Iai et al. (1990a, 1990b).

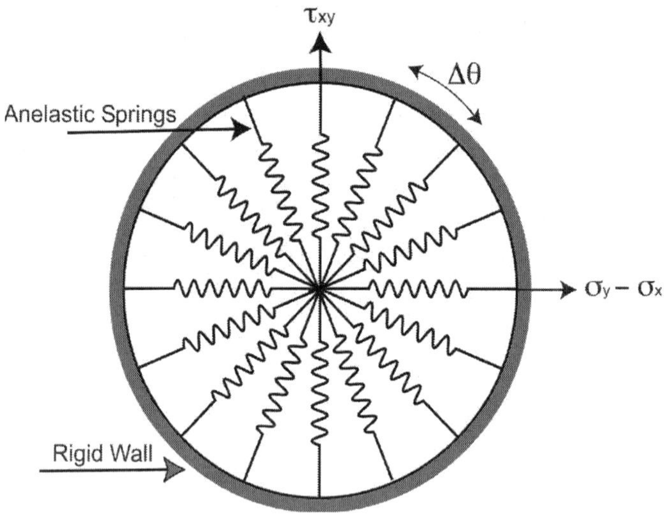

Fig. 4.4. Schematic figure for the multishear mechanism. The plane strain is the combination of pure shear (vertical axis) and shear by compression (horizontal axis) (after Towhata and Ishihara, 1985).

At this point, this formulation provides only the backbone curve. It is here that the hysteresis is now taken into account by using the generalized Masing rules. In fact, they are not simple rules but a state equation that describes hysteresis given a backbone curve (Bonilla, 2000). They are called generalized Masing rules because its formulation contains Pyke's (1979) and the original Masing models as special cases. Furthermore, this formulation allows, by controlling the hysteresis scale factor, the reshaping of the backbone curve as suggested by Ishihara et al. (1985) so that the hysteresis path follows a prescribed damping ratio.

4.4 The generalized Masing rules

In previous sections we use the hyperbolic model to describe the stress-strain space of soil materials subjected to cyclic loads. In the hyperbolic model, the nonlinear relation can be written as

$$\frac{G}{G_0} = \frac{1}{1 + |\gamma/\gamma_{ref}|}$$

where $\gamma_{ref} = \tau_0/G_0$ is the reference strain.
Introducing the equation above into $\tau_{xy} = G\,\gamma_{xy}$, where τ_{xy} is the shear stress and γ_{xy} is the shear strain; and adding the hysteresis operator, we have

$$\tau_{xy} = Hys(F_{bb})$$
$$F_{bb} = \tau_0 \frac{\gamma_{xy}/\gamma_{ref}}{1 + |\gamma_{xy}/\gamma_{ref}|}$$

where F_{bb} is the backbone curve, and $Hys(.)$ is the hysteresis operator (application of the generalized Masing rules).
Hysteresis behavior can be implemented in a phenomenological manner with the help of the Masing and extended Masing rules (Vucetic, 1990; Kramer, 1996). However, these rules are not enough to constrain the shear stress τ_{xy} to values not exceeding the strength τ_0. This happens when the time behavior of the shear strain departs from the simple cyclic behavior, and of course, noncyclic time behavior is common in seismic signals. Inadequacy of the Masing rules to describe the hysteretic behavior of complicated signals has been already pointed out and some remedies have been proposed (e.g. Pyke, 1979; Li and Liao, 1993).
The Masing rules consist of a translation and dilatation of the original law governing the strain-stress relationship. While the initial loading of the material is given by the backbone curve $F_{bb}(\gamma_{xy})$, the subsequent loading and unloading, the strain-stress relationship is given by:

$$\frac{\tau_{xy} - \tau_r}{\kappa_H} = F_{bb}\left(\frac{\gamma_{xy} - \gamma_r}{\kappa_H}\right)$$

where the coordinate (γ_r, τ_r) corresponds to the reversal points in the strain-stress space, and κ_H is the so-called hysteresis scale factor (Archuleta et al., 2000). In Masing's original formulation, the hysteresis scale factor κ_H is equal to 2. A first extension to the Masing rules can be obtained by releasing the constraint $\kappa_H = 2$. This parameter controls the shape of the loop in the stress-strain space (Bonilla et al., 1998). However, numerical simulations suggest spurious behavior of τ_{xy} for irregular loading and unloading processes even when extended Masing rules are used. A further generalization of Masing rules is obtained choosing the value of κ_H in such way to assure that the path τ_{xy}, at a given unloading or reloading, in the strain-stress space will cross the backbone curve, and becomes bounded by the maximum strength of the material τ_0. This can be achieved by having the following condition,

$$\lim_{\gamma_{xy} \to sign(\dot{\gamma}_{xy})|\gamma_f|} \kappa_{H_j} F_{bb}\left(\frac{\gamma_{xy} - \gamma_{r_j}}{\kappa_{H_j}}\right) \quad |\gamma_{r_j}| \le |\gamma_f| \le |\infty|$$

where γ_f is the specified finite or infinite strain condition, γ_{r_j} and κ_{H_j} correspond to the turning point and the hysteresis shape factor at the jth unloading or reloading; and $sign(\dot{\gamma}_{xy})$ is the sign of the strain rate. Thus,

$$sign(\dot{\gamma}_{xy})|\tau_f| = \lim_{\gamma_{xy} \to sign(\dot{\gamma}_{xy})|\gamma_f|} \kappa_{H_j} F_{bb}\left(\frac{\gamma_{xy} - \gamma_{r_j}}{\kappa_{H_j}}\right) + \tau_{r_j}$$

where $|\tau_f| = F_{bb}(|\gamma_f|)$, and $(\gamma_{r_j}, \tau_{r_j})$ is the turning point pair at the jth reversal. Replacing the functional form of the backbone (the hyperbolic model) and after some algebra we have,

$$\kappa_{H_j} = \frac{\left(sign(\dot{\gamma}_{xy})|\tau_f| - \tau_{r_j}\right)|sign(\dot{\gamma}_{xy})|\gamma_f| - \gamma_{r_j}|}{\tau_0\left(sign(\dot{\gamma}_{xy})|\gamma_f| - \gamma_{r_j}\right) - \gamma_{ref}(sign(\dot{\gamma}_{xy})|\tau_f| - \tau_{r_j})} \quad |\gamma_{r_j}| \leq |\gamma_f| \leq |\infty|$$

The equation above represents a general constraint on the hysteresis scale factor, so that the computed stress does not exceed τ_0 depending on the chosen maximum deformation γ_f that the material is thought to resist. The limit $\gamma_f \to \infty$ corresponds to the Cundall-Pyke hypothesis (Pyke, 1979), while $\gamma_f \to \gamma_{r_j}$ is similar to some extent to a method discussed in (Li and Liao, 1993).

In the following section, we will see an example of application of this soil constitutive model (Towhata and Ishihara, 1985; Iai et al., 1990a, 1990b) together with the Generalized Masing hysteresis operator (Bonilla, 2000).

4.5 Analysis of the 1987 Superstition Hills Earthquake

On 24 November 1987, the M_L 6.6 Superstition Hills earthquake was recorded at the Wildlife Refuge station. This site is located in southern California in the seismically active Imperial Valley. In 1982 it was instrumented by the U.S. Geological Survey with downhole and surface accelerometers and piezometers to record ground motions and pore water pressures during earthquakes (Holzer et al., 1989). The Wildlife site is located in the flood plain of the Alamo River, about 20 m from the river's western bank. *In situ* investigations have shown that the site stratigraphy consists of a shallow silt layer approximately 2.5 m thick underlain by a 4.3 m thick layer of loose silty sand, which is in turn underlain by a stiff to very stiff clay. The water table fluctuates around 2-m depth (Matasovic and Vucetic, 1993).

This site shows historically one direct *in situ* observation of nonlinearity in borehole data. The Wildlife Refuge liquefaction array recorded acceleration at the surface and 7.5-m depth, and pore pressure on six piezometers at various depths (Holzer et al., 1989). The acceleration time histories for the Superstition Hills events at GL-0 m and GL-7.5 m, respectively, are shown in Figure 4.5 (left). Note how the acceleration changes abruptly for the record at GL-0 m after the *S* wave. Several sharp peaks are observed; they are very close to the peak acceleration for the whole record. In addition, these peaks have lower frequency than the previous part of the record (the beginning of the *S* wave, for instance).

Zeghal and Elgamal (1994) used the Superstition Hills earthquakes to estimate the stress and strain from borehole acceleration recordings. They approximated the shear stress $\tau_{xy}(h,t)$ at depth h, and the mean shear strain $\bar{\gamma}_{xy}$ between the two sensors as follows,

$$\begin{aligned}
\tau_{xy}(h,t) &= \frac{1}{2}\rho h[\ddot{u}(0,t) + \ddot{u}(h,t)] \\
\ddot{u}(h,t) &= \ddot{u}(0,t) + \frac{h}{H}[\ddot{u}(H,t) - \ddot{u}(0,t)] \\
\bar{\gamma}_{xy}(t) &= \frac{u(h,t) - u(0,t)}{2}
\end{aligned}$$

where $\ddot{u}(0,t)$ is the horizontal acceleration at the ground surface, $\ddot{u}(h,t)$ is the acceleration at depth h (evaluated through linear interpolation); $\ddot{u}(H,t)$ is the acceleration at the bottom of the layer; $u(H,t)$ and $u(0,t)$ are the displacement histories obtained by integrating twice the corresponding acceleration histories; H is the thickness of the layer; and ρ is the density. Using this method, the stress and strain at GL-2.9 m were computed (Figure 4.5). This figure clearly shows the large nonlinearity developed during the Superstition Hills event. The stress-strain loops form an S-shape and the strains are as large as 1.5%. At this depth, there is a piezometer (P5 according to Holzer et al., 1989). With this information it is also possible to reconstruct the stress path (bottom right of Figure 4.5). Note that some of the pore pressure pulses are correlated with episodes of high shear stress development. The stress path shows a strong contractive phase followed by dilatancy when the effective mean stress is close to 15 kPa.

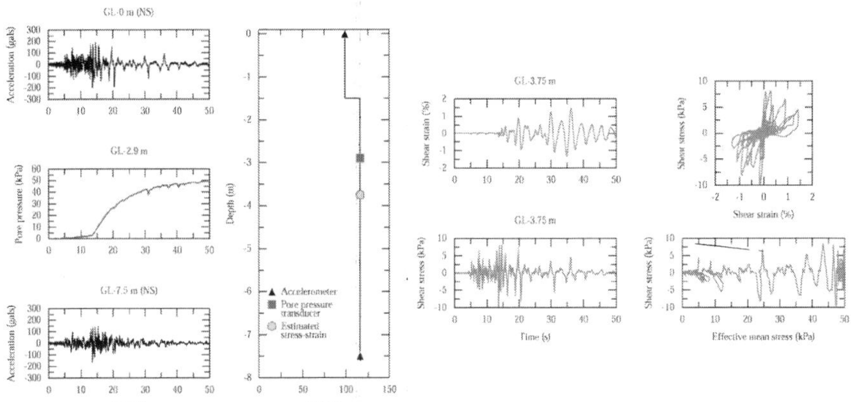

Fig. 4.5. Wildlife Refuge station that recorded the 1987 Superstition Hills earthquake both acceleration and pores pressure time histories (left). Computed stress and strain time histories according to Zeghal and Elgamal (1994), stress-strain loops and stress path history reconstitution (right).

Using the stress and stress time histories at GL-2.9 m computed earlier, Bonilla et al. (2005) performed a trial-and-error procedure in order to obtain the dilatancy parameters that best reproduce such observations. Figure 4.6 compares the computed shear stress time history with the observed shear strain at GL-2.9 m. The stress-strain hysteresis loops are also shown. We observe that the computed shear stress is well simulated; the stress-strain space also shows the same dilatant behavior (S-shape hysteresis loops) as the observed data.

Once the model parameters were determined, they proceed to compute the acceleration time history at GL-0 m using the north–south record at GL-7.5 m as input motion. Figure 4.7 shows the accelerograms (left) and the corresponding response spectra (right). The observed data are shown with no filtering, whereas the computed data are low-pass filtered at 10 Hz. The computed accelerogram shows the transition from high-frequency content between 0 and 15 sec to the intermittent spiky behavior after 15 sec. The response spectra show that the computed accelerogram accurately represents the long periods; yet, the short periods are still difficult to model accurately. This is the challenge of nonlinear simulations; the fit should be as broadband as possible.

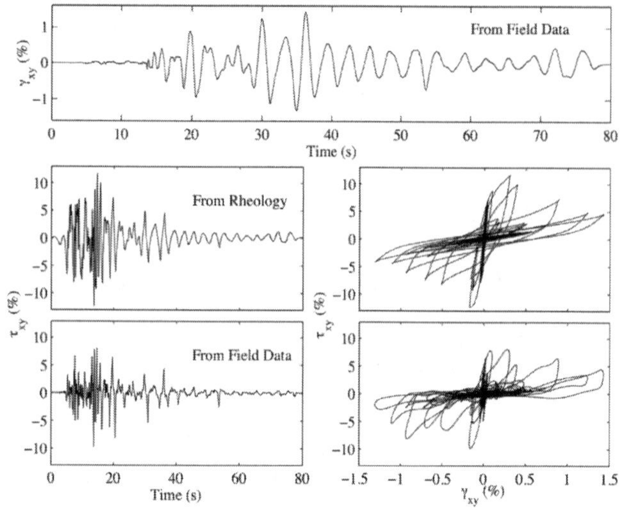

Fig. 4.6. The top panel shows the computed strain time history at the middle of the borehole. Middle panels show the computed stress by trial-and-error using the multispring model in order to find the best dilatancy parameters. Bottom panels indicate the computed stress time history from acceleration records (after Bonilla et al., 2005).

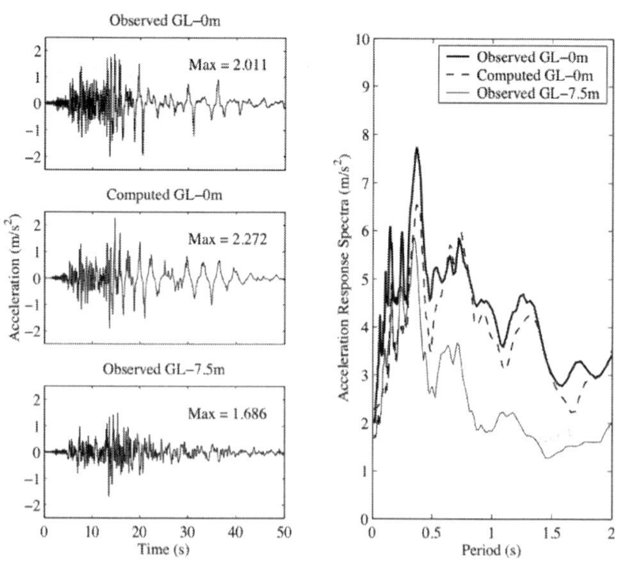

Fig. 4.7. Observed and computed acceleration time histories of the 1987 Superstition Hills earthquake (left) and their corresponding response spectra (right).

4.6 NOAH2D 2D P-SV analyses of the maximum observed peak acceleration

Current nonlinear formulations generally reproduce all first-order aspects of nonlinear soil responses. To illustrate this point, we present a nonlinear analyses of the largest peak ground acceleration recorded to date of > 4 g, that has a peak vertical acceleration of 3.8 g (Aoi et al., 2008). Aoi et al. (2008) analyzed ground motions recorded by the Kyoshin Network (Kik-net) during the M 6.9 2008 Iwate-Miyagi earthquake that included one soil-surface site that recorded a vertical acceleration of 3.8g (station IWTH25). The horizontal borehole and surface motions reported in Aoi et al. (2008) for station IWTH25 are generally consistent with the soil reducing surface horizontal accelerations at high frequencies, as is widely observed at soil sites (Field et al., 1997; 1998; Seed and Idriss, 1970; Beresnev et al., 1995, 2002) and reproducible using 1D nonlinear site response modeling (O'Connell, 2008). However, the surface vertical peak acceleration exceeded 3.8g, exceeding the maximum expected amplification, based on the site velocity profile between the borehole and the surface accelerometers, and current 1D linear or nonlinear theories of soil behavior (O'Connell, 2008). In particular, application of the nonlinear approach of shear-modulus reduction advocated and tested by Bersenev et al. (2002) to predict nonlinear vertical responses, failed to predict peak vertical accelerations in excess of 2g (O'Connell, 2008). Further, Aoi et al. (2008) observed largest upward accelerations at the surface that were 2.27 times larger than the largest downward accelerations, a result not reproduced using 1D approaches to approximate soil nonlinearity. The 2D nonlinear wave propagation implementation of Bonilla et al. (2006) uses a plane-strain model (Iai et al., 1990a, 1990b). In this section we show that this model could explain the first-order soil responses observed at station IWTH25 using fairly generic approximation to the site's nonlinear soil properties.

The P-SV nonlinear rheology developed by Iai et al. (1990a, 1990b) was used in the Bonilla et al. (2006) implementation of 2D nonlinear wave propagation. The constitutive equation implemented corresponds to the strain space multishear mechanism model developed by Towhata and Ishihara (1985) and Iai et al. (1990a, 1990b) with its backbone characterized by the hyperbolic equation (Hardin and Drnevich, 1972). The multishear mechanism model is a plane strain formulation to simulate cyclic mobility of sands under undrained conditions. In the calculations of this study, a total stress rheology (pore pressure was ignored) was used in the second order staggered grid P-SV plane-strain finite difference code. Perfectly matched layer (PML) absorbing boundary conditions were used to approximate elastic (transmitting) boundary conditions at the bottom and side edges using an implementation adapted for finite differences from Ma and Liu (2006). Linear hysteretic damping (Q) was implemented using the method of Liu and Archuleta (2006). The horizontal- and vertical component plane waves are inserted in the linear viscoelastic portion of the 2D with a user-selectable range of incident angles.

The Kyoshin Network, or Kik-net, in Japan (Fujiwara et al., 2005), has recorded numerous earthquakes with ground motion data recorded at surface and at depth in underlying rock and soil. We use the recording at Kik-net station IWTH25, where a 3.8g peak vertical acceleration was recorded (Aoi et al., 2008). Analyses of the combined downhole and surface ground motions from IWTH25 provide an opportunity to evaluate several strategies to estimate vertical ground motions since a P- and S-wave velocity profile is available to the bottom of the borehole at 260m (Aoi et al., 2008).

Station IWTH25 is located in a region of rugged topography adjacent to a high-gradient stream channel on a fluvial terrace. The ruggest topography reflects the station's hangingwall location relative to the reverse fault . Station IWTH25 is located near a region of large slip along strike and updip of the hypocenter. Consequently, IWTH25 is subjected to significant rupture directivity and near-fault radiation associated with strong gradients of slip and rupture velocity on the portions of the fault close to the station (Miyazaki et al., 2009). The IWTH25 ground motion has been of particular interest because of the extreme peak vertical acceleration (3.8g) and peculiar asymmetric amplitudes distribution of the vertical accelerations (Aoi et al., 2008; O'Connell, 2008; Hada et al., 2009; Miyazaki et al., 2009; Yamada et al, 2009a); the upward vertical acceleration is much larger than the downward direction, although in the borehole record at a depth of 260 m at the same site, the upward and downward accelerations have symmetric amplitudes (Aoi et al., 2008). Aoi et al. (2008) propose a conceptual model for this asymmetry. Their model uses a loose soil with nearly zero confining pressure near the surface. The soil particles separate under large downward acceleration, and in this quasi free-fall state, the downward accelerations at the surface only modestly exceed gravity. Conversely, large upward accelerations compact the soil and produce much larger upward accelerations. Aoi et al. (2008) report three cases of these anomalous large vertical acceleration amplifications in a search of 200,000 strong motion recordings. Hada et al. (2009) successfully reproduce the strong vertical asymmetric accelerations at IWTH25 with a simple 1D discrete element model, a model that is not a rigorous model of wave propagation. Yamada et al. (2009a) interpret the large upward spikes in acceleration as slapdown phases, which are also typically observed in near-field recordings of nuclear explosion tests. Our focus here is not the asymmetry of the IWTH25 vertical accelerations recorded at the surface, but showing that the simple total stress plane-strain model of soil nonlinearity in Bonilla et al. (2006) reproduces both the first-order peak horizontal and vertical velocities and accelerations and acceleration response spectra at station IWTH25 using the borehole motions at 260 m depth as inputs.

Yamada et al. (2009b) conducted geophysical investigations at the site and found lower velocities in the top several meters than reported in Aoi et al. (2008). Trail-and-error modeling was used to obtain the final refined velocity model consistent with the results of Yamada et al. (2009b); a lowest-velocity first layer of about 2 m thickness and shear-wave velocity on the order of 200 m/s was required to produce the maximum horizontal spectral responses observed near 10 Hz. The geologic environment at station IWTH25 will clearly produce lateral changes in shallow velocity structure. In particular, the hangingwall uplift associated with repeated faulting similar to the 2008 earthquake will produce a series of uplift terraces adjacent to the stream next to station IWTH25, with the lowest shallow velocities being found on the lowest terrace adjacent to the stream, where station IWTH25 is located. The width of the stream and lowest terrace is about 100m near station IWTH25.

We constructed a 2D velocity model by including a region 100 m wide with surface Vs=300 m/s layer 2-m deep and then extended Vs=500 m/s to the free surface in the region surrounding the 100-m-wide low-velocity surface layer. Station IWTH25 is assumed to be located relatively close (4-5 m) to the lateral velocity change within the lowst-velocity portion of the 2D velocity model because the geologic log from station IWTH25 indicates a only 1-2 m of young terrace deposits (Aoi et al., 2008), but the youngest terrace probably extends across and encompasses the stream channels and their margins. The dominant large-amplitude arrivals in the borehole motions are associated with large slip regions below and just south of station IWTH25. Consequently, a plane-wave incident at 80 degrees from the south was used to propagate the borehole motion to the surface in the 2D model.

The nonlinear properties were simplified to a depth-independent plasticity index (PI) of 20% for the NOAH2D calculations. Overall the 2D synthetic nonlinear horizontal motions provide a good fit to the acceleration response spectra (Figs. 4.8a and 4.8d) and acceleration seismograms (Figs. 4.8b and 4.8e). The 2D synthetic horizontal velocities match the observed velocity seismograms well, except in the early portion of the record where the translation ("fling") associated with permanent displacement that dominates early portions of the observed seismograms (Figs. 4.8c and 4.8f).

Synthetic vertical responses were calculated for each horizontal-vertical component pair which is a crude approximation to total 3D wavefield. The east component is nearly fault-normal and has the largest peak accelerations and velocities of the two horizontal components, so the east-vertical combination probably best corresponds to the dominant P-SV responses. Except for the obvious asymmetry in both the acceleration and velocity vertical seismograms, both the north-vertical and east-vertical 2D nonlinear synthetic vertical surface motions provide a good fit to the observed acceleration response spectra (Figures 4.9a and 4.9d), acceleration seismograms (Figures 4.9b and 4.9eb), and velocity seismograms (Figures 4.9c and 4.9f). Since station IWTH25 is located in the deformed hangingwall of a reverse fault in rugged topography, it is clear that even these 2D nonlinear calculations are a crude approximation to the field conditions and complex incident wavefield associated with the finite fault rupture. However, the 2D nonlinear calculations summarized in Figs. 4.8 and 4.9 for station IWTH25 clearly show that the 2D P-SV nonlinear approach of Bonilla et al. (2006) provide a sound basis to evaluate first-order nonlinear horizontal and vertical nonlinear responses, even for cases of extremely large incident accelerations and velocities.

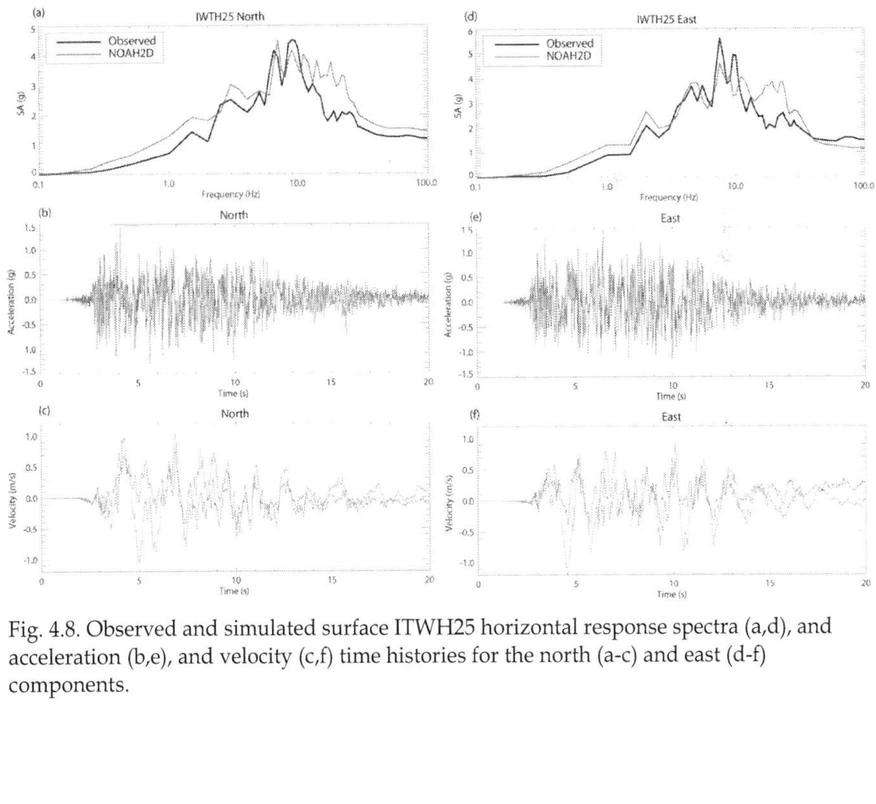

Fig. 4.8. Observed and simulated surface ITWH25 horizontal response spectra (a,d), and acceleration (b,e), and velocity (c,f) time histories for the north (a-c) and east (d-f) components.

It is important to mention some factors that are not explicitly accounted for in the approach of Bonilla et al. (2006). Goldberg (1960) was among the first to theoretically show the interaction between P and S waves in an elastic medium for large-amplitude seismic waves. His solution yielded the following results: (1) P- and S-waves couple, (2) S waves induce P waves, (3) the induced waves have a dominant frequency twice the S-wave frequency, (4) the induced P waves propagate ahead with P-wave velocity. Loukachev et al. (2002) develop a mechanism to explain the frequency dispersion in the P-wave spectra relative to S-wave spectra due to the interaction between compressional (P) and shear (S) waves in granular materials. Shear waves induce dilatancy and contractancy in granular materials which produces longitudinal dilatancy waves (so-called D waves) with approximately double frequency. The results of Goldberg (1960) are sufficient to partially explain the frequency doubling of the dominant vertical resonance frequency even if the IWTH25 material does not exhibit granular dilatancy and is dominated by relatively stiff clay responses. However, it appears more likely that the IWTH25 responses reflect a combination of large-amplitude P-S-wave linear coupling combined with the nonlinear dilatant material responses predicted by Loukachev et al. (2002) for granular soils. This may explain the broad frequency bandwidth (9-18 Hz) of large vertical spectral response. Even with these caveats, the synthetic results illustrated in Figures 9-12 show that the approach of Bonilla et al. (2006) clearly provides a sound basis to evaluate first-order nonlinear horizontal and vertical response at shallow soil sites and deeper soil basins.

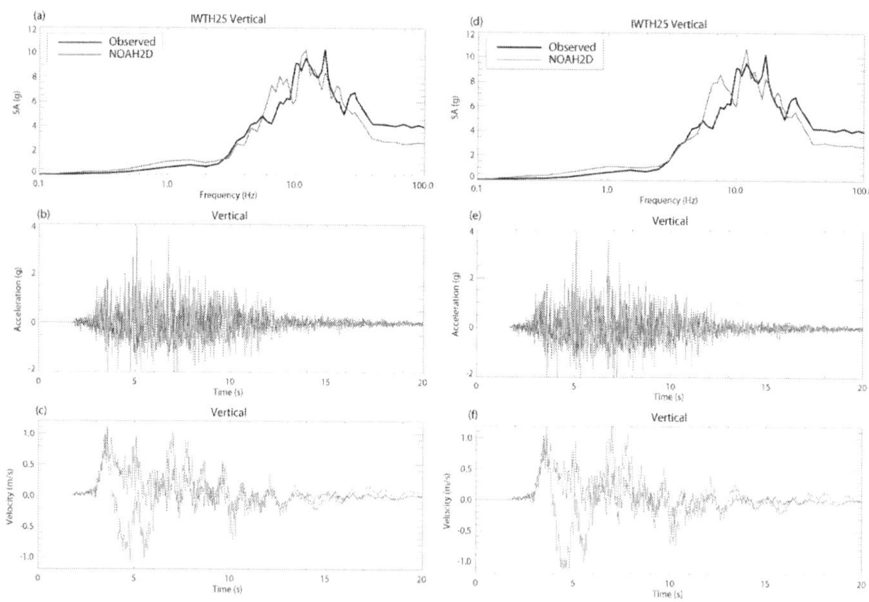

Fig. 4.9. Observed and simulated surface ITWH25 vertical response spectra (a,d), and acceleration (b,e), and velocity (c,f) time histories using the north-vertical (a-c) and east-vertical (d-f) components.

5. Ground motion prediction equations based on empirical data

Ground motion observations are the result of a long history of instrument development and deployment, instigated primarily by earthquake engineers, to acquire data to develop an empirical foundation to understand and predict earthquake ground motions for use in the design of engineered structures. Strong motion instruments usually record time histories of ground acceleration that can be post-processed to estimate ground velocities and displacements. A particularly useful derived quantity for engineering analyses are response spectra, which are the maximum amplitudes of modestly damped resonant responses of single-degree-of-freedom oscillators (an idealization of simple building responses) to a particular ground motion time history, as a function of natural period or natural frequency. While peak accelerations are always of concern for engineering analyses, peak ground velocity is now recognized as a better indicator of damage potential for large structures than is peak ground acceleration (EERI, 1994). Engineering analyses often consist of linear approaches to determine if structures reach their linear strength limits. Ground motion estimation quantities required for linear analyses are peak accelerations and velocities and associated response spectra. Nonlinear engineering analyses require estimates of future acceleration time histories. The discussion presented in this section focuses on empirical ground motion parameter estimation methods. Ground motion estimation methods required for nonlinear engineering analyses are presented in subsequent sections.

Historically the estimation of ground motion parameters such as peak acceleration, velocity, and displacement, and response spectral ordinates, and duration has been based on regression relationships developed using strong motion observations. These ground motion prediction equations strive to interpolate and extrapolate existing ground motion measurements to serve the needs to design for seismic loads.

5.1 Function form of GMPEs for regression

In their simplest form, these empirical GMPEs predict peak ground motions based on a limited parametric description of earthquake and site characteristics. Peak ground motion amplitudes generally increase with increasing magnitude up to a threshold magnitude range where peak accelerations saturate, i.e., only slightly increase or stay nearly constant above the threshold magnitude range (Campbell, 1981; Boore et al., 1994). Similarly, observed peak ground motion amplitudes decrease with increasing distance from the earthquake fault, but saturate at close distances to faults such that the decrease in amplitudes with increasing distance is small within several km of faults. These GMPEs relate specific ground motion parameters to earthquake magnitude, reduction (attenuation) of ground motion amplitudes with increasing distance from the fault (geometric spreading), and local site characteristics using either site classification schemes or a range of quantitative measures of shallow to deeper velocity averages or thresholds. The 30-m-average shear-wave velocity (Vs30) is most commonly used to account for first-order influences of shallow site conditions. Depths to shear-wave velocities of 1.0, 1.5, and 2.5 km/s (Z1.0 in Abrahamson and Silva (2008) and Chiou and Youngs (2008), Z1.5 in Choi et al. (2005) and Day et al. (2008), and Z2.5 in Campbell and Bozorgnia (2008), respectively) are sometimes used to account for influences of larger scale crustal velocity structure on ground motions.

The "Next Generation Attenuation" (NGA) Project was a collaborative research program with the objective of developing updated GMPEs (attenuation relationships) for the western

U.S. and other worldwide active shallow tectonic regions. These relationships have been widely reviewed and applied in a number of settings (Stafford et al., 2008; Shoja-Taheri et al., 2010). Five sets of updated GMPEs were developed by teams working independently but interacting throughout the NGA development process. The individual teams all had previous experience in the development of GMPEs. The individual teams all had access to a comprehensive, updated ground motion database that had been consistently processed (Chiou et al., 2008). Each team was free to identify portions of the database to either include or exclude from the development process. A total of 3551 recordings were included in the PEER-NGA database. The number of records actually used by the developers varied from 942 to 2754. The individual GMPEs are described in Abrahamson and Silva (2008), Boore and Atkinson (2008), Campbell and Bozorgnia (2008), Chiou and Youngs (2008), and Idriss (2008). These models are referred to as AS08, BA08, CB08, CY08, and I08, respectively, below. The NGA GMPEs developed equations for the orientation-independent average horizontal component of ground motions (Boore et. al., 2006).
The NGA account for these ground motion factors using the general form,

$$\begin{aligned}\ln Y = A_1 + A_2 \cdot M + A_3 (M - M_{REF}) + A_4 \cdot ln\left[\left(R^N + C_{Source}\right)^N\right] \\ + A_5 \cdot R + A_6 \cdot F^{source} + A_7 \cdot F^{site} + A_8 \cdot F^{HW} + A_9 \cdot F^{main}\end{aligned} \tag{9}$$

and,

$$\sigma_{lnY} = A_{10}(M, Vs30), \tag{10}$$

where Y is the ground motion parameter of interest (peak acceleration, velocity, displacement, response spectral ordinate, etc.), M is magnitude, R is a distance measure, M_{REF} and C_{SOURCE} are magnitude and distance terms that define the change in amplitude scaling and the $F^{[source,\ site,\ HW,\ main]}$ are indicator variables of source type, site type, hanging wall geometry and main shock discriminator. The A_i are coefficients to be determined by the regression. Not all of the five NGA GMPEs utilize all of these F indicator variables. The σ_{lnY} term represents the estimate of the period dependent standard deviation in the parameter lnY at the magnitude and distance of interest.
The NGA models use different source parameters and distance measures. Some of the models include the depth to top of rupture (TOR) as a source parameter. This choice was partially motivated by research (Somerville and Pitarka, 2006) that suggested a systematic difference in the ground motion for earthquakes with buried ruptures producing larger short period ground motions as compared to earthquakes with surface rupture. Large reverse-slip earthquakes tend to be buried ruptures more often than large strike-slip earthquakes so the effect of buried ruptures may be partially incorporated in the style-of-faulting factor. Not all the NGA developers found the inclusion of TOR to be a statistically significant factor.
All of the models except for I08 use the time-averaged S-wave velocity in the top 30 m of a site, Vs30, as the primary site response parameter. I08 is defined only for a reference rock outcrop with Vs30 = 450-900 m/s. Approximately two thirds of the recordings in the PEER-NGA database were obtained at sites without measured values of shear-wave velocity. Empirical correlations between the surface geology and *Vs30*, were developed (Chiou and others, 2008) and used with assessments of the surface geology to estimate values of *Vs30* at the sites without measured velocities. The implications of the use of estimated Vs30 on the standard deviation (σ_T) was evaluated and included by AS08.

All of the relationships that model site response incorporate nonlinear site effects. Two different metrics for the strength of the shaking are used to quantify nonlinear site response effects. AS08, BA08, and CB08 use the median estimate of PGA on a reference rock outcrop in the nonlinear site response term. CY08 uses the median estimate of spectral acceleration on a reference rock outcrop at the period of interest. The definition of "reference rock" varies from Vs30=535 m/s (I08) to Vs30=1130 m/s (CY08). A very small fraction of the strong-motion data in the PEER-NGA data set was obtained at sites with Vs30> 900m/s. Depths to shear-wave velocities of 1.0, 1.5, and 2.5 km/s (Z1.0 in Abrahamson and Silva (2008) and Chiou and Youngs (2008), Z1.5 in Choi et al. (2005) and Day et al. (2008), and Z2.5 in Campbell and Bozorgnia (2008), respectively) are sometimes used to account for influences of larger scale crustal velocity structure on ground motions. The implications of the methodology chosen to represent larger scale crustal velocity structure on ground motions is discussed in more detail below.

The standard deviation or aleatory variability, often denoted as sigma (σ_T), exerts a very strong influence on the results of probabilistic seismic hazard analysis (PSHA) (Bommer and Abrahamson 2006). For this reason it is important to note that the total aleatory uncertainties, as well as the intra- and inter-event uncertainties are systematically larger for the new NGA equations relative to previous relationships (Boore et al., 1997; Sadigh et al., 1997; Campbell, 1997). Three of the NGA models incorporate a magnitude dependence in the standard deviation. For magnitudes near 7, the five NGA models have similar standard deviations. However, for **M** < 5.5, there is a large difference in the standard deviations, with the three magnitude-dependent models exhibiting much larger standard deviations (σ_T > 0.7) than the magnitude-independent models (σ_T ~0.54). The three models that include a magnitude-dependent standard deviation (AS08, CY08 and I08) all included aftershocks, whereas the two models that used a magnitude-independent standard deviation (BA08 and CB08) excluded them. Including aftershocks greatly increases the number of small-magnitude earthquakes. However, there is a resulting trade-off of significantly larger variability in predicted ground motions than if only large magnitude mainshocks are used. Significant differences in the standard deviations are also noted for soil sites at short distances, this is most likely due to inclusion or exclusion of nonlinear site effects on the standard deviation.

In general, the NGA models predict similar median values (within a factor of ~1.5) for vertical strike-slip earthquakes with 5.5 < **M** < 7.5. The largest differences are for small magnitudes (**M** < 5.5), for very large magnitudes (**M** = 8), and for sites located over the hanging wall of dipping faults (Abrahamson et al., 2008). As more data has become available to the GMPE developers the number of coefficients in the relationships has increased significantly (>20 in some cases). However, the aleatory variability values (σ_T) have not decreased through time (J. Bommer, pers. comm.).

Since empirical GMPEs, including NGA GMPEs, are by necessity somewhat generic compared to the wide range of seismic source, crustal velocity structure, and site conditions encountered in engineering applications, there are cases when application of empirical GMPEs is difficult and most importantly, more uncertain. In the context of PSHA, these additional epistemic (knowledge) uncertainties, when quantified, are naturally incorporated into the probabilistic estimation of ground motion parameters. We present two situations of engineering interest, where the application of empirical GMPEs is challenging, to illustrate the difficulties and suggest a path forward in the ongoing process to update and improve empirical GMPEs.

5.2 Application of NGA GMPEs for near-fault Vs30 > 900 m/s sites

Independent analysis of the performance of the NGA GMPEs with post-NGA earthquake ground motion recordings demonstrate that use of measured site Vs30 characteristics leads to greatly improved ground motion predictions, with lower performance for sites where Vs30 is inferred instead of directly measured (Kaklamanos and Baise, 2011). Thus, the use of Vs30 represents a significant improvement over previous generations of GMPEs that use a simple qualitative site classification scheme. Kaklamanos and Baise (2011) suggest that development of of better site characteristics than Vs30 may also improve the prediction accuracy of GMPEs. In this section we illustrate the challenges presented in the use and application of Vs30 in the NGA GMPE regressions and application of the NGA GMPEs for "rock" sites.

It is becoming more common to need ground motion estimates for "rock" site conditions to specify inputs for engineering analyses that include both structures and shallow lower-velocity materials within the analysis model. In this section we consider the challenges in estimating ground motions for site conditions of Vs30 > 900 m/s close to strike slip faults. The problem is challenging for empirical GMPEs because most of the available recordings of near-fault strike-slip ground motions are from sites with Vs30 on the order of 300 m/s. The NGA GMPEs that implement Vs30 used empirical and/or synthetic amplification functions that involve modifying the observed ground motions prior to regression. In this section we discuss some of the challenges of this approach as it applies to estimating ground motions at rock (Vs30 > 900 m/s) sites that are typical of foundation conditions for many large and/or deeply embedded structures.

Earthquake (name)	**Date** (day,mon,yr)	**M**	**Station**	**Vs30** (m/s)	**JB Fault Distance** (km)
Parkfield	28 Jun. 1966	6.1	Cholame 2WA	185	3.5
Imperial Valley	15 Oct. 1979	6.5	El Centro Array #7	212	3.1
Superstition Hills	24 Nov. 1987	6.6	Parachute	349	1.0
Erzincan	13 Mar. 1992	6.9	95 Erzincan	275	2.0
Landers	28 Jun. 1993	7.3	Lucerne	665	1.1
Kobe, Japan	16 Jan. 1995	6.9	KJMA	312	0.6
Kocaeli, Turkey	17 Aug. 1999	7.4	Yarimca	297	2.6
Kocaeli, Turkey	17 Aug. 1999	7.4	Sakarya	297	3.1
Duzce, Turkey	12 Nov. 1999	7.1	Duzce	276	8.1
Geometric Mean		6.9		299	2.1

Table 5. NGA Near-Fault Strike-Slip Ground Motions

The four NGA GMPEs that implement Vs30 using deterministic ("constrained") amplification coefficients to remap the observed near-fault strike-slip strong motion data that have an average Vs30=299 m/s PSA (Table 5) prior to regression. In contrast, Boore et al. (1997) applied non-linear multi-stage regression using the observed data directly; the observed ground motion values were directly employed in their regression with no remapping of values due to site characteristics). Boore and Atkinson (2008) used the Choi and Stewart (2005) linear amplification coefficients to remap observed response spectra to a reference Vs30=760 m/s. Campbell and Bozorgnia (2008) used 1D nonlinear soil amplification simulation results of Walling et al. (2008) to deterministically fix nonlinear amplification and remap all response spectra with Vs30 < 400-1086 m/s, depending on period, to create the response spectral "data" input into the non-linear multi-stage regression. Abrahamson and Silva (2008) use an approach similar to Campbell and Bozorgnia (2008). Chiou and Youngs (2008) do not explicitly specify how the coefficients for linear and nonlinear amplification were constrained or obtained. Thus, Boore and Atkinson (2008) remap observed response spectra prior to regression using the linear coefficients from Choi and Stewart (2005), Campbell and Bozorgnia (2008) and Abrahamson and Silva (2008) remap observed response spectra prior to regression using the nonlinear coefficients from Walling et al. (2008), and it is not clear what Chiou and Youngs (2008) did. We use Boore and Atkinson (2008) and Campbell and Bozorgnia (2008) to illustrate how the observed response spectral data for sites with Vs30=300 m/s are changed to create the actual "data" used in the regression to estimate Vs30=915 m/s ground motions. It is instructive to compare the approaches and resulting near-fault ground motion predictions.

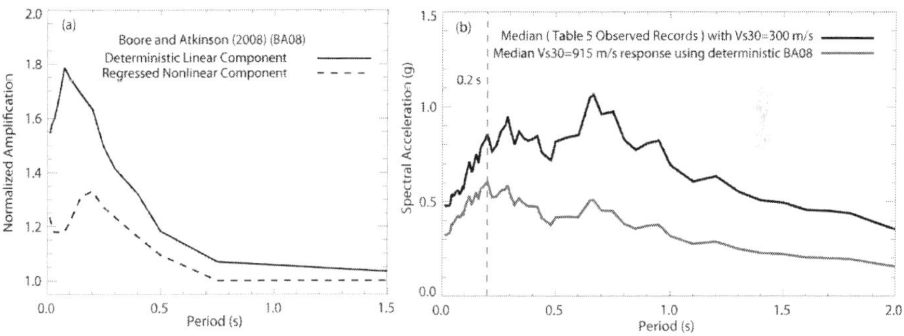

Fig. 5.1. Boore and Atkinson (2008) amplification functions (a) and original Vs30=300 m/s (black) and "Vs30=915 m/s remapped observed" responses spectra (red) (b) for M=7.0, distance of 2 km, and PGA=0.45 g.

For Boore and Atkinson (2008) amplification normalized by the GMPEs longest period amplification (10 s) is used in Fig. 5.1, to clearly illustrate the scale of the *a priori* deterministic linear amplification as a function of period. The *a priori* deterministic linear-amplification normalization (Fig. 5.1a) takes the original median near-fault response spectra that have a peak amplitude at about 0.65 s (Figure 5.1) and create response spectra with peak amplitude at 0.2 s that is used as the "observed data" (red curve in Figure 5.1b) in the nonlinear multi-stage GMPE regression. For Campbell and Bozorgnia (2008), the nonlinear Vs30 amplification coefficients are fixed and create the deterministic nonlinear amplification function (Figure 5.2a) that is always applied to Vs30 < 400 m/s PSA at all periods to create the "data" (red curve in Figure 5.2b) used in the non-linear multi-stage GMPE regression. In the case of nonlinear deterministic amplification it is necessary to specify a reference PGA. We use 0.45 g for the reference PGA for illustration since this is close to the median ground motion case for sites about 2 km from strike-slip faults and M > 6.; use of a higher reference PGA would increase the nonlinear amplification in Fig. 5.2a.

The use of a single deterministic amplification function for Vs30, whether linear or nonlinear, assumes that there is a one-to-one deterministic mapping of period-dependent amplification to Vs30, which Idriss (2008) suggests is not likely; a single Vs30 can be associated with a wide variety of amplification functions. Further, in the case of nonlinear amplification (Campbell and Bozorgnia, 2008 and Abrahamson and Silva, 2008), a single deterministic nonlinear amplification function used to account for modulus reduction and damping that vary widely as a function of soil materials, as discussed in Section 4 and Bonilla et al. (2005).

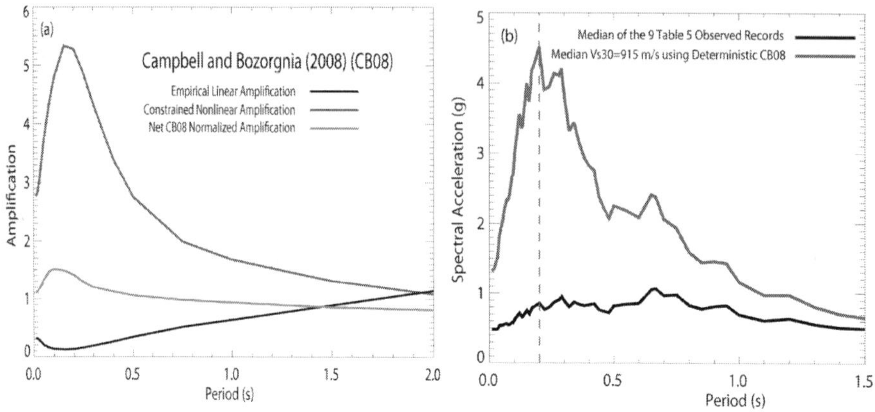

Fig. 5.2. Campbell and Bozorgnia (2008) amplification functions (a) and original Vs30=300 m/s (black) and "Vs30=915 m/s remapped observed" responses spectra (red) (b) for M=7.0, distance of 2 km, and PGA=0.45 g.

Earthquake (year-name)	**Focal Mechanism**	**M**	**Station**	**Vs30** (m/s)	**JB Fault Distance** (km)	**Period of Maximum GMRotI50** (s)
1989-Loma Prieta	reverse-oblique	7.0	Lexington Dam	1073	3.2	1.0
1989-Loma Prieta	reverse-oblique	7.0	Gilroy #1	1428	8.8	0.36
1994-Northridge	reverse	6.7	Pacoima	2016	4.9	0.38

Table 6. NGA Near-Fault Vs30 > 900 m/s Ground Motions

Figs. 5.1 and 5.2 show that the *a priori* deterministic amplification functions used in the NGA Vs30 GMPE regression procedures remap most of the observed data in the NGA database, which have Vs30 < 500 m/s (Chiou et al., 2008) to have a peak acceleration response at about 0.2 s, prior to regression. Thus, in hindsight it may not be a surprise that the NGA response spectra maintain a strong bias to peak at 0.2 s period that in large part is the result of the deterministic amplification modifications to the observed data prior to non-linear multi-stage regression. What is remarkable is that all four NGA GMPEs that implement Vs30 and Idriss (2008) predict that spectral accelerations normalized by peak ground acceleration always peak at about 0.2 s, virtually independent of magnitude for **M** > 6; the overall shape of Boore and Atkinson (2008) response spectra normalized by peak ground acceleration in Fig. 5.3a are representative of all NGA GMPE response spectral shapes in terms of overall spectra shape and the 0.2 s period of maximum response. Boore et al. (1997) obtained a quite different result, with the period of peak spectral amplitude shifting to longer periods as magnitude increases above **M** 6.6 (Fig. 5.3b). The few near-fault data from sites with Vs30 > 900 m/s (Table 6 and Fig. 5.3c) are more consistent with the Boore et al. (1997) results, with the median period of maximum spectral acceleration of 0.38 s being nearly twice that predicted by the four NGA GMPEs (Fig. 5.3c). Mostly, important for a wide class of larger buildings and critical structures, the broader longer-period spectral shift of Boore et al. (1997) suggests much larger (> 50% larger spectral responses at periods > 0.4 s (Fig. 5.3c) Based on current seismological knowledge and theory, what should we expect?

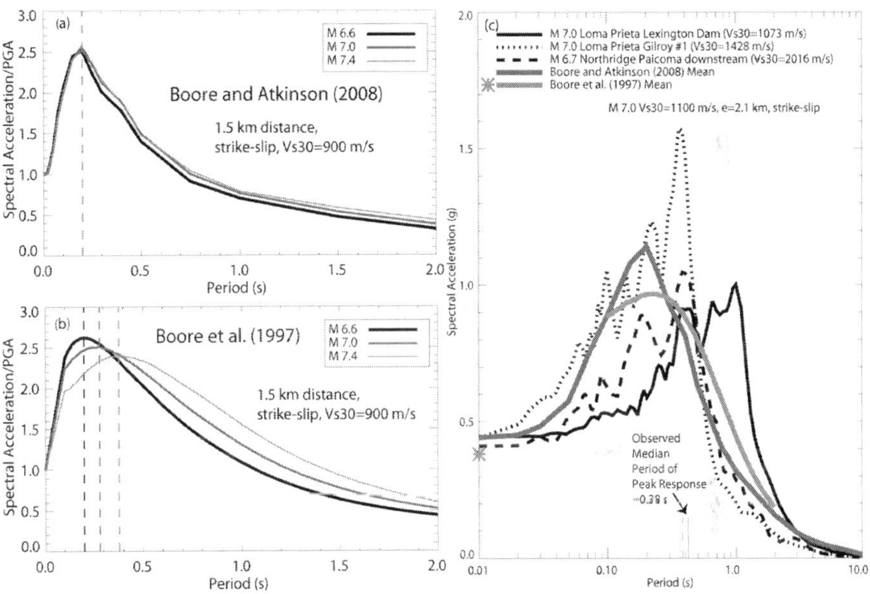

Fig. 5.3. Boore and Atkinson (2008) (a) and Boore et al. (1997) (b) response spectra normalized by peak ground acceleration for Vs30=900 m/s. The geometric mean spectral accelerations from the three observed Vs30 > 900 m/s ground motions in Table 6 is compared to the mean Boore and Atkinson (2008) and Boore et al. (1997) estimates in (c).

Ground motion acceleration at high frequency scales in proportion to dynamic stress drop (Boore, 1983). Average slip is proportional to the product of average dynamic stress drop and average rise time. Dynamic stress drop averaged over the entire fault plane is generally found to remain relatively constant with magnitude (Aki, 1983; Shaw, 2009). Thus, as average slip increases with magnitude (Somerville et al., 1999; Mai and Beroza, 2000; Mai et al., 2006) average rise time must also increase with increasing magnitude. Somerville (2003) notes that the period of the dominant amplitude near-fault motions is related to source parameters such as the rise time and the fault dimensions, which generally increase with magnitude. Mai et al. (2006) present an analysis of scaling of stress drop with seismic moment and find a strong increase of maximum stress drop on the fault plane as a function of increasing moment. In contrast, average stress drop over the entire fault plane at most only slightly increases with increasing moment; the substantial scatter of average stress drop values in Figure 1 of Mai et al. (2006) is consistent with average stress drop that is constant with moment. The Mai et al. (2006) results for maximum stress drop are consistent with first-order constraints on stochastic aspects of seismic source properties (Andrews, 1981; Boore, 1983; Frankel, 1991). As fault area increases, the probability of observing a larger stress drop somewhere on the fault plane increases since stress drop must exhibit correlated-random variability over the fault to explain the first-order observations of seismic source properties inferred from ground motion recordings, such as the ω^2 spectral shape (Andrews, 1981; Frankel, 1991). However, for the moment range (6.5 < **M** < 7.5) that dominates the hazard at many sites, the stress drop averaged over the entire fault plane is generally found to remain relatively constant with magnitude (Aki, 1983; Shaw, 2009), thus requiring average rise time to increase with increasing magnitude. These fundamental seismological constraints derived from analyses of many earthquakes require that the period that experiences peak response spectral amplitudes should increase with magnitude for some threshold magnitude. The results of the Boore et al. (1997) GMPEs suggest the threshold magnitude is about **M** 6.6 (Fig. 5.3b). That all five NGA GMPEs predict invariance of the period of peak spectral response amplitude for **M** > 6.6 to M 8.0 (example from **M** 6.6 to **M** 7.4 are shown in Fig. 5.3a) implies that stress drop increases strongly with increasing magnitude, which is inconsistent with current knowledge of seismic source properties. In contrast, the Boore et al. (1997) response-spectral-magnitude-period-dependent results are more consistent with available seismological constraints. It is important to understand why.

Boore et al. (1997) implements Vs30 site factors in a quite different manner than the four NGA GMPEs that use Vs30. Boore et al. (1997) applied non-linear multi-stage regression using the observed data directly, with no deterministic remapping of data by Vs30 prior to regression. Except for their deterministic treatment of Vs30, Boore and Atkinson (2008) use a similar regression approach for Boore et al. (1997). Since Boore and Atkinson (2008) regress period-by-period, the linear site-response remapping (Fig. 5.1a) effectively swamps any signal associated with a period shift with increasing magnitude observed by Boore et al. (1997); a non-linear regression will operate on the largest signals. The deterministic linear amplification function in Boore and Atkinson (2008) becomes a very large signal (Figure 5.1a) when operating on data from Vs30=300 m/s sites. The other NGA GMPE regressions normalize response spectra by peak ground acceleration prior to regression, which Boore et al. (1997) suggest tends to reduce resolution of the period-amplitude response-spectra variations in multi-stage regression. Figs 5.1 and 5.2 illustrate why the NGA GMPEs predict PSA shapes that barely change with magnitude (Fig. 5.3a) and why the NGA GMPEs do not match the first-order characteristics of M > 6.6 near-fault PSA (Fig. 5.3c). It simply might be true that once nonlinear amplification occurs it is impossible to resolve differences between period shifts associated with source processes and site responses. Yet, implicitly the NGA GMPE non-linear regressions assume resolution of all possible response-spectral shape changes as a function of magnitude using deterministic site response amplification functions, an assumption Idriss (2008) does not find credible. In contrast, Boore et al. (1997) used the actual unmodified response spectral data in their multi-stage regression and obtained results compatible with existing seismological constraints. Unfortunately, this leaves us in a bit of a conundrum based on GMPE grading criteria suggested by Bommer et al. (2010) and Kaklamanos and Baise (2011), which clearly establish that NGA is a significant improvement for a wide range of applications than previous generation GMPEs, including Boore et al. (1997). A primary contributor to this conundrum about appropriate spectral behaviour for near-fault Vs30 > 900 m/s sites is the lack of near-fault ground motion data for Vs30 > 900 m/s (Table 6 and Figure 5.3c), providing a vivid real-world example of epistemic uncertainty.

The site amplification approach used in NGA is discussed by Boore and Atkinson (2008), "The rationale for pre-specifying the site amplifications is that the NGA database may be insufficient to determine simultaneously all coefficients for the nonlinear soil equations and the magnitude-distance scaling, due to trade-offs that occur between parameters, particularly when soil nonlinearity is introduced. It was therefore deemed preferable to "hard-wire" the soil response based on the best-available empirical analysis in the literature, and allow the regression to determine the remaining magnitude and distance scaling factors. It is recognized that there are implicit trade-offs involved, and that a change in the prescribed soil response equations would lead to a change in the derived magnitude and distance scaling. Note, however, that our prescribed soil response terms are similar to those adopted by other NGA developers who used different approaches; thus there appears to be consensus as to the appropriate level for the soil response factors." This consensus is both a strength and weakness of the NGA results. The weakness is that if there is a flaw in the deterministic site response approach, then all the NGA GMPEs that use Vs30 are adversely impacted. Ultimately, three data points (Table 6 and Fig. 5.3c) are insufficient for the data to significantly speak for themselves in this particular case. Consequently, one can argue for one interpretation (invariant spectral shape) or the other (spectral peaks shift to longer periods at M > 6.6), and while a Bayesian evidence analysis shows that limited available data support a spectral shift with increasing magnitude, without data from more earthquakes, an honest result is that large epistemic uncertainty remains a real issue for Vs30 > 900 m/s near-fault sites.

Epistemic uncertainties can be rigorously accounted for in probabilistic ground motion analyses. However, it is necessary to develop a quantitative description of the epistemic uncertainties to accomplish this. Uncertainty in spectral shape as a function of magnitude, particular the period band of maximum acceleration response are important issues because many structures have fundamental modes of vibration at periods significantly longer than 0.2 s, the period the NGA GMPEs suggest that maximum acceleration responses will occur for M > 6.6 earthquakes at Vs30 > 900 m/s near-fault sites.

We can reduce these site uncertainties and improve ground motion prediction, with the ground motion data that currently exist by collecting more quantitative information about site characteristics that more directly and robustly determine site amplification, like Vs-depth profiles. Kaklamanos and Baise (2011) showed through empirical statistical analyses that actual Vs30 measurements produced better performance than occurred at sites where Vs30 is postulated based on geology or other proxy data.

Boore and Joyner (1997) suggested that quarter-wavelength approximation of Joyner et al. (1981) would likely be a better predictor of site responses than Vs30. For a particular frequency, the quarter-wavelength approximation for amplification is given by the square root of the ratio between the seismic impedance (velocity times density) averaged over a depth corresponding to a quarter wavelength and the seismic impedance at the depth of the source. The analyses of this section suggest that the combination of Vs30 and its deterministic implementation in NGA is not the best approach. Thompson et al. (2011) show that the quarter-wavelength approximation more accurately estimates amplification than amplification estimated using Vs30. Given the rapid growth in low-cost, verified passive measurement methods to quickly estimate robust Vs-depth to 50-100 m or more (Stephenson et al., 2005; Boore and Asten, 2008; Miller et al., 2010; O'Connell and Turner, 2011), it would greatly improve the prospects for substantial improvements in future GMPEs to acquire Vs-depth data for as much of the empirical ground motion database as possible to improve resolution of site amplification.

These results illustrate how difficult it is to formulate a GMPE formulation and regression strategy a-priori, for a "single" parameter like Vs30. This analysis does not show that the NGA GMPEs are incorrect. Instead, it demonstrates some of the trade-offs, dependencies, and uncertainties that occur in the NGA GMPEs between Vs30 and spectral shape. This near-fault high Vs30 example illustrates that it is important to conduct independent analyses to determine which GMPEs are best suited for a particular application and to use multiple GMPEs, preferably with some measure of independence in their development to account for realistic epistemic GMPE uncertainties.

5.3 Near-fault application of NGA GMPEs and site-specific 3D ground motion simulations: Source and site within the basin

In tectonically active regions near plate boundaries active faults are often located within or along the margins of sedimentary basins. Basins are defined by spatially persistent strong lateral and vertical velocity contrasts that trap seismic waves within the basin. Trapped seismic waves interact to amplify ground shaking and sometimes substantially increase the duration of strong shaking. Basin amplification effect is the result the combination of lateral and vertical variations in velocity that make the basin problem truly three-dimensional in nature and difficult to quantify empirically with currently available strong motion data. The basin problem is particularly challenging for estimating amplifications for periods longer than 1 s and sedimentary basin thicknesses exceeding about 3 km (Campbell and Bozorgnia 2008). Unfortunately, some of the largest urban populations in the world are located within basins containing active faults, including many parts of Japan (Kawase, 1996; Pitarka et al., 1998, NIED, 2011), the Los Angeles and other basins in southern California (Day et al., 2008), and Seattle, Washington, (Frankel et al., 2009). Consequently, estimation of long-period ground motions in sedimentary basins associated with near-fault faulting is an important practical need.

Choi et al. (2005) used empirical and synthetic analyses to consider the effects of two types of basin situations. They denoted sites located in a basin overlying the source as having coincident source and site basin locations (CBL) and differentiated them from distinct source and site basin locations (DBL). They used pre-NGA GMPEs for "stiff-soil/rock", but modified to account for Vs30 using Choi and Stewart (2005) to regress for additional basin amplification factors as a function a scalar measure of basin depth, Z1.5, the depth to a shear-wave velocity of 1.5 km/s. Using ground motion data from southern and northern California basins, Choi et al. (2005) found strong empirical evidence that ground-motion amplification in coincident source and site basin locations (CBL) is significantly depth-dependent at medium to long periods ($T > 0.3$ s). In contrast, They found that when the seismic source lies outside the basin margin (DBL), there is a much lower to negligible empirical evidence for depth-dependent basin amplification.

In support of NGA GMPE development Day et al. (2008) proposed a model for the effect of sedimentary basin depth on long-period response spectra. The model was based on the analysis of 3D numerical simulations (finite element and finite difference) of long-period 2–10 s ground motions for a suite of sixty scenario earthquakes (**M** 6.3 to **M** 7.1) within the Los Angeles basin region. Day et al. (2005) used a deterministic 3D velocity model for southern California (Magistrale et al., 2000) to calculate the wave responses on a grid and determine the amplification of basin sites as a function of Z1.5 in the 3D model. Being a purely synthetic model, but primarily concerned with ratios (amplification), it is relatively unimportant to consider the correlated-random effects on wave amplitude (Table 3) and phase (Table 4) to calculate to first-order amplification effects for shallow (< 2 km) and/or relatively fast basins. In shallow and/or fast basins the additional stochastic basin path length difference between the shallow basin and bedrock paths is less than a couple wavelengths at periods > 1 s, so the effects of differential correlated-random path lengths on S-wave amplification are negligible (O'Connell, 1999a). For typical southern California lower-velocity basins deeper than 3 km both the 3D viscoelastic finite-difference simulations of O'Connell (1999a) and phase-screen calculations of Hartzell et al. (2005) show correlated-random velocity variations will significantly reduce estimated basin amplification relative to deterministic 3D models. The primary purpose of O'Connell's (1999a) investigations was to determine the likely amplification of higher-velocity rock sites where few empirical data exist (see Section 5.2), relative to the abundant ground motion recordings obtained from stiff soil sites. O'Connell (1999a) showed that basin amplification in > 3 km deep basins is reduced relative to rock as the standard deviation of correlated-random velocity variations increases because the mean-free-path scattering in the basins significantly increases relative to rock at periods of 1-4 s. Consequently, because Day et al. (2008) use a deterministic 3D velocity model, we expect that their estimated basin amplifications will generally correspond to upper bounds of possible mean amplifications for southern California basins deeper than 3 km, but provide accurate first-order estimates of basin amplification for shallower (< 2 km) basins.

Several NGA GMPEs worked to empirically evaluate and incorporate "basin effects" in some way, but it is important to note that none of the empirical NGA GMPEs explicitly consider 3D basin effects by separately considering data in coincident source and site basin locations (CBL) from other data as Choi et al. (2005) showed is necessary to empirically estimate 3D basin effects for coincident source and site basin locations. NGA GMPEs lack sufficient parameterization to make this distinction, thus lumping all sites, CBL, DBL, and sites not located in basins into common Z1.0 or Z2.5 velocity-depth bins. All these sites, no matter what their actual location relative to basins and sources are apportioned some "basin-effect" through their Vs30 site velocity, Z1.0, and Z2.5 "basin-depth" terms (Day et al., 2008). It is important to understand that Z1.0 and Z2.5 are not empirically "basin-depth" terms, but "velocity-depth" terms. We use "velocity-depth" to refer to Z1.0 and Z2.5 instead of "basin-depth" because empirically, the NGA empirical GMPEs do not make the necessary distinctions in their GMPE formulations for these terms to actually apply to the problem of estimating 3D CBL amplification effects, the only basin case where a statistically significant empirical basin signal has been detected (Choi et al, 2005).

Campbell and Bozorgnia (2008) found empirical support for significant "velocity-depth" Z2.5 term after application of their Vs30 term, but only for sites where Z2.5 < 3 km, which roughly correspond to Z1.5 < 1.5. For Z2.5 > 3 km, Campbell and Bozorgnia (2008) used the parametric 3D synthetic basin-depth model from Day et al. (2008). Day et al. (2008) note that correlation between Vs30 and "basin" depth is sufficiently strong to complicate the identification of a basin effect in the residuals after having fit a regression model to Vs30. Chiou and Youngs (2008) found that to implement a velocity-depth term using Z1.5 would require removing the Vs30 site term from their GMPE because of the Z1.5-Vs30 correlation. Instead, Chiou and Youngs (2008) retained Vs30 at all periods and included a "velocity-depth" Z1.0 term to empirically capture the portion of velocity-depth amplification not fully accounted for by the correlation between Vs30 and Z1.0. Abrahamson and Silva (2008) used a similar Vs30 and Z1.0 parameterization approach for their GMPE. Since none of the NGA implementations of Z1.0 make the distinction whether a site is actually contained in a CBL or is not even in a basin, it is useful to evaluate the predictions of the four NGA GMPEs that implement Vs30, including the three NGA GMPEs that incorporate Z1.0 and Z2.5 velocity-depth terms, for four CBL sites along a portion of the North Anatolia Fault, where the fault is embedded below a series of connected 3D basins (Fig. 5.4).

The North Anatolia Fault is a major right-lateral strike slip fault forming the tectonic boundary between the Eurasian plate and Anatolian plate. Over much of its length in eastern and central Turkey, the NAF is a relatively simple, localized structure (Hubert-Ferrari et al. 2002). In the Marmara region, however, the North Anatolia Fault splays into several strands and forms a wide zone of deformation between the Anatolia block and Eurasian Plate. The northern strand traverses through Izmit Bay and forms the Northern Marmara Sea basin extending into the Aegean Sea through the Gulf of Saroz. The Sea of Marmara and Izmit Bay are primarily shaped by movement along the North Anatolia Fault (Fig. 5.4). Significant portions of the North Anatolia Fault, west of the **M** 7.5 1999 Izmit earthquake that involved primary rupture of the North Anatolia Fault are embedded within and below sedimentary basins in Izmit Bay and the eastern Marmara Sea (Fig. 5.4). We construct and use a 3D velocity model for the region indicated in Fig. 5.4 to simulate North Anatolia Fault ground motions for M 7.0 two-segment and M 7.1 three-segment rupture scenarios. Simulated ground motions are used to evaluate basin responses at five closely-spaced sites in Izmit Bay as a function of distance from the fault and basin depth (Fig. 5.4).

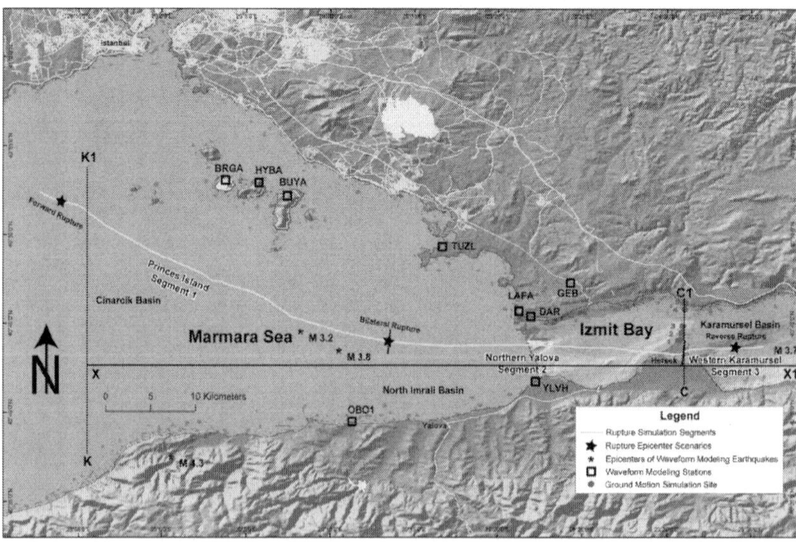

Fig. 5.4. North Anatolia Fault segments and sites for 3D ground motion modeling.

Three hypocenters positions are used to evaluate forward, bilateral, and reverse rupture directivity (Fig. 5.4). These simulated ground motions are compared to NGA GMPE response spectra predictions from the four NGA GMPEs with Vs30, including the three with velocity-depth terms (Abrahamson and Silva, 2008; Campbell and Bozorgnia, 2008; Chiou and Youngs, 2008), and Boore and Atkinson (2008) for periods of 1 s. The NGA results are modified to account for rupture directivity using Abrahamson (2001), Spudich and Chiou (2008), and Rowshandel (2010) to isolate residual 3D basin and directivity effects relative to the NGA-based empirical predictions.

A 3D velocity model encompassing the eastern Marmara Sea and Izmit Bay regions was constructed to span a region including the fault segments of interest, the ground motion estimation sites, and local earthquakes and recording stations (Fig. 5.5). Synthetic waveform modeling of local earthquake ground motions was used to iteratively improve and update the 3D model. The initial 3D velocity model was constructed using published 1-D velocity model data (Bécel et al., 2009; Bayrakci, 2009), tomographically assessed top of basement contours (Bayrakci, 2009), seismic reflection profiles (Carton et al., 2007; Kurt and Yucesoy, 2009), Boguer gravity profiles (Ates et al., 2003), geologic mapping (Okyar et al. 2008) and fault mapping (Armijo et al., 2002). Additional understanding of the basin-basement contact was gained by assessment of seismic reflection data collected by the SEISMARMARA cruise and made available at http://www.ipgp.fr/~singh/DATA-SEISMARMARA/.

The empirical wavespeed and density relations from Brocher (2005) were used to construct 3D shear-wave and density models based on the initial 3D acoustic-wave model. Shear-wave velocities were clipped so that they were not less than 600 m/s to ensure that simulated ground motions would be accurate for periods > 0.7 s for the 3D variable grid spacing used in the finite-difference calculations. This initial 3D velocity model was used to generate synthetic seismograms to compare with recordings of local **M** 3.2-4.3 earthquakes recorded on the margins of the Sea of Marmara, Izmit Bay, and inland locations north of Izmit Bay to assess the ground motion predictive performance of the initial 3D model. Several iterations of forward modeling were used to modify the 3D velocity model to obtain models that produce synthetic ground motions more consistent with locally-recorded local earthquake ground motions. The resulting shear-wave surface velocities mimic the pattern of acoustic-wave velocities that are consistent to first-order with the 3D acoustic-wave tomography results for the eastern Marmara Sea from Bayrakci (2009). Following O'Connell (1999a) and Hartzell et al. (2005) the final 3D model incorporates 5% standard deviation correlated-random velocity variations to produce more realistic peak ground motion amplitudes than a purely deterministic model. Since there are three distinct geologic volumes in the 3D model, three independent correlated-randomizations were used, one for the basin materials with a correlation length of 2.5 km, and one each for the basement north and south of the NAF that both used a correlation length of 5 km. Similar to Hartzell et al. (2010) we use a von Karman randomization with a Hurst coefficient close to zero and 5% standard deviation. Velocity variations are clipped to that shear-wave velocities are never smaller than 600 m/s to ensure a consistent dispersion limit for all calculations and randomized acoustic velocities were never larger than the maximum deterministic acoustic velocity to keep the same time step for all simulations.

Realistic ground motion simulations require accounting for first-order anelastic attenuation, even at long periods (Olsen et al., 2003). The fourth-order finite-difference code employs the efficient and accurate viscoelastic formulation of Liu and Archuleta (2006) that accurately

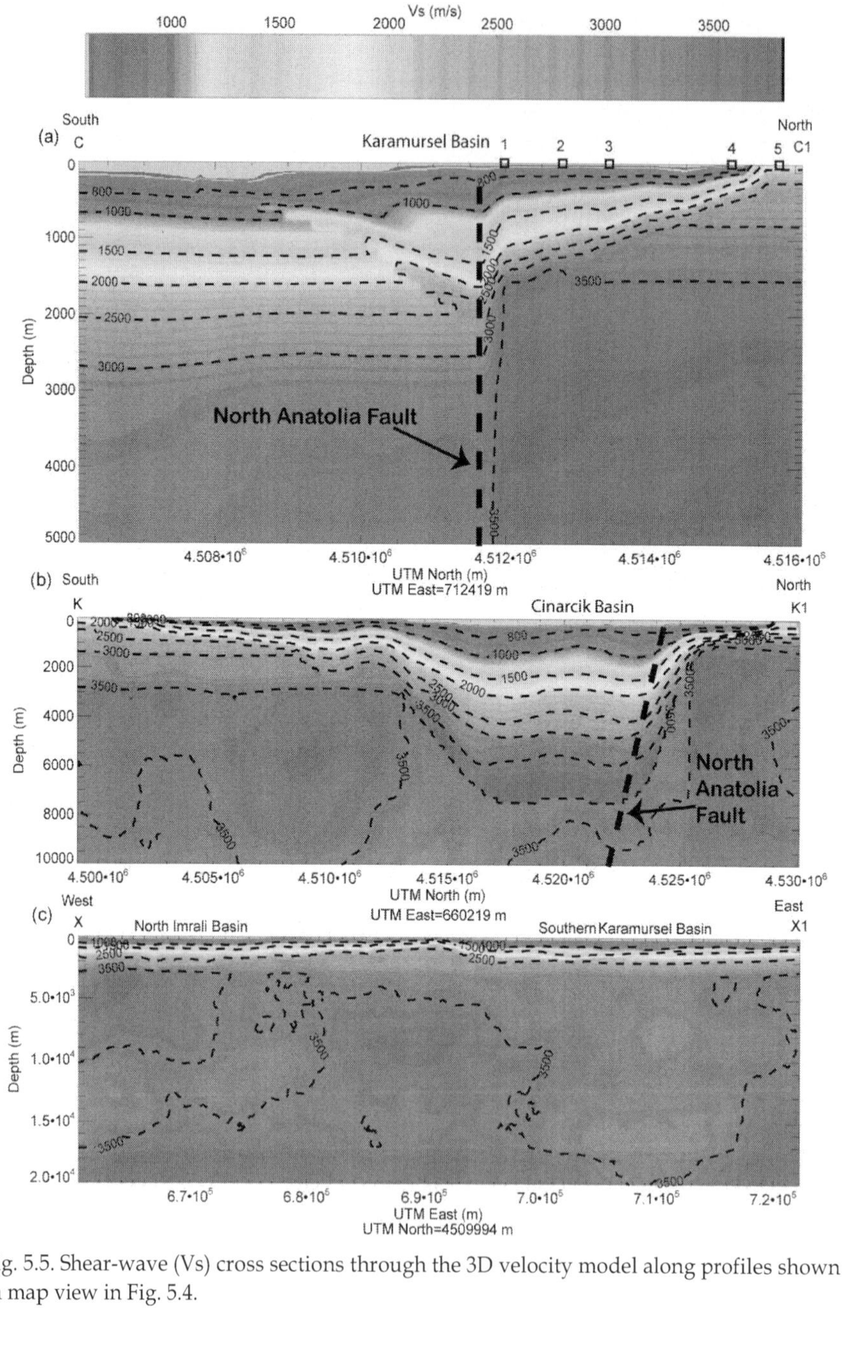

Fig. 5.5. Shear-wave (Vs) cross sections through the 3D velocity model along profiles shown in map view in Fig. 5.4.

accounts for Q. The Vs-Qs relationship of Olsten et al. (2003) developed for the Los Angeles basin in southern California was modified slightly to ensure that Qs was bounded at the lowest shear-wave velocity by Qs=20 and at the highest shear-wave velocity by Qs=200. The simple relationship Qp=1.5*Qs was used to specify acoustic wave attenuation. These Qs-Vs and Qs-Qp relationships produced synthetic responses consistent with observed ground shaking durations and overall amplitude-frequency responses observed in local recordings of moderate-sized local earthquakes.

The Karamürsel basin in Izmit Bay close to the ground motion sites is less than 2 km deep (Fig. 5.5a). The western Cinarcik basin is deep and broad (Fig. 5.5b). The basins in Izmit Bay are separated from the Cinarcik basin and other basins in the Marmara Sea by a relatively shallow bedrock sill (Fig. 5.5c) that is well constrained by seismic reflection and refraction data from Bayrakci (2009). Thus, in terms of longer period (> 2 seconds period) seismic waves, the basins in the Maramara Sea and Izmit Bay are separated by a relatively shallow bedrock and thin sedimentary basin deposits along the North Yalova segment (Figs. 5.4 and 5.5c).

A kinematic representation of finite fault rupture is used where fault slip (displacement), rupture time, and rise time are specified at each finite-difference grid node intersected by the fault. The 3D viscoelastic fourth-order finite-difference method of Liu and Archuleta (2002; 2006) was used to calculate ground motion responses from the kinematic finite fault rupture simulations. The kinematic rupture model mimics the spontaneous dynamic rupture behavior of a self-similar stress distribution model of Andrews and Boatwright (1998). The kinematic rupture model is also similar to the rupture model of Herrero and Benard (1994). Self-similar displacements are generated over the fault with rise times that are inversely proportional to effective stress. Peak rupture slip velocities evolve from ratios of 1:1 relative to the sliding (or healing peak) slip velocity at the hypocenter to a maximum ratio of 4:1. This form of slip velocity evolution is consistent with the dynamic rupture results of Andrews and Boatwright (1998) that show a subdued Kostrov-like growth of peak slip velocities as rupture grows over a fault. The kinematic model used here produces slip models with $1/k^2$ (k is wavenumber) distributions consistent with estimates of earthquake slip distributions (Somerville et al., 1999) and ω^2 (ω is angular frequency) displacement spectra in the far-field. Oglesby and Day (2002) and Schmedes et al. (2010) used numerical simulations of dynamic fault rupture to show that rupture velocity, rise time, and slip are correlated with fault strength and stress drop, as well as each other. The kinematic rupture model used here enforces correlations between these parameters by using a common fractal seed to specify relationships between all these fault rupture parameters. Oglesby and Day (2002), Guatteri et al. (2003), and Schmedes et al. (2010) used dynamic rupture simulations to demonstrate that rupture parameter correlation, as implemented in the stochastic kinematic rupture model outlined here, is necessary to produce realistic source parameters for ground motion estimation. The fault slip variability incorporates the natural log standard deviation of strike-slip displacement observed by Petersen et al. (2011) in their analyses of global measurements of strike-slip fault displacements. Consequently, although mean displacements are on the order of 1.5 m for the **M** 7.1 three-segment scenario earthquake, asperities within the overall rupture have displacements of up to 3-4 m. The Liu et al. (2006) slip velocity function is used with the specified fault slips and rise times to calculate slip-velocity time functions at each grid point.

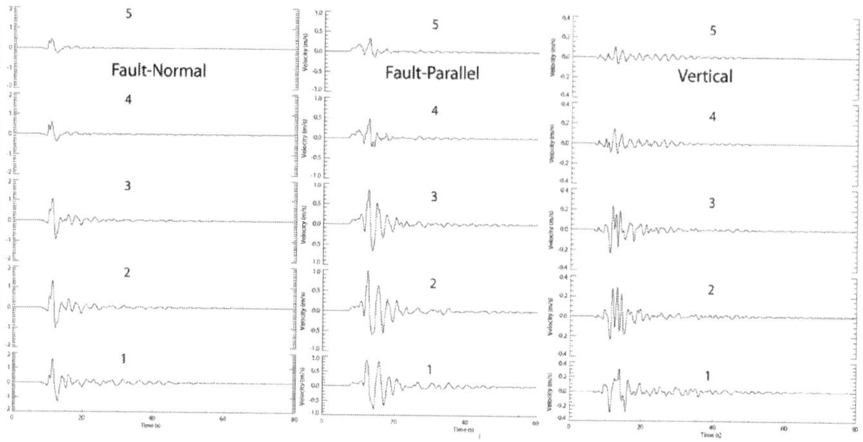

Fig. 5.6. Near-median bilateral three-segment rupture synthetic velocity seismograms for the five sites shown in Figs. 5.4 and 5.5a.

Three hypocenters were used to simulate forward, reverse, and bilateral ruptures relative to Izmit Bay sites (Fig. 5.4). To find an appropriate "median" randomization of the 3D velocity model, ten correlated-random 3D velocity models were created and then a single, three-segment randomized kinematic rupture model was used to simulate ten sets of ground motions. The randomized 3D model that most consistently produced nearly median motions across the five sites over the 1-10 s period band was used to calculate all the ground motion simulations for all two-segment and three-segment rupture scenarios. Ten kinematic randomizations were used for each case resulting in 60 rupture scenario ground motion simulations.

The simulated ground motions were post-processed to calculate acceleration response spectra for 5% damping. The geometric mean of Boore et al. (2006) (GMRotI50) was calculated from the two horizontal components to obtain GMRotI50 response spectra (SA). Response spectral results are interpreted for periods longer than 1 s, consistent with the fourth-order finite-difference accuracy for the variable grid spacing, minimum shear-wave velocity of 600 m/s, and broad period influence of oscillator response (Day et al., 2008). Fault-normal peak velocities decrease from sites 1-3 close to the fault and near the deeper portion of the basin (Fig. 5.5a) toward the shallow basin (site 4 in Figs. 5.4a and Fig. 5.6), and bedrock outside the basin (site 5 in Figs. 5.4a and Fig. 5.6).

The four NGA ground motion prediction equations (GMPE) that implement Vs30 were used to calculate ground motion estimates at all four sites using the Z1.0 and Z2.5 below each site in the 3D synthetic velocity model, Vs30=600 m/s, the directivity corrections of Abrahamson (2001), Spudich and Chiou (2008), and Rowshandel (2010) equally weighted, and the three rupture hypocenters (forward, bilateral, and reverse directivity in Fig. 5.3) equally weighted.

Site 4 was located away from basin-edge effects and in the shallow portion of the basin with the same Vs30=600 m/s, as sites 1-3 and consistent with a relatively linear site response making direct comparison of linear 3D simulated motions with empirical GMPE feasible.

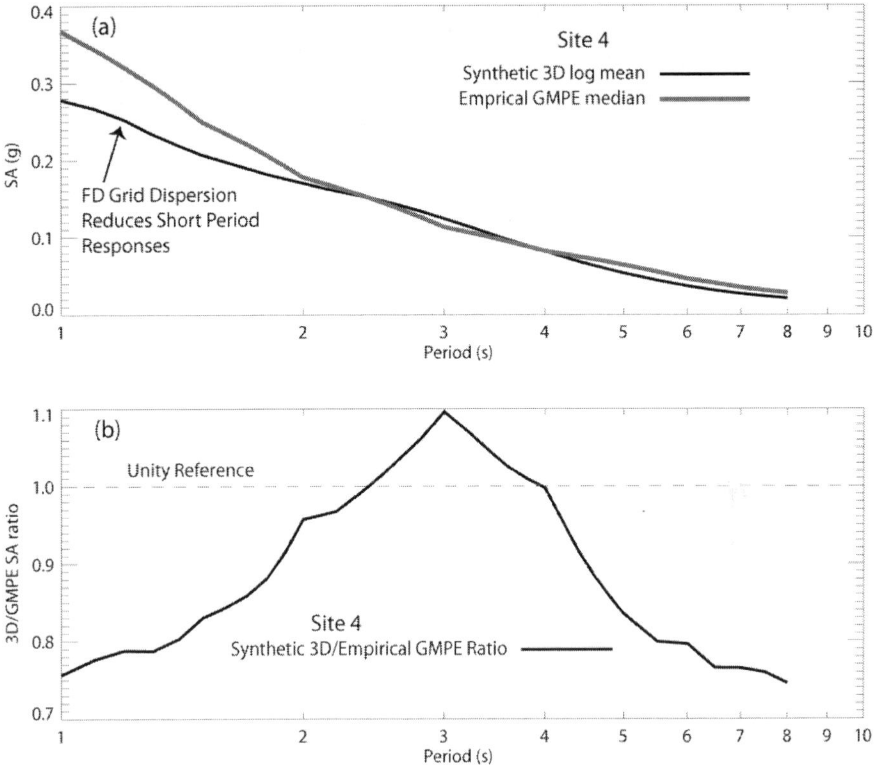

Fig. 5.7. Site 4 3D synthetic and NGA GMPE mean response spectra (a) and the 3D/NGA GMPE ratio (b).

Site 4 horizontal spectra where estimated as the log-mean average of the set of six earthquake rupture scenarios (two-segment and three-segment rupture, and forward, bilateral, and reverse directivity) used in the 3D ground motion simulations. To obtain robust estimates of mean synthetic spectra, we omitted the two largest and smallest amplitudes at each period to estimate log mean spectra for comparison. The Site 4 horizontal responses are comparable in amplitude to NGA predicted response spectra for periods > 1 s (Fig. 5.7a). The reduced synthetic responses between 1-2 s in Fig. 5.7 are an artifact of finite-difference grid dispersion similar to that noted by Day et al. (2008). The site 4 3D simulated responses are generally slightly less than the empirical GMPE median estimates over the 1-8 second period range, except for a small amplification at 3 seconds of < 10% (Fig. 5.7b). This confirmed that the 3D ground motion simulations and empirical NGA GMPE predict comparable spectral hazard at site 4 and establish site 4 as an appropriate reference point to compare to responses at sites 1-3 closer to the fault and within deeper portions of the basin.

We use the empirical NGA GMPEs to estimate the amplitude effects of differential source-site distances on amplitudes relative to site 4 and changes in Z1.0 and Z2.5 between sites 1-3 and reference site 4. The three empirical directivity relations are used with equal weight to remove the differential directivity effects for two separate sets of rupture cases designed to determine if 3D basin amplification is dependent on rupture directivity. We consider in the first case, the two rupture scenarios away from the sites, to determine 3D basin amplification in the absences of forward rupture directivity. In the second case, we average all six rupture scenarios, four of which have strong forward rupture directivity, to see if any of the sites shows significantly 3D basin amplification in the case of solely reverse rupture direction ground motions.

Empirical NGA distance, directivity, and Z1.0-Z2.5 sites 1-3 amplifications relative to reference site 4 are the lowest curves at the bottom of Fig. 5.8 and represent the sum total of the effects of all NGA GMPE terms related to differential distance, directivity, and Z1.0 and Z2.5 velocity-depth. Although sites 1-3 are much closer to the fault than site 4, the relative

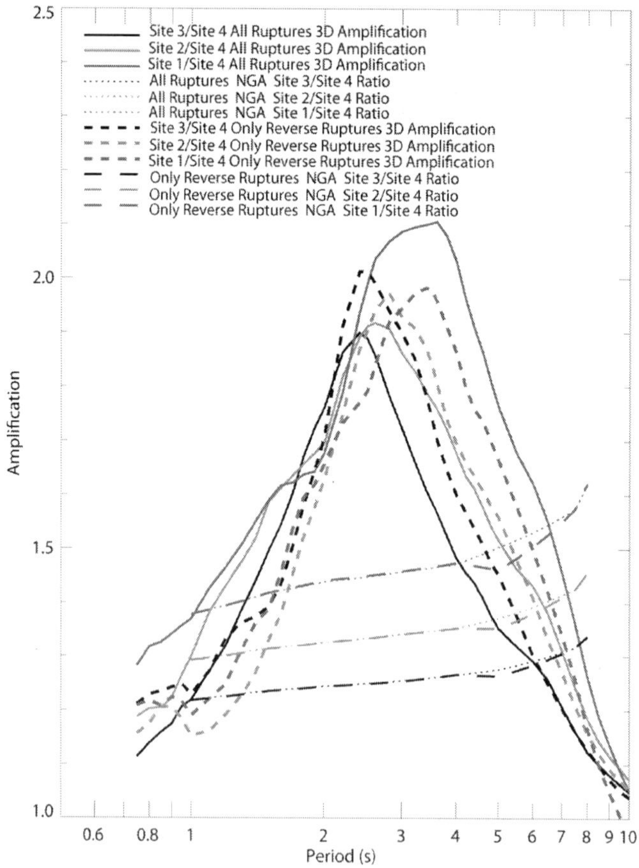

Fig. 5.8. Mean and reverse-rupture only residual 3D basin amplifications for sites 1-3 relative to reference site 4 with NGA differential site amplitude correction functions.

changes in amplitudes are much smaller than the proportional differences in site-source distances as a result of saturation, the condition enforced in NGA that ground motion amplitudes cease to increase as distance to the fault approaches zero. The directivity amplitude reduction from Rowshandel (2010) for reverse rupture accounts for the dip in longer-period NGA differential responses at periods of about 5 s in Fig. 5.8. The most striking aspect of the NGA transfer functions is that although three of the four GMPE include Z1.0 or Z2.5 "basin-depth" terms, there is no hint of an empirical resonant 3D basin response, just slight steady increases of "amplification" with increasing period. The non 3D-basin-like NGA differential amplification results are not surprising because the NGA basin-depth formulation pools ground motion observations from all scales of basins and non-basins in each Z1.0 and Z2.5 bin. Consequently, the NGA Vs30 and velocity-depth Z1.0 and Z2.5 basin terms do not capture any of the strongly period-dependent amplification associated with the site-specific basin of < 2 km total depth near the sites.

The residual site-specific synthetic 3D amplifications at sites 1-3 relative to reference site 4 are essentially independent of rupture direction (Fig. 5.8). Site 1 closest to the fault shows the largest amplification for case 2 with 2/3 forward rupture directivity, but the difference at site 1 between 2/3 forward rupture directivity basin amplification and reverse rupture basin amplification is < 10%. For sites 2 and 3 located slightly further from the fault, differences in case one and case two directivity 3D basin amplifications deviate < 4% from their mean peak amplifications. The remarkable result is that even in this case of a strike-slip fault embedded below the center of a basin and rupturing within basins continually along the entire rupture length, to first order 3D basin amplification is independent of rupture directivity/rupture direction. These 3D synthetic calculations shows that the three empirical directivity corrections applied with NGA GMPE effectively accounted for first-order directivity in this rather severe case of strike-slip fault rupture within a basin.

The Izmit Bay basins are quite similar in width, depth, and velocity characteristics to the San Fernando Basin, one of the basins included in the Day et al. (2008) 3D synthetic calculations to represent basin amplification in younger shallower basins. Thus, it is interesting to compare the Day et al. (2008) synthetic amplification predictions calculated across a spectrum of shallow and deeper basins using a deterministic 3D velocity model with these simulations using a site-specific weakly-randomized 3D velocity model. We calculate the 3D simulation and Day et al. (2008) response ratios of sites 1-3 to site 4 using the Z1.5 values from the 3D simulation model in the Day et al. (2008) Z1.5-amplification relationships (Fig. 5.9). Both 3D synthetic approaches predict comparable peak amplifications at comparable periods (Fig. 5.9), with the site-specific 3D model predicting a more rapid decrease with in increasing period that reflects the details of the site-specific 3D model; Day et al. (2008) have a wider period range of larger amplification because they pooled basin amplifications from a wider range of basin configurations than representative of the site-specific 3D velocity structure.

When soils are significantly less linear than clays with a plasticity index of 20, the fully nonlinear shear P-SV 2D investigations of O'Connell et al. (2010) suggest that combining the outputs of linear 3D simulations that omit the very-low-velocity basin with 1D nonlinear analyses to account for the very-low-velocity basins will produce comparable amplifications within the basin to full nonlinear 2D or 3D analyses. Linear 1D P-SV vertical analyses in the central portions of basins will typically provide appropriate vertical amplifications throughout most of the basin. Thus, it appears that it may be feasible in most of these cases to omit the shallow soft low-velocity regions from the top of basins from 3D linear or nonlinear analyses and use the outputs from linear 3D analyses with simplified 1D nonlinear SH and P-SV nonlinear amplification calculations to estimate realistic horizontal and vertical peak velocities and accelerations in the upper low-velocity soft soils.

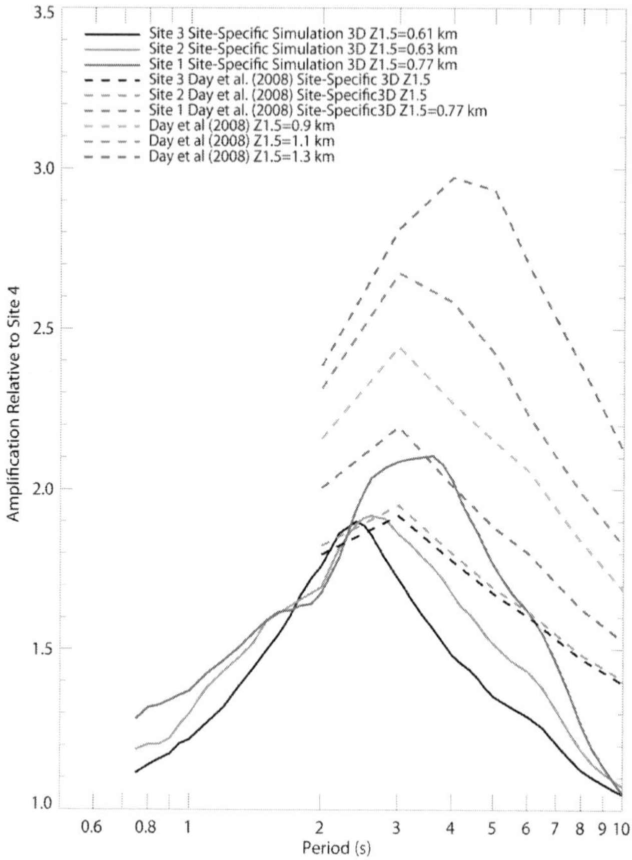

Fig. 5.9. Mean 3D site-specific simulation 3D amplification and Day et al. (2008) 3D amplifications for sites 1-3 relative to reference site 4.

These results illustrate that at present, the NGA GMPE do not effectively estimate site-specific 3D basin amplification for the most extreme case of a strike slip source and sites located within a closed basin. In such situations it is necessary to use site-specific 3D basin amplification calculations or compiled synthetic 3D generic basin amplification relations like Day et al. (2008) to estimate realistic site-specific 3D basin amplification effects. However, the NGA GMPEs and associated empirical directivity relations are shown to effectively account for geometric spreading and directivity in the demanding application of source and site located within a closed basin and provide a robust means to extract residual 3D basin amplification relative to NGA GMPE predictions. This approach requires a suitable

reference site in shallow portions of the basin that are not strongly influenced by basin effects or a site outside the basin.
In future GMPE development, the basin analyses of Choi et al. (2005), Day et al. (2008), and this analysis suggest separate consideration and analysis of data that is within closed basins with faults beneath or adjacent to the basin is warranted to evaluate empirical evidence for systematic basin responses. Such analyses need to be done separately for ground motion observations outside of this specific basin configuration to discern the relative effects of velocity-depth versus basin-depth on parameters like Vs30, Z1.0, and Z1.5. We suggest that it is more appropriate and prudent to refer to Z1.0 and Z1.5 as velocity-depth terms, not basin-terms, since they will fail to account for significant systematic period-dependent 3D basin amplification in the cases of sources and sites located within low-velocity basins.

6. Conclusion and recommendations

Geologic seismic source characterization is the fundamental first step in strong ground motion estimation. Many of the largest peak ground motion amplitudes observed over the past 30 years have occurred in regions where the source faults were either unknown or major source characteristics were not recognized prior to the occurrence of earthquakes on them. The continued development of geologic tools to discern and quantify fundamental characteristics of active faulting remains a key strong ground motion estimation research need.
As Jennings (1983) noted, by the early 1980s efforts to develop empirical ground motion prediction equations were hampered not only by the insufficient recordings of ground motions to constrain the relationships between magnitude, distance, and site conditions, but insufficient physical understanding of how to effectively formulate the problem. Strong ground motion estimation requires both strong motion observations and understanding of the physics of earthquake occurrence, earthquake rupture, seismic radiation, and linear and nonlinear wave propagation. In sections 2-4 we provided an overview of the physics of strong ground motions and forensic tools to understand their genesis. The physics are complex, requiring understanding of processes operating on scales of mm to thousands of km, most of the physical system is inaccessible, and the strong motion observations are sparse. As O'Connell (1999a) and Hartzell et al. (2005) showed, surface ground motion observations alone are insufficient to constrain linear and nonlinear amplification and seismic source properties. The observational requirements to understand the earthquake system and how ground motions are generated are immense, and require concurrent recording of ground motions at the surface and at depth. These observations have only recently been undertaken at a comprehensive large scale. In Japan, the National Research Institute for Earth Science and Disaster Prevention (NIED) operates K-NET (Kyoshin Network) with 660 strong motion stations. Each station records triaxial accelerations both at the surface and at sufficient depth in rock to understand the physics of earthquake fault rupture and to directly observe linear and nonlinear seismic wave propagation in the shallow crust. These borehole-surface data have provided fundamental new constraints on peak ground motions (Aoi et al., 2008), direct observation of nonlinear wave propagation, and new constraints on ground motion variability (Rodriguez-Marek et al., 2011). It will be necessary to expand the deployment of K-NET scale networks to other tectonically active regions like the western United States, to make real long-term progress understanding and significantly improving our ability to predict strong ground shaking. The synergy between

earthquake physics research and strong ground motion estimation is based on ground motion observations and geologic knowledge.

The need for new recordings of strong ground motions in new locations is clear, but there is immensely valuable information yet to be extracted from existing strong ground motion data. One of single biggest impediments to understanding strong ground motions is the lack of site velocity measurements for most of the current strong ground motion database (Chiou et al., 2008; Kaklamanos and Baise, 2011). The last 10 years has seen an explosion in the development and successful application of rapid, inexpensive, and non-invasive methods to measure site shear-wave velocities over depths of 50-1000 m that can provide site amplification estimates accurate to on the order of 10-20% (Stephenson et al., 2005; Boore and Asten, 2008). Using the large borehole-surface station network in Japan, Rodriguez-Marek et al. (2011) showed that the difference in the single-station standard deviation of surface and borehole data is consistently lower than the difference in ergodic standard deviations of surface and borehole data. This implies that the large difference in ergodic standard deviations can be attributed to a poor parameterization of site response.

Vs30 does not constrain frequency-dependent site amplification because literally, an infinite number of different site velocity-depth profiles can have the same Vs30. Even given geologic constraints on near-surface material variability, the scope of distinct velocity profiles and amplification characteristics that share a common Vs30 is vast. Ironically, the implementation of Vs30 in four of the NGA GMPE produced significant uncertainties in spectral shape as a function of magnitude as illustrated in Section 5.1. Vs30 also trades off with other velocity-depth factors (Section 5.2). We propose that one of the most valuable new strong ground motion datasets that can be obtained now is measurement of site shear-wave velocity profiles at the sites of existing strong ground motion recordings. These measurements would provide a sound quantitative basis to constrain frequency-dependent linear-site amplification prior to regression and reduce uncertainties in ground motion estimations, particularly spectral shape as a function of site conditions. As Rodriguez-Marek et al. (2011) note, reduction of exaggerated ground motion variability results in more realistic ground motion estimates across widely differing sites in probabilistic analyses.

The analyses of Choi et al. (2005) and section 5.2 suggest that accounting for positions of ground motion recordings and earthquake inside or outside of closed basins may provide a path forward to improve the ability of future empirical GMPE to accurately estimate responses within basins.

7. Acknowledgments

This paper is dedicated to the memory of William Joyner, who generously participated in discussions of directivity, wave propagation, site response, and nonlinear soil response, and who encouraged us to pursue many of the investigations presented here. David Boore kindly read the initial draft and provided suggestions that improved it. The authors benefited from helpful discussions with David Boore, Joe Andrews, Paul Spudich, Art Frankel, Dave Perkins, Chris Wood, Ralph Archuleta, David Oglesby, Steve Day, Bruce Bolt, Rob Graves, Roger Denlinger, Bill Foxall, Larry Hutchings, Ned Field, Hiro Kanamori, Dave Wald, and Walt Silva. Shawn Larson provided the e3d software, and Paul Spudich provided isochrone software. Supported by U.S. Bureau of Reclamation Dam Safety Research projects SPVGM and SEIGM and USGS award no. 08HQGR0068. The National Information Center for Earthquakes and Disasters, National Research Institute for Earth Science and Disaster Prevention in Japan provided the digital strong-motion data for station IWTH25.

8. References

Abrahamson, N. A. (2000), Effects of rupture directivity on probabilistic seismic hazard analysis, Earthquake Engineering Research Institute, Proceedings of the Sixth International Conference on Seismic Zonation (6ICSZ). Managing Earthquake Risk in the 21st Century. Palm Springs, CA. Nov. 12-15, 2000.

Abrahamson, N., Atkinson, G., Boore, D., Bozorgnia, Y., Campbell, K., Chiou, B., Idriss, I.M., Silva, W., and Youngs, R. ,2008. Comparison of the NGA Ground Motion Relations, *Earthq. Spectra,* 24, 45-66.

Abrahamson, N. A., and K. M. Shedlock, 1997. Overview, Seism. Res. Lett. 68, no. 1, 9-23.

Abrahamson, N. A. and W. J. Silva, 1997. Empirical response spectral attenuation relations for shallow crustal earthquakes, Seismol. Res. Lett. 68, 94-127.

Abrahamson, N.A. and Silva, W.J. (2008). Summary of the Abrahamson & Silva NGA Ground-Motion Relations, *Earthq. Spectra,* 24, 67-97.

Aguirre, J. and K. Irikura [1997]. "Nonlinearity, Liquefaction, and Velocity Variation of Soft Soil Layers in Port Island, Kobe, during the Hyogo-ken Nanbu Earthquake", Bull. Seism. Soc. Am., Vol. 87, pp. 1244-1258.

Aki, K., 1983. Strong-motion seismology, in "Earthquakes: Observation, Theory, and Interpretation", H. Kanamori and E. Boschi, eds., Proc. of the Int. School of Physics., Enrico Fermi, North-Holland Pub. Co., Amsterdam, pp. 223-250.

Alex, C. M., and K. B. Olsen, 1998. Lens-effect in Santa Monica?, Geophys. Res. Lett. 25, 3441-3444.

Anderson, J. G. and P. G. Richards, 1975. Comparison of strong ground motion from several dislocation models, Geophys. J. R. Astron. Soc. 42, 347-373.

Andrews, D. J., 1981. A stochastic fault model, 2, Time-dependent case, J. Geophys. Res. 86, 10,821-10,834.

Andrews, D. J. and J. Boatwright, 1998. Dynamic simulation of spontaneous rupture with heterogeneous stress drop, Seism. Res. Lett. 69, 143.

Archuleta, R. J., 1984. A faulting model for the 1979 Imperial Valley earthquake, J. Geophys. Res. 89, 4559-4585.

Archuleta, R. J. [1998]. "Direct observation of nonlinear soil response in acceleration time histories", Seism. Res. Lett., Vol. 69, p. 149.

Archuleta, R.J., L. F., Bonilla, and D. Lavallée, 2000. Nonlinearity in observed and computed accelerograms, in Proceeding of the 12 the World Conference on Earthquake Engineering, Aston Koedyk Ltd (pub). Paper reference number 0338.

Armijo, R., Meyer, B., Navarro, S., King, G., and Barka, A., 2002. Asymmetric slip partitioning in the Sea of Marmara pull-apart: a clue to propagation processes of the North Anatolian Fault?, Terra Nova, 14, 80-86.

Assimaki, D., and Steidl, J. H. (2007), Inverse analysis of weak and strong motion borehole array data from the Mw7.0 Sanriku-Minami earthquake, Soil Dyn. Earthquake Eng. 27:73–92.

Assimaki, D., W. Li, and A. Kalos (2010). A Wavelet-based Seismogram Inversion Algorithm for the In Situ Characterization of Nonlinear Soil Behavior, Pure Appl. Geophys., DOI 10.1007/s00024-010-0198-6.

Ates, A., Kayırana, T., Sincer, I., 2003, Structural interpretation of the Marmara region, NW Turkey, from aeromagnetic, seismic and gravity data. Tectonophysics, 367, 41-99.

Bayrakci, G, 2009. Approach to the complex 3D upper-crustal seismic structure of the Sea of Marmara by artificial source tomography on a 2D grid of OBS, PhD Thesis, Institut de Physique du Globe de Paris.

Bécel, A., Laigle, M., de Voogd, B., Hirn, A., Taymaz, T., Galve, A., Shimamura, H., Murai, Y., Lepine, J.C., Sapin, M., Ozalaybey, S., 2009. Moho, crustal architecture and deep deformation under the North Marmara Trough, from the SEISMARMARA Leg 1 offshore-onshore reflection-refraction survey. Tectonophysics, 467, 1-21.

Belardinelli, M. E., M. Cocco, O. Coutant, and F. Cotton, 1999. Redistribution of dynamic stress during coseismic ruptures: Evidence for fault interaction and earthquake triggering, J Geophys. Res. 104, 14,925-14945.

Bent, A. L., and D. V. Helmberger, 1989. Source complexity of the October 1, 1987, Whittier Narrows earthquake, J. Geophys. Res. 94, 9548-9556.

Beresnev, I.A. (2002). *Nonlinearity at California generic soil sites from modeling recent strong-motion data*, Bull. Seism. Soc. Am., 92, 863-870.

Beresnev, I. A., K.-L. Wen, and Y. T. Yeh (1995). Nonlinear soil amplification: Its corroboration in Taiwan, Bull. Seism. Soc. Am. 85, 496-515

Beresnev, I. A., and K. L. Wen (1996). "Nonlinear site response - a reality?", Bull. Seism. Soc. Am., Vol. 86, pp. 1964-1978.

Beresnev, I. A., and G. M. Atkinson, 1997. Modeling finite-fault radiation from the □n spectrum, Bull. Seism. Soc. Am. 87, 67-84.

Beresnev, I. A., A. M. Nightengale, and W. J. Silva (2002). Properties of vertical ground motions, Bull. Seism. Soc. Am. 92, 3152-3164.

Bernard, P., A. Herrero, and C. Berge, 1996. Modeling directivity of heterogeneous earthquake ruptures, Bull. Seism. Soc. Am. 86, 1149-1160.

Beroza, G. C., and Spudich, P., 1988. Linearized inversion for fault rupture behavior: Application to the 1984 Morgan Hill, California, earthquake, J. Geophys. Res. 93, 6275-6296.

Boatwright, J., 1982. A dynamic model for far-field acceleration, Bull. Seism. Soc. Am. 72, 1049-1068.

Bonilla, L. F., J. H. Steidl, G. T. Lindley, A. G. Tumarkin, and R. J. Archuleta, 1997. Site amplification in the San Fernando Valley, California: Variability of site-effect estimation using the S-wave, coda, and H/V methods, Bull. Seism. Soc. Am. 87, 710-730.

Bonilla, L. F., 2000. Computation of linear and nonlinear response for near field ground motion, Ph.D. Dissertation, University of California, Santa Barbara.

Bonilla, L. F., J. H. Steidl, G. T. Lindley, A. G. Tumarkin, and R. J. Archuleta (1997). Site amplification in the San Fernando Valley, CA: Variability of site effect estimation using the S-wave, coda, and H/V methods, Bull. Seism. Soc. Am., 87, 710-730.

Bonilla, L.F., R.J. Archuleta, and D. Lavallee [2005]. "Hysteretic and Dilatant Behavior of Cohesionless Soils and Their Effects on Nonlinear Site Response: Field Data Observations and Modeling", Bull. Seism. Soc. Am., Vol. 95, pp. 2373-2395.

Bonilla L.F., Liu P.C., Nielsen S. (2006). 1D and 2D linear and non linear site response in the Grenoble area, Proc. 3rd Int. Symp. on the Effects of Surface Geology on Seismic Motion (ESG2006), University of Grenoble, LCPC, France.

Bonilla, L.F., K. Tsuda, N. Pulido, J. Regnier, and A. Laurendeau [2011]. "Nonlinear site response evidence of K-NET and KiK-net records from the 2011 off the Pacific coast of Tohoku Earthquake", Earth Planets Space, Vol. 63, pp. 1-5.

Bommer, J. J., and N. A. Abrahamson, 2006. Why do modern probabilistic seismic-hazard analyses often lead to increased hazard estimates? *Bull. Seismol. Soc. Amer.*, 96, 1967–1977.

Bommer, J. J., Douglas, J., Scherbaum, F., Cotton, F., Bungum, Hi., Fah, D. (2010), On the selection of ground-motion prediction equations for seismic hazard analysis, Seism. Rese. Lett., 81, 783-793.

Boore, D. M., 1972. A note of the effect of simple topography on seismic SH waves, Bull. Seism. Soc. Am. 62, 275-284.

Boore, D. M., 1973. The effect of simple topography on seismic waves: Implications for the acceleration recorded at Pacoima Dam, San Fernando Valley, California, Bull. Seism. Soc. Am. 63, 1603-1609.

Boore, D. M., 1983. Stochastic simulation of high-frequency ground motions based on seismological models of the radiated spectra, Bull. Seism. Soc. Am. 73, 1865-1884.

Boore, D. M. and W. B. Joyner, 1989. The effect of directivity on the stress parameter determined from ground motion observations, Bull. Seism. Soc. Am. 79, 1984-1988.

Boore, D. M. and M. W. Asten (2008). Comparisons of shear-wave slowness in the Santa Clara Valley, California, using blind interpretations of data from invasive and non-invasive methods, Bull. Seism. Soc. Am. 98, 1983-2003.

Boore, D. M., W. B. Joyner, and T. E. Fumal (1994). Estimation of response spectra and peak accelerations from western North American earthquakes: An interim report, Part 2, U.S. Geological Survey Open-File Report 94-127.

Boore, D. M. and W. B. Joyner, 1997. Site amplifications for generic rock sites, Bull. Seism. Soc. Am. 87, 327-341.

Boore, D. M., W. B. Joyner, and T. E. Fumal, 1997. Equations for estimating horizontal response spectra and peak acceleration from western North America earthquakes: a summary of recent work, Seism. Res. Lett. 68, No. 1, 128-153.

Boore, D.M., Watson-Lamprey, J., and Abrahamson, N.A. (2006). GMRotD and GMRotI: Orientation-Independent Measures of Ground Motion, *Bull. Seismol. Soc. Am.* 96, 1502-1511.

Boore, D.M. and Atkinson, G..M. (2008). Ground-Motion Prediction Equations for the Average Horizontal Component of PGA, PGV, and 5%-Damped PSA at Spectral Periods Between 0.01s and 10.0 s, *Earthq. Spectra,* 24, 99-138.

Borcherdt, R. D., J. F. Gibbs, and T. E. Fumal, 1979. Progress on ground motion predictions for the San Francisco Bay region, California, in Progress on Seismic Zonation in the San Francisco Bay Region, E. E. Brabb (Editor), U. S. Geol. Surv. Circular 807, 13-25.

Bouchon, M. C. A. Schultz, and M. N. Toksoz, 1996. Effect of three-dimensional topography on seismic motion, J. Geophys. Res. 101, 5835-5846.

Bouchon, M., M. Campillo, and F. Cotton, 1998. Stress field associated with the rupture of the 1992 Landers, California, earthquake and its implications concerning the fault strenght at the onset of the earthquake, J. Geophys. Res. 103, 21091-21097.

Bouchon, M., M. P. Bouin, H. Karabulut, M. N. ToksÎz, M. Dietrich, A. Rosakis, 2001. How Fast is Rupture during an Earthquake? New Insights from the 1999 Turkey Earthquakes, Geophys. Res. Lett. 28 , 2723-2726.

Brune, J. N., 1970. Tectonic stress and the spectra of seismic shear waves from earthquakes, J. Geophys. Res. 75, 4997-5009.

Campbell, K. W. (1981). Near-source attenuation of peak horizontal acceleration. *Bull. Seism. Soc. Am.* 71, 2039–70.

Campbell, K.W., 1997. Empirical near-source attenuation relationships for horizontal and vertical components of peak ground acceleration, peak ground velocity, and pseudo-absolute acceleration response spectra, Seism. Res. Lett. 68, 154-179.

Campbell, K.W. and Bozorgnia, Y. (2008). NGA Ground Motion Model for the Geometric Mean Horizontal Component of PGA, PGV, PGD and 5% Damped Linear Elastic Response Spectra for Periods Ranging from 0.01 to 10 s, *Earthq. Spectra,* 24, 139-171.

Carton, H., S. C. Singh, A. Hirn, S. Bazin, B. de Voogd, A. Vigner, A. Ricolleau, S. Cetin, N. Oçakoğlu, F. Karakoç, and V. Sevilgen (2007), Seismic imaging of the three-dimensional architecture of the Çınarcık Basin along the North Anatolian Fault, J. Geophys. Res., 112, B06101, doi:10.1029/2006JB004548.

Castro, R. R., M. Mucciarelli, F. Pacor, and C. Petrungaro, 1997. S-wave site-response estimates using horizontal-to-vertical spectra ratios, Bull. Seism. Soc. Am. 87, 256-260.

Catchings, R. D., and W. D. Mooney, 1992. New Madrid and central California apparent Q values as determined from seismic refraction data, Seism. Res. Lett. 63, 49.

Catchings, R. D., and W. M. Kohler, 1996. Reflected seismic waves and their effect on strong shaking during the 1989 Loma Prieta, California, earthquake, Bull. Seism. Soc. Am. 86, 1401-1416.

Çelebi, M., J. Prince, C. Dietel, M. Oñate, and G. Chavez (1987). *The culprit in Mexico city – amplification of motions,* Earthquake spectra, 3, 315-328.

Chang, C. Y., C. M. Mok, M. S. Power, Y. K. Tang, H. T. Tang, and J. C. Stepp, 1991. Development of shear modulus reduction curves based on Lotung downhole ground motion data, in Proc. of the 2nd Internation Conf. on Recent Advances in Geotechnical Earthquake Engineering and Soil Dynamics, St. Louis, Missouri, 111-118.

Chapman, C. H., and P. M. Shearer, 1989. Ray tracing in azimuthally anisotropic media-II. Quasi-shear wave coupling, Geophys. J. 96, 65-83.

Chavez-Garcia, F. J., W. R. Stephenson, M. Rodriguez, 1999. Lateral propagation effects observed at Parkway, New Zealand. A case history to compare 1D versus 2D site effects, Bull. Seism. Soc. Am. 89, 718-372.

Chiou, B.S.J. and Youngs, R.R. (2008). An NGA Model for the Average Horizontal Component of Peak Ground Motion and Response Spectra, *Earthq. Spectra,* 24, 173-215.

Chiou, B., Darragh, R., Gregor, N., and Silva, W. (2008). NGA Project Strong-Motion Database, *Earthq. Spectra* 24, 23-44.

Choi, Y., and Stewart, J. P., 2005. Nonlinear site amplification as function of 30 m shear wave velocity, Earthquake Spectra, 21, 1–30.

Choi, Y., Stewart, J. P., and Graves, R. W., 2005. Empirical model for basin effects accounts for basin depth and source location, Bull. Seism. Soc. Am. 95, 1412–1427.

Coutant, O., 1996. Observations of shallow anisotropy on local earthquake records at the Garner Valley, Southern California, downhole array, Bull. Seism. Soc. Am. 86, 477-488.

Cultera, G. C., D. M. Boore, W. B. Joyner, and C. M. Dietel, 1999. Nonlinear soil response in the vicinity of the Van Norman Complex following the 1994 Northridge, California, earthquake, Bull. Seism. Soc. Am. 89, 1214-1231.

Darragh, R.B., and A.F. Shakal [1991]. "The site response of two rock and soil station pairs to strong and weak ground motion", Bull. Seism. Soc. Am., Vol. 81, pp. 1885-1899.

Das, S. and K. Aki, 1977. A numerical study of two-dimensional spontaneous rupture propagation, Geophys. J. Roy. astr. Soc. 50, 643-668.

Das, S. and D. Kostrov, 1983. Breaking of a single asperity: Rupture process and seismic radiation. J. Geophys. Res. 88, 4277-4288.

Davis, P. M., J. L. Rubinstein, K. H. Liu, S. S. Gao, L. Knopoff, 2000. Northridge earthquake damage caused by geologic focusing of seismic waves, Science 289, 1746-1750.

Day, S. M., 1982. Three-dimensional simulation of spontaneous rupture: The effect of nonuniform prestress, Bull. Seism. Soc. Am. 72, 1881-1902.

Day, S. M., Bielak, J., Dreger, D., Graves, R. W., Larsen, S., Olsen, K. B., and Pitarka, A., 2005. 3-D ground motion simulation in basins: Final report for Lifelines Project 1A03, Pacific Earthquake Engineering Research Center.

Day, S. M., R. Graves, Bielak, J., Dreger, D., Larsen, S., Olsen, K. B., and Pitarka, A., and L. Ramirez-Guzman, 2008. Model for basin effects on long-period response spectra in southern California, Earthquake Spectra, 24, 257–277.

Earthquake Engineering Reserach Institute, 1994. EERI Special Earthquake report, August 1994, EERI Newsletter 26, 1-12.

Farra, V., P. Bernard, and R. Madariaga, 1986. Fast near source evaluation of strong ground motion for complex source models, in "Fifth Maurice Ewing Symposium on Earthquake Source Mechanics," Series 6 (S. Das, J. Boatwright, and C. Scholz, eds.), Am. Geophys. Union, pp. 121-130.

Field, E. H. and K. H. Jacob, 1995. A comparison and test of various site response estimation techniques including threee that are not reference site dependent, Bull. Seism. Soc. Am. 85, 1127-1143.

Field, E. H., 2000. A modified ground-motion attenaution relationship for southern California that accounts for detailed site classification and a basin-depth effect, Bull. Seism. Soc. Am. 90, S209-S221.

Field, E. H., P. A. Johnson, I. A. Beresnev, and Y. Zeng, 1997. Nonlinear ground-motion amplification by sediments during the 1994 Northridge earthquake, Nature, 390, 599-602.

Field, E. H., Y. Zeng, P. A. Johnson, and I. A. Beresnev [1998]. "Nonlinear sediment response during the 1994 Northridge earthquake: observations and finite source simulations", J. Geophys. Res., Vol. 103, pp. 869-883.

Frankel, A., 1991. High-frequency spectral falloff of earthquakes, fractal dimension of complex rupture, b value, and the scaling strength on faults, J. Geophys. Res. 96, 6291-6302.

Frankel, A., 1993. Three-dimensional simulations of ground motions in the San Bernardino Valley, California, for hypothetical earthquakes on the San Andreas fault, Bull. Seism. Soc. Am. 83, 1020-1041.

Frankel, A., 1999. How does the ground shake, Science 283, 2032-2033.

Frankel, A. and R. W. Clayton, 1986, Finite difference simulations of seismic scattering: implications for the propagation of short-period seismic waves in the crust and models of crustal heterogeneity, J. Geophys. Res. 91, 6465-6489.

Frankel, A., and J. Vidale, 1992. A three-dimensional simulation of seimic waves in the Santa Clara Valley, California, from a Loma Prieta aftershock, Bull. Seism. Soc. Am. 82, 2045-2074.

Frankel, A., and W. Stephenson, 2000. Three-dimensional simulations of ground motins in the Seattle region for earthquakes in the Seattle fault zone, Bull. Seism. Soc. Am. 90, 1251-1267.

Frankel, A., D. Carver, E. Cranswick, T. Bice, R. Sell, and S. Hanson, 2001. Observations of basin ground motions from a dense seismic array in San Jose, California, Bull. Seism. Soc. Am. 91, 1-12.

Frankel, A. D., W. J. Stephenson, and D. L. Carver, 2009. Sedimentary basin effects in Seattle, Washington: Ground-motion observations and 3D simulations. Bull. Seism. Soc. Am. 99, 1,579–1,611

Fukuyama, E. and R. Madariaga, 1995. Integral equation method for plane crack with arbitary shape in 3D elastic medium, Bull. Seism. Soc. Am. 85, 614-628.

Ghayamghamain, M. R., and H. Kawakami, 1996. On the characteristics of non-linear soil response and dynamic soil properties using vertical array data in Japan, Earthquake Eng. Struct. Dyn. 25, 857-870.

Graves, R. W., A. Pitarka, and P. G. Somerville, 1998. Ground motion amplification in the Santa Monica area: Effects of shallow basin structure, Bull. Seism. Soc. Am. 88, 1224-1242.

Guatteri, M., and P. Spudich, 1998. Coseismic temporal changes of slip direction: The effect of absolute stress on dynamic rupture, Bull. Seism. Soc. Am. 88, 777-789.

Guatteri, M., P. M. Mai, G. C. Beroza, and J. Boatwright (2003). Strong ground motion prediction from stochastic-dynamic source models, Bull. Seism. Soc. Am. 93, 301-313.

Hanks, T. C., and K. Krawinkler, 1991. The 1989 Loma Prieta, California, earthquake and its effects: Introduction to the Special Issue, Bull. Seism. Soc. Am. 81, 1415-1423.

Harris, R. A., and S. M. Day, 1997. Effects of a low-velocity zone on dynamic rupture, Bull. Seism. Soc. Am. 87, 1267-1280.

Hartzell, S., D. Carver, E. Cranswick, and A. Frankel, 2000. Variability of site response in Seattle, Washington, Bull. Seism. Soc. Am. 90, 1237-1250.

Hartzell, S., S. Harmsen, and A. Frankel, 2010. Effects of 3D random correlated velocity perturbations on predicted ground motions, Bull. Seism. Soc. Am. 100, 1415-1426.

Hernandez, B., F. Cotton, and M. Campillo, 1999. Contribution of radar interfermetry to a two-step inversion of the kinematic process of the 1992 Landers earthquake, J. Geophys. Res. 104, 13,082-13,099.

Herrero A. and P. Bernard, 1994. A kinematic self similar rupture process for earthquakes, Bull. Seism. Soc. Am. 84, 1216-1228.

Holzer, T.L., T.L. Youd, and T.C. Hanks [1989]. "Dynamics of liquefaction during the 1987 Superstition Hills, California, earthquake", Science, Vol. 244, pp. 56-59.

Hubert-Ferrari, A., A. Barka, E. Jacques, S. Nalbant, B. Meyer, R. Armijo, P. Tapponier and G.C.P. King, 2000. Seismic hazard in the Marmara Sea following the 17 August 1999 İzmit earthquake, Nature, 404, 269-272.

Hutchings, L., 1994. Kinematic earthquake models and synthesized ground motions using empirical Green's functions, Bull. Seism. Soc. Am. 84, 1028-1050.

Hutchings, L. and F. Wu, 1990. Empirical Green's functions from small earthquakes - A waveform study of locally recorded aftershocks of the San Fernando earthquakes, J. Geophys. Res. 95, 1187-1214.

Hutchings, L., and S. Jarpe, 1996. Ground-motion variability at the Highway 14 and I-5 interchange in the northern San Fernando Valley, Bull. Seism. Soc. Am. 86, S289-S299.

Iai, S., Y. Matsunaga, and T. Kameoka (1990a). Strain space plasticity model for cyclic mobility, Report of the Port and Harbour Research Institute, 29, 27–56.

Iai, S., Y. Matsunaga, and T. Kameoka (1990b). Parameter identification for cyclic mobility model, Report of the Port and Harbour Research Institute, 29, 57–83.

Iai, S., T. Morita, T. Kameoka, Y. Matsunaga, and K. Abiko (1995). Response of a dense sand deposit during 1993 Kushiro-Oki Earthquake, Soils Found. 35, 115–131.

Idriss, I.M. (2008). An NGA Empirical Model for Estimating the Horizontal Spectral Values Generated by Shallow Crustal Earthquakes, *Earthq. Spectra,* 24, 217-242.

Idriss, I. M., and Seed, H. B., 1968a. Seismic response of horizontal soil layers, Proc. Am. Soc. Civil Eng. J. soil Mech. Found. Div. 94, SM4, 1003-1031.

Idriss, I. M., and Seed, H. B., 1968b. An analysis of ground motions during the 1957 San Francisco earthquake, Bull. Seism. Soc. Am. 58, 2013-2032.

Ishihara, K., 1996. Soil Behavior in Earthquake Geotechnics, Oxford Univ. Press, New York.

Ishihara, K., N. Yoshida, and S. Tsujino (1985). Modelling of stress-strain relations of soils in cyclic loading, Fifth International Conference on Numerical Methods in Geomechanics, Nagoya, 373-380.

Jennings, P.C., 1983. Engineering seismology, In: "Earthquakes: Observation, Theory, and Interpretation" (H. Kanamori and E. Boschi, Eds.), pp. 138-173, North Holland, Amsterdam-New Year—Holland.

Joyner, W. B., 1991. Directivity for nonuniform ruptures, Bull. Seism. Soc. Am. 81, 1391-1395.

Joyner, W. B., and A. T. F. Chen, 1975. Calculation of nonlinear ground response in earthquakes, Bull. Seism. Soc. Am. 65, 1315-1336.

Joyner, W. B., R. E. Warrick, and T. E. Funal, 1981. The effect of Quaternary alluvium on strong ground motion in the Coyote Lake, California earthquake of 1979, Bull. Seism. Soc. Am. 71, 1333-1349.

Kaklamanos, J., and L. G. Baise (2011). Model validations and comparisons of the next generation attenuation of ground motions (NGA-West) project, Bull. Seismol. Soc. Am. 101, 160-175.

Kawasaki, I., and T. Tanimoto, 1981. Radiation pattern of body waves due to the seismic dislocation occurring in an anisotropic source medium, Bull. Seism. Soc. Am. 71, 37-50.

Kawase, H., 1996. The cause of the damage belt in Kobe: "The basin-edge effect," constructive interference of the direct S-wave with the basin-induced diffracted/Rayleigh waves, Seism. Res. Lett. 67, no. 5, 25-34.

Koketsu, K., and M. Kikuchi, 2000. Propagation of seismic ground motion in the Kanto Basin, Japan, Science 288, 1237-1239.

Konder, R.L., and J.S. Zelasko (1963). A hyperbolic stress-strain formulation for sands, in Proc. of 2nd Pan American Conference on Soil Mechanics and Foundation Engineering, Brazil, 289-324.

Kramer, S.L. (1996). *"Geothechnical Earthquake Engineering"*, Prentice Hall, New Jersey.

Kurt, H., and Yücesoy, E., 2009. Submarine structures in the Gulf of _Izmit, based on multichannel seismic reflection and multibeam bathymetry. Marine Geophysical Researches, 30, 73-84.

Langston, C. A., 1979. Structure under Mount Ranier, Washington, inferred from teleseismic body waves, J. Geophys. Res. 84, 4749-4762.

Lermo, J. and F. Chavez-Garcia, 1993. Site effect evaluation using spectral ratios with only one station, Bull. Seism. Soc. Am. 83, 1574-1594.

Lettis, W. R., D. L. Wells, and J. N. Baldwin, 1997. Empirical observations regarding reverse earthquakes, blind thrust faults, and Quaternary deformation: Are blind thrust faults truly blind, Bull. Seism. Soc. Am. 87, 1171-1198.

Li, X., and Z. Liao (1993). Dynamic skeleton curve of soil stress-strain relation under irregular cyclic loading Earthquake research in China, 7, 469-477.

Li, Y. G., J. E. Vidale, K. Aki, C. J. Marone, W. H. K. Lee, 1994. Fine structure of the Landers fault zone; segmentation and the rupture process, Science 256, 367-370.

Liu, H.-L., and T. Heaton, 1984. Array analysis of the ground velocities and accelerations from the 1971 San Fernando, California, earthquake, Bull. Seism. Soc. Am. 74, 1951-1968.

Liu, H. L., and D. V. Helmberger, 1985. The 23:19 aftershock of the 15 October 1979 Imperial Valley earthquake: More evidence for an asperity, Bull. Seism. Soc. Am. 75, 689-709.

Liu, P-C., R. J. Archuleta (2006). Efficient modeling of Q for 3D numerical simulation of wave propagation, Bull. Seis. Soc. Am. 96, 1352 - 1358.

Liu, P-C and R. J. Archuleta (2002). The effect of a low-velocity surface layer on simulated ground motion, Seism. Res. Lett., 73, 267.

Lomnitz, C., J. Flores, O. Novaro, T. H. Seligman, and R. Esquivel, 1999. Seismic coupling of interface modes in sedimentary bains: A recipe for distaster, Bull. Seism. Soc. Am. 89, 14-21.

Madariaga, R., 1977. High-frequency radiation from crack (stress-drop) models of earthquake faulting, Geophys. J. 51, 625-652.

Magistrale, H. and S. Day, 1999. 3D simulations of multi-segment thrust fault rupture, Geophys. Res. Lett. 26, 2093-2096.

Mai, P. M., and G. C. Beroza, 2000. Source scaling properties from finite-fault-rupture models, Bull. Seism. Soc. Am. 90, 604-615.

Mai, P. M., P. Somerville, A. Pitarka, L. Dalguer, S. G. Song, G. Beroza, H. Miyake, and K. Irikura (2006), On scaling of fracture energy and stress drop in dynamic rupture models: Consequences for near-source ground motions, Earthquakes: Radiated Energy and the Physics of Faulting, Geophysical Monograph Series 170, AGU, pp. 283-294.

Matasovic, J., and M. Vucetic [1993]. "Analysis of seismic records obtained on November 24, 1987 at the Wildlife Liquefaction Array", *Research Report, Civil Engineering Department, University of California, Los Angeles.*

Meremonte, M., A. Frankel, E. Cranswick, D. Carver, and D. Worley, 1996. Urban seismology: Northridge aftershocks recorded by multiscale arrays of portable digital seismographs, Bull. Seism. Soc. Am. 86, 1350-1363.

Miller R.D., J. H. Bradford, and K. Holliger, 2010. Advances in near-surface seismology and ground-penetrating radar: Introduction, Society of Exploration Geophysicists, Tulsa, OK pp. xiii-xvii.

NIED, 2011, 2011 Off the Pacific Coast of Tohoku Earthquake, Strong Ground Motion, http://www.reluis.it/images/stories/Rapporto_Kik-Net.pdf

O'Connell, D. R. H., 1999a. Replication of apparent nonlinear seismic response with linear wave propagation models, Science 283, 2045-2050.

O'Connell, D. R. H., 1999b. Possible super-shear rupture velocities during the 1994 Northridge earthquake, EOS 80, F941.

O'Connell, D. R. H., 1999c. Ground motion evaluation for Casitas Dam, Ventura River Project, California, U. S. Bureau of Reclamation Seismotectonic Report 99-4, Denver, CO.

O'Connell, D.R.H., 2008, Assessing Ground Shaking, Science, 322, 686-687.

O'Connell, D.R.H., S. Ma, and R.J. Archuleta, 2007, Influence of dip and velocity heterogeneity on reverse- and normal-faulting rupture dynamics and near-fault ground motions, , Bull. Seis. Soc. Am., 97, 1970-1989; DOI: 10.1785/0120070042.

O'Connell, D.R.H., P.C. Liu, and L. F. Bonilla, 2010. Influence of 2D Soil Nonlinearity on Basin and Site Responses, U. S. Geol. Surv. Technical Report 08HQGR0068: http://earthquake.usgs.gov/research/external/reports/08HQGR0068.pdf

O'Connell, D.R.H., and J. P. Turner, 2011, Interferometric multichannel analysis of surface waves (IMASW), Bull. Seis. Soc. Am.

Oglesby, D. D. and R. J. Archuleta, 1997. A faulting model for the 1992 Petrolia earthquake: Can extreme ground acceleration be a source effect?, J. Geophys. Res. 102, 11,877-11,897.

Oglesby, D. D., R. J. Archuleta, and S. B. Nielsen, 1998. Earthquakes on dipping faults:The effects of broken symmetry, Science 280, 1055-1059.

Oglesby, D. D., R. J. Archuleta, and S. B. Nielsen, 2000. The three-dimensional dynamics of dipping faults, Bull. Seism. Soc. Am. 901, 616-628.

Oglesby, D. D., and S. M. Day 2002, Stochastic fault stress: Implications for fault dynamics and ground motion, Bull. Seism. Soc. Am. 92 3006-3021.

Okyar, M., Pinar; A., Tezcan, D., Kamaci, D., 2008. Late quaternary seismic stratigraphy and active faults of the Gulf of İzmit (NE Marmara Sea). Mar. Geophys. Res. 29, 89–107.

Olsen, A. H. and R. J. Apsel, 1982. Finite faults and inverse theory with applications to the 1979 Imperial Valley earthquake, Bull. Seism. Soc. Am. 72, 1969-2001.

Olsen, K. B. and R. J. Archuleta, 1996. Three-dimensional simulation of earthquakes on the Los Angeles fault system, Bull. Seism. Soc. Am. 86, 575-596.

Olsen, K.B., R. Madariaga, R.J. Archuleta, 1997. Three-dimensional dynamic simulation of the 1992 Landers earthquake, Science 278, 834-838.

Olsen, K. B., S. M. Day, and C. R. Bradley (2003). Estimation of Q for long period(>2 sec) waves in the Los Angeles basin, Bull. Seismol. Soc. Am. 93, 627-638.

Petersen, M.D., T.E. Dawson, R. Chen, T. Cao, C.J. Wills, D.P. Schwartz, and A.D. Frankel, 2011. Fault displacement hazard for strike-slip faults, , Bull. Seismol. Soc. Am. 101, 805-825.

Phillips, W. S., S. Kinoshita, and H. Fujiwara, 1993. Basin-induced Love waves observed using the strong motion array at Fuchu, Japan, Bull. Seism. Soc. Am. 83, 65-84.

Pitarka, A., K. Irikura, T. Iwata, H. Sekiguchi, 1998. Three-dimensional simulation of the near-fault ground motions for the 1995 Hyogo-Nanbu (Kobe), Japan, earthquake, Bull. Seism. Soc. Am. 88, 428-440.

Pitarka, A., P. Somerville, Y. Fukushima, T. Uetake, K. Irikura, 2000. Simulation of near-fault strong-ground motion using hybrid Green's functions, Bull. Seism. Soc. Am. 90, 566-586.

Pyke, R. (1979). Nonlinear soil model for irregular cyclic loadings. J. of the Geotech. Engrg. Div., 105, 715-726.

Rodriguez-Marek, A., J. D. Bray, and N. A. Abrahamson, 2001. An empirical geotechnical seismic site response procedure, Earthquake Spectra 65, 65-87.

Rowshandel, B. (2010), Directivity correction for the next generation attenuation (NGA) relations, Earthquake Spectra, 26, 525-599.

Sadigh, K., C.-Y. Chang, J. A. Egan, F. Makdisi, and R. R. Youngs, 1997. Atenuation relationships for shallow crustal earthquakes based on California strong motion data, Seism. Res. Lett. 68, no. 1, 180-189.

Sato, H., and M. C. Fehler, 1998. Seismic Wave Propagation and Scattering in the Heterogenous Earth, Springer-Verlag, New York, 308 pp.

Satoh, T., H. Kawase, and T. Sato, 1995. Nonlinear behavior of soil sediments identified by using borehole records observed at the Ashigara Valley, Japan, Bull. Seism. Soc. Am. 85, 1821-1834.

Satoh, T., M. Horike, Y. Takeuchi, T. Uetake, and H. Suzuki, 1997. Nonlinear behavior of scoria evaluated from borehole records in eastern Shizuoka prefecture, Japan, Earthquake Eng. Struc. Dyn. 26, 781-795.

Satoh, T., M. Fushimi, and Y. Tatsumi, 2001. Inversion of strain-dependent nonlinear characteristics of soils using weak and strong motions observed by borehole sites in Japan, Bull. Seism. Soc. Am. 91, 365-380.

Schmedes, J., and R. J. Archuleta (2008). Near-source ground motion along strike-slip faults: Insights into magnitude saturation of PGV and PGA, Bull. Seismol. Soc. Am. 98, 2278-2290.

Schmedes, J., R. J. Archuleta, and D. Lavallé (2010), Correlation of earthquake source parameters inferred from dynamic rupture simulations, J. Geophys. Res., 115, B03304.

Schneider, J. F., W. J. Silva, and C. Stark, 1993. Ground motion model for the 1989 M 6.9 Loma Prieta earthquake including effects of source, path, and site, Earthquake Spectra, 9, 251-287.

Schultz, C. A., 1994. Enhanced backscattering of seismic waves from irregular interfaces, Ph.D. Thesis, Massachusetts Institue of Technology, Cambridge.

Seed, H.B. and I.M. Idriss (1969). Influence of soil conditions on ground motions during earthquakes, J. Soil Mech. and Found. Div., ASCE, Vol. 95 (SM1).

Seed, H. B., and I. M. Idriss (1970). Analyses of ground motions at Union Bay, Seattle, during earthquakes and distant nuclear blasts, Bull. Seism. Soc. Am. 60, 125-136.

Seed, H. B., M. P. Romo, J. I. Sun, A. Jaime, and J. Lysmer, 1988. The Mexico earthquake of September 19, 1985 - relationships between soil conditions and earthquake ground motions, Earthquake Spectra, 4, 687-729.

Shaw, B. E. (2009), Constant stress drop from small to great earthquakes in magnitude-area scaling, Bull. Seism. Soc. Am. 99, 871-875.

Shin, T. C., K. W. Kuo, W. H. K. Lee, T. L. Teng, and Y. B. Tsai, 2000. A preliminary report of the 1999 Chi-Chi (Taiwan) earthquake, Seism. Res. Lett., 71, 24-30.

Shoja-Taheri, J., Naserieh, S., and Hadi, G. (2010), A Test of the Applicability of the NGA Models to the Strong-Motion Data in the Iranian Plateau, Jour. Earthquake Eng., 14, 278-292.

Somerville, P.G. (2003), Magnitude scaling of the near fault rupture directivity pulse, Physics Earth Planet. Interiors, 137, 201-212.

Somerville, P., and J. Yoshimura, 1990. The influence of critical Moho reflections on strong ground motions recorded in San Francisco and Oakland during the 1989 Loma Prieta earthquake, Geophys. Res. Lett. 17, 1203-1206.

Somerville, P., M. Sen, and B. Cohee, 1991. Simulations of strong ground motions recorded during the 1985 Michoacan, Mexico and Valparaiso, Chile, earthquakes, Bull. Seism. Soc. Am. 81, 1-27.

Somerville, P., K. Irikura, R. Graves, S. Sawada, D. Wald, N. Abrahamson, Y. Iwasaki, T. Kagawa, N. Smith, and A. Kowada, 1999. Characterizing crustal earthquake slip models for the prediction of strong ground motion, Seism. Res. Lett. 70, 59-80.

Somerville, P., and A. Pitarka. (2006). Differences in earthquake source and ground motion characteristics between surface and buried earthquakes. In *Proceedings, Eighth National Conference on Earthquake Engineering,* Proceedings CD-ROM, Paper No. 977. Oakland, Calif.: Earthquake Engineering Research Institute.

Spudich, P., and B. S. J. Chiou (2008), Directivity in NGA earthquake ground motions: Analysis using isochrone theory, Earthquake Spectra, 24,279–298.

Spudich, P., and E. Cranswick, 1984. Direct observation of rupture propagation during the 1979 Imperial Valley earthquake using a short baseline accelerometer array, Bull. Seism. Soc. Am. 74, 2083-2114.

Spudich, P., and L. N. Frazer, 1984. Use of ray theory to calculate high-frequency radiation from earthquake sources having spatially variable rupture velocity and stress drop, Bull. Seism. Soc. Am. 74, 2061-2082.

Spudich, P., and L. N. Frazer, 1987. Errata for "Use of ray theory to calculate high-frequency radiation from earthquake sources having spatially variable rupture velocity and stress drop", Bull. Seism. Soc. Am. 77, 2245-2246.

Spudich, P., and D. Oppenheimer, 1986. Dense seismograph array observations of earthquake rupture dynamics, in "Fifth Maurice Ewing Symposium on Earthquake Source Mechanics," Series 6 (S. Das, J. Boatwright, and C. Scholz, eds.), Am. Geophys. Union, Washington, DC, pp. 285-296.

Spudich, P., and R. J. Archuleta, 1987. Techniques for earthquake ground-motion calculation with applications to source parameterization to finite faults, in "Seismic Strong Motion Synthesis," (B. A. Bolt, ed.), Academic Press, Orlando, Florida, pp. 205-265.

Spudich, P., and M. Iida, 1993. The seismic coda, site effects, and scattering in alluvial basins studied using aftershocks of the 1986 North Palm Springs, California, earthquakes as source arrays, Bull. Seism. Soc. Am. 83, 1721-1743.

Spudich, P., M. Hellweg, and W. H. K. Lee, 1996. Directional topographic site response at Tarzana observed in aftershocks of the 1994 Northrige, Calfornia, earthquake: Implications for mainshock motions, Bull. Seism. Soc. Am. 86, S193-S208.

Spudich, P., M. Guatteri, K. Otsuki, and J. Minagawa, 1998. Use of fault striations and dislocation models to infer tectonic shear stress during the 1995 Hyogo-ken Nanbu (Kobe) earthquake, Bull. Seism. Soc. Am. 88, 413-427.

Stafford, P.J., Strasser, F.O., and Bommer, J.J. (2008). An Evaluation of the Applicability of the NGA Models to Ground Motion Prediction in the Euro-Mediterranean Region, *Bull. Earthq. Eng.* 6, p. 144-177.

Steidl, J.H., A.G. Tumarkin, and R.J. Archuleta (1996). *What is A Reference Site?* Bull. Seism. Soc. Am., 86, 1733 - 1748.

Stein, R.S. and G. Ekström, 1992. Seismicity and geometry of a 110-km-long blind thrust fault: 2. Synthesis of the 1982-1985 California earthquake sequence, J. Geophys. Res. 97, 4865-4883.

Stephenson, W. J., J. N. Louie, S. Pullammanappallil, R. A. Williams, and J. K. Odum (2005). Blind shear-wave velocity comparison of ReMi and MASW results with boreholes to 200 m in Santa Clara Valley: Implications for earthquake ground-motion assessment, Bull. Seism. Soc. Am. 95, 2506 - 2516.

Thompson, E. M., L. G. Baise, R. E. Kayen, E. C. Morgan, and J. Kaklamanos, 2011, Multiscale site-response mapping: a case study of Parkfield, California, Bull. Seism. Soc. Am. 101, 1081-1100.

Towhata, I., and K. Ishihara [1985]. "Modeling Soil Behavior Under Principal Axes Rotation", *Fifth International Conference on Numerical Methods in Geomechanics*, Nagoya, pp. 523-530.

Tumarkin, A. G., R. J. Archuleta, and R. Madariaga, 1994. Scaling relations for composite earthquake models, Bull. Seism. Soc. Am. 84, 1279-1283.

Tsubio, S., M. Saito, and Y. Ishihara, 2001. Verification of horizontal-to-vertical spectral-ratio technique for estimate of site response using borehole seismographs, Bull. Seism. Soc. Am. 91, 499-510.

Vidale, J. E., 1989. Influence of focal mechanism on peak accelerations of strong motions of the Whittier Narrows, California, earthquake and an aftershock, J. Geophys. Res. 94, 9607-9613.

Vucetic, M. [1990]. Normalized behavior of clay under irregular cylic loading, *Can. Geotech. J.* 27, 29–46.

Wald, D. J., and R. W. Graves, 1998. The seismic response of the Los Angeles Basin, California, Bull. Seism. Soc. Am. 88, 337-356.

Walling, M., W. Silva, and N. Abrahamson (2008), Nonlinear site amplification factors for constraining the NGA models, Earthquake Spectra, 24, 243-255.

Wells, D.L. and K. J. Coppersmith, 1994. New empirical relationships amoung magnitude, rupture length, rupture width, rupture area, and surface displacement, Bull. Seism. Soc. Am. 84, 974-1002.

Wen, K. L., I. A. Beresnev, and Y. T. Yeh, 1995. Investigation of nonlinear site amplification at two downhole strong ground motion arrays in Taiwan, Earthquake Eng. Struct. Dyn. 24, 313-324.

Williams, R. A., W. J. Stephenson, A. D. Frankel, and J. K. Odum, 1999. Surface seismic measurements of near-surface P- and S-wave seismic velocities at earthquake recording stations, Earthquake Spectra 15, 565-584.

Zeghal, M., and A.-W. Elgamal [1994]. "Analysis of Site Liquefaction Using Earthquake Records", Jour. Geotech. Engrg., Vol. 120, No. 6, pp. 996-1017.

Zeng, Y., J. G. Anderson, and G. Yu, 1994. A composite source model for computing realistic strong ground motions, Geophys. Res. Lett. 21, 725-728.

Zeng, Y., and J. Anderson, 2000. Evaluation of numerical procedures for simulating near-fault long-period ground motions using the Zeng method, Pacific Earthquake Engineering Res. Center Utilities Program Report 2000/01.

2

Prediction of High-Frequency Ground Motion Parameters Based on Weak Motion Data

Sebastiano D'Amico[1], Aybige Akinci[2], Luca Malagnini[2] and Pauline Galea[1]
[1]Physics Department, University of Malta, Msida,
[2]Istituto Nazionale di Geofisica e Vulcanologia, Rome,
[1]Malta
[2]Italy

1. Introduction

Large earthquakes that have occurred in recent years in densely populated areas of the world (e.g. Izmit, Turkey, 17 August 1999; Duzce, Turkey, 12 November 1999; Chi-Chi, Taiwan 20 September 1999, Bhuj, India, 26 January 2001; Sumatra 26 December 2004; Wenchuan, China, May 12, 2008; L'Aquila, Italy, April 6, 2009; Haiti, January 2010 Turkey 2011) have dramatically highlighted the inadequacy of a massive portion of the buildings erected in and around the epicentral areas. For example, the Izmit event was particularly destructive because a large number of buildings were unable to withstand even moderate levels of ground shaking, demonstrating poor construction criteria and, more generally, the inadequacy of the application of building codes for the region. During the L'Aquila earthquake (April, 06, 2009; Mw=6.3) about 300 persons were killed and over 65,000 were left homeless (Akinci and Malagnini, 2009). It was the deadliest Italian earthquake since the 1980, Irpinia earthquake, and initial estimates place the total economic loss at over several billion Euros. Many studies have already been carried out describing the rupture process and the characteristics of local site effects for this earthquake (e.g. D'Amico et al., 2010a, Akinci et al., 2010). It has been observed that many houses were unable to withstand the ground shaking.

Building earthquake-resistant structures and retrofitting old buildings on a national scale may be extremely costly and may represent an economic challenge even for developed western countries, but it is still a very important issue (Rapolla et al., 2008). Planning and design should be based on available national hazard maps, which, in turn, must be produced after a careful calibration of ground motion predictive relationships (Kramer, 1996) for the region.

Consequently, the assessment of seismic hazard is probably the most important contribution of seismology to society. The prediction of the earthquake ground motion has always been of primary interest for seismologists and structural engineers. For engineering purposes it is necessary to describe the ground motion according to certain number of ground motion parameters such as: amplitude, frequency content and duration of the motion. However it is necessary to use more than one of these parameters to adequately characterize a particular ground motion.

Updating existing hazard maps represents one of the highest priorities for seismologists, who contribute by recomputing the ground motion and reducing the related uncertainties. The quantitative estimate of the ground motion is usually obtained through the use of the so-called predictive relationships (Kramer, 1996), which allow the computation of specific ground-motion parameter as a function of magnitude, distance from the source, and frequency and they should be calibrated in the region of interest. However this is only possible if seismic records of large earthquakes are available for the specific region in order to derive a valid attenuation relationship regressing a large number of strong-motion data (e.g. Campbell and Bozorgnia, 1994; Boore *et al.*, 1993; Ambraseys *et al.*, 1996, Ambraseys and Simpson, 1996; Sabetta and Pugliese, 1987, 1996; Akkar and Bommer 2010). For the Italian region the most used attenuation relationships are those obtained by Sabetta and Pugliese (1987, 1996) regressing a few data recorded for earthquakes in different tectonic and geological environments. It has been shown in several cases that it is often not adequate to reproduce the ground motion in each region of the country using a single model. Furthermore the different crustal properties from region to region play a key role in this kind of studies. However, the attenuation properties of the crust can be evaluated using the background seismicity as suggested by Chouet et al. (1978) and later demonstrated by Raoff et al. (1999) and Malagnini et al (2000a, 2007). In other words, it becomes possible to develop regionally-calibrated attenuation relationships even where strong-motion data are not available. One of the purposes of this work is to describe quantitatively the regional attenuation and source characteristics for constraining the amplitude of strong motion expected from future earthquakes in the area. In this work we describe how to use the background seismicity to perform the analysis (details in Malagnini et. 2000a, 2007). In particular, this chapter describes the procedures and techniques to study the ground motion and will focus on describing both strong motion attenuation relationships and the techniques used to derive the ground motion parameters even when strong ground motion data are not available. We will present the results obtained for different regions of the Italian peninsula, showing that the attenuation property of the crust and of the source can significantly influence the ground motion. In addition, we will show that stochastic finite-fault modeling based on a dynamic frequency approach, coupled with field investigations, confirms to be a reliable and practical method to simulate ground motion records of moderate and large earthquakes especially in regions prone to widespread structural damage.

2. Weak motion data set processing

In general, a large dataset of weak motion data are processed by following the approach described in details by Malagnini *et al.* (2000a, 2007). The advantage of the procedure is that no hypothesis on the functional form of the scaling laws needs to be formulated before the analysis. The method takes into account the duration parameter, as a function of frequency and distance, through the statistical tool called Random Vibration Theory (RVT, see Cartwright & Longuet-Higgins 1956). The latter is used to estimate the peak ground motion of a random time history, given the empirical attenuation parameter, source spectrum and its duration in time. A detailed description of the method and data processing technique are provided in Malagnini *et al.* (2000a, 2007).

Each waveform is visually inspected to eliminate recordings with low signal-to-noise ratios, anomalous glitches, and calibration issues., and seismograms are corrected for the instrument response. The pickings of the P- and S-wave arrivals are also reviewed. Each corrected time series is filtered using bandpass filter at every f_0 constructed as the contribution of two 8-pole Butterworth filters: a low-pass filter and a high-pass filter with corner frequency, respectively, at $\sqrt{2}\, f_0$ and $1/\sqrt{2}\, f_0$.

A general form for a predictive relationship for observed ground motion is (Fig. 1):

$$A_{ij}(\mathrm{f},\mathrm{r_{ij}}) = EXC_i(\mathrm{r_{ref}}, f) + SITE_j(\mathrm{f}) + D(r_{ij}, r_{ref}, \mathrm{f}) \tag{1}$$

where $A_{ij}(f)$ represents the logarithm of peak amplitude of ground-motion velocity at site "j" for the earthquake "i" on each filtered seismogram recorded at the hypocentral distance r_{ij}, $EXC_i(\mathrm{r_{ref}}, f)$ is the excitation term; $SITE_j(f)$ represents the site term and describes the site modification effects; and $D(r_{ref}, r, f)$ is the crustal propagation term and represents an estimate of the average frequency-dependent crustal response for the region. Usually the value of 40 km is chosen for r_{ref}. A piece-wise linear function defined by fixed-distance nodes was used to model the $D(r,\mathrm{r_{ref}},f)$ function. The number of nodes of this piece-wise linear function, and the spacing between the nodes are selected according to the distance distribution of our data. Equation (1) is solved in the time domain, from multiple narrow band-passed signals.

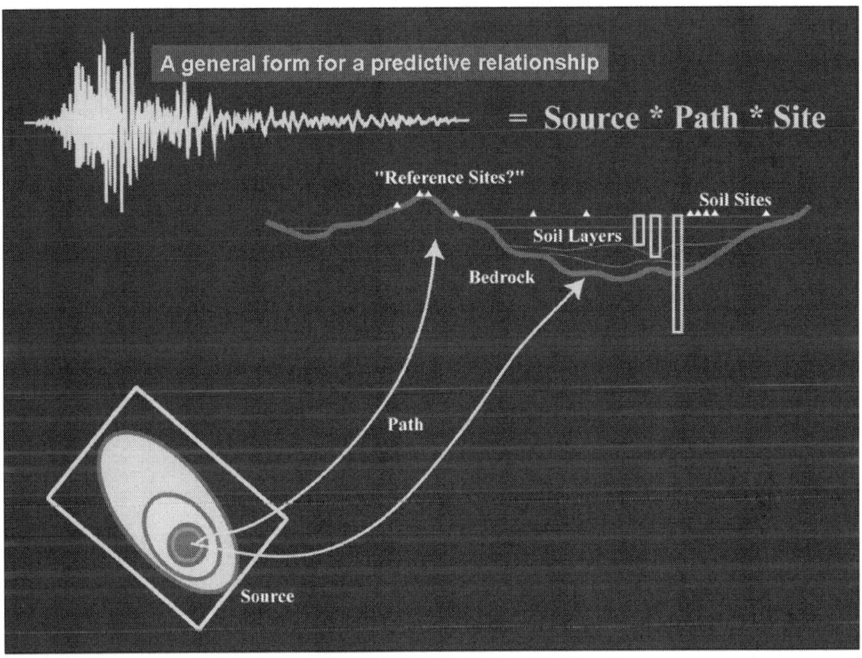

Fig. 1. Schematic representation of a general form for a predictive attenuation relationship.

By using the model equation (1), we can arrange all our observations in a large matrix and then invert to obtain source, path and site terms. According to Malagnini *et al.* (2000a) the regression requires two constraints for stable inversion:

$$D(r_{ij}=r, r_{ref}, f)=0 \tag{2a}$$

$$\Sigma SITE_i\ (f)=0 \tag{2b}$$

The first constraint defines the distance to which the excitation term is projected. The effect of the second one is that common site effects are mapped on the excitation term. During the inversion the sum of all site terms is forced to zero (relation 2b) for each frequency, so that the source terms represent what would be recorded at the reference hypocentral distance by the average network site. Each individual site term measures the deviation from the mean seismic spectra for each station, which is due to the physical properties of the shallow geology at the recording site or, in some cases, to instrumental calibration.

The duration of ground motion is a function of the fault size and of the dispersion of elastic waves along the path between the source and the seismic station. To make specific predictions for large earthquakes, the duration of the event must be added in the form of a constant term equal to the inverse of the corner frequency. Since no unique definition of the effective duration of the ground motion is possible, we define the effective duration for the ground motion as given by Raoof *et al.* (1999). Then for each seismogram the duration T is determined as the width of the time window that delimits the 5%-75% portion of the seismic energy following the S-wave arrivals.

The functional form for the propagation terms expresses the effects of frequency-dependent geometrical spreading and anelastic attenuation (Aki, 1980), depending, for each frequency, on the average velocity structure along the propagation path and rock's physical properties. Therefore, we obtain the results of the regression at a set of sampling frequencies, and model them by using the following functional form:

$$D(r,r_{ref},f)=\log[g(r)]-\log[g(r_{ref})]-\frac{\pi f(r-r_{ref})}{\beta Q(f)} \tag{3}$$

where $g(r)$ is the apparent geometrical spreading, β is the shear-wave velocity, and the frequency-dependent attenuation is defined through the quality factor, $Q(f)$, which is defined as

$$Q(f)=Q_0\left(\frac{f}{f_{ref}}\right)^{\eta} \tag{4}$$

with the reference frequency f_{ref} chosen at 1 Hz. The parameter η defines the frequency dependence of $Q(f)$.

The term: $\log_{10}(D(r, r_{ref}, f))$ is modeled as a piecewise linear function between a number of distance nodes, and assumed to be zero at a reference distance r_{ref}, which defines the excitation term $\log_{10}(EXC_i(r_{ref}, f))$. The r_{ref} is chosen such that mislocations in source depth would not significantly change the reference hypocentral distance. A body-wave-like geometric attenuation and a surface-wave-like decay is used to model the decay of the Fourier amplitudes as a function of distances (Malagnini et al., 2000a)

The source excitation, $EXC_i(r_{ref}, f)$ in equation (1), may be thought of as the average expected level of ground motion at r_{ref} for each earthquake. The observed excitation is related to the actual source spectrum through the expression

$$EXC_i(r_{ref}, f) = s(f, M_W)\, g(r_{ref}) \exp[-\pi f r_{ref} / Q(f)\beta]\{V(f) \exp(-\pi f \kappa_o)\}_{avg} \tag{5}$$

where $s(f,M_w)$ is the source excitation as a function of moment magnitude, $V(f)$ is a frequency-dependent 'regional' site amplification (the average site term of the stations included in Eq. 2b) and κ_0 controls near-surface attenuation at high frequency. The term $[g(r_{ref}) \exp(-\pi f r / Q(f)\beta)]$ represents the effect of the propagation at the reference distance due to geometrical spreading and crustal attenuation. The term $\{V(f) \exp(-\pi f \kappa_o)\}_{avg}$ term controls the average site modification of the signal spectrum. *V(f)* represents the average site amplification term relative to hard rock (in the manner of Atkinson & Silva, 1997), and can be calculated from the shallow shear-wave velocity structure near the site (Boore, 1996). κ_0 describes the depletion of the high-frequency motion at the site, which may be caused by the local Q(z) structure. The form of $s(f,M_w)$ representing the Fourier velocity spectra is,

$$s(f, M_w) = K\,[M_o/4\pi\rho\beta^3]\,(2\pi f)\,S(f) \tag{6}$$

where $\log M_o = 1.5(M_w + 6.03)$ (M_0 in Nm) (from Hanks & Kanamori 1979), $K = (0.55 \cdot 2.0 \cdot 0.707)$ is a coefficient composed of the effects of the average radiation pattern, free-surface amplifications for vertically incident *S* waves and the energy partition of initial shear wave amplitude into two horizontal components, M_o is the seismic moment, ρ is the mass density at the source and β is the shear wave velocity at the source (D'Amico et al., 2011b). *S(f)* is the single-corner source term:

$$S(f) = 1/\,[1 + (f/f_c)^2] \tag{7}$$

where $f_c = 0.49\beta(\Delta\sigma/M_o)^{1/3}$. Here the corner frequency f_c is determined from the spectra and is related to the stress drop, $\Delta\sigma$ (Pa), seismic moment, M_o (N-m) and the shear wave velocity β (m s^{-1}). The constant 0.49 depends on the type of model spectra and the geometry of the source.

In order to model the excitation term, we use Brune's (1970, 1971) ω^2 source model, which describes the spectrum of shear radiation in terms of stress drop and moment magnitude. The average stress parameter required to best reproduce the observed propagation-corrected source spectra for the larger earthquakes in the data is adjusted using a trial-and-error procedure. In order to properly calibrate the $\Delta\sigma$, and k_0 the scaling relations must be constrained by using moment magnitudes derived from regional moment tensor inversion, both for small and moderate earthquakes (e.g. D'Amico et al. 2010b, D'Amico et al 2011a). In general, a tradeoff exists between the stress parameter and κ_0. κ_0 governs the high-frequency decay of the theoretical excitation terms, as well as the $\Delta\sigma$, which affects the radiated spectra beyond their corner frequencies. In the frequency band of our interest, however, the effect of the stress parameter is strongest for the largest earthquakes, while κ_0 completely controls the behavior of the small earthquakes at high frequency. For this reason, it is necessary first to find estimates of the high frequency parameter, k_0, by examining the spectra of small events, and then, knowing k_0, of the stress parameter of the larger events. During the inversion the sum of all site terms is forced to zero for each frequency. This constraint represents what would be recorded at the reference hypocentral distance by the average network site. The site term measures the deviation from the mean seismic spectra for each station, which is due to the physical properties of the shallow geology at the recording site.

The methodology described above for determining the attenuation properties has been successfully applied in different part of the world: California (Raoof et al. 1999; Malagnini et al. 2007), northwestern United States (Herrmann and Dutt, 1999; Jeon and Herrmann 2004), central United States (Herrmann and Malagnini, 1996), Mexico (Ortega et al., 2003), Greece and Crete (Pino et al., 2001), Italy (Malagnini et al., 2000a, 2002; Malagnini and Herrmann 2000; Morasca et al. 2006; Scognamiglio et al. 2005), Central Europe (Malagnini et al, 2000b, Bay et al. 2003), Turkey (Akinci et al., 2001, 2006), India (Bodin et al., 2004). For example, comparisons can be done among different Italian regions in which this kind of studies have been conducted. It has been found that the western Alps (Morasca *et al.*, 2006), eastern Alps (Malagnini *et al.*, 2002), Southern Appenines, central Italy (Malagnini et al 2000a) and eastern Sicily (Scognaniglio et al 2005) have different characteristics for the attenuation parameters. For instance the crustal wave propagation in eastern Sicily is more efficient than in other Italian regions. The combination of the geometrical spreading function and the parameter $Q(f)$ is strictly related to the crustal characteristics. Figure 2 reports the attenuation values for each region of Italy for a comparison.

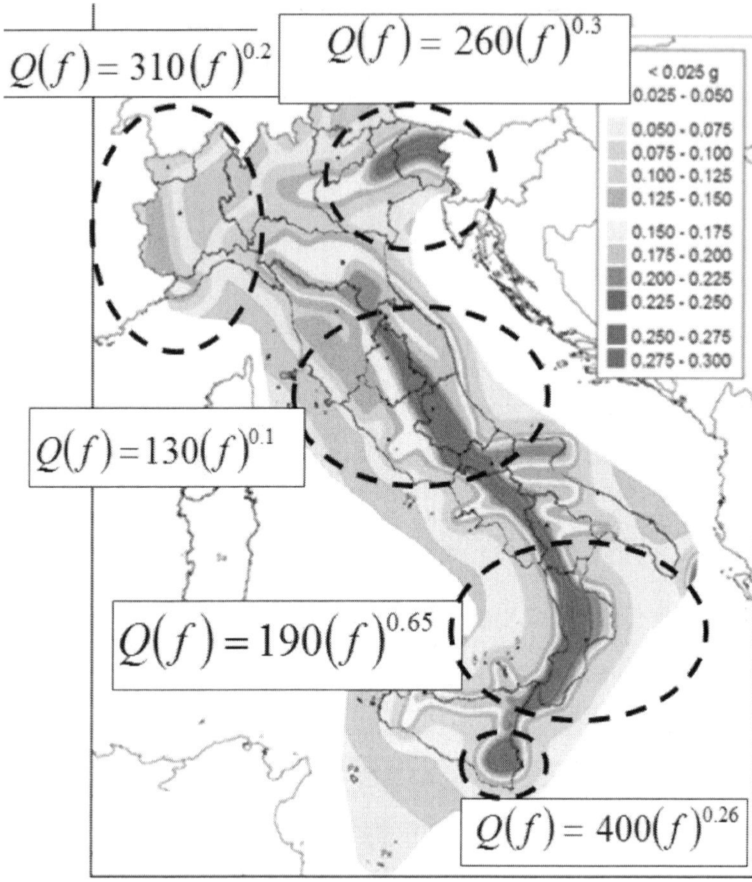

Fig. 2. Attenuation values for different Italian regions: western Alps (Morasca *et al.*, 2006), eastern Alps (Malagnini *et al.*, 2002), Southern Appenines (D'Amico et al. 2011d), central Italy (Malagnini et al 2000a) and eastern Sicily (Scognaniglio et al 2005). The values are plotted on the Italian seismic hazard map please refer to the following website for further details http://zonesismiche.mi.ingv.it ("Mappa di pericolosità sismica del territorio nazionale".

3. Prediction of ground motion parameters

The estimation of ground motion for a particular region and also site-specific investigation is essential for the design of engineered structures. Estimates of expected ground motion at a given distance from an earthquake of a given magnitude are fundamental inputs to earthquake hazard assessments. The determination of seismic design criteria for engineered structures depends upon reproducible estimates of the expected lifetime of the structures. There are also several site classification schemes used in different papers ranging from qualitative description of the near surface material to very quantitative definitions based on shear wave velocities. In order to predict the expected ground motion parameters (e.g. in terms of peak ground acceleration (PGA) and peak ground velocity (PGV) as a function of distance and magnitude we used a stochastic approach. We performed the task by using two widely used computer codes; for point-source model, SMSIM and for extended-source model, EXSIM that is originally developed by Boore (1996, 2003) and Motazedian & Atkinson (2005) & Boore (2009), respectively. During the SMSIM point-source simulations we did not use any information about fault geometry and rupture properties. In general, the simulations are carried out by using the regional propagation parameters derived as described above and, for the EXSIM simulations, having rectangular faults having length and width proportional to the moment magnitude according to the relationship proposed by Wells and Coppersmith (1994). Each fault is assigned with a random slip distribution, if known, otherwise it is reasonable to assume a random slip distribution if we do not have any constraint on the slip distribution. It has been shown that only the gross features of slip distribution on a fault plane that do not diverge significantly from the average value of slip may be reliable; all other complexities could be extremely uncertain (Beresnev and Atkinson, 2002). During each simulation the fault plane is discretized into several subfaults. Site effects at a specific station are very important and may be used for engineering purpose to define the regional predictive law and the seismic hazard. In the simulations in order to consider different site conditions, we will refer to the NEHRP classification (BSSC, 1994). Figures 3, 4, 5, and 6 shows several examples of stochastic simulations (both using SMSIM and EXSIM) carried out in different parts of the globe. Figure 3 shows a ccomparison between observed and simulated (a) PGA and (b) PGV obtained for the large and destructive Chi-Chi (Taiwan) event that occurred on 20th September 1999. The simulation values are obtained using the EXSIM computer codes. The PGA and PGV simulated values were estimated for an earthquake of $\mathbf{M_w}$ 7.6 using the parameters derived by D'Amico et al. (2011b), and the proposed fault geometry by Ma et al. (2001). The grey symbols represent the observed data of the Chi-Chi earthquake (http://www.cwb.gov.tw). Predictions obtained using Akkar & Bommer (2010), Boore & Atkinson (2008) and Campbell & Bozorgnia (2008) are also shown.

Figure 4 shows the observed PGA and PGV for the L'Aquila (6 April 2009) main shock (Mw= 6.2, D'Amico et al. 2011d). The figures show the theoretical predictions based on the results by Malagnini et al . 2011 for the Generic Rock Site (Boore & Joyner 1997), and from the strongmotion relationships by Sabetta & Pugliese (1996). Predictions are given in terms of medians values $\pm 1\sigma$. Predictions by Akinci *et al.* (2010), computed using the ground motion model developed by Malagnini *et al.* (2008), are only slightly different from the ones obtained in this study, and fit quite well the recorded strong motion data up to a distance of 200 km (Malagnini et al. 2011).

Figure 5 and 6 report the ground motion simulated parameter in terms of PGA and PGV, for the island of Malta due to the possible activation of two active faults located (i) on the Hyblean-Maltese Escarpment, at an epicentral distance of 140 km and (ii) about 20 km south of Malta (D'amico et al. 2011c). In particular for the fault on the Hyblean-Maltese Escarpment we simulated the possible ground motion parameters for four earthquakes of different moment magnitude. It is worth noting on these figures that peak ground accelerations close to 0.1g would be generated on Class B and C sites from a magnitude 7.6 event on the Malta escarpment. Such an event is comparable to the 1693 and the 1169 events

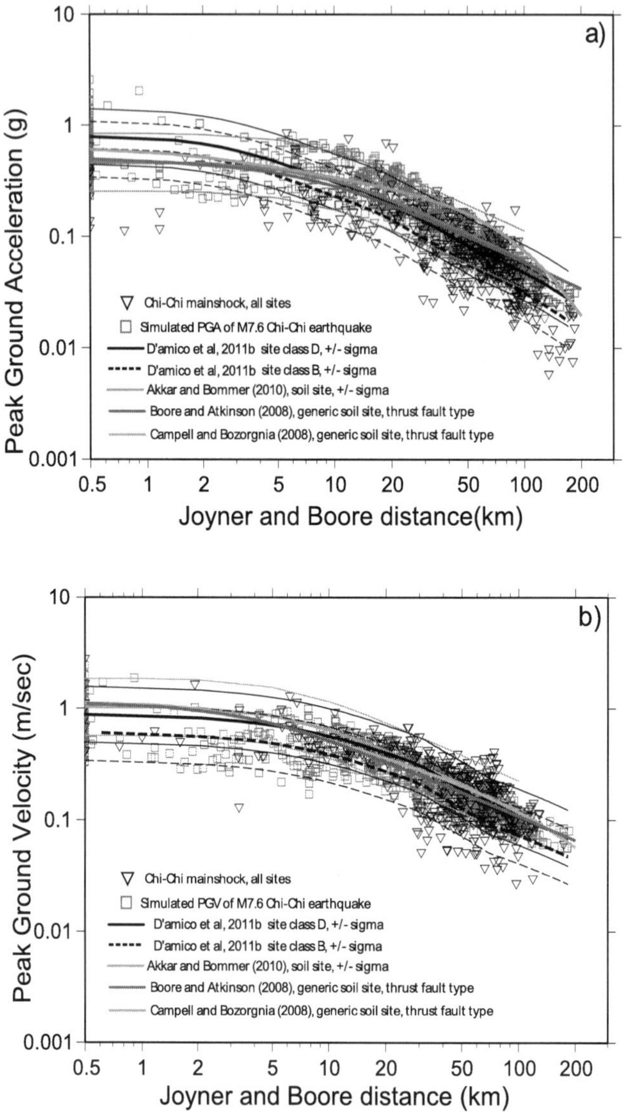

Fig. 3. Comparisons between observed and simulated (a) PGA and (b) PGV. The simulation values are obtained using the EXSIM computer codes. The PGA and PGV simulated values were estimated for an earthquake of $\mathbf{M_w}$ 7.6 using the parameters derived by D'Amico et al. 2011b, and the proposed fault geometry by Ma et al. (2001). The grey symbols represent the observed data of the Chi-Chi earthquake (http://www.cwb.gov.tw). Predictions obtained using Akkar & Bommer (2010), Boore & Atkinson (2008) and Campbell & Bozorgnia (2008) are also shown.

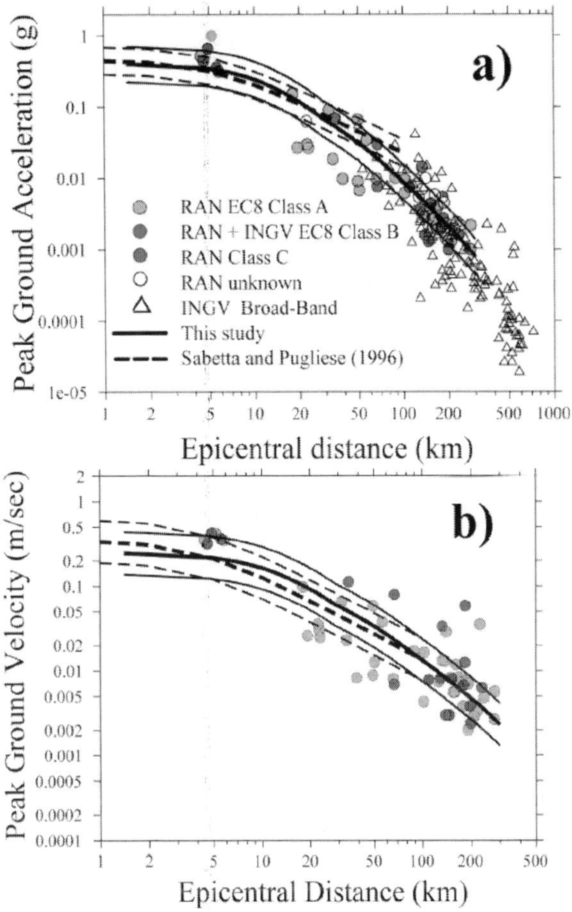

Fig. 4. (a) Predictions of peak ground acceleration (PGA) for the L'Aquila main shock (*Mw* 6.15) and comparison with the observed data (Malagnini et al. 2011). Solid lines refer to the predictions calculated in this study (median $\pm$ 1σ), whereas dashed lines refer to Sabetta Pugliese (1996, SP96 in the figure). (b) Same as in frame (a) but for peak ground velocity (PGV). The agreement of the ground motion model presented in his study with the observations gathered during the main shock allow an objective judgment of the model's performance. Observed PGAs and PGVs are characterized, when available, by their Eurocode 8 class (symbols in colours).

on the same fault (Azzaro and Barbano, 2000). Because of the inherent brittleness, lack of ductility and lack of tensile strength of unreinforced masonry buildings (URM), it is expected that even such accelerations could cause significant damage in these buildings (Hess, 2008). It is also observed that a magnitude 5 event on the active fault zone immediately south of Malta would produce similar PGA/PGV values as the above event, albeit the frequency content would be different. Such a magnitude is rare on the Sicily

Channel faults but approximate calculations based on radius of perceptibility indicate that they have occurred on at least one occasion (1911) in the past century.

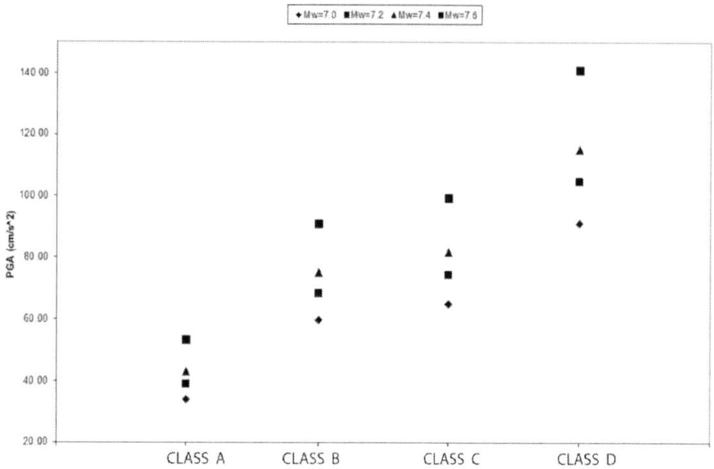

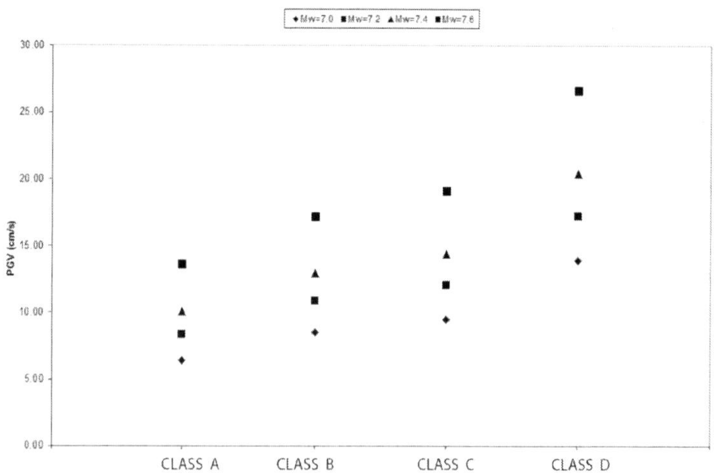

Fig. 5. Predictions of Peak Ground Acceleration (PGA) and Peak Ground Velocity (PGV) for earthquakes of different magnitudes located on the Malta escarpment, at 140km distance. The simulation values are obtained using the EXSIM computer codes. The PGA and PGV simulated values were estimated for an earthquake of Mw 7.0, 7.2, 7.4, and 7.6 (see D'Amico et al. 2011c for details. Each site class is characterized by the average Shear-Wave Velocity over the Upper 30 m (Vs30). In particular, for the class A Vs30~2900m/s; class B Vs30~620m/s; class C Vs30~520m/s; class D Vs30=255m/s (Boore and Joyner, 1997). Preliminary investigations of shallow crustal properties in the Malta area indicate a high variability of soil categories (Panzera *et al* 2011) and therefore the simulations were carried out for each site class.

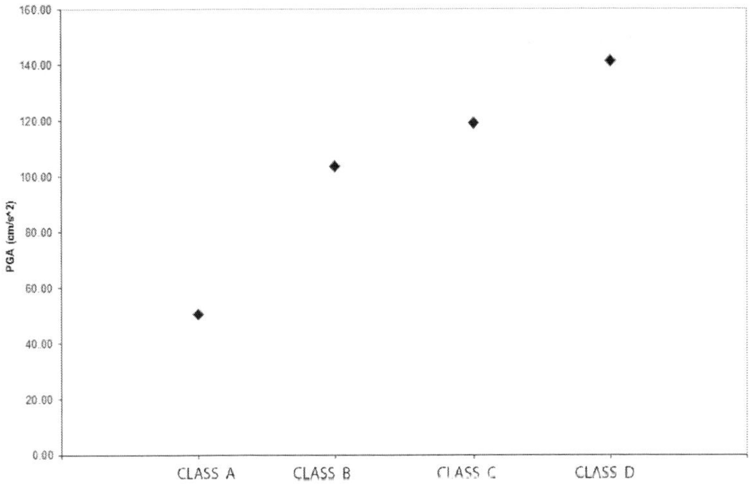

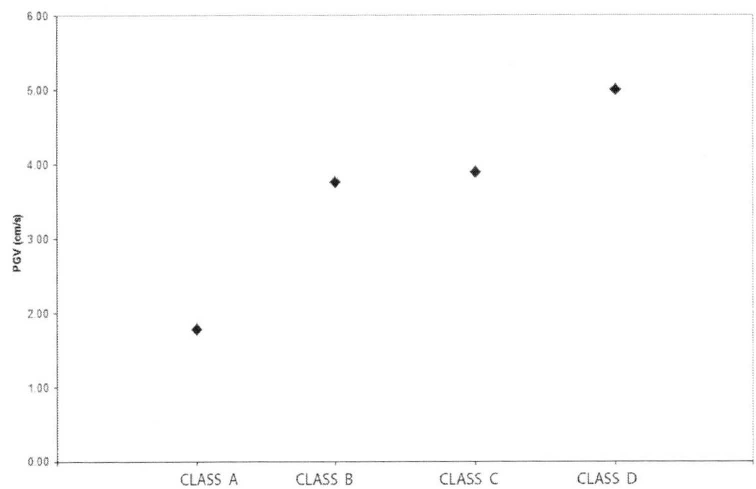

Fig. 6. Predictions of Peak Ground Acceleration (PGA) and Peak Ground Velocity (PGV) for a magnitude 5 earthquake located at about 20 km south of Malta. See caption of Figure 5 for details.

4. Comparison between simulated ground motions for different moment magnitude events from different regions of the world

Because the technique used here for the investigation of the scaling properties of the high-frequency parts of the ground motion is virtually identical to those used in studies

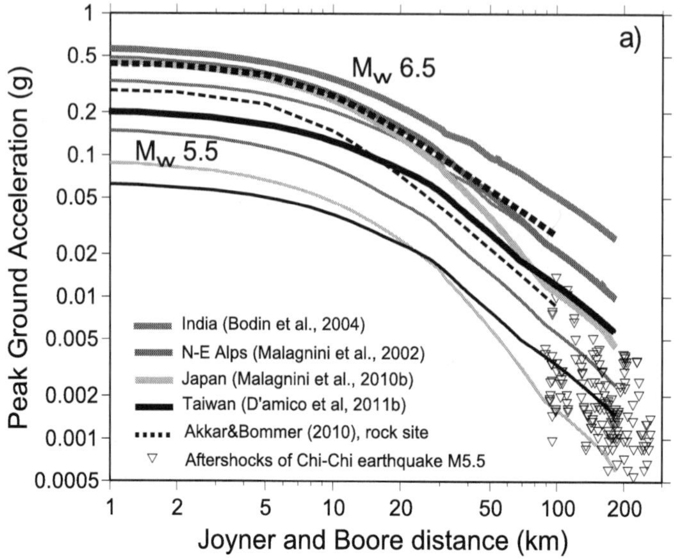

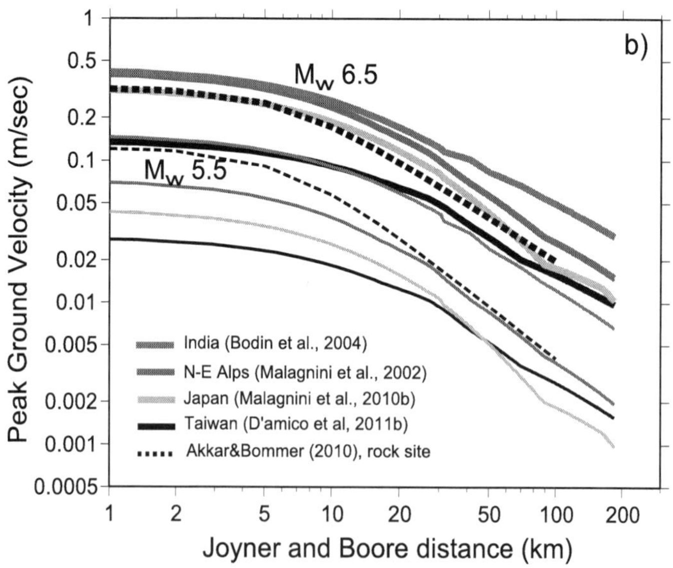

Fig. 7. Predictions of (a) PGA, in g and (b) PGV, in m/sec for different regions of the world: India (red lines, Bodin *et al* ., 2004); N-E Alps (blue lines, Malagnini *et al* ., 2002); Japan (green lines, Malagnini *et al* ., 2010b); Taiwan (black lines, this study). Predictions obtained using Akkar & Bommer (2010) are also shown by dashed black lines. The thick and thin lines are the prediction of PGA and PGV for moment magnitude of 5.5 and 6.5, respectively

conducted in a wide variety of environments (e.g., Bodin *et al.*, 2004; Malagnini *et al.*, 2010), we can compare our results with those studies to reveal systematic similarities and differences with other areas. Figure 7 shows the predicted ground motions computed for moment magnitude of $\mathbf{M}_w$ 6.5 and 5.5 from different regions of the world using the generic rock sites by Boore & Joyner (1997), and provide a comparison with the attenuation relationship developed by Akkar and Boommer (2010) (hereafter AB10).

For example in Figure 7 it is clear that weak-motion based predictive relationships fit quite well with the observed data from $\mathbf{M}_w$ 6.5 and 5.5 events in Taiwan. The predictive relationships of AB10 overestimate the PGA and PGV with compare to both our RVT-generated ground motion predictions and the observed data at lower magnitude, $\mathbf{M}_w$ 5.5. The predicted decay of PGA and PGV with distance clearly indicates that Taiwanese ground motions may be similar to that observed in tectonically active regions, rather than in stable

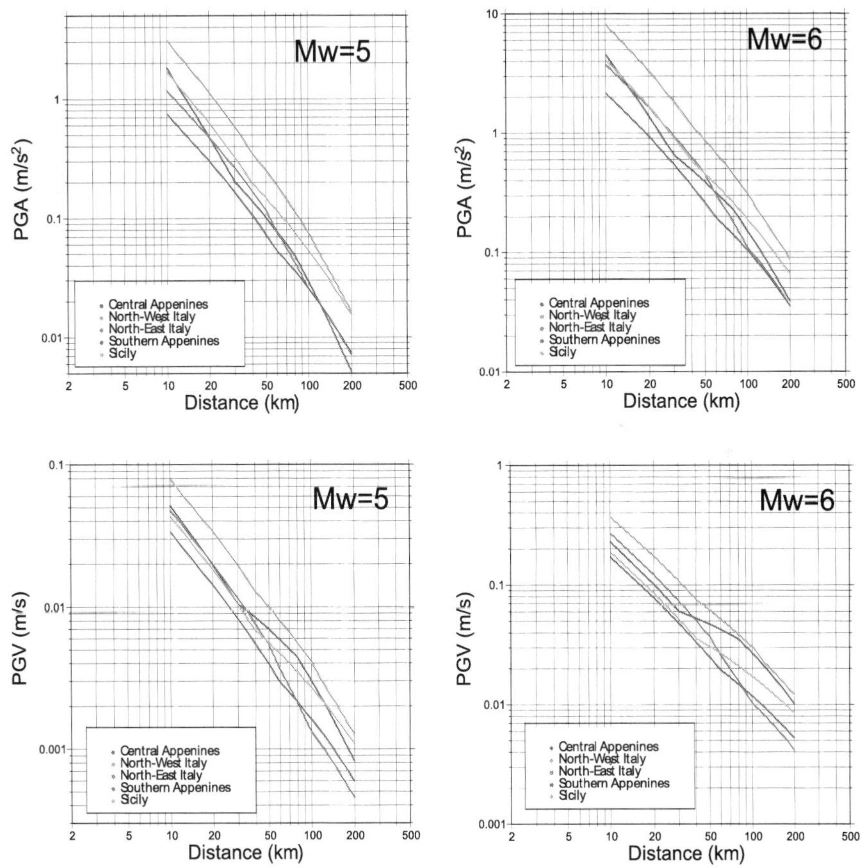

Fig. 8. Comparison of different estimates of PGA and PGV for different Italian regions as a function of moment magnitude Mw.

areas like the Indian plate. The differences in the predicted peak amplitudes between regions get larger at lower magnitudes. The stronger attenuation is experienced by the high-frequency part of the spectrum since peak ground motions are carried by a dominant frequency band that is magnitude-dependent. Most of the seismic energy is radiated around the earthquake corner frequency, so that at larger magnitudes, peak ground motions are dominated by lower frequencies and are not very sensitive to crustal attenuation. Therefore, peak ground motions decay faster for the $\mathbf{M_w}$ 5.5 event than for the $\mathbf{M_w}$ 6.5 event. Moreover, the level of ground shaking increases as the quality factor Q_o increases, and as the slope of the effective geometrical spreading function gets more gentle. The effect gets more severe for distant records. Due to the heterogeneities that are present (mainly vertically) in the crustal structure, different slopes may be presented in the predicted curves. Unfortunately, many of the existing predictive relationships (e.g., Ambraseys, 1995; Sabetta & Pugliese, 1996; Akkar & Bommer, 2010) were obtained by forcing a body-wave geometrical spreading to a distance range where supercritical reflections at the Moho appear to be quite important. Another important parameter is the stress drop, which may cause the differences in the ground motion levels at short distances and must be carefully calibrated in the region of interest. For example, Malagnini et al. (2010) observed that the absolute levels of the stress drop parameters are region-dependent that results in different estimation of gorund motion paramters. As an example, Figure 8 shows the comparison of simulated PGA and PGV as a function of distance for moment magnitude of 5 and 6 obtained using the SMSIM programs (Boore, 2003; http://www.daveboore.com/software_online.htm). It is clear how the predictions of the ground motion are different in the different areas, therefore the regional calibration of attenuation properties and source scaling is a really important task even considering a quite small area.

5. Conclusions

Such kind of studies clearly demonstrate that the weak-motion based ground-motion predictive equations successfully predict the ground motions induced by larger events, once the source term is adequately calibrated (D'Amico et al. 2011b; Malagnini *et al.*, 2010).

The results obtained applying the technique described in this chapter could be used for upgrading the most recent hazard map of Italy and for engineering designs as well. The results are also useful to implement tools like *Shake Map*® (Wald et al. 2005) which use this kind of information to generate a rapid earthquake response. *ShakeMap*® is a tool used to portray the extent of potentially damaging shaking following an earthquake. It can be used for emergency response, loss estimation, and public information. Shake maps show the distribution of ground shaking in the region, information that can be really critical for emergency management decision making. In fact, it is the distribution of peak ground motion and intensity rather than the magnitude that provides useful information about areas prone to damage. Having this information in real time will result in lives saved and reduction in property damage. After a damaging earthquake, emergency managers must quickly find answers to important questions such as the localization of the most serious damage, and the areas with less; the resources that must be mobilized and in what quantity. Usually government response organizations answer these questions after a preliminary survey of the damaged area. This reconnaissance can require several hours or sometimes some days to be completed. As a result, decisions regarding search and rescue, medical emergency response, care and shelter for the damaged and displaced persons, and other critical response needs must often be made while information is still incomplete. In this context a rapid and automatic response for the injured area is really important; hence predictive attenuation relationships play a key and really important role in similar tools.

6. Acknowledgements

This research was carried out using computational facilities procured through the European Regional Development Fund, Project ERDF-080 'A supercomputing laboratory for the University of Malta (http://www.um.edu.mt/research/scienceeng/erdf_080). Some of the figures were made using the Generic Mapping Tools (GMT) by Wessel and Smith (1991). The authors are also very thankful to Mr. Igor Babic and Ms Ivana Lorkovic for their support.

7. References

Aki, K., 1980. Attenuation of shear waves in the lithosphere for frequencies from 0.05 to 25 Hz, *Phys. Earth Planet. Inter.*, 21, 50-60.

Akkar, S., & Bommer, J. J., 2010, Empirical equations for the prediction of PGA, PGV, and spectral accelerations in Europe, the Mediterranean, and the Middle East. *Seismological Research Letters* 81 (2), 195–206.

Akinci, A., Malagnini, L., Pino, N. A. , Scognamiglio, L., Hermann, R. B. & Eyidogan H. , 2001. High-frequency ground motion in the Erzincan region. Turkey: inferences from small earthquakes, *Bull. Seism. Soc. Am.* 91, 1446–1455.

Akinci A., Malagnini, L., Herrmann, R. B., Gok R., & Sorensen M. B., 2006. Ground motion scaling in Marmara region, Turkey. *Geophys. Journ Int.*, 166, 635-651

Akinci, A., Malagnini, L., 2009. The 2009 Abruzzo earthquake, Italy. IRIS Newsletter, 1, 3.

Akinci, A., Malagnini, L. & Sabetta, F., 2010. Characteristics of the strong ground motions from the 6 April 2009 L'Aquila earthquake, Italy, *Soil Dyn. Earthq. Eng.*, 30, 320-335, doi:10.1016/j.soildyn.2009.12.006

Ambraseys, N. N., 1995. The Prediction of Earthquake Peak Ground Acceleration in Europe, *Earthquake Engineering and Structural Dynamics*, 24, 467-490

Ambraseys, N. N., K. A. Simpson, and J. J. Bommer 1996. Prediction of horizontal response spectra in Europe, *Earthquake Eng. Struct. Dyn.* 25, 371–400

Ambraseys, N. N., and Simpson K. A., 1996. Prediction of vertical response spectra in Europe, Earthquake Eng. Struct. Dyn., 25, 401–412

Atkinson, G.M., & Silva, W. 1997. An empirical study of earthquake source spectra for California earthquakes. *Bull. Seism. Soc. Am.*, 87, 97-113.

Azzaro, R. and Barbano, M.S., 2000. Analysis of the seismicity of Eastern Sicily: a proposed tectonic interpretation. Ann. Geofis. 43(1), 171 - 188

Bay, F., Fah, D., Malagnini, L., Giardini, D., 2003. Spectral shear-wave ground motion scaling for Switzerland, Bull. Seism. Soc. Am., 93, 414–429.

Beresnev I., & Atkinson G., 1998. FINSIM: a FORTRAN program for simulating stochastic acceleration time histories from finite faults, *Seism. Res. Lett.*, 69, 27-32

Bodin, P., Malagnini, L. & Akinci A., 2004. Ground motion scaling in the Kachchh basin, India, deduced from aftershocks of the 2001 Mw7.6 Bhuj earthquake, *Bull. Seism. Soc. Am.* 94, 1658–1669

Boore, D.M., 1996. SMSIM - Fortran programs for simulating ground motion from earthquakes: version 1.0, U.S. Geological Survey open-file report 96-80-A, 73 pp.

Boore, D.M., 2003. Simulation of ground motion using the stochastic method. *Pure and Applied Geophysics*, 160, 635-676

Boore, D. M., 2009. Comparing stochastic point-source and finite-source ground-motion simulations: SMSIM and EXSIM, *Bull. Seism. Soc. Am.* 99, 3202-3216

Boore, D. M. & Joyner, W. B., 1997. Site amplification for generic rock sites, *Bull. Seism. Soc. Am.* 87, 327-341.

Boore, D. M.,W. B. Joyner, and T. E. Fumal, 1993. Estimation of response spectra and peak acceleration from western North America earthquakes: an interim report, U.S. Geol. Surv. Open-File Rept. 93-509, 72 pp

Boore, D. M. & Atkinson, G. M., 2008. Ground-motion prediction equations for the average horizontal component of PGA, PGV, and 5%-damped PSA at spectral periods between 0.01 s and 10.0 s, *Earthquake Spectra* 24, 99-138.

Brune, J.N., 1970. Tectonic stress and the spectra of seismic shear waves from earthquakes. *J.Geophys. Res.*, 75, 4997-5009.

Brune, J.N., 1971. Correction. J. *Geophys. Res.*, 76, 5002.

Campbell, K. W., and Y. Bozorgnia (1994). Near-source attenuation of peak horizontal acceleration from worldwide accelerograms recorded from 1957 to 1993, in *Proc. Fifth U.S. National Conference on Earthquake Engineering*, EERI, Berkeley, California, Vol. 1, 283–292.

Campbell, K. W. & Bozorgnia, Y. , 2008. NGA ground motion model for the geometric mean horizontal component of PGA, PGV, PGD and 5%-damped linear elastic response spectra at periods ranging from 0.1 s to 10.0 s. *Earthquake Spectra,* 24 (1), 139–171.

Cartwright D.E., &Longuet-Higgins, M.S., 1956. The statistical distribution of the maxima of a random function, *Proc. R. Soc. London*, 237, 212-232.

Chouet B. K., & Tsujiurs M., 1978. Regional variation of the scaling law of earthquake source spectra. *Bull. Seism. Soc. Am.* 68, 49-79.

D' Amico, S.,K. D. Koper, R. B. Herrmann, A. Akinci, L. Malagnini, 2010a, Imaging the rupture of the Mw 6.3 April 6, 2009 L' Aquila, Italy earthquake using back-projection of teleseismic P-waves, Geophys. Res. Lett., 37, L03301, doi:10.1029/2009GL042156

D'Amico S., Orecchio B., Presti D., Zhu L., Herrmann R. B., Neri G., 2010b. Broadband waveform inversion of moderate earthquakes in the Messina straits, Southern Italy, Physics of Earth and Planetary Interiors, 179, 97-106, doi: 10.1016/j.pepi.2010.01.012

D'Amico S., Orecchio B., Presti D., Gervasi A., Guerra I., Neri G., Zhu L., Herrmann R. B., 2011a. Testing the stability of moment tensor solutions for small and moderate earthquakes in the Calabrian-Peloritan arc region. Boll. Geo. Teor. Appl. doi:10.4430/bgta0009 (in press)

D'Amico S., Akinci A., Malagnini L., 2011b. Predictions of high-freqeuncy ground-motion in Taiwan based on weak motion data, submitted to Geophysical Journal International

D'Amico S., Galea P., Borg P. R., Lotteri A., 2011c. Earthquake ground-motion scenario: case study for the Xemjia Bay (Malta) area. Proceedings of the International conference of Europen Council of Civil Engineers, 149-160

D'Amico S., Akinci A., Malagnini L., Herrmann R. B., 2011d. Ground motion scaling in southern Apennines (Italy). In preparation

Hanks, T.C. & Kanamori H., 1979. A moment magnitude scale. *J. Geophys. Res.*, 84, B5, 2348-2350.

Herrmann, R. B., and Malagnini, L., 1996. Absolute ground motion scaling in the New Madrid Seismic Zone, Seism. Res. Lett. 67, 40.

Herrmann, R. B., and Dutt, J., 1999. High frequency vertical ground motion in the Pacific Northwest using PNSN data, Seism. Res. Lett. 70, 216

Hess, R. L., 2008, Unreinforced Masonry (URM) Buildings, in The ShakeOut Scenario, USGS Open File Report 2008-1150

Jeon Y.S., and Herrmann, R. B., 2004. High-frequency ground-motion scaling in Utah and Yellowstone. Bull. Seism. Soc. Am. 94, 1644-1657.

Kramer S.L. (1996). Geotechnical Earthquake Engineering. Prentice Hall, Upper Saddle River, NJ.

Ma, K.F., Mori, J., Lee, S.-J., & Yu, S.B., 2001. Spatial & temporal distribution of slip for the 1999 Chi-Chi, Taiwan, Earthquake, *Bull. Seism. Soc. Am.* 91, 1069–1087

Malagnini, L. & Herrmann, R.B., 2000. Ground motion scaling in the region of the Umbria-Marche earthquake of 1997, *Bull. seism. Soc. Am.*, 90,1041–1051.

Malagnini, L., Herrmann, R.B., & M. Di Bona 2000a. Ground motion scaling in the Apennines (Italy). *Bull. Seism. Soc. Am.*, 90, 1062-1081.

Malagnini, L., Herrmann, R.B., & Koch, K., 2000b. Ground motion scaling in Central Europe, *Bull. Seism. Soc. Am.*, 90, 1052-1061.

Malagnini, L., Akinci, A., Herrmann, R.B., Pino, N.A. & Scognamiglio, L, 2002. Characteristics of the ground motion in Northeastern Italy, *Bull. seism. Soc. Am.*, 92, 2186–2204.

Malagnini, L., Scognamiglio, L., Mercuri, A., Akinci, A. & Mayeda, K., 2008. Strong evidence for non-similar earthquake source scaling in central Italy, *Geophys. Res. Lett.*, 35, L17303, doi:10.1029/2008GL034310.

Malagnini, L., Nielsen, S., Mayeda, K. & Boschi, E., 2010. Energy radiation from intermediate to large magnitude earthquakes: implications for dynamic fault weakening, *J. geophys. Res.*, doi:10.1029/2009JB006786.

Malagnini, L., Akinci, A., Mayeda, K., Herrmann, R.B. & Munafo', I., 2010b. Accurate predictions of strong ground motion based on weak motion data: case studies from Italy and Japan, in *Proceedings of the 2010 SSA Annual Meeting*, Portland, OR, 21–23 April 2010

Morasca, P., Malagnini, L., Akinci, A., Spallarossa, D., Herrmann, R.B., 2006. Ground motion scaling in the western Alps. J. Seismol., 10, 315-333.

Motezedian, D., & Atkinson, G. M., 2005. Stochastic finite-fault modeling based on a dynamic corner frequency, *Bull. Seism. Soc. Am.* 95, 995-1010.

Ortega, R., Herrmann, R. B., Quintanar L., 2003. Earthquake ground-motion scaling in Central Mexico between 0.7 and 7 Hz, Bull. Seism. Soc. Am. 93, 397–413

Panzera F., D'Amico S., Galea P., Lombardo G., Pace S., 2011. Vs30 estimates in Malta. In preparation

Pino, N. A., Malagnini, L., Akinci, A., Scognamiglio, L., Herrmann, R. B., Stavrakakis, G., Chouliaras, G., 2001. Ground motion scaling relationships for mainland Greece and Crete, Seism. Res. Lett. 72, 258.

Raoof, M., Herrmann, R.B. & Malagnini, L., 1999. Attenuation and excitation of three component ground motion in Southern California, *Bull. Seism. Soc. Am.* 89, 888-902

Rapolla A., Akinci A., Bruno P. P., D'Amico S., Di Fiore V., Malagnini L., Maschio L., Paoletti V., Secomandi M., Vietri A., 2008. La pericolosità sismica: dalla classificazione sismica alla microzonazione dei territori comunali, alla risposta sismica del sito. Liguori Editore

Sabetta, F., and Pugliese A., 1987. Attenuation of peak horizontal acceleration and velocity from Italian strong-motion records, Bull. Seism. Soc. Am. 77, 1491–1511.

Sabetta, F., and Pugliese A. 1996. Estimation of response spectra and simulation of nonstationary earthquake ground motion, Bull. Seism. Soc. Am. 86, 337–352.

Scognamiglio, L., Malagnini, L., Akinci, A., 2005. Ground motion scaling in eastern Sicily, Italy. Bull. Seism. Soc. Am. 95, 568-578

Wald, D., Worden, B., Quitoriano, V., Pankow, K. L., 2005. Shake Map manual: technical manual, user's guide, and software guide. U.S. Geological Survey Techiniques and Methods, book 12, section A, 132 pp.

Wells, D. L., and Coppersmith K. J., 1994. New empirical relationships among magnitude, rupture length, rupture width, rupture area, and surface displacement, Bull. Seism. Soc. Am. 84, 974–1002.

Wessel, P. & Smith, W. H. F., 1991. Free software helps map and display data, *Eos Trans.*, AGU, 72, 441

3

Application of Empirical Green's Functions in Earthquake Source, Wave Propagation and Strong Ground Motion Studies

Lawrence Hutchings and Gisela Viegas
Lawrence Berkeley National Laboratory
USA

1. Introduction

Over the last couple of decades, empirical Green's functions [EGFs] have been increasingly used in earthquake source studies, crustal attenuation studies, strong ground-motion prediction, finite rupture modeling, and site-response studies. Theoretically, Green's functions are the impulse response of the medium, and EGFs are recordings used to provide this impulse response. In this chapter, we review the theoretical and observational basis for identifying and using small earthquakes as empirical Green's functions—and their application. We generally refer to analyses that use EGFs as "the EGF method". The seismic record of an earthquake contains information on the earthquake source, the path the seismic waves propagated through, the site response of the geology beneath the recording site, and the response of the instrument that recorded the ground motion. In seismology, we are interested in isolating either source, path, or site information, depending on the study focus. The instrument response is usually known and easy to remove from recordings, but the same is not true for the other factors. Seismic waves propagate in the earth in a complex way. They are reflected and refracted at interfaces between rocks with different properties, attenuated throughout the path between the earthquake source and the station by scattering and anelastic effects, subjected to energy focusing and defocussing due to lateral changes in the refractive properties of the rock, and amplified and highly attenuated near the recording station as waves pass through unconsolidated near-surface material.

These propagation complexities are not well captured by crustal models, which provide the basis for calculating Green's functions, especially for high frequencies (> 1 Hz). At higher frequencies, wave propagation is very sensitive to small crustal heterogeneities, which are generally not well known; at low frequencies (< 1 Hz), wave propagation can be modeled fairly accurately. EGFs can be used instead of mathematical calculations to more accurately represent seismic wave propagation in the geologically heterogeneous crust. The EGF method is the best available method because it empirically corrects for unknown path and site effects, for which a short wavelength resolution is needed. However, true EGFs contain the source rupture process of the small earthquakes in the recorded seismograms. No earthquake has a true impulsive source. Therefore, one must be careful using EGFs.

There has not been uniformity in defining EGF's. Recordings of earthquakes with magnitudes less than 1.0 to 7.0 have been used as EGFs. Here, we apply a strict mathematical definition to EGFs and discuss different uses of EGFs with respect to this definition. We define EGFs as recordings of sources that satisfy impulsive-point dislocation criteria within the frequency band the EGFs are used. The word "satisfy" is included in the definition of EGFs because there is no true impulsive point source in nature. Therefore, the definition of an EGF can be dependent upon the frequency range of interest. Sources in nature may be explosions, rock bursts, earthquakes, or any impulsive source. Under certain constraints, EGFs can be used as point sources. The impulsive point source definition is consistent with the mathematical definition of an elastodynamic Green's function, which is the response at a particular location to a uni-directional, unit-impulsive, point source at another location.

EGFs were initially used in attenuation and earthquake-source studies (e.g., Bakun and Bufe, 1975, and references within). Bakun and Bufe noted that for small earthquakes (<M4), for which corner frequencies are within the frequency band most affected by attenuation and site effects, deconvolution could isolate source and propagation-path effects. Initially, the method was applied in the frequency domain by a spectral ratio method. The underlying reasoning for doing this was that common station recordings of closely located earthquakes shared the same propagation path and could be used to form "ratios of spectra," and thus "cancel the common propagation path effects" (Bakum and Bufe, 1975).

Subsequent studies further developed EGF methods. Frankel (1982), Mueller (1985), Hough et al., (1991), Abercrombie and Rice (2005), and Viegas et al. (2010) performed analysis of EGFs in the frequency domain to obtain source parameters of larger earthquakes and attenuation prosperities of the medium. Frankel and Kanamori (1983), Frankel et al. (1986), Mori and Frankel (1990), Mori et al. (2003) performed similar analysis in the time domain. Hough (1997), Hough et al. (1999), Prejean and Ellsworth (2001), Ide et al. (2003), Prieto et al. (2004), Shearer et al. (2006) analyzed multiple earthquakes simultaneously to obtain source properties and kappa (near site attenuation). Mayeda et al. (2007) and Viegas (2009) used source spectra derived from coda waves instead of direct waves to obtain source properities.

Hartzell (1978) and Wu (1978) first suggested using EGFs to calculate strong ground motion. Using small earthquakes to provide EGF's for synthesizing larger earthquakes is very practical; small earthquakes occur hundreds of times more frequently than larger earthquakes and EGFs can be readily obtained in a short period of time before a large earthquake occurs. Hartzell and Wu suggested using EGFs as the Green's function in the representation relation along with synthetic rupture processes for calculating (synthesizing) the resulting ground motion. That is, the fault of the large earthquake is represented as a summation of subfaults, or elemental point sources, for which EGFs are available. Figure 1 dramatizes the wave propagation effects that occur from different portions of a fault, so that as an earthquake ruptures, EGFs account for the different travel paths through the heterogeneous geology. Hutchings (1991) and Hartzell et al. (1999) designed particular rupture models to represent what actual earthquake might do. This followed Boatwright (1981), who designed quasi-dynamic sources. Guatteri et al. (2003) further developed source modeling by including actual dynamic rupture process in the calculations. Irikura (1984) used a relatively large earthquake as an EGF and modified it to represent the strong ground motion from an even larger earthquake (usually 1 or 2 magnitude units higher). In this approach the EGF was not isolated as an impulsive point source, and the larger earthquake was made up of fairly large sub-events. Joyner and Boore (1986) examined the statistics of how to add up sub-events to create a larger earthquake. Frankel (1991) proposed a fractal summation of scaled EGFs to represent the source process as a statistical process. Tumarkin (1994) and Abrahamson and

Bolt (1997) modified recordings of relatively large earthquakes to fit a target spectra of an even larger earthquake. Somerville et al. (1991) used the recordings near a large earthquake as an empirical source function and calculated the wave propagation effects. All these source models have different implications for the synthesis process, and are discussed below. We also examine validation procedures to verify the usefulness of methods.

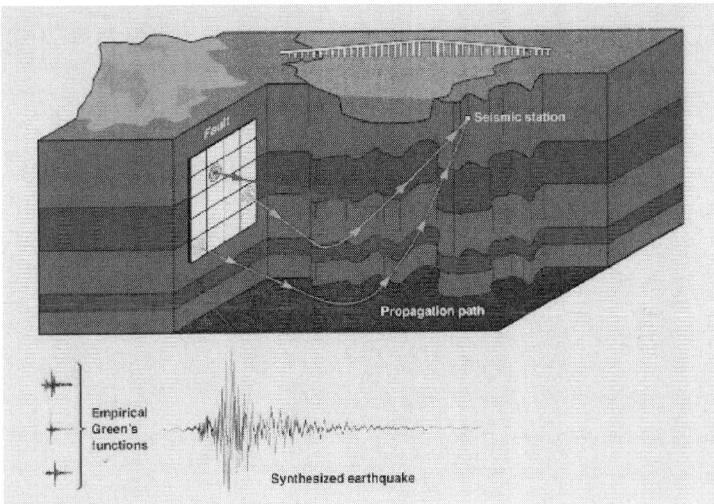

Fig. 1. Illustration of wave propagation effects that occur from different portions of a fault as an earthquake ruptures. Each EGF accounts for a different travel path through the heterogeneous geology.

There are several limitations to EGFs. The selection of an EGF for use in source studies follows strict criteria that are not always possible to fulfill. Thus, a suitable EGF function may not be available, limiting the number of earthquakes that can be analyzed. The bandwidth of the EGF for which enough signal above noise is available is another limiting factor in source studies. For very small EGF earthquakes or noisy surface stations, the available bandwidth may not be enough to perform the source analysis, even though there is sufficient signal from the earthquake being analyzed. Limited instrument bandwidth, another potentially serious limitation, may bias the source parameters. However, the EGF method profits from having a good distribution of recording stations, providing a good azimuthal coverage so that source directivity effects may be accounted for.

There are also a number of limitations in applying EGF methods to ground-motion modeling:

1. In virtually all practical applications, there are insufficient number of EGFs to provide an impulse response for all portions of a fault rupture to be modeled.
2. EGFs cannot accurately model variations in focal mechanism solutions.
3. Noise levels in recordings limit their usable frequency band (usually between 0.2 Hz to 25.0 Hz or narrower in seismically active regions such as the Western U.S).
4. A distribution of stations with good azimuthal coverage is necessary to ascertain the locations and source parameters of the small earthquakes that would provide EGFs.

5. Good-quality recordings of seismograms have to be captured at the locations of interest for ground-motion synthesis.

In this chapter, we describe the theoretical and observational basis for identifying and using small earthquakes as EGFs and review some of the typically used methods in earthquake source and strong-ground-motion studies. We also discuss the advances in the scientific knowledge of earthquake sources made possible by EGFs. Finally, we discuss using the application of the EGF method (in strong-ground-motion prediction) to probabilistic seismic-hazard analysis.

2. Theoretical considerations

The representation relation (Aki and Richards, 2002) is the fundamental, elastodynamic, mathematical description of an earthquake and the resulting ground motion. It is expressed as

$$u_n(X,t) = \int_A m_{pq}(X',t') * G_{np,q}(X',t';X,t)dA\,, \tag{1}$$

where u_n is the ground displacement in the direction $\hat{x}_n$, at location X and time t, resulting from the integral over the fault surface of the convolution of the source function $m_{pq}(X',t')$ with the derivative of the Green's function tensor $G_{np}(X',t';X,t)$, with respect to the $\hat{x}_q$ direction, at location X' and time t' on the source; $*$ is the convolution operator, and A is the fault surface. The Green's function tensor is the contribution to the displacement in the $\hat{x}_n$ direction from a unidirectional unit-impulse in direction $\hat{x}_p$. A complete description of Equation (1) can be found in Aki and Richards (2002, Chapter 3), whose notation we follow. Every EGF synthesis approach utilizes this representation relation. The differences between methods are in how the equation is solved, what the source function is, and what is used for the Green's functions.

The source function is defined as the moment density tensor m_{pq},

$$m_{pq}(X',t') = \left[s_i(X',t')\right]\hat{n}_j(X')c_{ijpq}\,, \tag{2}$$

which is the strength of the double couple with dimensions of moment per unit area. Here, $\hat{n}$ is the unit normal vector to the fault in source coordinates, and $s(X',t')$ is the source-time function. The source-time function $s_i(X',t')$ specifies the spatial and time distribution of slip along the fault in the $\hat{x}_i$ direction. [ ] symbolizes directional displacement between two sides of the fault. c_{ijpq} is the fourth-order elasticity tensor, which relates stress to strain.

For a fault surface embedded in a linearly elastic and isotropic body, and located many wavelengths away from the recording site, the moment density tensor m_{pq} is defined as

$$m_{pq}(X',t') = \lambda\,\hat{n}_k(X')\left[s_k(X',t')\right]\delta_{pq} + \mu\left\{\left[s_p(X',t')\right]\hat{n}_q(X') + \left[s_q(X',t')\right]\hat{n}_p(X')\right\}. \tag{3a}$$

Here, λ and μ are the Lamé constants at the source, and δ_{pq} is the Kronecker delta function (equal to 1 for p=q, equal to 0 for p≠q). Assuming that the fault slip is parallel to the fault surface, so that the product between the fault normal and the slip is zero, Equation (2) can be simplified to

$$m_{pq}(X',t') = \mu\{[s_p(X',t')]\hat{n}_q(X') + [s_q(X',t')]\hat{n}_p(X')\}. \tag{3b}$$

The source-time function can be described in scalar form if we assume that at each point along the fault, the slip function has the same time dependence in all directions (Johnson, 1975). Then, the moment density tensor can be simplified to

$$m_{pq}(X',t') = \mu S(X',t')\{\hat{s}_p(X')\hat{n}_q(X') + \hat{s}_q(X')\hat{n}_p(X')\}. \tag{4}$$

$S(X',t')$ is the scalar-source time function, and the symbol for differential slip is dropped. The term $(\hat{s}_v\hat{n}_q + \hat{s}_q\hat{n}_p)$ is the focal mechanism with slip in direction $\hat{s}$ and fault normal in direction $\hat{n}$. To express this relation for an arbitrary fault orientation requires general coordinates that describe the fault orientation. We use the convention of Aki and Richards (2002, page 109).

The impulsive point-source solution describes what we will call the EGF. For observed wavelengths much greater than the linear dimensions of the fault, the Green's functions are effectively in phase from all points along the fault and independent of the X' coordinates. This is the condition for a point source, and Equation (1) can then be written as

$$u_n(X,t) = \int_A m_{pq}(X',t')dA * G_{np,q}(X',t';X,t). \tag{5}$$

Substituting Equation (4) into Equation (5), we obtain

$$u_n(X,t) = \int_A \mu S(X',t')\{\hat{s}_p(X')\hat{n}_q(X') + \hat{s}_q(X')\hat{n}_p(X')\}dA * G_{np,q}(X',t';X,t). \tag{6}$$

If the periods of observed waves are also much greater than the duration of faulting, the time dependence of the source cannot be observed, and the source-time function at the source $S(X',t')$ can be replaced by a unit Heaviside step function scaled by the average slip over the fault surface, $\bar{s}\cdot H(t')$. In addition, replacing the integral over the fault surface by the total area A, and using e instead of u to describe the ground displacement associated of the EGF, Equation (6) can be rewritten as

$$e_n(X,t) = \mu A\bar{s}H(t')(\hat{s}_p\hat{n}_q + \hat{s}_q\hat{n}_p) * G_{np,q}(X',t';X,t), \tag{7}$$

in which the focal mechanism solution is that of the overall fault displacement. Defining the seismic moment of the empirical Green's function as

$$M_0^e = \mu A\bar{s}, \tag{8}$$

and substituting it into Equation (7), the observed ground motion from a point, step-impulsive source-time function can be expressed in the formal notation of a Green's function as

$$e_n(X_0',t_0';X,t) = M_0^e(\hat{s}_p\hat{n}_q + \hat{s}_q\hat{n}_p)H(t' - t_0') * G_{np,q}(X_0',t_0';X,t). \tag{9}$$

Here $e_n(X_0',t_0';X,t)$ is the solution for an EGF at position X and time t due to an impulsive-point dislocation source at the hypocenter X_0', and origin time t_0'; and $H(t'-t_0')$ is the unit step source-time function. Since a true point-source dislocation does not occur in nature, this equation only applies to recordings of earthquakes in frequencies below the source corner frequency. So, EGFs are recordings that essentially have impulsive point dislocation sources, i.e. in the frequency band of observation the spectra is flat and they cannot be distinguished from true impulsive point sources.

EGFs can be used to solve the representation relation exactly, under certain conditions. If we represent the fault area as a discrete summation of subfaults, the integral in Equation (6) can be replaced by a discrete sum over N elemental areas A_i such that ΣA_i equals the total rupture area,

$$u_n(X,t) = \sum_{i=1}^{N} \mu_i A_i S(t')_i (\hat{s}_p \hat{n}_q + \hat{s}_q \hat{n}_p)_i * G_{np,q}(X',t';X,t)_i \;. \tag{10}$$

We can use the EGF analytical solution described by Equation (9) to provide the focal mechanism $(\hat{s}_p \hat{n}_q + \hat{s}_q \hat{n}_p)$ and the derivative of the elastodynamic Green's functions $G_{np,\,q}$ for each elemental area, and substitute it into Equation (10) to get

$$u_n(X,t) = \sum_{i=1}^{N} \frac{\mu_i A_i S'(t')_i}{M_{0\,i}^{e}} * e_n\left(X',t';X,t\right)_i \;. \tag{11}$$

Here, the analytical solution for the slip function of the larger event, $S'(t')_i$ is the desired slip function for element i deconvolved by the EGF step function, $S'(t')_i = S(t') \otimes H(t'-t'_0)_i$, where $\otimes$ represents deconvolution. Both large earthquakes and EGF earthquakes are assumed to share the same focal mechanism, so that the dislocation vectors have the same orientations, and the wave phases have the same polarity and relative amplitudes. $M_{0\,i}^{e}$ is the scalar seismic moment of the source event for the EGF, and $e_n(X,t)_i$ is the EGF for the ith element obtained from recordings of small earthquakes with an effective step-source time function; u_n has the same units as e_n, and any differentiation with respect to t operates only on u_n and e_n. Each element can have a different recording as an EGF. EGFs are adjusted so that each has an origin time at the arrival of a rupture front. Because EGFs cannot be recorded from each point on the fault, they must be interpolated. (Interpolation schemes are not shown as part of Equation (11) and are discussed later in the chapter.) Unfortunately, it is not possible to decompose the EGF into a tensor component solution so that different focal-mechanism solutions could be used in Equation (9). Hutchings and Wu (1990) provide a detailed description of this problem, and we discuss its implications later in this chapter. If larger earthquakes are used to provide EGFs, then e_n has a source-time function that is not necessarily known. Then, the entire Equation (9) can be identified as the solution for the composite event, where $S'(t')_i$ is not known but part of the empirical recording.

An important form of the representation relation can be derived if the slip function is discretized into a series of step functions,

$$S(t') = \sum_{j=1}^{L} \bar{s} H(t' - \tau_j^{s}) \,, \tag{12}$$

where τ_j^s are phase delays due to the slip function, and their sum is the rise time. If the step function of the EGF is used, then the deconvolution with a step function as in Equation (11) is not necessary, and Equation (10) can be written as

$$u_n(X,t) = \sum_{i=1}^{N}\sum_{j=1}^{L} \frac{\mu_i A_i \bar{s}_{ij}}{M_{oi}^e} e_n(X,t_i^{'} - \tau_i^r - \tau_j^s)_i \ , \quad (13)$$

where s_{ij} is calculated from the slip function and τ_i^r are phase delays due to rupture. To develop this relationship further and into a form often used in ground-motion synthesis, we combine time delays and summations into a single summation,

$$u_n(X,t) = \sum_{j=1}^{\eta} \kappa_j e_n(X,t_i^{'} - \tau_j)_j \ , \quad (14)$$

where, τ_j includes all time delays, and

$$\kappa_j = \frac{\mu_j A_j \bar{s}_j}{M_{oj}^e}, \ \text{and} \ \eta = \sum_{i=1}^{N} L_i \ . \quad (15)$$

Equations (13) and (14) are often used as the form for composite earthquake ground-motion-synthesis models. In this approach, relatively large events are used as EGFs, with a large EGF source-time function employed as part of the source-time function of the larger event. Equation (14) would be expressed as

$$u_n(X,t) = \sum_{j=1}^{\eta} \kappa_j \left[\int_A m_{pq}(X^{'},t^{'}) * G_{np,q}(X^{'},t^{'};X,t) dA \right]_j \ , \quad (16)$$

where Equation (1) describes the recording of a large event used as an EGF. Then, a solution with point sources would be

$$u_n(X,t) = \sum_{j=1}^{\eta} \kappa_j \sum_{l=1}^{\varsigma} \kappa_l e_n(X,t_i^{'} - \tau_l - \tau_j)_l \ . \quad (17)$$

To examine the Fourier amplitude spectra of synthesized earthquakes with EGFs, we can express the modulus of the power spectrum of Equation (14) as

$$\left| U(\omega)^2 \right| = \sum_{j=1}^{\eta} \kappa_j^2 E_j(\omega) E_j^*(\omega) + \sum_{j=1}^{\eta} \sum_{\substack{k=1 \\ j \neq k}}^{\eta} \kappa_j \kappa_k E_j(\omega) E_k^*(\omega) e^{-i\omega\tau_j} e^{i\omega\tau_k} \ . \quad (18)$$

$E(\omega)$ is the Fourier transform of the EGF and * indicates a complex conjugate. The phase spectrum defined by the rupture process is

$$\left| P(\omega) \right| = \sum_{j=1}^{\eta} \sum_{\substack{k=1 \\ j \neq k}}^{\eta} e^{-i\omega\tau_j} e^{i\omega\tau_k} \ . \quad (19)$$

A simple form of the Fourier amplitude spectrum can be developed if it is assumed that EGF Fourier amplitude spectra are similar at a particular recording site, even if their phase spectra can be quite different. Then, the Fourier amplitude spectrum of the synthesized seismogram can be expressed as

$$|U(\omega)| = |E(\omega)| \sqrt{\sum_{j=1}^{\eta} \kappa_j^2 + \sum_{j=1}^{\eta} \sum_{\substack{k=1 \\ j \neq k}}^{\eta} \kappa_j \kappa_k \cos(\varphi_j - \varphi_k + \omega\tau_k - \omega\tau_j)}\,, \tag{20}$$

where $\varphi(\omega)$ is the phase spectrum of the EGF.
Scaling effects and the effects of different rupture parameters on the Fourier amplitude spectra are fairly easy to observe. At low frequencies, $\omega \to 0$, the phase spectrum of different Green's functions is the same, and the spectral amplitudes are expressed as

$$|U(\omega)| = |E(\omega)| \frac{M_0}{M_0^e}\,. \tag{21}$$

Equation (21) is an expression for the largest spectral amplitudes possible for the synthesized seismograms. Spectra at higher frequencies are modulated by the phase effects of different EGFs and by the delay times caused by rupture velocity and the slip function (Equation 18). The summation of these modulated spectra gives the shape of the spectra at high frequencies. It is clear, then, that the frequency content of synthesized seismograms (and actual recordings as well) is dependent upon the constructive and destructive interference effects caused by the rupture parameters and spatial variation of the Green's functions. No scaling relation is needed to describe the large earthquake other than the sum of the moments of the elemental areas, which need to add up to that of the large earthquake.
Determining earthquake source parameters typically involves the deconvolution of the target earthquake with a single EGF earthquake. This assumes that the Green's function is the same for all points along the rupture of the larger event, so this approach works well only if both earthquakes are recorded in the far field. To successfully deconvolve the large earthquake by a single EGF earthquake, further constraints have to be introduced that guarantee the suitability of the EGF earthquake. Both earthquakes need to be recorded at the same station. The larger and smaller earthquakes have to be co-located so that the deconvolution process removes all propagation-path and recording-site effects from the records of the large event. There should be at least a one-magnitude difference between the two earthquakes, so that the EGF and larger-earthquake source effects are distinguishable. Then, from Equation (1), we have

$$\begin{aligned} u_n(X,t) \otimes e_n(X,t) &= \int_A m_{pq}^u(X',t') * G_{np,q}(X',t';X,t)dA \otimes \int_A m_{pq}^e(X',t') * G_{np,q}(X',t';X,t)dA \\ &= \left[\mu A^u \bar{s}^u d^u(t'-t_0')(\hat{s}_p\hat{n}_q + \hat{s}_q\hat{n}_p)_i * G_{np,q}(X',t';X,t)\right] \otimes \\ &\qquad \left[\mu A^e \bar{s}^e d^e(t'-t_0')(\hat{s}_p\hat{n}_q + \hat{s}_q\hat{n}_p)_i * G_{np,q}(X',t';X,t)\right] \\ &= M_0^u(\hat{s}_p\hat{n}_q + \hat{s}_q\hat{n}_p)d^u(t'-t_0') \otimes M_0^e(\hat{s}_p\hat{n}_q + \hat{s}_q\hat{n}_p)d^e(t'-t_0') \end{aligned} \tag{22}$$

Here, $d(t'-t'_0)$ is a unit function that describes the shape of the source slip function, and the superscript indices u and e refer to the large and egf earthquakes. $d(t'-t'_0)$ is sometimes represented as a ramp-like function, set equal to zero before the onset of slip at time t'_0, and for which the time interval when slip occurs is called the rise time. For an impulsive point source, the rise time is instantaneous, and $d(t'-t'_0)$ takes the shape of a Heaviside unit-step function. It is common to apply the deconvolution method in the frequency domain, where the deconvolution transforms to division, thus simplifying the calculations. Equation (22) can then be written in the frequency domain as

$$\frac{u_n(X,\omega)}{e_n(X,\omega)}=\frac{M_0^u(\hat{s}_p\hat{n}_q+\hat{s}_q\hat{n}_p)D^u(\omega)}{M_0^e(\hat{s}_p\hat{n}_q+\hat{s}_q\hat{n}_p)D^e(\omega)}, \tag{23}$$

where $D(\omega)$ is the Fourier transform of $d(t'-t'_0)$. If we further assume that both events have the same focal mechanism solution, we have

$$\frac{u_n(X,\omega)}{e_n(X,\omega)}=\frac{M_0^u\,D^u(\omega)}{M_0^e\,D^e(\omega)}. \tag{24a}$$

It should be pointed out that this representation holds even when there is a component of slip perpendicular to the fault surface (Equation 3a)—that is, even if the faulting mechanism is not purely a shear mechanism—as long as both the large earthquake and the EGF earthquake have the same focal mechanism, so that it will cancel out when the deconvolution or the spectral division is performed.

The displacement field solved for a general second-order symmetric moment tensor (point source) solution in an infinite homogeneous medium shows three relative-magnitude terms that depend on the source-receiver distance: near-field, intermediate-field, and far-field (Aki and Richards, 2002, eq. 4.29). In the far-field term, the shape of the displacement pulse is proportional to the particle velocity at the source averaged over the fault plane. The displacement field can then be approximated by its far-field representation, and the particle slip-time function at the source $d(t'\ t'_0)$ can be approximated by the particle velocity-time function at the source $\dot{d}(t'-t'_0)$. If the source slip function has a ramp or Heaviside function shape, the displacement pulse, or far-field source-time function, will have a boxcar or a delta Dirac function shape, respectively. Equation (24a) can then by expressed as

$$\frac{u_n(X,\omega)}{e_n(X,\omega)}\approx\frac{M_0^u\,\dot{D}^u(\omega)}{M_0^e\,\dot{D}^e(\omega)}=\frac{M_0^u}{M_0^e}\dot{D}'(\omega). \tag{24b}$$

$\dot{D}'(\omega)$ is the far-field unit-normalized relative source-time function; its amplitude is proportional to the ratio of the moments. If we inverse-transform to the time domain, we get $\dot{d}'(t')$, the far-field relative source-time function. Often, small earthquakes have moments greater than the threshold necessary to be effective impulsive point sources. However, if both earthquakes share the same focal mechanism, the deconvolution method can still be applied. Care must be taken, however, when measuring source durations from the source pulses, because it is a relative duration. Figure 2 shows a deconvolution of a M = 3.5 earthquake with a M = 2.3 earthquake, and the inverse transform to the time domain to obtain $\dot{d}'(t')$. If the smaller event is not effectively an impulsive point source, then we can only achieve relative values for the larger event.

Kinematic Earthquake Models and Synthesized Ground Motion Using Empirical Green's Functions

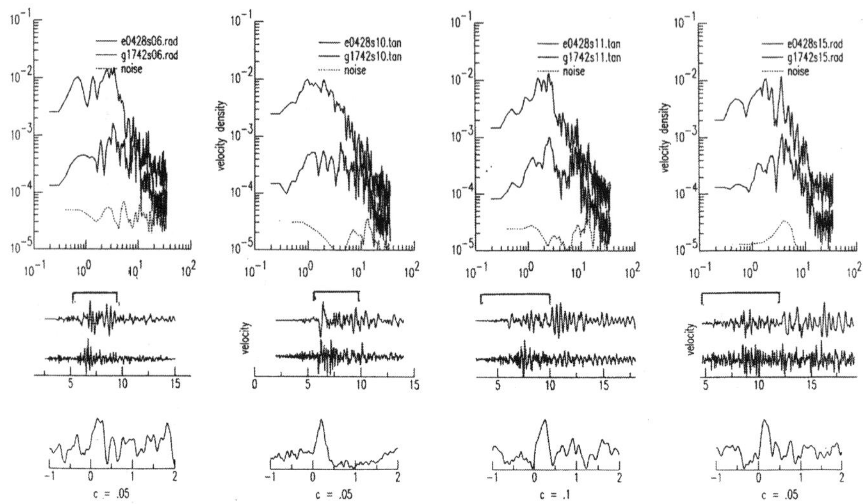

Fig. 2. Deconvolution at four stations of a M3.5 larger earthquake with a M2.3 earthquake, which is assumed to be effectively an EGF. *Top:* velocity spectra of both earthquakes. *Middle*: recorded velocity seismograms and time window used in the analysis. *Bottom*: relative source time function rate (from Hutchings, 1991).

3. Observations of Empirical Green's Functions (EGFs)

By the definition stated previously in section 2.0, Equation (9), EGFs are recordings of earthquakes for which the effects of rupture and size cannot be observed in the frequency range of usage. This definition provides a basis for understanding the influence of frequency band, geology, and the source rupture on recorded seismograms used as EGFs. Theoretically, the Fourier amplitude spectra of displacement will be flat up to a source corner frequency, which is inversely proportional to the duration of the earthquake rupture. The duration of an earthquake is dependent on the fault dimensions and the earthquake rupture process. A flat Fourier amplitude spectrum is the result of a zero duration source, or source with shorter duration than the period of the highest frequency in the spectrum. Therefore, seismograms filtered to include frequencies below their source corner frequency exclusively do not include the effects of duration, and they will effectively be impulsive point sources for that frequency range. If they are also located close to one another and have the same focal mechanism solution, they will have nearly identical waveforms over that frequency range. In fact, waveform similarity is the best selection criteria for a good EGF earthquake.

Figure 3 shows several recordings of earthquakes along the Hayward fault in California with very closely located hypocenters and nearly identical focal mechanisms, but varying moments. The earthquakes range in magnitude from 1.8 to 4.1 and the recorder is 10 km distant. The dimension of the sources ranges from about 20 m to 500 m. Also shown are the Fourier amplitude spectra of the recordings (Figure 3c). It is apparent that the source corner frequency decreases for larger magnitude earthquakes, and that all spectra are approximately flat below 2 Hz. Unfiltered (Figure 3a), the records appear quite different. However, when filtered to include only frequencies less than the corner frequency (Figure 3b) of the largest earthquake (2 Hz), they appear nearly identical, and are effectively from point sources.

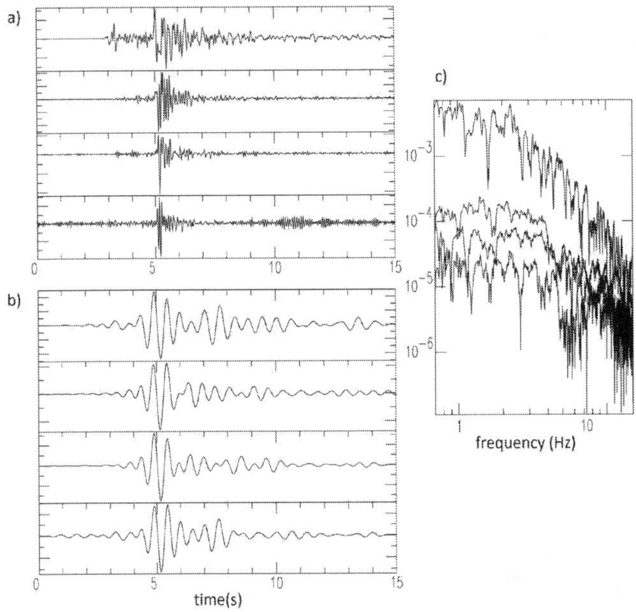

Fig. 3. (a) Unfiltered records of events with different magnitudes that occurred at approximately the same location along the Hayward fault and were recorded at the same station (YBI). (b) Band-pass filtered between 1.0 Hz and 2.0 Hz, for which there is no source effect and signal to noise is good. (c) Fourier amplitude spectra of the recordings.

The use of EGFs has to be quantified for the low frequency end of the spectrum as well. Most recordings of small earthquakes are band limited by cultural noise or instrument response, and do not include frequencies below about 0.5 Hz, so that full broadband use of EGFs is not generally possible. This is the frequency range wherein near-field terms are important for source studies and strong ground-motion synthesis. However, it has long been recognized that synthetic seismograms can be accurately calculated for frequencies less than 0.5 Hz in areas where the geology is well known. Often, tomographic inversion for velocity structure provides the basic information necessary to accurately synthesize low-frequency seismograms (Gök and Hutchings, 2004). Thus, there is a convenient match between the synthetic Green's functions that can be accurately calculated for frequencies commensurate with the scale in which the geology can be well known, and can include the near-field wave propagation effects that are difficult to record—and the EGFs that record the high frequencies for which the geology is not well known and for which primarily far-field arrivals are important. Merging the two frequency bands provides a broadband simulation, so that synthetic Green's functions can be used for frequencies 0.0 to about 0.5 Hz, and EGFs can generally be used for frequencies from about 0.5 to 25.0 Hz (or higher for stable continental interiors). This covers the frequency range of engineering interest.

3.1 Scaling relations

A significant subject in seismology is how earthquakes scale—that is, whether earthquakes are self-similar and follow a constant scaling relationship (constant stress drop) between the earthquake moment and its size, or whether there is a non-constant scaling relationship. Constant stress drop is widely accepted as a first-order approximation for large and moderate earthquakes (Aki, 1967), although stress-drop values show a wide variation from about 0.1 MPa to 100 MPa (e.g., Hanks, 1977). However, a non-constant scaling relationship has been proposed for small earthquakes, based on the observation of a small systematic shift to lower apparent stress (scaled energy multiplied by the medium rigidity) at lower magnitudes (Figure 4). Energy scaling relationships remain unresolved. There are contradictory studies arguing for self-similarity where scaled energy and apparent stress are constant (e.g., McGarr, 1999; Ide and Beroza, 2001; Ide et al., 2003) and for non-self-similarity where scaled energy decreases with earthquake size (e.g., Abercrombie, 1995; Prejean and Ellsworth, 2001; Mayeda and Walter, 1996). This decrease in scaled energy suggests differences (with earthquake size) in frictional behavior during rupture (Kanamori and Heaton, 2000).

The scatter in stress-drop estimates reflects the different tectonic regimes in which the earthquake occur, variations within a regime, the diversity of methods used to retrieve source parameters, data quality, and the available bandwidth. Uncertainties in stress-drop estimates propagate from seismic moment and source-dimension estimates, since stress drop is directly proportional to seismic moment and inversely proportional to the source dimension cubed (see Equation 33). Today, most methods estimate seismic moment within a factor of 2, but there are still huge uncertainties in fault-dimension estimates. Fault dimension is determined from the corner frequency in spectral analyses or, alternatively, from the source duration in time analyses (see equations 31 and 32, respectively). Retrieving accurate small-earthquake corner frequencies is particularly challenging, because it requires path and site effects corrections, which are not easily determined. Also, good-quality data with high-frequency content is required to estimate corner frequencies of small to moderate earthquakes. Limited bandwidth introduces a false cut-off at high frequencies and a bias in earthquake selection, leading to the underestimation of corner frequency. In addition, attenuation-corrected corner frequencies are higher than the observed corner frequencies (Hough et al., 1989; Anderson, 1986; Ide et al., 2003). Consequently, uncorrected corner frequencies lead to an underestimation of stress drop and could be responsible for an apparent non-constant scaling relationship. Historically, the bias in stress drop and thus the non-constant scaling relationship introduced by an inaccurate corner frequency (or rupture duration) estimate was argued by many to be a source effect and not a path effect. To date, the EGF deconvolution method is the best available method for retrieving accurate earthquake source properties, because it empirically corrects for path and site effects. Nevertheless, it depends on the quality of the EGF, which many studies to not adequately test.

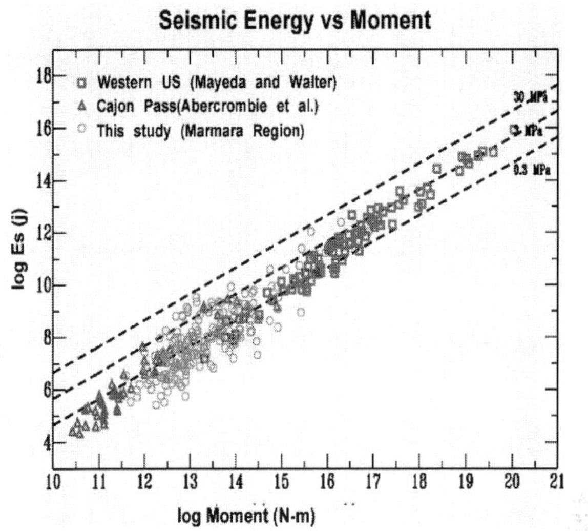

Fig. 4. Moment and energy estimation from several studies, and line of constant stress drop. A decrease in stress drop with decreasing earthquake size is observed within each dataset. (Abercrombie, 1995; Mayeda and Walter, 1996; Gök et al., 2003) These studies used a similar approach to calculate energy, so that if there is a bias in the approach, then the possibility exists that small earthquakes are plotted at lower energy values than they should be, thus causing the apparent shift. (figure from Gök et al., 2003).

3.2 *fmax*, κ, and t^*

Parameters *fmax*, κ (kappa), and t^* (t-star) are important parameters used in understanding EGFs. The corner frequency of spectra is usually observed on log-log plots, and *fmax* is an observed corner frequency of spectra that is constant for different sizes of earthquakes. A Brune source model with 10 MPa stress drop predicts a source corner frequency of 10 Hz and 25 Hz for events with moment of 1.0×10^{14} Nm and 1.0×10^{13} Nm, respectively. However, such events generally are observed to have the same corner frequency near 10 Hz. Recently it has been shown that this constant corner frequency is due to attenuation, where the source corner frequency is much higher than can be observed in the frequency range recorded, so only the effect of attenuation is observed. If the spectra is plotted on a log-linear scale, the plot is a straight line and the slope is called κ. So, below a certain size earthquake, when the source corner frequency is very high, *fmax* and κ are measures of the filtering effect of the site geology.

Parameters κ and *fmax* are not a physical model for attenuation. A common interpretation is that attenuation is caused by absorption of energy expressed by the well-known seismic quality factor Q, such that $e^{-\omega(R/Qv)}$ is the attenuation operator in the frequency domain, where R is travel distance, v the seismic velocity, and ω the angular frequency. The terms in parentheses have units of time and are often referred to as t^*. Therefore,

$$e^{-\omega(R/Qv)} = e^{-\omega\kappa} = e^{-\omega t^*}. \tag{25}$$

However, t^* generally refers to any path attenuation operator between the earthquake and the site, and κ generally refers to the attenuation just beneath a site, and is often equated with a site characterization (Anderson and Hough, 1984) ; *fmax* is the resulting effect of attenuation and is generally assumed to be caused by the same near-site attenuation as κ. However, in reality, *fmax* can result from any whole-path attenuation effect that acts as a filter and limits the frequency band of observations. Baise et al. (2001) observed, for example, that the whole-path attenuation band limited recordings in a 61 m borehole at Yerba Buena Island in San Francisco Bay to about 3 Hz, such that events as large as magnitude 4.0 have essentially the same observed corner frequency as the smallest events that could be detected (M2.0). They attributed this to a regional effect of a heterogeneous and highly attenuating Franciscan formation.

Hanks (1982) and Anderson and Hough (1984) found the threshold moment for constant κ and *fmax* to be about 1.5×10^{14} Nm. Interestingly, Hutchings and Wu (1990) plotted pulse widths of the first arriving P-waves and found the same effect, a constant pulse width below a moment threshold of 1.5×10^{14}. Once a source pulse of an earthquake is generated, it broadens as it propagates due to attenuation. This threshold is dependent on the geology, and for boreholes the threshold can be substantially lower.

Abercrombie (1998) addresses the apparent paradox that *fmax* is near 10 Hz, no matter what the geology of the site region. Her borehole and literature analysis concludes that micro-cracks in the geology that occur due to the diminishing lithostatic load near the surface causes intrinsic attenuation, which is essentially the same regardless of the geology. This results in a Q of about 10 to 100 near the surface. Inelastic attenuation, attributed to energy absorption, is very low at greater depths, and intrinsic attenuation is minimal as well, so the combined Q approaches 1000. Very near the surface, inelastic attenuation in soft soils causes significant loss of energy.

Some authors have attributed this constant maximum corner frequency to be the result of a minimum source dimension for earthquakes, which would result in a decrease in stress drop for smaller events (Archuleta et al., 1982; Archuleta 1986; Papageorgiou and Aki, 1983). However, a wide body of literature supports the interpretation that the constant maximum corner frequency (or minimum source dimension) results from whole-path or near-site attenuation. First, several authors have noticed that this apparent maximum corner frequency was different at different recording sites, so that it had a site dependence and not a source dependence (Anderson and Hough, 1984; Frankel and Wennerberg, 1989; Hutchings and Wu, 1990; Blakeslee and Malin, 1991; Aster and Shearer, 1991). Further, Abercrombie (1995) estimated source corner frequencies from events recorded in granite at a depth of 2.5 km in a Cajon Pass scientific drill hole, avoiding the highly attenuating surface layers, and observed corner frequencies about a factor of 10 higher than observed at the surface. They attributed this to near-surface attenuation. Abercrombie also observed that corner frequencies of recordings at depth continued to increase with decreasing seismic moment and found to that they follow self-similarity. Prejean and Ellsworth (2001) made the same observations about near-surface attenuation and increasing corner frequencies with decreasing moments, from a 2 km deep borehole in the Long Valley Caldera. These observations are supported by previous, less definitive studies (Malin and Walter, 1985; Hauksson et al., 1987; Aster and Shearer, 1991). In sum, the overriding evidence is that attenuation near the surface of the crust acts as a filter on recordings of seismograms, limiting the observable maximum corner frequency—so that this apparent non-constant stress-drop scaling was caused by path effects. However, the non-constant scaling of scaled energy referred to in the previous section is apparent after the (more or less successful) removal of all path and bandwidth effects, so that the controversy regarding source scaling still continues. Until uncertainties in energy measurements can be significantly reduced, the issue will not be resolved.

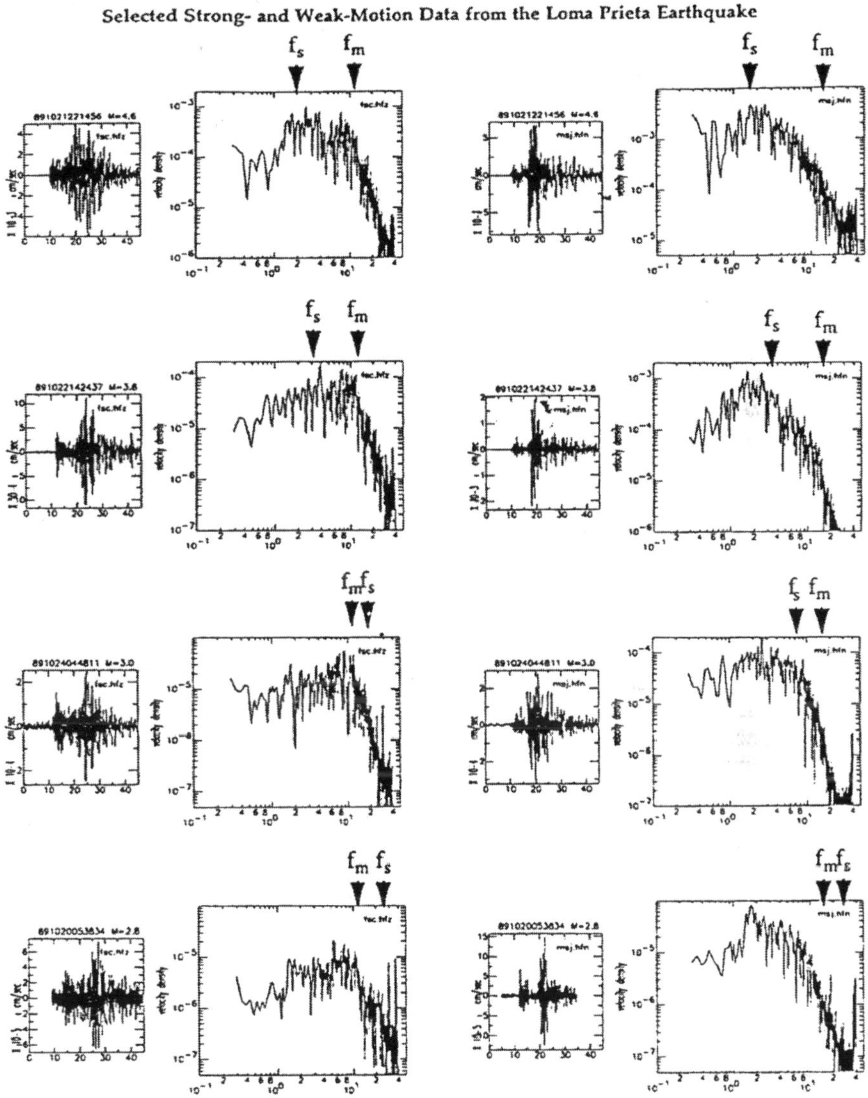

Fig. 5. Events of different magnitudes recorded at two different locations. *fmax* (f_m) is different for each site, but remains the same as events change in magnitude. The source corner frequency (f_s) gets smaller as the magnitudes increase.

The parameter *fmax* provides a convenient threshold for EGFs. Recordings of events below the threshold moment do not have significant effects due to the source; i.e., the true source corner frequency is higher than *fmax*. Hutchings and Wu (1990) confirmed this by showing that events located very close to each other, with the same focal mechanism solution and with moments below the threshold for *fmax*, had virtually identical waveforms that scaled linearly in amplitude only with differences in their moments. Figure 5 shows several events at the same location with moments above and below the threshold. The source corner frequency and *fmax* corner frequencies are plotted in the figure. As the events get smaller and smaller, the source corner frequency moves to higher frequencies, but the *fmax* corner remains the same. An interesting question regarding *fmax* is that if there is a spectral fall-off caused by attenuation and an apparent source corner frequency in the spectra, then there should be an additional fall-off at frequencies above the actual source corner frequency (Jarpe et al., 1989). This is apparent in Figure 5.

3.3 Finite bandwidth

The fact that earthquake records are limited by finite bandwidth can introduce significant bias when solving for scaling relationships. The sampling rate constrains the dataset resolution (Nyquist frequency) of the maximum corner frequency and smallest pulse duration that can be observed. The recording instrument characteristics also constrain the lowest frequency for which signals can be recorded. Therefore, when choosing the small events in a data set, we are limited to observing corner frequencies less than the maximum frequency exclusively, and we are thus biased because higher corner frequency events are excluded. The opposite will happen for the selection of the largest magnitude earthquakes in the dataset. In that case, the selection bias is choosing only lower corner frequencies. Also, in subsequent interpretation, very high and very low stress drop events will be excluded. Figure 6 shows the effect of the high frequency bias. The red dots that plot in a straight line show the upper limit of stress drop that can be interpreted with a Nyquist frequency of 80 Hz. The "true" corner frequency of these events is above the line. The other red dots are the result of an EGF study. They show a scatter with a slope trending parallel to the upper limit straight line. Viegas et al. (2010) pointed out that the apparent slope of the stress values as a function of moment is due to the elimination of higher stress drop values imposed by the Nyquist frequency.

The blue dots in Figure 6 show stress drop values obtained with non-EGF approaches. In this case, researchers have to model attenuation which is very often done incorrectly and subsequently cause an apparent corner frequency. There is a large trade-off between attenuation and apparent corner frequency (Anderson and Hough, 1984), and many possible combinations of these two variables provide equally good fits to the source spectra data (Hough et al., 1991). The EGF method uses the actual attenuation, so it does not suffer from this effect. The incorrect estimation of attenuation can result in miss identification of the source corner frequency. Such is apparent the case of the blue dots as they show a parallel line to upper limit imposed by the Nyquist frequency. In this case the calculated attenuation value systematically forced an apparent corner frequency just below the Nyquist frequency. So, the apparent stress drop variation as a function of moment is due to the upper limit of imposed by the Nyquist frequency and the choice of attenuation and not a source effect. This may be one of the causes of interpretations of stress-drop scaling with magnitude (see section 3.1). Ide and Beroza (2001) showed that most of the dependence on size interpreted for small earthquakes in apparent stress versus seismic moment relationships results from limited bandwidth.

The finite nature of the bandwidth has also to be taken into account when applying the EGF method in the time domain. The duration of the delta function has to be estimated based on the sampling rate and number of points needed to represent it, so that only earthquakes with source durations above the delta function are considered, and a bias toward decreasing stress drop with decreasing moment is avoided.

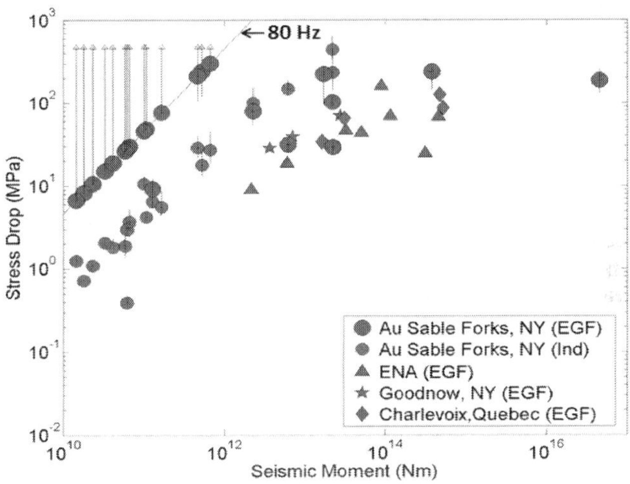

Fig. 6. Stress drop versus Seismic moment. Comparison of stress drop results obtained with the EGF method (red symbols) with the ones obtained with an individual fitting method (blue circles). All results were converted to a Madariaga (1976) source model for comparison purposes. Arrows indicate a minimum stress drop estimate. Data points are: Au Sable Forks, NY (EGF and Ind, Viegas et al., 2010); ENA (Shi et al., 1998); Goodnow, NY (Xie et al., 1991); Charlevoix, Quebec (Li et al., 1995). (Adapted from Viegas et al., 2010).

4. Earthquake source parameters

Earthquakes are characterized by their source parameters, such as hypocentral location, equivalent fault radius, stress drop, apparent stress, radiated energy, and focal mechanism. These parameters are routinely determined for most moderate to large events (>M4) in a straightforward way. Events are usually well recorded by regional seismographs and often with broadband recording capabilities, with frequencies as low as 0.1 Hz or even 0.01 Hz. Moments are easily calculated from the long period asymptote on displacement spectra (Aki and Richards, 2002, sec. 3.4) or by moment-tensor inversion programs (Dreger, 2003; Pasyanos et al., 1996). Corner frequencies, from which source dimensions are calculated, are mostly less than 0.5 Hz (depending on the tectonic setting), below frequencies for which there are significant site response and attenuation effects on their spectra. For these events, corner frequencies can be identified by the intercept of the flat, low-frequency portion of displacement spectra with the near ω^{-2} fall-off of the higher frequencies. A second corner frequency (*fmax*, discussed above) is separated enough in the spectra from the source corner frequency that misidentification or interference is unlikely. Calculating hypocentral location and focal mechanism solutions are not discussed here. Typically, earthquakes are recorded by local or regional networks and these parameters can be well constrained. Estimating these parameters is well described in many textbooks (e.g., Lay and Wallace, 1995).

Determining the moment and source corner frequency of smaller earthquakes is much more difficult. Source corner frequencies of small events are higher than 1 Hz and must be determined from the portion of the spectra that are greatly affected by site response. The spectra of small events are usually band-limited by cultural noise, so only frequencies above ~0.2 to 0.5 Hz and below about 15 to 25 Hz are usually recorded for surface sites. This is the frequency band for site response and near-surface attenuation effects on spectra can be significant. Even frequencies as low as 0.2 Hz for soil sites are effected by site response (Jarpe and Kasameyer, 1996; and Bonilla et al., 1997). Further some rock sites can also have a significant site response effect (Steidl et al., 1996). It is in determining the source parameters of small earthquakes that EGF methods prove to be particularly useful, because propagation-path and site-response effects are empirically removed.

4.1 EGF methods for determining source parameters

Here we discuss the EGF method for determining source parameters, whereby a small earthquake located in close proximity to a larger earthquake, and with the same focal mechanism, is used to remove the propagation path and site response from the records of the larger earthquake. Although all EGF methods share the same basic principles, there are variations in the way the method is applied. It can be applied in the frequency domain, time domain, or both, using direct, surface, or coda waves, to a single pair of earthquakes at a time or to multiple earthquakes simultaneously. In this section, we briefly review the various EGF methods, indicating their main advantages and disadvantages.

4.1.1 Frequency domain EGF

The EGF method was initially applied in the frequency domain and was designated by a spectral ratio method. Bakum and Bufe (1975) were one of the first to use the method to determine attenuation characteristics for propagation paths along the San Andreas Fault trace. The method was further developed and applied to source studies (e.g., Frankel 1982; Mueller, 1985; Hutchings and Wu, 1990; Hough et al., 1991; Abercrombie and Rice 2005; Viegas et al., 2010) commonly used today.

The source spectral ratio of P or S wave radiation between two earthquakes is calculated assuming a point-source approximation. First, the complex Fourier spectra of the displacement records are calculated, after correcting for instrument response. The theoretical far-field P and S wave radiation from a point source is calculated by solving Equation (7) for the specific wave type motion (P, SH, SV). The Fourier amplitude displacement spectra of P or S wave displacement records can be modeled using the theoretical spectrum equation (Brune, 1970)

$$\Omega(f) = \frac{\Omega_0 \, e^{-\frac{\pi f t}{Q}}}{\left[1+\left(\frac{f}{f_c}\right)^{\gamma n}\right]^{\frac{1}{\gamma}}}, \tag{26}$$

where $\Omega(f)$ is the displacement amplitude spectrum, Ω_0 is the long-period level (or asymptote) of the spectrum, f_c is the corner frequency, t is the travel time of the seismic wave of interest (P or S), Q is the frequency-independent whole-path attenuation quality factor, and γ and n are constants that control the shape of the spectrum curvature around the corner frequency and the high frequency fall-off, respectively. For γ =1 and n = 2, Equation (26) corresponds to the Brune (1970) source model, and $\gamma = n = 2$ corresponds to the modified Boatwright (1980) source model. Boatwright's spectrum has a sharper corner frequency than Brune's. There is no rationale for which spectral model to chose; usually, the spectral shape that best fits the observed spectrum is the one used. Although an approximation to a one-corner theoretical spectra is usually assumed, theoretical source spectra should have two corner frequencies, one representing the particle dislocation history or rise time, the other representing the fault finiteness. Depending on the fault geometry and slip history, the two corners will be closer in frequency or further apart, thus being better approximated by a sharper or smoother one-corner frequency model, respectively. $e^{(-\pi f t/Q)}$ is the attenuation operator and is equal to 1 for null attenuation ($Q \to \infty$). After calculating the complex spectra of both earthquakes, the complex spectral division is performed. The corresponding amplitude spectral ratio, $\Omega_r(f)$, can be modeled using (Abercrombie and Rice, 2005)

$$\Omega_r(f)=\Omega_{0r}\left[\frac{1+\left(\frac{f}{fc_2}\right)^{\gamma n}}{1+\left(\frac{f}{fc_1}\right)^{\gamma n}}\right]^{\frac{1}{\gamma}} . \tag{27}$$

Equation (27) represents the theoretical spectral ratio and corresponds to the division of the large earthquake by the EGF earthquake theoretical source spectra, as described by Equation (23), assuming that the attenuation operator is the same for both spectra. fc_1 and fc_2 are the corner frequencies of the large and EGF earthquakes, respectively, and Ω_{0r} is the relative long-period level (or asymptote). The spectral ratios are modeled in a bandwidth within which there are good signal-to-noise ratios for both earthquakes, which usually corresponds to the usable bandwidth of the EGF earthquake, since the larger earthquake has a stronger signal over a wider frequency band. The spectral ratio is modeled so that the corner frequencies of both earthquakes and the ratio between the two long-period levels are obtained. The spectral ratio is usually fit with Equation (27) using some type of nonlinear regression, so that the objective function that measures the differences between the model and the data is minimized. Viegas et al. (2010) minimized the function

$$res=\sum_{i=1}^{N}\left(\log\Omega_{rM}(f_i)-\log\Omega_{rD}(f_i)\right)^2 , \tag{28}$$

where the indices M and D refer to model and data, respectively, and N is the number of points in the spectral ratio, using a Nelder-Mead simplex method.

Some studies additionally perform an inverse Fourier transform of the complex spectral ratio back to the time domain, to obtain the earthquake's far-field relative source-time function (e.g., Tomic et al., 2009; Viegas et al., 2010; Viegas 2011). A clear source pulse would indicate that the EGF is sufficient in both amplitude and phase. If the corner frequencies of the two earthquakes are different by a sufficient value, and the signal-to-noise ratio is adequate over a broad-enough frequency range, this method offers the opportunity to simultaneously extract the corner frequencies of both earthquakes, after having the path effects removed. However, Abercrombie et al. (2012) found that the corner frequency of the smaller earthquake in the ratio (EGF) is often underestimated. A good approach is to use successive smaller earthquakes as EGFs of succesivelly smaller earthquakes until the resolution limit is reached. Figure 7 exemplifies the EGF spectral-ratio-fit analysis applied to two pairs of aftershocks of the M5 2002 Au Sable Forks, NY, earthquake. Two examples are shown, one in which the corner frequency of the EGF earthquake is outside the bandwidth and cannot be determined, and another in which the corner frequencies of both earthquakes are within the usable bandwidth and can be determined.

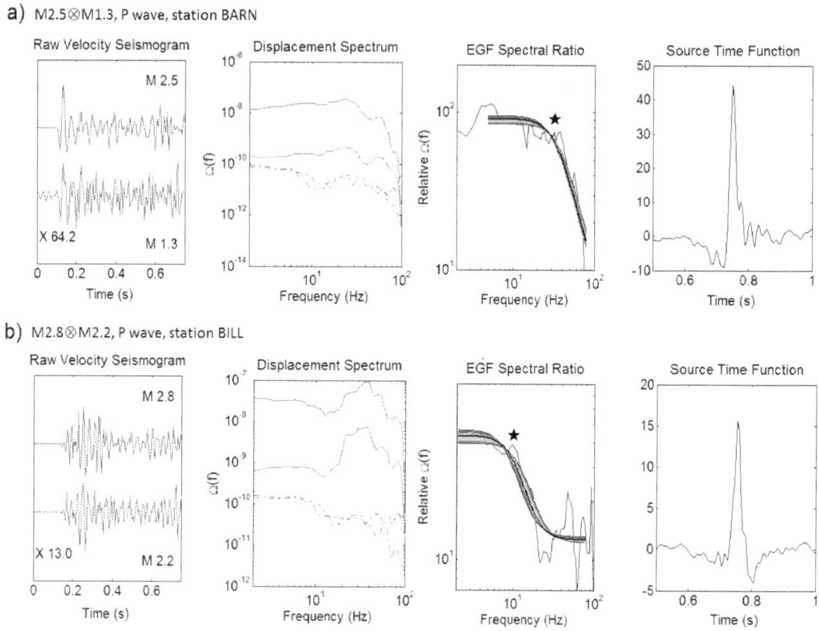

Fig. 7. EGF spectral-ratio-fit analysis. Example of the EGF spectral-ratio-fit analysis of two small earthquakes (P waves). *Left panel:* P wave raw velocity seismograms of the two aftershocks of the M5 2002 Au Sable Forks, New York, USA, earthquake. The number on the bottom left corner of the plot indicates the multiplication factor applied to the EGF record to match the amplitude of the larger earthquake Magnitudes are indicated on the right. *Central left panel:* Calculated displacement spectra of the seismograms on the left panel, after integration and instrument correction. The full lines correspond to wave signal and the dotted lines to pre-P noise. *Central right panel:* Best fit (thick black line) model to the spectral ratio (thin black line). The star indicates the best-fit corner frequency of the larger earthquake. *Right panel:* Obtained Earthquake source time function (adapted from Viegas et al., 2010).

4.1.2 Time domain EGF

The EGF time-domain deconvolution method is also commonly used in the time domain (e.g., Frankel and Kanamori, 1983; Frankel *et al.*, 1986; Mori and Frankel, 1990; Mori et al.,

2003). Figure 2 shows source pulses obtained through the time deconvolution of two similar seismograms. Once the pulse duration is measured, the source dimension can be estimated. Care must be taken in measuring the source pulse, since it is a relative source pulse. Even if the EGF is an effective impulsive point source, because discrete time series are being used and are subjected to a certain sampling rate (and Nyquist frequency) and filtering techniques, the delta function will have a minimum duration greater than three data points. This effect will decrease the relative duration of the source pulse. A large difference in magnitude between the two earthquakes, such that the duration of the target earthquake is many times that of the EGF earthquake (or delta function), will help minimize these effects. Some attention must also be paid to measuring the onset of the pulse, because noise introduces ambiguity in the deconvolved trace. Some studies use the zero crossing as the onset, others the change in gradient. Mori et al. (2003) picked the onset of the pulse on the original data, which is usually clearer, and then calculated the pulse start time.

4.1.3 Multiple EGFs

When dealing with large datasets, investigators find it preferable to apply the EGF method simultaneously to a large number of earthquakes. Hough (1997) first proposed the method called Multiple EGF (see Section 4.3 below), which was adopted by later studies with slight changes (e.g., Hough et al., 1999; Prejean and Ellsworth 2001; Ide et al., 2003). The method is the same as the standard spectral-ratio EGF method for determining source parameters, but takes advantage of multiple records in the inversion for attenuation. Here, it simultaneously inverts for kappa using a cluster of closely located earthquakes constrained by the calculated EGF source parameters. Using large numbers of same-station records for clustered earthquakes decreases the uncertainty in the source-station attenuation estimates.

Prieto et al. (2004) used a stacking approach to estimate source parameters of over 400 M0.5-to-M3.4 earthquakes recorded in southern California, following the method developed by Warren and Shearer (2002) to calculate upper mantle attenuation. The method takes advantage of a large dataset in which there are multiple receivers for each source and multiple sources for each receiver. Prieto et al. uses several events recorded at one station to calculate the site response. They do this for all stations. Then, source parameter for a single earthquake recoded at many stations is determined by simultaneously fitting all corrected spectra (Andrews, 1986; Boatwright et al., 1991). The process is repeated iteratively until a best solution is found. The obtained source spectra of all the events are relative stacked spectra and need to be scaled to absolute values using magnitude information. So, the spectra are averaged in magnitude bin intervals and corrected for attenuation by using the smallest bin spectrum as an EGF for the larger moment bin spectra.

The stack method has been applied to a few large datasets. Shearer et al. (2006) used it to determine the source parameters of 68,803 events of M_L0.5 to M_L 5.2 recorded in Southern California between 1989 and 2001. In that study, the authors used a theoretical spectral shape to fit the stacked spectra and solve for the EGF, instead of using the smallest bin spectrum as an EGF, so that lower-magnitude bin spectra close to the minimum bin magnitude could be used. Allmann and Shearer (2007) applied this method to determine temporal and spatial stress-drop variations in 42,367 earthquakes recorded between 1984 and 2005 within the Parkfield segment of the San Andreas fault, and Allmann and Shearer (2009) applied it to determine global variations in stress drop using ~2,000 >M5.5 earthquakes recorded globally between 1990 and 2007. Because it deals with such huge datasets through staking, grouping, and averaging, this method can only resolve average source properties, limited by theoretical source models and average velocity models. However, the relative variations in stress drop between earthquakes are robust, and the method is efficient when looking for spatial and temporal changes within large datasets.

4.1.4 Coda EGF

Mayeda et al. (2003), Baltay et al. (2010), and Viegas et al. (2011), among others, have used a version of the EGF deconvolution method to obtain earthquake source parameters using coda waves. Coda waves are waves that were backscattered multiple times from numerous heterogeneities in the earth and arrive later in the seismogram. Because coda waves sample the crustal volume multiple times they can provide an average model of the geology and the earthquake source, since all source directivity and radiation effects were averaged out. The Fourier amplitude source spectra derived from coda waves is dimensionless and independent for each frequency band and needs to be tied to some form of a theoretical model to get the shape of a source spectra and seismic moment units. First, "calibration events" are selected for which independent moment estimates are available. These usually are the larger events in the dataset. Then, a source spectral shape is estimated for the calibration events by choosing a stress drop, and thus a corner frequency. The Fourier amplitude spectral values estimated from the coda that don't fall on the theoretical plot are assumed to be representative of the site response. This term is subtracted from all the other spectra. If the original estimation of stress drop is correct, then the spectral shape from the smallest event will be flat up to its high corner frequency. If it is not flat, then a different stress drop is estimated, and the process is repeated until all the events are flat below their corner frequencies (Walter and Mayeda, 1996). This can be seen as an EGF approach, where the spectral shape of the smallest earthquake, which is mostly flat, is used to correct the spectra of the larger events. For an effective point source, the EGF spectra must be

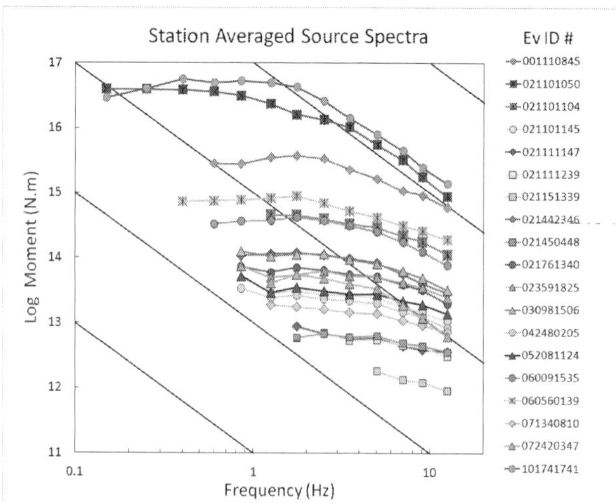

Fig. 8. Coda-derived earthquake source spectra for a series of M2 to M5 Eastern North America earthquakes. The earthquakes used as magnitude calibration events are plotted with a square plus a cross symbol. An apparent stress drop of 10 MPa was used for the theoretical spectra (from Viegas et al., 2011).

completely flat within the available bandwidth, but this is not the typical case, and corner frequencies are usually modeled by assuming a certain stress drop (Figure 8). Although minimal, this assumption can propagate to the other spectra and bias the corner frequency estimates. To avoid this, Mayeda et al. (2007) and Viegas (2009) started using spectral ratios from coda-derived individual source spectra. Figure 9 shows coda-derived spectral ratios for Parkfield, California, USA, earthquakes using the same M1 EGF with two M2 earthquakes. The advantage of using coda waves is that because of its averaging properties, source parameters can be estimated from a single station, the equivalent to a 9- to 25-station network average using direct waves (Mayeda et al., 2003). The big disadvantage of using coda waves is that precisely because of its averaging properties, source directivity cannot be determined, and the fault plane cannot be identified.

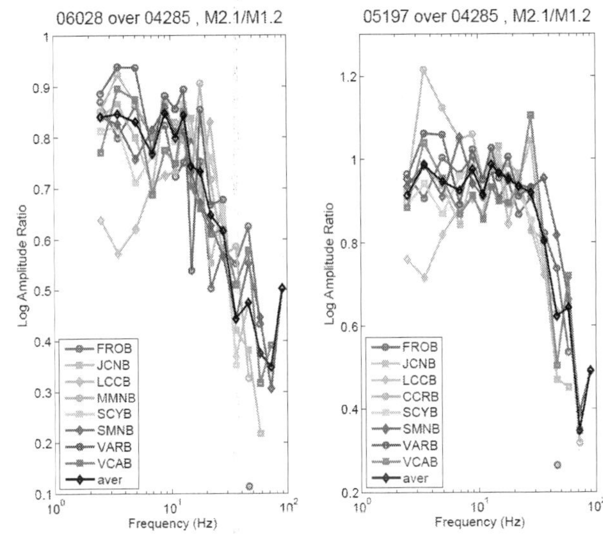

Fig. 9. Amplitude spectral ratio for two Parkfield target earthquake pairs (M2.1/M1.2) at several stations. The black line with diamonds is the average spectral ratio (from Viegas, 2009).

4.1.5 EGF earthquake selection criteria

The selection of a suitable EGF earthquake is fundamental to obtaining reliable source parameters. The basic criteria of using as an EGF (1) a collocated earthquake with (2) a similar focal mechanism that (3) is at least one order of magnitude lower than the target earthquake assure well-constrained source parameters. The fulfillment of criteria requirements (1) and (2) can be verified through waveform similarity, confirming that the waveforms have comparable shapes in terms of direct wave polarities and amplitudes at all three components and close S-P times. Often earthquake catalogues have improved epicentral locations obtained through relocation techniques that help to ensure accurate selection of EGFs. The best pairs are chosen based on close location, preferable with overlapping error location ellipses, and on high (≥0.6) waveform cross-correlation coefficients (Viegas et al., 2010). Mori et al. (2003) noted that the EGF time deconvolution method started to break down when using earthquake pairs with hypocentral separation larger than ~1.400 km, for M4–M5 earthquakes in California recorded at 10 km to 50 km. Hutchings and Wu (1990) found high waveform cross-correlation for earthquake pairs with hypocentral separation less than 1 km to 3 km, depending on the recording site geology, for M2–M3 Californian earthquakes with similar focal mechanisms, recorded at 3 km to 33 km, for frequencies up to 10 Hz (see Figure 13). Hutchings (1987) investigated spectral similarity for the same events as Hutchings and Wu (1990) and found high coherency (>0.8) for event separations less than 1.5 km. Note that the location uncertainty also plays a role in determining the maximum event separation for which waveform cross-correlation is high. Data with larger location uncertainty will show larger event separation. The event-separation distances can be larger when using coda waves instead of direct waves, because of the averaging properties of coda waves. Mayeda et al. (2007) estimated a limit inter-event separation of 25 km for regional ~M4 earthquakes in California recorded at ~1 km to 220 km. Viegas (2009) estimated an event separation of 750 m for local M2 earthquakes recorded at 4 km to 10 km using coda waves. Comparing the difference in dimension between the two coda-wave studies, we see that the scaling relationship holds.

After the initial selection of EGF earthquake pairs, further constraints can be imposed to assure the adequacy of the EGF earthquake. Viegas et al. (2010) suggested several constraints to improve accuracy obtained with the EGF method. Viegas et al. found that: a clear source pulse in the time domain should be obtained; the corner frequency of the larger earthquake should be well within the frequency band limits of the spectral ratio, so that the fit converges; the corner frequency of the EGF earthquake should be well within the frequency band limits of the spectral ratio to be considered; the high-frequency plateau of the spectral ratio should be well defined; the difference between the low and high frequency plateaus of the spectral ratio should be large enough to be distinguished from spectral modulations; and the normalized-variance of the fit has preferably a parabola shape for a sequence of fixed f_{c1} values, with the minimum variance corresponding to the best fit.

Selecting an EGF that is well recorded at several stations, in order to provide a good azimuthal coverage of the source, is also preferable, so that no bias is introduced by source propagation and directivity effects.

4.2 Source parameters

Once the long-period level of the displacement spectra and the corner frequency or source pulse duration (or rise time) are estimated, the source parameters can be estimated.

4.2.1 Seismic moment and moment magnitude

The scalar seismic moment, M_0, is calculated from the long-period part of the displacement spectrum, following Aki and Richards (2002),

$$M_0 = \frac{4\pi\rho(x)^{1/2}\rho(x')^{1/2}c(x)^{1/2}c(x')^{5/2}R\,\Omega_0}{FU_{\phi\theta}}, \tag{29}$$

where c is the velocity of the wave of interest (α for P wave; and β for S wave), R the hypocentral distance, F the free-surface parameter (F = 2), and $U_{\phi\theta}$ the mean radiation pattern, with $U_{\phi\theta}$ = 0.52 for P waves and $U_{\phi\theta}$ = 0.63 for S waves (Aki and Richards, 2002).

Moment magnitude, M_w, is frequently estimated from the seismic moment using the Hanks and Kanamori (1979) relationship

$$M_w = \frac{2}{3}\log_{10} M_0 - 6.04 \qquad \text{(in N.m)}\,. \tag{30}$$

4.2.2 Fault radius and stress drop

In earthquake-source parameters studies, it is common to assume a circular fault model. Such a model seems to work well for small to moderate earthquakes, for which the fault dimension correlates well with the rise time. However, the model seems to break down for large earthquakes, in which the fault width is limited by the seismogenic depth and the fault length becomes several times larger. There are several circular source models from which to estimate fault radius, *r*, from the corner frequency, f_c. The most commonly used are Brune (1970), Sato and Hirasawa (1973), and Madariaga (1976) and are represented by

$$r_B = \frac{2.34c}{2\pi f_c}, \qquad r_{SH} = \frac{Cc}{2\pi f_c}, \qquad r_M = \frac{k\beta}{f_c}, \tag{31}$$

where the index letters in the radius correspond to the model's authors. *C* and *k* are constants related to the rupture velocity v_r (C = 1.5 for P waves and C = 1.9 for S waves, k = 0.32 for P waves and k = 0.21 for S waves, assuming $v_r = 0.9\beta$). All three source models implicitly assume a constant subsonic rupture velocity. Brune's static solution model was developed for S waves and later adapted to P waves. Sato and Hirasawa (1973) and Madariaga (1976) models are dynamic solutions using stopping phases. Brune's model radii are approximately 1.76 times larger than Madariaga's radii ,with Sato and Hirasawa radii lying in between. Depending on the source model used, stress drop estimates will differ up to 5.5 times.

Boatwright (1980) proposed a circular rupture model that relates source radius to the source pulse duration, τ, using P waves,

$$r_{Bt} = \frac{v_r \tau}{2\left(1 - \dfrac{v_r \sin\theta}{\alpha}\right)}, \tag{32}$$

where θ is the take-off angle, assumed to have an average value of 0.64. The dependence on the take-off angle comes from rupture directivity effects. Tomic et al. (2009) found that this model is suitable for P waves but is not consistent with a Madariaga source model for S waves. Viegas et al. (2010) proposed to convert S waves source pulse duration, τ, to corner frequency, using $f_c = 2/\pi\tau$ (Lay and Wallace, 1995) and then estimate the fault radius using Madariaga's solution.

Rupture directivity makes the relative duration and amplitude of the source pulse dependent on the recording-station location relative to the fault-plane orientation. The source pulse increases in amplitude and decreases in duration if the rupture propagates towards that station, but it will decrease in amplitude and increase in duration if the rupture propagates away from that station. Sometimes, fault rupture can happen bilaterally, adding complexity to the solution and making it more difficult to distinguish the rupture direction and identify the fault plane.

The static stress released by faulting – that is, the static stress drop, $\Delta\sigma$ – is usually determined from the seismic moment and source radius using Eshelby's (1957) circular static crack solution,

$$\Delta\sigma = \frac{7}{16}\frac{M_0}{r^3}. \tag{33}$$

Depending on the source model used, stress-drop estimates can vary by a factor of 5.5, with the Madariaga's (1976) source model 5.5 larger than Brune's (1970), and Sato and Hirasawa's (1973) located between the two.

4.2.3 Radiated energy

The radiated energy of small earthquakes is a particularly difficult parameter to calculate, because it uses the high-frequency portion of the source spectrum. Radiated energy, E_S, is commonly calculated from the integral over the velocity squared source spectra, following Boatwright and Fletcher (1984),

$$E_S = \frac{4\pi\rho\, cR^2 \langle U_{\phi\theta}\rangle^2}{F^2 U_{\phi\theta}^2} \times 2\int_{f_1}^{f_2} |\dot{\Omega}(f)|^2\, df \tag{34}$$

where $|\dot{\Omega}(f)|$ is the Fourier velocity amplitude spectrum, f_1 and f_2 are the integration limits, and $\langle U_{\phi\theta}\rangle$ is the rms radiation pattern. When the earthquake focal mechanism is not known, which is often the case, $\langle U_{\phi\theta}\rangle = U_{\phi\theta}$ is assumed. Taking advantage of the EGF method, the spectral ratio can be used to correct for propagation path and site response effects. The velocity spectrum is obtained by differentiation of the EGF spectral ratio, after setting the long-period level of the spectral ratio equal to the long-period level of the displacement spectrum. The energy is calculated from $f_1 = f_c/10$ to $f_2 = 10f_c$ to obtain at least 90% of the total radiated energy (Ide and

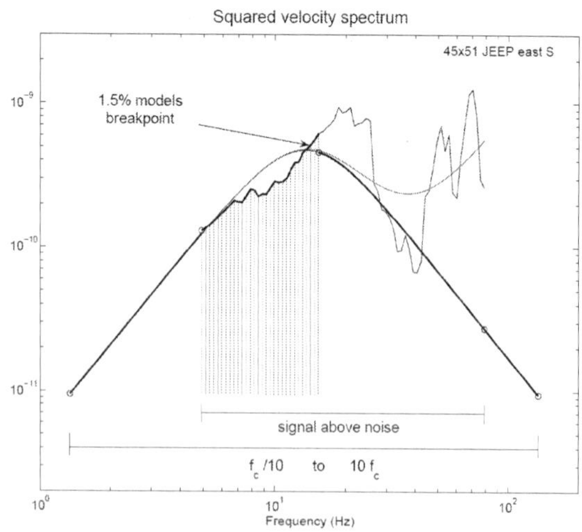

Fig. 10. Illustration of radiated energy calculation. Energy is calculated from the squared spectral ratio. Real data is used between two cut-off frequencies, and model data is used for the remaining frequency band up until a decade above and below the corner frequency. Low frequency cut-off is set for signal to noise ratio above 3. The high frequency cut-off is set when the ratio of the EGF fit model to the individual fit model is above 1.5% (from Viegas et al., 2010).

Beroza, 2001). The available bandwidth does not always reach one or both of those limiting frequencies, so a modeled spectral ratio is used to fill the gaps. The EGF spectral ratio contains two corner frequencies and starts diverging from a ω^2 source model (Brune, 1970) in the vicinity of the smaller earthquake corner frequency. To avoid this distortion at high frequencies, Viegas et al. (2010) use the data spectral ratio exclusively until a certain cut-off frequency is reached, and then use a ω^2 source model for the remaining frequency interval. They set the cut-off frequency to the point where the modeled spectral ratio starts diverging from the modeled ω^2 source by more than 1.5% (Figure 10). Lower-magnitude earthquakes, with corner frequencies closer to the limiting maximum frequency, have a larger energy contribution from the model to fill the high-frequency gap. Earthquakes deconvolved by EGF earthquakes with close magnitudes, and thus close corner frequencies, also have a large model contribution to the energy estimate. The total energy radiated by an earthquake is calculated by summing the contributions of P and S wave energies (Ide et al., 2003).

4.3 Solving for attenuation

Attenuation is often characterized by Q, the frequency-independent attenuation quality factor. Q is usually calculated from modeling the source displacement spectrum with a theoretical source spectrum such as Brune's (1970) (equation 26) and solving for Ω_0, f_c and Q. Ω_0, is very stable and well determined within a factor of 2, but there are several combinations of f_c and Q that fit the spectra equally well. That is, there is a tradeoff between f_c and Q, and any of the combinations provide an optimal fit. Hough et al. (1991) found a way to avoid this caveat by using the corner frequencies calculated from the EGF deconvolution to constrain the source spectrum model when solving for Q. Then, using the obtained Q, Hough et al. (1991) estimated the corner frequencies of other earthquakes that shared common source-station paths, but could not be analyzed using the EGF method. They found the fits were just slightly poorer than the ones obtained solving for Q and f_c simultaneously. Viegas (2011) also used a fixed corner frequency derived from the EGF method and fitted the individual instrument-corrected displacement-spectra using Equation (24a), solving for Q and Ω_0 for all source-station paths. Figure 11 shows an example of Viegas' (2011) EGF analysis applied to the 2010 Germantown, Maryland, USA, M3.4 mainshock and M2.1 aftershock (EGF) pair, and subsequent individual spectrum fitting to obtain Q constrained by the EGF results. It can be seen that the recording site-response observable in the individual spectra was removed by the EGF deconvolution and is no longer observable in the spectral ratio. f_c estimates from fitting the individual spectrum would be biased by the site effects at this station.

Hough (1997) expanded Hough et al.'s (1991) method to determine Q using EGF constrained f_c's, by determining the common source-station attenuation parameter through simultaneously inverting several clustered EGF pairs. This expanded method was called Multiple Empirical Green's Function (MEGF) and has been adopted by other authors with minimal changes (e.g., Hough et al., 1999; Prejean and Ellsworth 2001; Ide et al., 2003; Imanishi and Ellsworth, 2005). Because it simultaneously uses several pairs, it better constrains the attenuation estimate, since the outliers are down-weighed by the most common average properties.

Another common approach to estimating attenuation is to estimate the necessary correction needed to flatten the individual source spectra of an effective impulsive point source EGF, for which the corner frequency is outside the frequency band (e.g., Viegas et al., 2010).

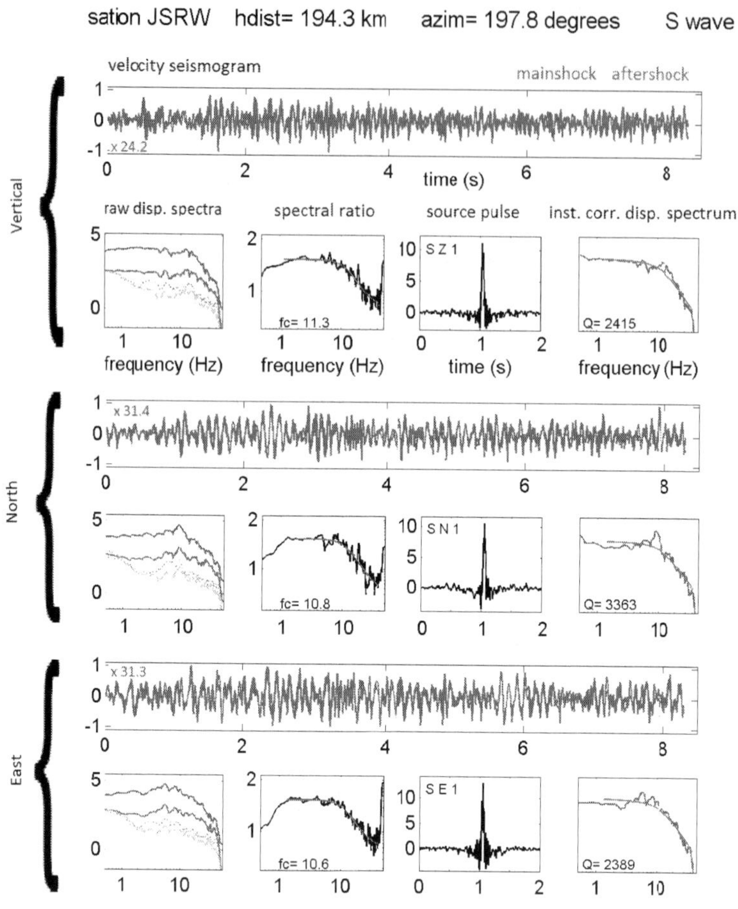

Fig. 11. EGF spectral-ratio-fit analysis. Example of the EGF spectral-ratio-fit analysis at station JSRW (S waves). *Top horizontal panel:* S wave raw velocity seismograms of the 2010 Germantown, Maryland, USA, M3.4 mainshock (blue) and M2.1 aftershock (red). The red number on the plot indicates the multiplication factor applied to the aftershock record to match the amplitude of the mainshock. *Left panel:* Calculated displacement spectra of the seismograms in bottom panel, after integration and instrument correction. The full lines correspond to wave signal, the dotted lines to pre-P noise. *Second panel from left:* Best fit (red) of the EGF spectral ratio (black). The bottom value indicates the best-fit corner frequency. *Third panel from left:* Obtained earthquake source time function. Top left characters correspond to (1) P or S wave, (2) vertical (Z), North (N) or East (E) components, and (3) good (1) or bad (0) fit quality. *Right panel:* Best fit (magenta) of the instrumented corrected displacement spectrum (blue). The bottom value indicates the best-fit attenuation quality factor obtained with the fixed best-fit corner frequency. The panels sequence is repeated three times, one for each component (from Viegas, 2011).

As mentioned in a previous section, Andrews (1986), Boatwright et al. (1991), Prieto et al. (2004), and others used an iterative process to simultaneously solve for source and attenuation terms when dealing with large datasets.

4.4 Increased source resolution with better data and better methods

In the last two decades, we have witnessed a vast improvement in data quality, with the development of borehole acquisition systems and the exponential decrease in computer memory cost, which have enabled the storing of high sampling-rate data. Simultaneously, the use of EGF methods to estimate source parameters of small-to-moderate earthquakes has become more popular, and better attenuation corrections have been conducted. All in all, better data associated with better methods has resulted in better constrained source

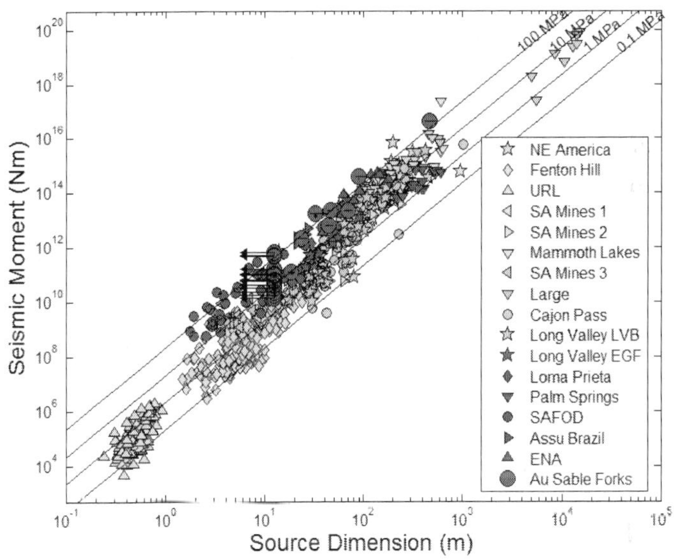

Fig. 12. Global compilation of seismic moment versus source radius, color coded by method used or data type. Dark blue and red indicates EGF methods, peach indicates slip inversion methods or deep borehole data, and light blue indicates individual fitting methods. Viegas et al. (2010) results are shown in red circles. Arrows indicate radii smaller than the maximum resolution threshold. Diagonal lines indicate constant stress drop. (from Viegas et al., 2010, adapted from Abercrombie and Leary, 1993). Data points are: NE America (Boatwright 1994), Fenton Hill (Fehler & Phillips, 1991), URL (Gibowicz et al., 1991), S. A Mines 1 (McGarr et al., 1990), S. A Mines 2 (Spottiswoode and McGarr, 1975), Mammoth Lakes (Archuleta et al., 1982), S. A. Mines 3 (Yamada et al., 2007), Large (compilation of Abercrombie and Rice, 2005, Table 5: Dreger and Helmberger 1991, Mori 1996, Hough and Dreger 1995, Ji et al. 2002, Venkataraman et al. 2000, Wald 1995, Wald and Heaton 1994, Wald et al. 1996, Wald et al. 1991), Cajon Pass (Abercrombie, 1995), Long Valley (Ide et al., 2003, individual and EGF studies), Loma Prieta (Hough et al., 1991), Palm Springs (Mori and Frankel 1990), SAFOD (Imanishi and Ellsworth, 2005), Açu Brazil (Tomic et al., 2009), ENA (Xie et al., 1991; Li et al., 1995; Shi et al., 1998), Au Sable Forks (Viegas et al., 2010).

parameters. An overview of the improvement in source resolution with improved methods and data is illustrated in Figure 12, which is a compilation of earthquake source dimension estimates obtained with various methods. We see in Figure 12 that all stress drop values from EGF methods are above 1 MPa. It seems that with increasing data quality and more accurate methods, stress drop estimates are getting higher—a trend particularly apparent for small earthquakes (<M3). A good example is the Long Valley, CA (Ide et al., 2003) study, showing EGF stress drops that are higher than the individually estimated stress drops.

5. Strong ground-motion synthesis

5.1 EGFs for synthesizing strong ground motion

The use of EGFs in strong ground-motion synthesis has to be qualified not only for high frequencies but for low frequencies as well. Most recordings of small earthquakes are band limited by cultural noise or instrument response, and do not include frequencies below about 0.5 Hz, so that full broadband use of EGFs is not generally possible. This is the frequency range where near-field terms are important for source studies and strong ground-motion synthesis. However, it has long been recognized that synthetic seismograms can be accurately calculated for frequencies < 0.5 Hz in areas where the geology is well known. Often, tomographic inversion for velocity structure provides the basic information necessary to accurately synthesize low-frequency seismograms (Gök and Hutchings, 2004). Thus, there is a convenient match between synthetic Green's functions, which can be accurately calculated for frequencies at which geology can be well known (and can include the near-field wave-propagation effects that are difficult to record)—and EGFs, which record the high frequencies for which the geology is not well known (and for which far-field arrivals are more important). Merging the two frequency bands provides a broadband simulation: synthetic Green's functions can be used for frequencies 0.0 to about 0.5 Hz, and EGFs can generally be used for frequencies from about 0.5 to 25.0 Hz (or higher for stable continental interiors). This covers the frequency range of engineering interest.

5.1.1 Source-event parameters

Earthquakes provide the recordings that are typically used as EGFs. Knowing the source parameters of these earthquakes is critical for correctly including their recordings in mathematical computations. Important source parameters include the hypocentral location, moment, source corner frequency, stress drop, and focal mechanism solution. Moment and source corner frequency, which are particularly important parameters, can be the most difficult to estimate. Virtually all EGF synthesis methods add up or scale EGFs based upon their moment, and the amplitudes of synthesized records are linearly dependent upon these estimates. Similarly, many EGF methods scale records with stress drops or deconvolve out the source event spectra, and these computations are directly dependent upon the source corner-frequency estimate. Typically, a source model (e.g., a Brune source or Boatwright source) is used to characterize the source corner frequency and moment,, and from this the source dimensions and stress drop is derived. (Calculating hypocentral location and focal mechanism solutions are not discussed here. Earthquakes are typically recorded by local or regional networks, and these parameters can be well constrained).

5.1.2 Interpolation of EGFs

Typically, EGFs from recordings of small earthquakes are not necessarily on the faults to be modeled. Moreover, if they are on the fault to be modeled, they are not necessarily recorded from all locations along a fault, or share the same focal mechanism solution. As a consequence, to represent all the fault elements, it is necessary to interpolate EGFs and modify their focal mechanism solution. Hutchings and Wu (1990), researching the spatial dependence of EGFs, found that the variability in ground motion caused by differences in source location and/or focal mechanism solutions are much less than those caused by the site response. Hutchings (1991), Hutchings (1994), and Jarpe and Kasameyer (1996) also found that interpolation for different source locations along a fault works quite well. Wossner et al. (2002) found that the number of EGFs affects the amplitude of synthesized seismograms. They found that amplitudes in the synthesized records for frequencies higher than the source corner frequency generally tend to be higher for records that use a single EGF interpolated over the fault, and they achieved more stable and reliable results if at least five EGFs were used. In synthesis, we have the option of correcting for different focal mechanism solutions, but Hutchings and Wu (1990) and Jarpe and Kasameyer (1996) found that for high frequencies, this correction does not improve the synthesis. Interpolation is performed by correcting for attenuation, geometrical spreading, and P-wave and S-wave arrival times resulting from differences in source distances, as discussed in Hutchings and Wu (1990). Figure 13 shows cross-correlation values for events with spatial separations up to 5 km and with similar focal mechanism solutions. It can be seen that the best cross-correlation coefficients are obtained when both small event separation and common focal mechanisms are verified.

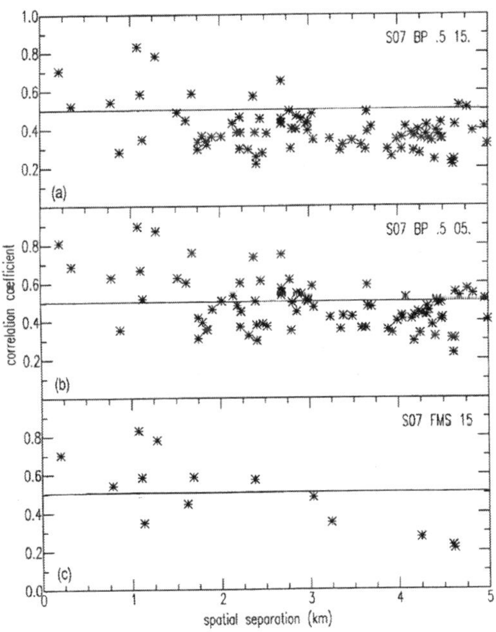

Fig. 13. Cross-correlation values of events with spatial separations up to 5 km and with similar focal mechanism solutions, at several frequency bands: (a) 0.5 Hz to 15.0 Hz; (b) 0.5 Hz to 5.0 Hz; **(c)** 0.5 Hz to 15.0 Hz for which variations in the focal solution angles strike, slip, and dip less than 15 degrees. (from Hutchings and Wu, 1991).

5.1.3 Scaling relations

A significant issue for strong ground-motion synthesis with EGFs is how to properly scale small earthquakes to model a large earthquake. Most strong ground-motion synthesis methods depend upon a scaling relation to determine how many EGFs to use in synthesizing a larger earthquake. If constant stress-drop scaling is assumed between small and larger earthquakes, and the earthquakes follow a non-constant scaling relationship, small earthquakes will systematically have a larger moment for the same apparent stress drop as large earthquakes, and thus fewer EGFs will be needed to add up to the moment of the larger event. Therefore, the assumed scaling relation directly affects the amplitude of the synthesized seismograms. Hutchings (1991 and 1994) proposed that if one follows the representation relation in synthesis and uses EGFs that are effectively impulsive point sources, then one only needs to use the number of EGS that add up to the moment of the larger earthquake. This avoids scaling relations (see equations 20 and 21).

5.1.4 Deconvolution to obtain EGFs

Often, small earthquakes have moments greater than the threshold necessary to effectively be impulsive point sources. One can use the output of source parameter studies to deconvolve out the finite source from the recordings to create EGFs. Generally, investigators confine this approach to events with magnitude < 4.0 to minimize finite source effects and keep the basic assumption of the Brune source model. The Brune source has zero phase shifts, such that in the deconvolution, only the amplitude spectra are affected. There is no apparent noncausal effect in the time series due to the deconvolution. Figure 14 shows several recordings that have been corrected to effectively provide impulsive point, earthquake source-event recordings (EGFs). The time series are in acceleration, and their displacement spectra are shown. That the spectra are not flat after the deconvolution is assumed to result from the effects of attenuation and site response (*fmax* effect).

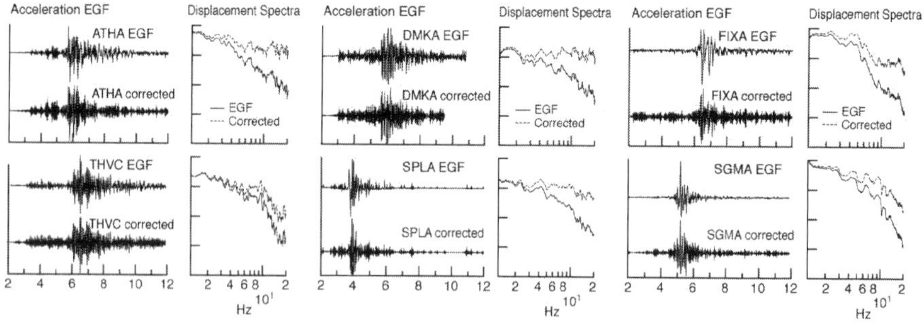

Fig. 14. Several recordings that have been corrected to effectively provide impulsive point earthquake sources. The time series are in acceleration, and their displacement spectra are shown. The fall-off at higher frequencies of the corrected spectra is due to attenuation and site effects.

5.1.5 Empirical Green's Functions merged with synthetic Green's functions

Since EGFs are recordings of actual earthquakes, they are band limited by cultural noise and instrument response. One way to overcome this band limitation is to augment these recordings by synthetic Green's functions from ground-motion synthesis. The basic premise in combining empirical and synthetic Green's functions is that each offers the best accuracy over particular frequency bands. EGF's, as noted above, are more accurate at high frequencies, at which wherein geologic inhomogeneities are not well modeled, and the synthetic Green's functions are more accurate at lower frequencies, at which EGFs do not have sufficient energy. The overlap is in the range from 0.2 to 1.0 Hz. In this range, the geology can be modeled with some accuracy, and EGFs have sufficient energy to be well recorded.

There are limitations to the mathematical solution for Green's functions. First, analytical Green's functions cannot account for small-scale geologic inhomogeneities, and as a result cannot reasonably model ground motion for frequencies much greater than about ~1 Hz. Given this limitation, geologic inhomogeneities in the spatial dimension of ~3 km or smaller cannot be modeled. Second, inaccurate estimates of attenuation and scattering properties can significantly affect synthesized frequency content. Hutchings and Wu (1990) presented a means for providing site response functions to synthetic Green's functions that improves the reliability of synthetic Green's functions.

5.1.6 Empirically constrained stochastic EGFs

We can create synthetic high-frequency Green's functions with Gaussian white noise shaped by an envelope function and modified by attenuation, geometrical spreading, and kappa functions. This approach has been described by Boore (1983 and 2003). Regional propagation information obtained by regression on recorded seismicity (Malagnini and Herrmann, 2000) and site-specific K obtained from geotechnical information (Akinci et al., 2001) can be used to modify the shape of the source spectra—so that the stochastic Green's functions have the averaged wave propagation characteristics of the region, and site-specific information when possible. In this sense, Green's functions are empirically derived. These Green's functions with impulsive point dislocation sources can be used in the representation relation to calculate seismograms using equations (11) or (13). (Low frequencies (< ~1.0 Hz) might better be computed by synthetic seismograms and merged, as discussed in the previous section.) One approach proposed by Hutchings et al. (2005) is to use synthetic Green's function for low frequencies, EGFs for middle frequencies, and stochastic Green's functions for very high frequencies. Stochastic Green's functions can be expressed as

$$e_n(X,t) = P(t) * D(R,t) * R(t)\,, \tag{35}$$

where *P(t)* is a stochastic time series with a specified envelope shape and duration, and with a Fourier amplitude spectrum from an impulsive point shear dislocation. D(R,t) is the propagation effect resulting from geometrical spreading and attenuation determined by inversion on background seismicity, such as developed by Herrmann and Malagnini (1996). *R(t)* is the site response from either recorded data or site-specific kappa obtained from geotechnical data (Akinci et al., 2001).

The Fourier amplitude spectrum of the Green's function displacement can be expressed as

$$\left|E_n(M_0,R,\omega)\right| = \left|P_n(M_0,R,\omega)\right| \cdot D(R,\omega) \cdot R(\omega)\,. \tag{36}$$

The far-field displacement spectrum for seismic P-waves from a point shear dislocation source in a homogeneous whole space (Aki and Richards, 2003, eqs. 4.93 - 4.95) is

$$|P_P(M_0, R, \omega)| = \frac{\mu A \bar{s}(x')U^P_{\phi\theta}}{4\pi R^{\gamma}\rho(x)^{1/2}\rho(x')^{1/2}\alpha(x)^{1/2}\alpha(x')^{5/2}}, \tag{37}$$

where the primed terms are values at the source and the unprimed are values at the receiver. $\bar{s}(x')$ is the average displacement of the point dislocation, A is the fault area, and μ is the rigidity. These three terms together are the seismic moment of the stochastic Green's function. We only specify the moment of the point source and not these individual values. The spectrum is flat. Similarly, for S-waves,

$$|P_S(M_0, R, \omega)| = \frac{M^e_0 U^S_{\phi\theta}}{4\pi R^{\gamma}\rho(x)^{1/2}\rho(x')^{1/2}\beta(x)^{1/2}\beta(x')^{5/2}}, \tag{38}$$

The average radiation values ($U^P_{\phi\theta}$ and $U^S_{\phi\theta}$) of P, SH, and SV waves are used, and α, β, and ρ are the P- and S-wave velocities and density of the medium at the surface or on the fault. R^{γ} is the geometrical spreading, where R is the hypocentral distance and γ is a geometrical spreading factor, generally between 0.9 and 1.1, and is determined from regression of regional seismicity as discussed below.

5.2 Source models

Strong ground-motion synthesis with EGFs has been the subject of research and application since 1978. It has intrigued researchers because such a synthesis provides seismograms that appear very realistic in frequency content, duration, and energy distribution. Our primary interest in this section is in using the EGF method for engineering purposes. For that reason, the frequency band of interest is about 0.5 to 25.0 Hz. Basically, the idea of using a recording of an earthquake for the wave propagation of the synthesis model is the same for all approaches, though there are differences regarding the number and size of EGFs used, as well as regarding the scaling relation between the EGFs and the larger event. Some investigators use only one EGF for the entire synthesis and assume that the wave propagation characteristics are the same from all portions of a fault. Others use several EGFs to try to account for differences in wave propagation. There are never enough small earthquakes to provide EGFs for all portions of a fault, so the assumption of slowly varying Green's functions for source locations along the fault is typically made. Most methods use an interpolation scheme to fill out the fault with EGFs. To be mathematically accurate, the EGFs must have the same focal mechanism as the larger event to be modeled. Since this is not generally true, an interpolation for differences in focal mechanism solutions and an estimation of the error involved should be included in the results—as discussed in the previous section.

The greatest difference between methods is in the source function assumed. There are five basic classes of source functions: point source, stochastic, kinematic, quasi-dynamic, and dynamic. There are further classifications of synthesized models, such as target spectrum, composite source, and continuous source summation—but these are secondary classifications, since they can have a stochastic, kinematic, or quasi-dynamic source. Point-source models generally are forced to fit a target spectrum without any phase information. Stochastic methods generally have time delays that are randomly chosen from a distribution, so they represent a finite fault in that sense, but there is no coherency that might result from a rupture process. Kinematic modeling is generally interpreted to mean that the contribution from each portion of a fault is specifically prescribed. Generally, this includes some form of rupture propagation, rise time, and slip distribution. Quasi-dynamic modeling generally involves earthquake models that are consistent with the physics of the rupture process and attempt to simulate the dynamic character of rupture. Thus, they must be consistent with the elastodynamic equations of seismology and fracture energy, and be consistent with a physical understanding of how earthquakes rupture, as well as the laboratory experiments, numerical modeling results, and field observations involved in earthquake processes. Dynamic modeling is a full calculation of the state of stress as a function of time and displacements, and stress release depends upon a friction law and material prosperities. Dynamic modeling has not previously been applied with EGFs and is not discussed here.

Size differences in the EGFs that are used affect the source time function. Those that use events larger than point sources to obtain EGFs are basically obtaining part of their source function from the EGF itself. Some methods discussed here use synthetic Green's functions, but EGFs could just as well have been used, so they are also examined. Similarly, any of the EGF synthesis methods could equally as well use synthetic records as Green's functions instead of EGFs. All the approaches follow the representation relation in some form (equations 11 and 13), and this relation is used as a common basis for comparison between methods.

5.2.1 Point source

Point-source approaches make no assumption about how earthquakes rupture. Rather, they add up EGFs with no phase information or force them to fit a target spectra. This is the basic approach first presented by Joyner and Boore (1986), where an original seismogram (EGF) spectra is modified to fit the target spectra by iteratively shaping in the frequency and time domain.

5.2.2 Composite source modeling

Most EGF methods utilize a composite source summation. This approach has its roots in research that described large events as composite of smaller "subevents" (Wyss and Brune, 1967; Trifunac and Brune, 1970; Kanamori and Stewart, 1976). However, these are generally confined to three or four subevents. Inversion studies indicate that subevents may be large patches of high-energy release along the rupture surface (Hartzell and Heaton, 1983, 1986; Hartzell and Iida, 1990), and deconvolution studies reveal multiple rupture events within a larger event's source time functions (Lay et al., 1982; Lynnes and Lay, 1988). These studies generally indicate that there are three or four subevents in most ruptures, which does not support a composite earthquake model in which tens or even hundreds of events are repeatedly summed to create a synthesis of a larger event. Nevertheless, many authors have been successful in synthesizing observed seismograms with the composite earthquake model. This approach can be summarized as (equation 14)

$$u_n(X,t) = \sum_{j=1}^{\eta} \kappa_j e_n\left(X, t' - \tau_j\right)_j, \tag{39}$$

where κ_j is a scaling factor for Green's functions, η is the number of times small events are added to create the synthetic seismogram, and τ_j are determined from the rupture model.

Joyner and Boore (1986) proposed one form of the composite model: they assumed κ was constant and that only one EGF was used, and described EGFs that have corner frequencies in the observable frequency range, i.e., they used a composite source model. They then developed constraints on the numbers of composite sources necessary to create the desired time series of a larger earthquake and showed that if constant stress-drop scaling and Brune spectra were applied,

$$\eta = \left(\frac{M_0}{M_0^e} \right)^{\frac{4}{3}} \quad \text{and} \quad \kappa = \left(\frac{M_0}{M_0^e} \right)^{\frac{1}{3}} . \tag{40}$$

This results from requiring the moment of the small earthquake to be scaled and added to be that of the large earthquake, and requiring the high frequency fall-off of the small earthquake, above its corner frequency, to match the amplitude of the larger earthquake. They showed that if τ_j were evenly distributed, this approach results in amplitude deficiencies in the spectrum near the corner frequency of the simulated event, and holes in the spectrum. If we examine Equation (39) in view of the representation relation (equations 13 and 14), the Joyner and Boore approach is basically a Haskell (1964) rupture, which assumes equal amplitude of slip with constant rise time and constant rupture velocity throughout the rupture surface. Note that the Haskell rupture model has been shown to be deficient in frequencies near the corner frequency and have spectral holes in the high frequency (Hutchings, 1994).

5.2.3 Stochastic source modeling

Many investigators take the approach of modeling a large earthquake as the sum of smaller earthquakes that are added together stochastically. That is, there is a probability distribution that describes the time delays in the summation. For example Joyner and Boore (1986), as discussed in the previous section, assumed an even probability distribution for time delays in the summation. That is, τ_j in Equation (39) is evenly distributed between 0.0 and the duration of the earthquake. The distribution replicates rupture in time and space, but there is no coherency in the rupture that would occur during a real earthquake. This approach has been termed the single-stage approach, because there is only one time-delay probability distribution.

Wennerberg (1990) showed that the deficiencies in the spectra observed by Joyner and Boore (1986) were caused by the transform of a uniform probability density having a periodic set of zeros. From Equation (20), the scaling can also be described by the cosine function's periodicity, where the τ's are equal. To overcome these problems, Wennerberg proposed a summation scheme in two stages. The first is a probability distribution of the event summation, and the second is a distribution about each of the first,

$$u_n(X,t) = \sum_{i=1}^{\eta} \sum_{j=1}^{\gamma} \kappa_j e_n(X, t_i^{'} - \tau_i^r - \tau_k^s)_i \; . \tag{41}$$

Wennerberg also kept κ constant. It is apparent that Equation (41) is in the same form as Equation (13). Equation (41) has the characteristics of a more complicated rupture, with the first stage representing rupture along the fault and the second stage representing rise time. However, there is no coherency to the rupture. Also, Wennerberg assumed that the spectra of the large event fits the omega-squared Brune source model, which is not necessarily true for the near-source region, where earthquake damage usually occurs.

It is instructive to compare the frequency-domain solutions of equations (18) and (19) to equations (39) and (41). Equation (18) is the frequency domain solution to Equation (14). Equation (18) is derived assuming an impulsive point-source event, so comparisons are only precisely valid below the corner frequency of source events used by Joyner and Boore or Wennerberg, but higher frequencies essentially have the same result. Had the phase spectrum in Equation (19) been a probability distribution in τ, only one EGF been used, and the scaling factors κ_j been constant, then Equation (14) would be identical to Wennerberg's Equation (1). If the probability distribution were described as a uniform probability, it would be identical to Equation (A2) in Joyner and Boore's (1986) Appendix. Thus, the main difference between stochastic modeling and actual rupture of earthquakes is that the phase delays associated with rupture are approximated by probability distributions. However, it is critical what distributions are used, and their implications need to be identified. Also, using only one EGF assumes that the amplitude and phase spectrum of the EGF are constant over the fault rupture area. Although there is substantial evidence that the amplitude spectrum is fairly constant for different locations of EGFs, assuming that the phase spectrum remains constant is fairly dubious.

5.2.4 Kinematic source modeling

Kinematic modeling involves simulating the actual rupture process by prescribing what happens at all points along a simulated fault rupture. Typically, rupture velocity, rise time, and slip distribution are prescribed. Irikura (1986) has developed one of the most widely used methods for synthesizing ground motion with EGFs—a kinematic, composite source model approach. He applied the relation

$$u_n(X,t) = \sum_{i=1}^{M} \left(\frac{r}{r_i}\right) F\left(t - t_i^{'}\right) * e_n(X,t), \tag{42}$$

where

$$F(t) = \delta(t) + \frac{1}{n} \sum_{j=1}^{n\left(\sqrt{M}-1\right)} \delta\left[t - \frac{(j-1)\tau}{n\left(\sqrt{M}-1\right)}\right], \tag{43}$$

and where M describes a finite fault with N rows times N columns, M=N×N, with the number of subevents determined by the ratio of the source parameters of the large event being synthesized to the small event being used as an EGF. Only one EGF is used. r and r_i are the hypocentral distances of the EGF earthquake and of the ith subfault element to the recording site, respectively. This is the interpolation scheme used by Irikura to provide an effective EGF at each source area. t'_i is the sum time delay from the rupture starting point to the ith element and that from the ith element to the site. τ is the rise time of the source rupture process of the large event. n is an appropriate integer number to shift fictitious periodicity $\tau/(N-1)$ into a high frequency out of the range of interest. The primed terms refer to descriptions on the source, and * is the convolution operator. *F(t)*, addressed as a filtering function, is a series of delays in summation of the Green's function over the rise time of the larger event. Substituting Equation (43) into (42), we have

$$u_n(X,t)=\sum_{i=1}^{M}\frac{r}{r_i}e_n(X,t-t_i^{'})+\sum_{i=1}^{M}\sum_{j=1}^{N}\frac{r}{nr_i}e_n(X,t-t_i^{'}-t_j^{'})\,, \tag{44}$$

where

$$t_j^{'}=\frac{(j-1)\tau}{N} \quad \text{and} \quad N=n\left(\sqrt{M}-1\right). \tag{45}$$

Equation (44) is in the form of a single-stage approach added to a two-stage approach, as discussed in the composite source section above (equations 39 and 41, respectively). Equation (44) is also in the same form as Equation (11), so it is a solution of the representation relation. In the spirit of Equation (10) through Equation (15), where the Green's function is separate from the source function and the interpolation scheme for EGFs is not shown as part of the equation, $\hat{S}^{'}(t^{'})$ of Equation (11) can be expressed with Irikura's filter function as

$$\hat{S}^{'}(t^{'})=\frac{M_{0i}^{e}}{\mu_i A_i}\left[\delta(t-t_i)+\frac{1}{n}\sum_{j=1}^{N}\delta\left(t-t_i-t_j\right)\right], \tag{46}$$

where M_{0i}^{e}, μ_i, and A_i are as described for equations (11) through (15). To examine the slip function of the representation relation expressed in Equation (10), it is important to keep in mind that the retrieved source function is a relative source obtained through deconvolution, so that

$$\hat{S}(t^{'})=\hat{S}^{'}(t^{'})*S_e(t^{'})\,, \tag{47}$$

where $S_e(t')$ is the source function of the EGF event and $*$ is the convolution operator. If the EGF is from an impulsive point-source event, i.e., at frequencies below its source corner frequency (so that the EGF source function is a step function), then $\hat{S}(t^{'})$ of Equation (10) can be expressed as

$$\hat{S}(t^{'})=\frac{M_{0i}^{e}}{\mu_i A_i}\left[H(t-t_i)+\frac{1}{n}\sum_{j=1}^{N}H\left(t-t_i-t_j\right)\right] \tag{48}$$

where $H(t')$ is the step function and is the form of Equation (12). As discussed below, the key issue for the slip function is how one describes the phase delays t_i and t_j. Irikura describes these as constant throughout the fault, and consequently they describe a constant velocity rupture front with a step at the arrival time of the rupture front plus a ramp slip over the rise time. Except for the initial step, this is a Haskell rupture.

Hartzell et al. (1999) proposed a kinematic model based upon a fractal distribution of subevents,

$$u_n(X,t)=\left[\sum_{k=1}^{L}S_{km}(t)*\sum_{j=1}^{M_k}S_{ok}(t)*\sum_{i=1}^{N_k}\frac{r_0}{r_{ij}}\left(\frac{\Delta\sigma_{ij}}{\Delta\sigma_0}\right)\delta\left(t-t_{sij}-t_{rij}\right)\right]*e(x,y,t)_{ij} \tag{49}$$

where $e(x,y,t)$ is the displacement for a small event as discussed above (EGF). It could be from an EGF or a calculated subevent. L is the number of different subevent spatial sizes, M_k is the number of subevents with size k, and N_k is the number of the smallest subevents needed to fill the fault area of subevent jk. r_0 is the distance from the subevent to the station, and r_{ij} scales this for geometrical spreading due to actual distance from actual subevent. $(\Delta\sigma_{ij}/\Delta\sigma_0)$ scales the stress drop at each subevent relative to that of the Green's function. The spectrum of the source function $S_{ij}(f)$ is that of the ratio of a Brune spectra between the EGF to that of the subevent. This source model has no theoretical relationship between rupture parameters and the elastodynamic equation of motion. The value at each element is arbitrarily applied. However, the similarity to the representation relation is apparent: the terms in the square bracket are the source time function of the representation relation (Equation 11 and following).

5.2.5 Quasi-dynamic modeling

Quasi-dynamic modeling generally involves earthquake models that are consistent with the physics of the rupture process and attempt to simulate the dynamic character of rupture. Several investigators have approached strong ground-motion synthesis with quasi-dynamic rupture models, including Boatwright (1981), Hartzell (1982), Heaton (1982), Papageiorgiou and Aki (1983), Spudich and Frazier (1984), Cohee et al. (1991), Hutchings (1991 and 1994), and Guatteri et al. (2003). Quasi-dynamic source models allow for model combinations of rupture parameters that have not yet occurred in nature. Bounds on input parameters are based on physical understanding and naturally bound the synthesized ground motions.

5.2.5.1 Hutchings' method

Hutchings (1991, 1994, and 2007) used the Kostrov (1964) slip function to describe slip at a point. The Kostrov slip function was derived from the analytical solution for rupture of a circular crack in a homogeneous medium (Buridge and Willis, 1969),

$$K(X,t')=\frac{0.81\Delta\sigma\beta}{\mu}\sqrt{t'^2-\frac{X_i^2}{v_r^2}}\,, \qquad (50)$$

where X_i^2 is the position vector from the hypocenter, t' is relative to the origin time of the earthquake, v_r is the rupture velocity, β is the shear wave velocity, 0.81 is a constant used for v_r equal to 0.9β, μ is rigidity, and $\Delta\sigma$ is stress drop. This expression does not include healing or termination of the slip. Hutchings transformed the Kostrov slip function to be relative to element time and added healing to the slip process. The Kostrov slip function at an element becomes

$$S(t')_i=\frac{0.81\Delta\sigma_i\beta_i}{\mu_i}\sqrt{t_i'^2+2t_r^i t_i'}\otimes H(t') \qquad t_i'\le\tau_i\,, \qquad (51)$$

where t'_i is zero at the arrival time of a rupture front, so that $t'_i = t' - t'_r$, and t'_r is the rupture time from the hypocenter, χ/v_r, in Equation (50). τ_i is the rise time at the ith element, and $\otimes H(t')$ is the deconvolution with the step function. The slip at an element follows Equation (51) from the arrival time of the rupture front to its rise time. Hutchings modeled the rise time as the time for termination of slip due to a stress pulse propagating back into the fault area after rupture has terminated at an edge. This is the result determined by Kostrov and Das (1990) from dynamic modeling. Now, t'_r places a rupture distance dependence on Equation (49), and for long faults it causes $S(t')_i$ to increase with spatial separation from the hypocenter. To constrain this, t'_r is limited to be equal to or less than the rupture time to the nearest edge but not less than 1.0 sec. Hutchings equates this limiting value to an "asperity termination" of fault rupture and a departure from a pure crack solution for an extended rupture. This is consistent with the observation of a slip pulse as identified by Heaton (1990).

Other features of this quasi-dynamic model include rigidity diminishing at the same rate as the stress drop near the surface. This has two effects: first, reducing the rigidity results in very little moment contribution for rupture near the surface; second, the commensurate diminishing of stress drop results in significant displacements (although not significantly seismogenic) at the surface. Further, as a result of these near-surface rupture effects, Hutchings does not allow a slip healing-phase to be generated for the termination of rupture at or near the surface.

Investigators often use knowledge about how earthquakes rupture to estimate bounds of possible rupture parameters that may have been identified prior to an earthquake. Rupture parameters are selected by a Monte Carlo method using a triangular distribution between limits. The limits of input parameters will naturally bound the range of synthesized ground motions. Also, because input parameters are correlated through a physical model, combinations that cannot occur in nature are excluded. Certainly, a major area of research in the future will be identifying the physical limits to input parameters in order to get the maximum physical range of ground motion possible (Bommer et al., 2004).

Fault shapes, slip distributions, and asperities are constrained to have the character of previously occurring earthquakes, but are not determined by regression on these parameters. Slip distributions of previous earthquakes show that rectangular ruptures generally occur on faults that rupture through the entire crust (Scholz, 1980), whereas earthquakes that terminate before reaching the surface of the entire crustal thickness have an elliptical shape (Hartzell and Iida 1990; Hartzell and Heaton 1983, 1986; Wald and Graves 2001). The definition of an asperity as defined by Somerville et al. (1999) provides guidelines for asperities. Somerville et al. used inversion results, primarily from Hartzell or Wald and Graves, to characterize the slip distribution of an asperity as having a slip amplitude greater than 1.5 times the overall average slip amplitude. From regression with all their data and results from Wells and Coppersmith (1994), Somerville calculates asperities to occupy 22% of the total rupture area and account for 44% of the total slip.

Rupture velocity is allowed to vary from 0.75 to 1.0 times the shear-wave velocity, as derived from dynamic rupture modeling (Das and Kostrov, 1990, and many others). Healing velocity is 0.8 and 1.0 times the rupture velocity, lying roughly between the Raleigh wave velocity and the shear wave velocity, as observed in dynamic rupture modeling. Stress drop in Equation (49) is a dependent variable derived from the Kostrov slip function. In this derivation, stress drop is that which results in a strain discontinuity and displacement on the fault, and results in seismic radiation. It is equivalent to the Orowan stress drop (Orowan, 1960). Stress drop is constrained to diminish near the surface of the earth due to the lithostatic load (about 30 MPa at 1.7 km depth). Figure 15 shows an example of synthetic waveforms and derived spectra obtained with the Hutchings' method that best model the rupture of the M6 1997 Colfiorito, Italy, earthquake (Scognamiglio and Hutchings, 2009).

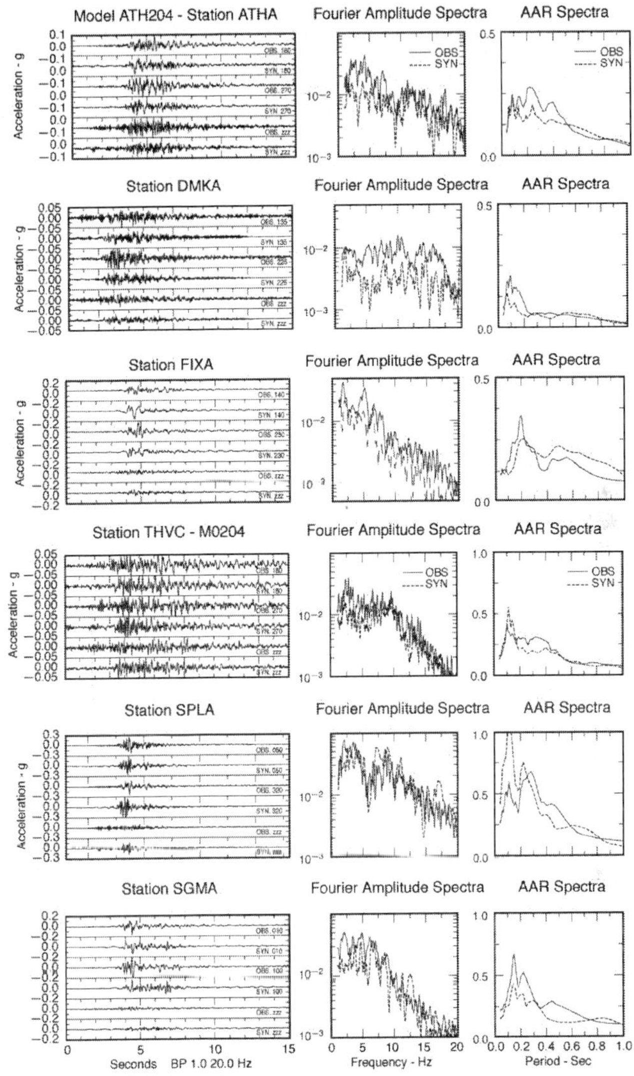

Fig. 15. Example of modeling the rupture of the M6 1997 Colfiorito, Italy, earthquake using the Hutchings' method. The fit between the observed (solid lines) and synthesized (dashed lines) seismograms (*left panels*), displacement Fourier spectra (*center panels*) and acceleration response spectra (*right panels*) are shown for all available stations for the preferred rupture model. Notice that the basic characteristics of the synthesized seismograms (SYN) closely match the observed seismograms (OBS). In particular, the basic waveforms, durations, and frequency content match well. The frequency content is evident from the Fourier amplitude and ARR spectra (from Scognamiglio and Hutchings, 2009).

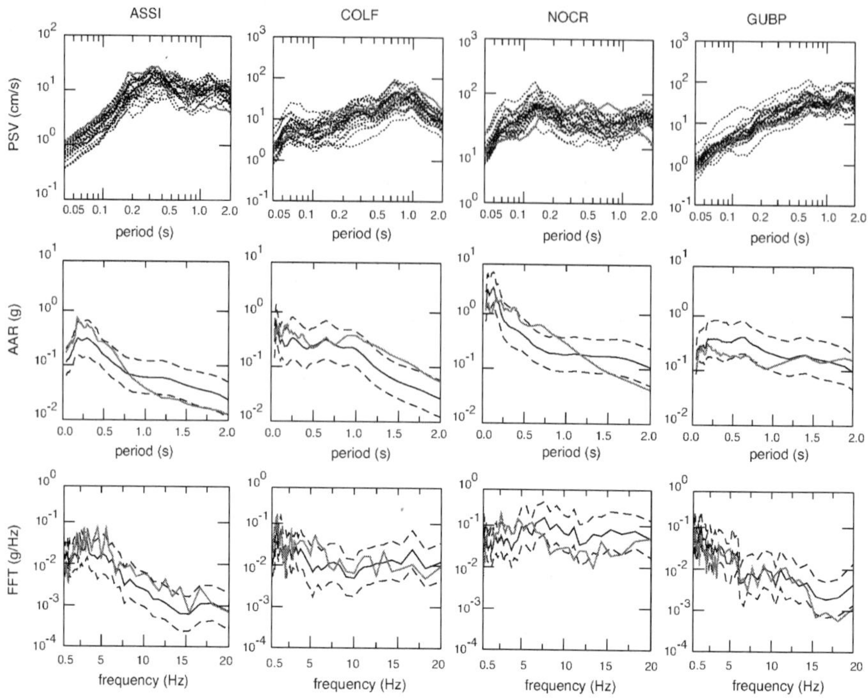

Fig. 16. Calculated twenty-three scenarios using "tight" constraints (dotted lines) in the smaller source volume where the M6 1997 Colfiorito, Italy earthquake is thought to have occurred at the available recording stations, as well as those observed from the actual earthquake (red curves). *Top panels:* PVR calculated for all models. *Middle and bottom pannels*: mean (thick solid line) and plus and minus standard deviation of the distribution (thick dashed lines) of acceleration response spectra (AAR) and Fourier spectra (FFT) for all models that includes epistemic and aleatory uncertainty. (from Scognamiglio and Hutchings, 2009)

Obviously, when rupture parameters are varied, they combine to give a distribution of ground-motion amplitudes. In studying the effects of source parameter changes on acceleration time series and Fourier amplitude spectra, Wossner et al. (2002) found that hypocenter location greatly affects near-source stations due to a directivity effect, that different fault plane geometries significantly affect rise-time distribution and thus amplitude and frequency content of signal, and that variations in rupture and healing velocities can cause up to a factor-of-five difference in time-series amplitudes. These variations are caused by parameters that generally cannot be known prior to an earthquake. If parameters can be constrained, then the distribution should narrow. Scognamiglio and Hutchings (2007) showed that if some parameters are known in advance, then the distribution can be narrowed. The test for a physically based approach to ground-motion prediction is that when the parameters are narrowed, the distribution should still contain what actually happened. Figure 16 shows that for the M=6.0 Colfiorito, Italy earthquake, this is true. The recorded data are within the one-standard-deviation value of the synthesized suite of spectra, with the shapes of the synthesized spectra matching (in general) the shapes of the observed spectra, thus indicating the prediction capability of the method.

6. Strong ground motion prediction and application to seismic hazard analysis

The main goal of strong ground-motion prediction is to model ground motion near sources of potentially damaging earthquakes. It is this near-source region, within about two-fault-lengths' distance, wherein damage from earthquakes principally occurs. It is also within this distance range that finite rupture can significantly influence the character of ground motion. Therefore, a principal concern of strong ground-motion synthesis is to model the rupture process of large earthquakes. In addition, near-source ground-motion synthesis should also accurately account for the complexity of wave propagation. Unfortunately, EGF's do not include the effects of material nonlinearity caused either by modular damping and reduction or by loss of effective stress (liquefaction), and therefore the synthesized ground motion is for linear response. Typically, non-linear soil modeling is used to convert this to potential non-linear response (Heuze et al., 1994).

Making accurate earthquake predictions requires a physical model for earthquakes. One approach is to use quasi-dynamic source models as we have discussed above to generate rupture scenarios that span the variability of potential ground motion in a predictive situation. The basic premise is that fifty years of strong motion records is insufficient to capture all possible ranges of site and propagation path conditions, rupture processes, and spatial geometric relationships between source and site. Predicting future earthquake scenarios is necessary; models that have little or no physical base but have been tested and adjusted to fit available observations can only "predict" what happened in the past, which is "describing" not "predicting" at all. The ultimate solution for modeling earthquakes would be dynamic solutions that satisfy both elastodynamic equations and fracture mechanics, and that have known elastic constants and constituent relations for faulting processes. Estimates of these parameters for the fault zone carry large uncertainties and require several poorly bounded assumptions. The resultant uncertainties in computations limit their usefulness in better understanding of the earthquake process and in providing bounds for kinematic rupture models.

The basic premises of quasi-dynamic modeling for purposes of ground-motion prediction are:

1. Accurate synthesis of recorded ground motions for a particular fault rupture scenario, sufficient for engineering purposes, is possible;
2. A general description of the rupture is sufficient for synthesizing realistic ground motions;
3. The rupture characteristics of a fault can be constrained in advance by interpreting physical properties such as geologic structure, seismicity, and regional tectonics;
4. The range of possible fault-rupture scenarios is narrow enough to functionally constrain the range of strong ground motion predictions; and
5. A discrete set of rupture scenarios is sufficient, for engineering purposes, to span the infinite combinations possible from a given range of rupture parameters.

An important corollary for testing this approach is that if a scenario earthquake matches the seismograms from an actual earthquake, then the rupture parameters of that scenario are close to what actually happened. Pavic et al. (2000), for example, made an assessment of uncertainties and confidence level in the selection of rupture parameters.

Historically, strong ground-motion prediction has generally taken one of two paths: probabilistic or deterministic. Following Cornell (1968), probabilistic seismic hazard analyses (PSHA) require: (1) an interpretation of seismic sources that constitute a hazard to a particular site, from which the distance of an earthquake to the site can be determined; (2) an interpretation of earthquake recurrence for each source; and (3) an evaluation of ground-motion attenuation for the region. The ground-motion attenuation relationships are simple functions of earthquake magnitude and source-site distance (and in some cases a few additional source parameters), and are empirically derived from the strong-motion database recorded from past earthquakes worldwide. Finally (4), given the input evaluations, the PSHA method integrates over all values of the variables and produces an estimate of the mean yearly frequency of exceedance of ground-motion amplitude at the site (i.e., a hazard curve). Figure 17 illustrates the components of a probabilistic hazard study.

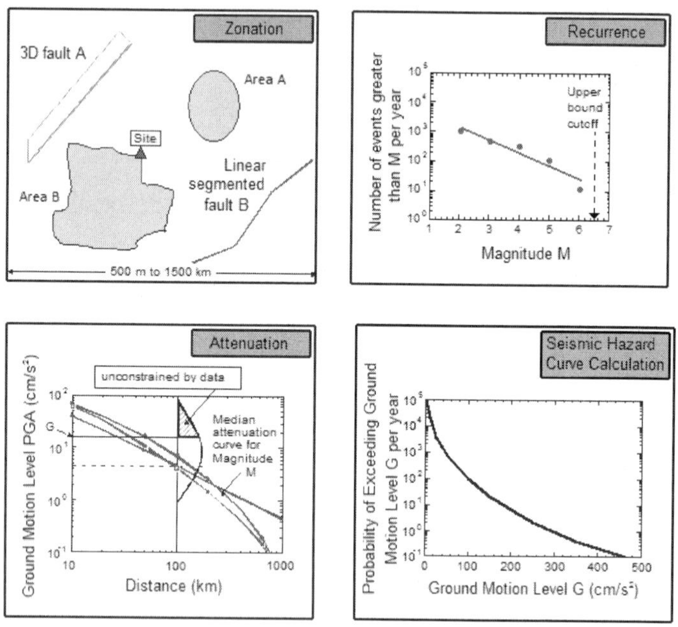

Fig. 17. Components of a probabilistic seismic hazard (PSHA) study. *Top left*: Representation of the source region for PSHA. *Top right*: Typical example of a Gutenberg-Richter b-value relationship. *Bottom left*: Distribution of attenuation relations along with the log normal distribution. *Bottom right*: Seismic hazard curve.

Both earthquake source models and ground-motion attenuation relationships are subject to significant uncertainties, which are expressed as probability distributions (giving an estimate of the median and standard deviation) on earthquake occurrence rates and on the ground-motion relationships. The uncertainty in a ground-motion relationship arises from the variability in source characteristics among events of the same size in the strong-motion database, and from the different earth structures through which the seismic waves from the events propagated. In PSHA studies, this is considered aleatory uncertainty, the uncertainty due to inherent randomness of the process. Current PSHA studies are based upon the ergodic assumption that the randomness in space from several sources is the same as the randomness in time from the same source (Anderson and Brune, 1999). With this ergodic assumption, correlation between the ground motion and the specific source, path, and site is lost, thereby leading to potentially higher total uncertainty in hazard estimates than if each earthquake-source release of energy were individually propagated to the site of interest. There is also an attempt to model epistemic uncertainty, the uncertainty in knowledge about earthquake processes. This refers to factors such as strike, dip, slip vector, etc., that could further reduce aleatory uncertainty if they were known and included as regression parameters.

Alternatively, a deterministic approach identifies significant faults or source zones and establishes the Maximum Possible Earthquake (MPE) for PSHA; Design Operating Earthquake (DOE), essentially the mean of PSHA; and the Operating Basis Earthquake (OBE), maximum during lifetime of facility. Deterministic hazard studies have had the problem of identifying the appropriate source for each earthquake and their likelihood of occurrence, and PSHA has had the problem of insufficient historical data and accurately accounting for epistemic uncertainty with respect to the source and propagation of strong ground motion.

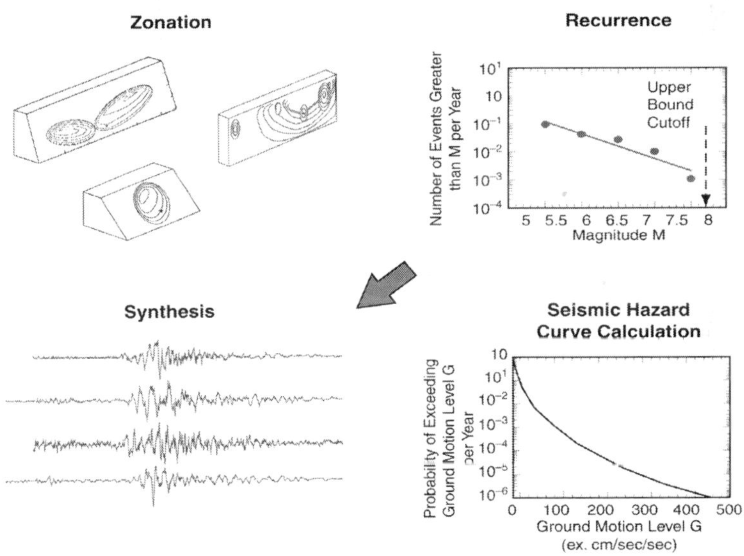

Fig. 18. Components of a physics-based PSHA study. *Top left*: Fault rupture scenarios represent the source region. *Top right*: Typical example of a Gutenberg-Richter b-value relationship. *Bottom left*: Synthesized ground motions. *Bottom right*: Seismic hazard curve for a particular synthesized ground motion.

We can incorporate the combination of deterministic studies by calculating the actual earthquake rupture and recorded ground motion relevant to a particular site, and incorporating this into PSHA studies by replacing the use of attenuation relations. The output from this PSHA approach is a library of source- and site-specific ground-motion time series that would comprise a sample of all the earthquakes that could affect a site during its design life. The uncertainty of the PSHA is defined by bounds on the physical parameters that go into the computation of ground motion rather than having an unbounded PSHA developed from unbound shapes to probability distributions. The library of seismograms is used to either develop hazard curves of traditional engineering parameters, in the form of the annual probability of exceedance, or used directly in developing risk estimates. This alternative approach is illustrated in Figure 18. It replaces the aleatory uncertainty that current PSHA studies estimate by regression of empirical parameters from the worldwide database, with epistemic uncertainty as to what specific sources may actually do at specific sites. Figure 19 illustrates the hazard curve that would result from such an approach. This approach is essentially that recommended by the Southern California Earthquake Center, Phase III study (Filed et al., 2000), which concluded that complex propagation effects unique to earthquake-rupture/site combinations result in uncertainties that require replacing the standard empirical-regression method with one based on a more physics-based approach to ground-motion modeling. This approach has further been endorsed by the National Research Council (2003). A sound framework with which to implement this approach on a large scale has been proposed by Field et al. (2003).

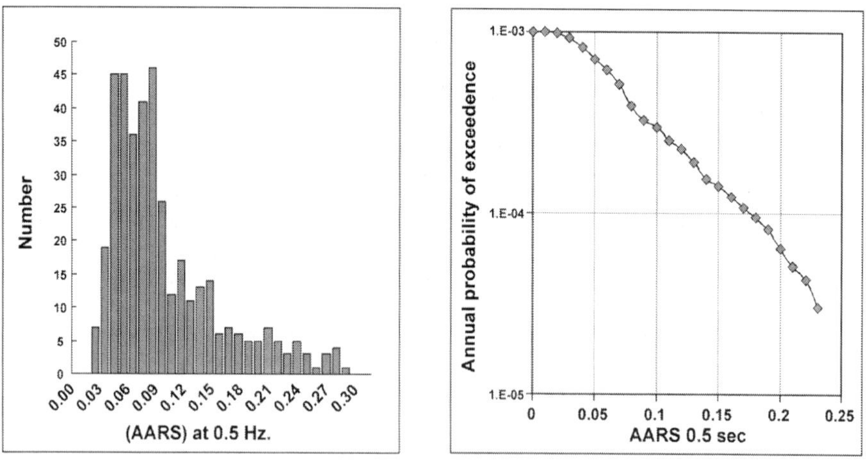

Fig. 19. Hazard Curve .*Left*: Histogram of the number of occurrence of each value of the hazard parameter. *Right*: Cumulative number of occurrences exceeding each hazard value normalized by year.

7. References

Abercrombie, R. E. (1995). Earthquake source scaling relationship from –1 to 5 ML using seismograms recorded at 2.5-km depth, *J. Geophys. Res.*, Vol. 100, pp. 24015-24036.

Abercrombie, R. E. (1998). A summary of attenuation measurements from borehole recordings of earthquakes: The 10 Hz transition problem. *Pure and Appl. Geophys.*, Vol. 153, pp. 475-487.

Abercrombie, R. E., and P. C. Leary (1993). Source parameters of small earthquakes recorded at 2.5 km depth, Cajon Pass, southern California: implications for earthquake scaling, *Geophys. Res. Lett.*, Vol. 20, pp. 1511-1514.

Abercrombie, R. E., and J. R. Rice (2005). Can observations of earthquake scaling constrain slip weakening? *Geophys. J. Int.*, Vol. 162, pp. 406-424, doi: 10.1111/j.1365-246X.2005.02579.x.

Abercrombie, R. E., W. Walter, and R. Gök (2012). Comparison of Direct and Coda Wave stress drop measurements for the Wells, Nevada, earthquake sequence, (in prep).

Abrahamson, N.A., and B. A. Bolt (1987). Array Analysis and Synthesis Mapping of Strong Ground Motion, in *Strong Motion Synthetics*, edited by B.A. Bolt, pp. 55-90, Academic Press, Orlando, Florida, U.S.A.

Aki, K. (1967). Scaling law of seismic spectrum, *J. Geophys. Res.*, Vol. 72, pp. 1217–1231.

Aki, K. and P. G. Richards (2002). *Quantitative Seismology* (2nd Edition), University Science Books, ISBN: ISBN0935702962, 9780935702965, Sausalito, California, U.S.A.

Akinci, A., L. Malagnini, R. B. Herrmann, N. A. Pino, L. Scognamiglio, and H. Eyidogen (2001). High-frequency ground motion in the Erzincan region, Turkey; inferences from small earthquakes, *Bull. Seism. Soc. Am.*, Vol. 91, No. 6, pp. 1446-1455.

Allmann, B. P., and P. M. Shearer (2007). Spatial and temporal stress drop variations in small earthquakes near Parkfield, California, *J. Geophys. Res.*, Vol. 112, No. B4, B04305, doi: 10.1029/2006JB004395.

Allmann, B. P., and P. M. Shearer (2009). Global variations of stress drop for moderate to large earthquakes, *J. Geophys. Res.*, Vol. 114, No. B01310, doi: 10.1029/2008JB005821.

Anderson, J. G. (1986). Implication of attenuation for studies of the earthquake source, in *Earthquake Source Mechanics, Geophysical Monograph 37, Maurice Ewing Vol. 6,* S. Das, J. Boatwright, and C. H. Scholz (Editors), pp. 311-318, American Geophysical Union, Washington, D.C., U.S.A.

Anderson, J. G. and J. Brune (1999). Probabilistic seismic hazard analysis without the ergodic assumption. *Seism. Res. Lett.*, Vol. 71, No. 1, pp. 19-28.

Anderson, J. G. and S. Hough (1984). A model for the shape of the Fourier amplitude spectra of accelerograms at high frequencies, *Bull. Seism. Soc. Am.*, Vol. 74, pp. 1969-1994.

Andrews, D. J. (1986). Objective determination of source parameters and similarity of earthquakes of different size, in *Earthquake Source Mechanics, Geophysical Monograph 37, Maurice Ewing Vol. 6,* S. Das, J. Boatwright, and C. H. Scholz (Editors), pp. 259-268, American Geophysical Union, Washington, D.C., U.S.A.

Archuleta, R. J. (1986). Downhole recordings of seismic radiation, in *Earthquake Source Mechanics, Geophysical Monograph 37, Maurice Ewing Vol. 6,* S. Das, J. Boatwright and C. H. Scholz (Editors), pp. 319-329, American Geophysical Union, Washington, D.C., U.S.A.

Archuleta, R. J., E. Cranswick, C. Mueller, and P. Spudich (1982). Source parameters of the 1980 Mammoth Lakes, California, earthquake sequence, *J. Geophys. Res.*, Vol. 87, pp. 4595-4607.

Aster, R.C. and P.M. Shearer (1991). High-frequency borehole seismograms recorded in the San Jacinto Fault Zone, Southern California, Part 2. Attenuation and site effects, *Bull. Seismol. Soc. Am.*, Vol. 81, pp. 1081-1100.

Baise, L. G., L. J. Hutchings, and S. D. Glaser (2001). Site Response at Yerba Buena Island, San Francisco Bay from Weak Motion Recordings. *Bollettino Di Geofisica Teorica ed Applicata*, Special Issue on "Site Response Estimate from Observed Ground Motion Data.", Vol. 42, No. 3-4, pp. 219-244.

Bakun, W. H. and C. G. Bufe (1975). Shear-wave attenuation along the San Andreas Fault zone in Central California, *Bull. Seism. Soc. Am.*, Vol. 65, pp. 439-460.

Baltay, A., G. Prieto, and G. Beroza (2010). Radiated seismic energy from coda measurements indicates no scaling in apparent stress with seismic moment, *J. Geophys. Res.*, Vol. 115, No. B8, doi: 10.1029/2009JB006736.

Boatwright, J. (1980). A spectral theory for circular seismic sources: simple estimates of source duration, dynamic stress drop, and radiated energy, *Bull. Seismol. Soc. Am.*, Vol. 70, pp. 1-28.

Boatwright, J. (1981). Quasi-dynamic models of simple earthquake: an application to an aftershock of the 1975 Oroville, California earthquake. *Bull. Seis. Soc. Am.*, Vol. 71, pp. 69-94.

Boatwright, J. (1994). Regional propagation characteristics and source parameters of earthquakes in Eastern North America, *Bull. Seism. Soc. Am.*, Vol. 84, pp. 1-15.

Boatwright, J., and J. B. Fletcher (1984). The partition of radiated energy between *P* and *S* waves, *Bull. Seism. Soc. Am.*, Vol. 74, pp. 361-376.

Boatwright, J., J. B. Fletcher, and T. E. Fumal (1991). A general inversion scheme for source, site, and propagation characteristics using multiply recorded sets of moderate-sized earthquakes, *Bull. Seism. Soc. Am.*, Vol. 81, No. 5, pp. 1754-1782.

Bommer, J. J., N. A. Abrahamson, F. O. Strasser, A. Pecker, P.-Y. Bard, H. Bungum, F. Cotton, D. Fah, F. Sabetta, F. Scherbaum, and J. Studer (2004). The Challenge of Defining Upper Bounds on Earthquake Ground Motion. *Seis. Res. Lett.*, Vol. 75, No. 1, pp. 82-95, doi: 10.1785/gssrl.75.1.82.

Bonilla, L. F., J. H. Steidl, G. T. Lindley, A. G. Tumarkin, and R. J. Archuleta (1997). Site amplification in the San Fernando valley, CA: variability of the site effect estimation using the S-wave coda, and H/V methods, *Bull. Seism. Soc. Am.* Vol. 87, pp. 710-730.

Boore, D. M. (1983). Stochastic simulation of high-frequency ground motions based on seismological models of the radiated spectra, *Bull. Seism. Soc. Am.*, Vol. 73, pp. 1865--1894.

Boore, D. M. (2003). Simulation of ground motion using the stochastic method, *Pure and Applied Geophysics*, Vol. 160, pp. 635-676.

Brune, J. (1970). Tectonic stress and the spectra of seismic shear waves from earthquakes, *J. Geophys. Res.*, Vol. 75, pp. 4997-5009.

Burridge, R. and J. Willis (1969). The self-similar problem of the expanding elliptical crack in an anisotropic solid, *Proc. Cambridge Phil. Soc.*, Vol. 66, pp. 443-468.

Cohee, B. P., P. G. Somerville, and N. A. Abrahamson (1991). Simulated Ground Motions for Hypothesized Mw= 8 Subduction Earthquakes in Washington and Oregon. *Bull. Seism. Soc. Am.*, Vol. 81, pp. 28-56.

Cornell, C. A. (1968). Engineering Seismic Risk Analysis, *Bull. Seism. Soc. Am.*, Vol. 58, pp. 1583-1606.

Das, S., and B. V. Kostrov (1990). Inversion for seismic slip rate history and distribution with stabilizing constraints: Application to the 1986 Andreanof Islands Earthquakes. *Journal of Geophysical Research,* Vol. 95, No. B5, pp. 6899–6913, doi: 10.1029/90JB00701, ISSN: 0148-0227.

Dreger, D., and D. Helmberger (1991). Source parameters of the Sierra Madre earthquake from regional and local body waves, *Geophys. Res. Lett.*, Vol. 18, pp. 2015–2018.

Dreger, D. S. (2003). TDMT_INV: Time Domain Seismic Moment Tensor INVersion, *International Handbook of Earthquake and Engineering Seismology,* Vol. 81B, p 1627.

Eshelby, J. D. (1957). The determination of the elastic field of an ellipsoidal inclusion and related problems, *Proc. Roy. Soc. Lond., A,* Vol. 241, pp. 376-396.

Fehler, M. and W. S. Phillips (1991). Simultaneous inversion for *Q* and source parameters of microearthquakes accompanying hydraulic fracturing in granitic rock, *Bull. Seism. Soc. Am,* Vol. 81, pp. 553–575.

Field, E. H. and K. H. Jacob (1993). The theoretical response of sedimentary layers to ambient seismic noise. *Geophys. Res. Lett.*, Vol. 20, No. 24, pp. 2925-2928.

Field, E.H. and the Phase II Working Group (2000). Accounting for site effects in probabilistic seismic hazard analyses of southern California: Overview of the SCEC Phase II Reeport, *Bull. Seism. Soc. Am.*, Vol. 90, pp. S1-S31.

Field, E.H., T. H. Jordan, and C. A. Cornell (2003). OpenSHA: a developing community-modeling environment for seismic hazard analysis, *Seism. Res. Lett.*, Vol. 74, No. 4, pp. 406–419.

Frankel, A. (1982). Precursors to a magnitude 4.8 earthquake in the Virgin Islands: spatial clustering of small earthquakes, anomalous focal mechanisms, and earthquake doublets, *Bull. Seism. Soc. Am.*, Vol. 72, pp. 1277-1294.

Frankel, A. (1991). High-frequency spectral falloff of earthquakes, fractal dimension of complex rupture, b value, and the scaling of strength on faults, *J. Geophys. Res.*, Vol. 96, pp. 6291 - 6302.

Frankel, A. and H. Kanamori (1983). Determination of rupture duration and stress drop from earthquakes in southern California, *Bull. Seism. Soc. Am.* Vol. 73, pp. 1527-1551.

Frankel, A., J. Fletcher, F. L. Vernon, L. Haar, J. Berger, T. Hanks, and J. Brune (1986). Rupture characteristics and tomographic source imaging of *M L* ~ 3 earthquakes near Anza, southern California, *J. Geophys. Res.*, Vol. 91, pp. 12633-12650.

Frankel, A. and L. Wennerberg (1989). Microearthquake spectra from the Anza, California, seismic network: site response and source scaling, Bull. Seis. Soc. Am. 79, 581-609.

Gibowicz, S.J., R. P. Young, S. Talebi and D. J. Rawlence (1991). Source parameters of seismic events at the Underground Research Laboratory in Manitoba, Canada: scaling relations for events with moment magnitude smaller than -2, *Bull. seism. Soc. Am.*, Vol. 81, pp. 1157 1182.

Gök, R., K. Doan and L. Hutchings (2003). Source parameters and scaling relations for aftershocks of the 1999 Marmara earthquakes, *EOS, Trans. Am. Geophy. Union.*

Gök, R. and L. Hutchings (2004). Broadband Green's Functions for the 1999 Marmara Earthquake Region Based on Local Earthquake Recordings, in *American Geophysical Union, Western Pacific Geophysics Meeting*, Honolulu, Hawaii, 16-20 August 2004.

Guatteri, M. P., M. Mai, G. C. Beroza and J. Boatwright (2003). Strong Ground-motion Prediction from Stochastic-Dynamic Source Models. *Bull. Seis. Soc. Am.*, Vol. 93, pp. 301-313.

Hanks, T. (1977). Earthquake stress drops, ambient tectonic stresses and stresses that drive plate motions, *Pure and Applied Geophysics,* Vol. 115, No. 1-2, pp. 441-458, doi: 10.1007/BF01637120.

Hanks, T. (1982). *fmax, Bull. Seism. Soc. Am.*, Vol. 72, pp. 1869-1879.

Hanks, T. and H. Kanamori (1979). A moment magnitude scale. *J. Geophys. Res.*, Vol. 84, pp. 2348-2350.

Hanksson, E., T.-L. Teng, and T. L. Henyey (1987). Results from a 1500 m deep, three-level downhole seismometer array: site response, low Q values and *fmax, Bull. Seism. Soc. Am.*, Vol. 77, pp. 1883-1904.

Hartzell, S. (1978). Earthquake aftershocks as Green's functions, *Geophys. Res. Lett.*, Vol. 5, pp. 1-4.

Hartzell, S. (1982). Simulation of ground accelerations for May 1980 Mammoth Lakes, California, earthquakes, *Bull. Seis. Soc. Am.*, Vol. 72, pp. 2381-2387.

Hartzell, S. and T. H. Heaton (1983). Inversion of strong ground motion and teleseismic waveform data for the fault rupture history of the 1979 Imperial Valley, California earthquake, *Bull. Seismol. Soc Am.*, Vol. 73, pp. 1553-1583.

Hartzell, S. and T. H. Heaton (1986). Rupture history of the 1984 Morgan Hill, California, earthquake from the inversion of strong motion records, *Bull. Seism. Soc Am.*, Vol. 76, pp. 649-674.

Hartzell, S. and M. Iida (1990). Source complexity of the 1987 Whittier Narrows, California, Earthquake from the inversion of strong motion records, *J. Geophys. Res.*, Vol. 95, pp. 12475-12485.

Hartzell, S., S. Harmsen, A. Frankel and S. Larsen (1999). Calculation of Broadband Time Histories of Ground Motion: Comparison of Methods and Validation using Strong-Ground Motion from the 1994 Northridge Earthquake. *Bull. Seism. Soc. Am.*, Vol. 89, pp. 1484-1504.

Haskell, N. A. (1964). Total energy spectral density elastic of wave radiation from propagating faults, *Bull. Seismo. Soc. Am.*, Vol. 54, pp. 1811-184I.

Haskell, N. A. (1966). Total energy and energy spectral density of elastic wave radiation from propagating faults, II, *Bull. Seism. Soc. Am.*, Vol. 56, pp. 125–140.

Heaton, T. H. (1982). The 1971 San Fernando earthquake: a double event? *Bull. Seism. Soc. Am.*, Vol. 72, part A, pp. 2037-2063.

Heaton, T. H. (1990). Evidence for and implications of self-healing pulses of slip in earthquake rupture, *Phys. Earth Planet Int.*, Vol. 64, pp. 1-20.

Herrmann, R. B., and L. Malagnini (1996). Absolute ground motion scaling in the New Madrid Seismic Zone, *Seism. Res. Lett.*, Vol. 67, pp. 40.

Heuze, F. E., T. S. Ueng, L. J. Hutchings, S. P. Jarpe, and P. W. Kasameyer (1994). A coupled seismic-geotechnical approach to site-specific strong motion, Soil Dynamics and Earthquake Engineering. 16 (4): 259-272 JUN 1997.

Hough, S. E. (1997). Empirical Green's function analysis: taking the next step, *J. Geophys. Res.*, Vol. 102, pp. 5369-5384.

Hough, S. E. , K. Jacob, and P. Filberg (1989). The 11/25/88, M = 6 Saguenay earthquake near Chicontimi, Quebec: evidence for anisotropic wave propagation in northeastern North America, *Geophys. Res. Lett.*, Vol. 16, pp. 645-648.

Hough, S. E., L. Seeber, A. Lerner-Lam, G. Armbruster, and H. Guo (1991). Empirical Green's function analysis of Loma Prieta aftershocks, *Bull. Seism. Soc. Am.*, Vol. 81, pp. 1737-1753.

Hough, S. E., and D. S. Dreger (1995). Source parameters of the 23 April 1992 M 6.1 Joshua Tree, California, earthquake and its aftershocks; empirical Green's function analysis of GEOS and TERRAscope data, *Bull. Seism. Soc. Am.*, Vol. 85, pp. 1576–1590.

Hough, S. E., J. M. Lees, and F. Monastero (1999). Attenuation and source properties at the Coso Geothermal Area, California, *Bull. Seism. Soc. Am.*, 89(6), 1606-1619.

Hutchings, L. (1987). Modeling near-source earthquake ground motion with empirical Green's functions, Ph.D. Thesis, State University of New York.

Hutchings, L. (1991). Prediction of strong ground motion for the 1989 Loma Prieta earthquake using empirical Green's functions, *Bull. Seism. Soc. Am.*, Vol. 81, pp. 88-121.

Hutchings, L. (1994). Kinematic Earthquake Models and Synthesized Ground Motion Using Empirical Green's Functions, *Bull. Seism. Soc. Am.*, Vol. 84, pp. 1028-1050.

Hutchings, L. and F. Wu (1990). Empirical Green's functions from small earthquakes: waveform study of locally recorded aftershocks of the San Fernando earthquake, *J. Geophys. Res.*, Vol. 95, pp. 1187-1214.

Hutchings, L., B. Foxall, P. Kasameyer, S. Larsen, C. Hayek, C. Tyler-Turpin, J. Aquilino, and L. Long (2005). Deep Borehole Instrumentation along San Francisco Bay Bridges: 1996 - 2003 and Strong Ground Motion Synthesis along the San Francisco/Oakland Bay Bridge. Final Report July 15, 2005. Also, annual progress reports. Caltrans contract no. 59A0238. UCRL-TR-2117303.

Hutchings, L., Ioannidou, E., Kalogeras, I., Voulgaris, N., Savy, J., Foxall, W., Scognamiglio, L., and Stavrakakis, G. (2007). A strong motion prediction methodology; application PSHA and the 7 September 1999, M = 6.0 Athens Earthquake, *Geophys. J. Int.*, Vol. 168, pp. 659–680, doi: 10.1111/j.1365-246X.2006.03178.x.

Ide, S., and G. C. Beroza (2001). Does apparent stress vary with earthquake size? *Geophys. Res. Lett.*, Vol. 28, pp. 3349-3352.

Ide, S., G. C. Beroza, S. G. Prejean, and W. L. Ellsworth (2003). Apparent break in earthquake scaling due to path and site effects on deep borehole recordings, *J. Geophys. Res.*, Vol. 108, No. B5, doi:10.1029/2001JB001617.

Imanishi, K., and W. L. Ellsworth, (2006). Source scaling relationships of microearthquakes at Parkfield, CA, determined using the SAFOD pilot hole seismic array, in *Earthquakes: Radiated Energy and the Physics of Faulting, Geophysical Monograph 170*, edited by R.E. Abercrombie, A. McGarr, H. Kanamori and G. Di Toro, pp. 81–90, American Geophysical Union, Washington, D.C., U.S.A.

Irikura, K. (1984). Prediction of strong ground motions using observed seismograms from small events, *Proc. 8th World Conf. on Earthq. Eng.*, Vol. 2, pp. 465-472.

Irikura, K. (1986). Prediction of strong acceleration motion using empirical Green's function. *Proc. 7th Japan Earthq. Eng. Synp.*, pp. 151-156.

Jarpe, S., L. J. Hutchings, T. F. Hauk, and A. F. Shakel (1989). Selected Strong and Weak Motion Data from the Loma Prieta Earthquake Sequence, *Seism. Res. Letters*, Vol. 60, pp. 167-176.

Jarpe, S., and P. K. Kasameyer (1996). Validation of a Methodology for Predicting Broadband Strong Motion Time Histories using Kinematic Rupture Models and Empirical Green's Functions. *Bull. Seism. Soc. Am.*, Vol. 86, pp. 1116-1129.

Ji, C., D. J. Wald, and D. V. Helmberger (2002). Source description of the 1999 Hector Mine, California, earthquake, Part II: Complexity of slip history, *Bull. Seism. Soc. Am.*, Vol. 92, pp. 1208–1226.

Joyner, W. B. and D. M. Boore (1986). On simulating large earthquakes by Green's-function addition of smaller earthquakes, in *Earthquake Source Mechanics, Geophysical Monograph 37, Maurice Ewing Vol. 6*, S. Das, J. Boatwright, and C. H. Scholz (Editors), pp. 269-274, American Geophysical Union, Washington, D.C., U.S.A.

Kanamori, H., and G. S. Stewart (1976). Mode of the strain release along the Gibbs fracture zone, Mid-Atlantic ridge, *Phys. Earth Planet. Int.*, Vol. 11, pp. 312-332.

Kanamori, H., and T. H. Heaton (2000). Microscopic and macroscopic physics of earthquakes, in *Geocomplexity and the Physics of Earthquakes, Geophysical Monograph Series* Vol. 120, edited by J. Rundle, D. L. Turcotte and W. Kein, pp. 147-155, American Geophysical Union, Washington, D.C., U.S.A.

Kostrov, B. V. (1964). Self similar problems of propagation of shear cracks, *J. Appl. Math. Mech. (PMM)*, Vol. 28, pp. 1077-1087.

Kostrov, B. V., and S. Das (1988). *Principles of earthquake source mechanics*, in Cambridge Monographs on Mechanics and Applied Mathematics, Cambridge University Press, Cambridge, UK.

Lay, T., H. Kanamori and L. Ruff (1982). The asperity model and the nature of large subduction zone earthquake occurrence, *Earthquake Prediction Research*, Vol. 1, pp. 3-71.

Lay, T., and T. C. Wallace (1995). *Modern Global Seismology*, Academic Press, ISBN: 012732870X, 9780127328706, San Diego, California, U.S.A.

Li, Y., C. Jr. Doll, and M. N. Toksöz (1995). Source characterization and fault plane determination for M_{bLg} = 1.2 to 4.4 earthquakes in the Charlevoix Seismic Zone, Quebec, Canada, *Bull. Seism. Soc. Am.*, Vol. 85, pp. 1604-1621.

Lynnes, C. S. and T. Lay (1988). Source process of the great 1977 Sumba earthquake, *J. Geophys. Res.*, Vol. 93, pp. 13407-13420.

Madariaga, R. (1976). Dynamics of an expanding circular crack, *Bull. Seism. Soc. Am.*, Vol. 66, pp. 639-666.

Malagnini, L., and R. B. Herrmann (2000). Ground motion scaling in the region of the 1997 Umbria-Marche earthquake (Italy), *Bull. Seism. Soc. Am.*, Vol. 90, pp. 1041-1051.

Malin, P.,E., and J. A. Waller (1985). Preliminary results from vertical seismic profiling of Oroville microearthquake S-waves, *Geophys. Res. Lett.*, Vol. 12, No. 3, pp. 137-140

Mayeda, K., and W.R. Walter (1996). Moment, energy, stress drop and source spectra of western United States earthquakes from regional coda envelopes, *J. Geophys. Res.*, Vol. 101, pp. 11195-11208.

Mayeda, K., A. Hofstetter, J. O'Boyle, and W.R. Walter (2003). Stable and transportable regional magnitudes based on coda-derived moment-rate spectra, *Bull. Seism. Soc. Am.*, Vol. 93, pp. 224-239.

Mayeda, K., L. Malagnini, W.R. Walter (2007). A new spectral ratio method using narrow band coda envelopes: Evidence for non-self-similarity in the Hector Mine sequence, *Geophys. Res. Lett.*, Vol. 34, No. L11303, doi:10.1029/2007GL030041.

McGarr, A., (1999). On relating apparent stress to the stress causing earthquake fault slip, *J. Geophys. Res.*, Vol. 104, pp. 3003-3011.

McGarr, A., J. Bicknell, J. Churcher and S. Spottiswoode (1990). Comparison of ground motion from tremors and explosions in deep gold mines, *J. Geophys. Res.*, Vol. 95, pp. 21777–21792.

Mori, J., (1996). Rupture directivity and slip distribution of the M4.3 foreshock to the 1992 Joshua Tree earthquake, southern California, *Bull. Seism. Soc. Am.*, Vol. 86, pp. 805–810.

Mori, J., and A. Frankel (1990). Source parameters for small events associated with the 1986 North Palm Springs, California, earthquake determined using empirical Green functions, *Bull. Seism. Soc. Am.*, Vol. 80, pp. 278-295.

Mori, J., R. E. Abercrombie, and H. Kanamori (2003). Stress drops and radiated energies of aftershocks of the 1994 Northridge, California, earthquake, *J. Geophys. Res.*, Vol. 108, No. B11, 2545, doi: 10.1029/2001JB000474.

Mueller, C. S. (1985). Source pulse enhancement by deconvolution of an empirical Green's function, *Geophys. Res. Let.*, Vol. 12, pp. 33-36.

National Research Council (2003). *Living on an Active Earth: Perspectives on Earthquake Science*, National Academic Press, Washington, D.C

Orowan, E. (1960). Mechanism of seismic faulting, *Geol. Soc. Am. Bull.*, 79, 323–345.

Papageorgiou, A.S. and K. Aki (1983). A Specific Barrier Model for the Quantitative Description of Inhomogeneous Faulting and the Prediction of Strong Ground Motion. Part I: Description of the Model, *Bull. Seism. Soc. Am.*, Vol. 73, No. 3, pp. 693-722.

Pasyanos, M.E., D.S. Dreger, and B. Romanowicz (1996). Toward Real-Time Estimation of Regional Moment Tensors, *Bull. Seism. Soc. Am.*, Vol. 86, pp. 1255-1269.

Pavic, R., M. G. Koller, P.-Y. Bard, and C. Lacave-Lachet (2000). Ground Motion prediction with the empirical Green's function technique: an assessment of uncertainties and confidence level, *Journal of Seismology*, Vol. 4, pp. 59-77.

Prejean, S. G., and W. L. Ellsworth (2001). Observations of earthquake source parameters at 2 km depth in the Long Valley Caldera, eastern California, *Bull. Seismol. Soc. Am.*, Vol. 91, pp. 165-177.

Prieto, G., P. M. Shearer, F. L. Vernon, and D. Kilb (2004). Earthquake source scaling and self-similarity estimation from stacking *P* and *S* spectra, *J. Geophys. Res.*, Vol. 109, No. B08310, doi: 10.1029/2004JB003084.

Sato, T., and T. Hirasawa (1973). Body wave spectra from propagating shear cracks, *J. Phys. Earth*, Vol. 21, pp. 415-431.

Scognamiglio, L. and L. Hutchings (2009). A test of a physically-based strong ground motion prediction methodology with the 27 September 1997, Mw = 6.0 Colfiorito (Umbria-Marcha sequence), Italy earthquake, *Tectonophysics*, Vol. 476, pp. 145-158.

Scholz, C. H. (1982). Scaling laws for large earthquakes: Consequences for physical models, *Bull. Seism. Soc. Am.*, Vol. 72, pp. 1-14.

Shearer, P. M., G. A. Prieto, and E. Hauksson (2006). Comprehensive analysis of earthquake source spectra in southern California, *J. Geophys. Res.*, Vol. 111, No. B06303, doi: 10.1029/2005JB003979.

Shi, J., W.-Y. Kim, and P. Richards (1998). The corner frequency and stress drops of intraplate earthquakes in the northeastern United States, *Bull. Seism. Soc. Am.*, Vol. 88, pp. 531-542.

Somerville, P. G., M. Sen, and B. Cohee (1991). Simulation of strong ground motion recorded during the 1985 Michoacan Mexico and Valparaiso, Chile earthquakes. *Bull. Seism. Soc. Am.*, Vol. 81, pp. 1–28.

Somerville, P., K. Irikura, R. Graves, S. Sawada, D. Wald, N. Abrahamson, Y. Iwasaki, T. Kagawa, N. Smith, and A. Kowada (1999). Characterizing Crustal Earthquake Slip Models for the Prediction of Strong Ground Motion, *Seis. Res. Lett.*, Vol. 70, No. 1, pp. 59-80.

Spottiswoode, S. M. and A. McGarr (1975). Source parameters of tremor in a deep-level gold mine, *Bull. Seism. Soc. Am.*, Vol. 65, pp. 93–112.

Spudich, P. and L. N. Frazier (1984). Use of ray theory to calculate high-frequency radiation from earthquake sources having spatially variable rupture velocity and stress drop. *Bull. Seism. Soc. Am.*, Vol. 74, pp. 2061-2082.

Steidl, J. H., A. G. Tumarkin and R. J. Archuleta (1996). What is a reference site? *Bull. seism. Soc. Am.*, Vol. 86, pp. 1733–1748.

Tomic, J., R. E. Abercrombie, and A. F. do Nascimento (2009). Source parameters and rupture velocity of small M≤2.1 reservoir induced earthquakes, *Geophys. J. Int.*, doi:10.1111/j.1365-246X.2009.04233.x.

Trifunac, M. D. and J. N. Brune (1970). Complexity of energy release during the Imperial Valley, California, earthquake of 1940, *Bull Seism. Soc. Am.*, Vol. 60, pp. 137-160.

Tumarkin, A.G., R.J. Archuleta, and R. Madariaga (1994). Scaling relations for composite models. *Bull. Seism. Soc. Am.*, Vol. 84, No. 4, pp 1279-1283.

Venkataraman, A., J. Mori, H. Kanamori, and L. Zhu (2000). Fine structure of the rupture zone of the April 26 and 27, 1997, Northridge aftershocks, *J. geophys. Res.*, Vol. 105, pp. 19085–19093.

Viegas, G. (2009). Earthquake Source Properties and Wave Propagation in Eastern North America, *Ph.D. Thesis, Boston University.*

Viegas, G. (2011). Source parameters of the July 16 2010 M3.4 Germantown, MD, Earthquake, *Seism. Res. Lett.*, (in review).

Viegas, G., R. E. Abercrombie, and W.-Y. Kim (2010). The 2002 *M*5 Au Sable Forks, NY, earthquake sequence: Source scaling relationships and energy budget, *J. Geophys. Res.*, Vol. 115, No. B07310, doi: 10.1029/2009JB006799.

Viegas, G., R. E. Abercrombie, and K. Mayeda (2011). Coda derived source parameters in Eastern North America, *Seism. Res. Lett.*, (in review).

Wald, D. J., D. V. Helmberger, and T. H. Heaton (1991). Rupture model of the 1989 Loma Prieta earthquake from the inversion of strong-motion and broadband teleseismic data, *Bull. seism. Soc. Am.*, Vol. 81, pp. 1540–1572.

Wald, D. J., and T. H. Heaton (1994). Spatial and temporal distribution of slip for the 1992 Landers, California, earthquake, *Bull. seism. Soc. Am.*, Vol. 84, pp. 668–691.

Wald, D. J., (1995). A preliminary dislocation model for the 1995 Kobe (Hyogo-ken Nanbu), Japan, earthquake determined from strong motion and teleseismic waveforms, *Seism. Res. Lett.*, Vol. 66, pp. 22–28.

Wald, D. J., T. H. Heaton, and K. W. Hudnut (1996). The slip history of the 1994 Northridge, California, earthquake determined from strong-motion, teleseismic, GPS, and leveling data, *Bull. seism. Soc. Am.*, Vol. 86, No. 1, Part B Suppl., pp. 49–70.

Wald, D. J., and R. W. Graves (2001). Resolution Analysis of Finite Fault Source Inversion Using 1D and 3D Green's Functions, Part II: Combining Seismic and Geodetic Data, *J. Geophys. Res.*, Vol. 106, No. B5, pp. 8767.

Wells, D. L. and K. J. Coppersmith (1994). New Empirical Relationships among Magnitude, Rupture Length, Rupture Width, Rupture Area, and Surface Displacement. *Bull. Seism. Soc. Am.*, Vol. 84, pp. 974-1002.

Wennerberg, L. (1990). Stochastic summation of empirical Green''s functions, *Bull. Seismol. Soc. Am.*, Vol. 80, pp. 1418-1432.

Wossner, J., M. Treml and F. Wenzel (2002). Simulation of Mw = 6.0 earthquakes in the Upper Rhinegraben using empirical Green functions, *Geophys. J. Int.*, 151, 487-500.

Wu, F. (1978). Prediction of strong ground motion using small earyhquakes, in *Proceedings of 2nd International Conference on Microzonation Nati. Sci. Found.*, San Francisco, Calif., 701-704.

Xie, J., Z. Liu, R. B. Herrmann, and E. Cranswick (1991). Source processes of three aftershocks of the 1983 Goodnow, New York, earthquake: High-resolution images of small, symmetric ruptures, *Bull. Seism. Soc. Am.*, Vol. 81, pp. 818-843.

Yamada, T., J. Mori, S. Ide, R. E. Abercrombie, H. Kawakata, M. Nakatani, Y. Iio, and H. Ogasawara (2007). Stress drops and radiated seismic energies of microearthquakes in a South African gold mine, *J. Geophys. Res.*, Vol. 112, No. B03305, doi: 10.1029/2006JB004553.

4

Seismic Source Characterization for Future Earthquakes

Jorge Aguirre Gonzales[1], Alejandro Ramirez-Gaytán[2], Carlos I. Huerta-López[3] and Cecilia Rosado-Trillo[2]
[1]*Engineering Institute, Mexico National Autonomous University, Interna Circuito, Universitary City, Coyoacan Delegation, D.F.*
[2]*University of Guadalajara, Civil Engineering Department, Guadalajara, Jalisco,*
[3]*Scientific Investigation Center and Superior Education of Ensenada, BC, Ensenada, Baja California*
Mexico

1. Introduction

Although the great advance in science and technology, currently there is no available methodology to predict an earthquake. One of the most important task in seismology is the develop of methodologies that allow predict and simulate strong ground motions. High accelerations are produced by earthquakes of large magnitude in urban areas located in relative close proximity to seismic sources. Strong ground motions allows to generate models that are necessary to understand the seismic source and to generate response spectra, both useful information in structural engineering.

One of the methodologies to simulate strong ground motion produced for big earthquakes is the empirical Green's function method (EGFM). This method developed by Irikura (1986), requires a small magnitude earthquake with hypocenter near to the main earthquake. An important characteristic of the EGFM is that information of structure and site effects are included in the simulations, since records of the element event used as seeds already include them. This means that instrumentation and detailed studies to know the cortical structure and site effect are not necessary. Another important characteristic of the EGFM is that allow model in the frequency interval of 1-10 Hz, in this range many buildings, bridges and civil constructions have their dominant vibration periods.

EGFM is a methodology applied in Mexico to simulated different big earthquakes. Applying this methodology, Garduño (2006) generate a source model for July 15, 2006 earthquake in the state of Guerrero Mexico, Aguirre Gonzalez (phone inquire) generate a source model for April 25, 1989 San Marcos Mexico earthquake, and Ramirez-Gaytán *et al.* (2010) applying this methodology generate a source model for January 22, 2003 Tecoman earthquake.

In these three different studies the relationships proposed by Somerville *et al.* (2002) for subduction earthquakes was applied. Somerville *et al.* (2002) relationships related the seismic moment with inner and outer source parameters. The comparison between the inner and outer parameters generated in these 3 different simulations, show poor adjust with Somerville *et al.* (2002) relations. The fit in some cases are minor to 27% respect to proposed by Somerville *et al.* (2002). The results obtained in the studies mentioned above might suggest that not all of the relationships proposed by Somerville *et al.* (2002) are applicable to the subduction zone in Mexico.

Under this premise, an immediate doubt arises: Somerville *et al.* (2002) relationships are appropriate to be applied in the simulation of strong ground motion for Mexico subduction zone?.

In the first part of this document we show an interesting application of the application of EGFM in Mexico conducted by Ramirez-Gaytan *et al.* (2010). In this study we take the strong ground motions generated in the model with best fit. Although of good fit the principal values of inner and outer parameters of source are shorter comparing with Somerville *et al.* (2002) relations. As consequence of these results, in the second part of this document we show the results of the investigation conducted to estimate a new relationships between seismic moment versus inner and outer seismic source parameters, but with the particularity that in this case we use only data from Mexican subduction earthquakes.

2. Tecomán earthquake: Physical implications of seismic source modeling, applying the empirical green's function method

In the study of Ramirez-Gaytan *et al.* (2010) a source model for Tecomán Mexico earthquake (21 January 2003, 20:06) was generated. The presence of soft soils and the location of 8 of the

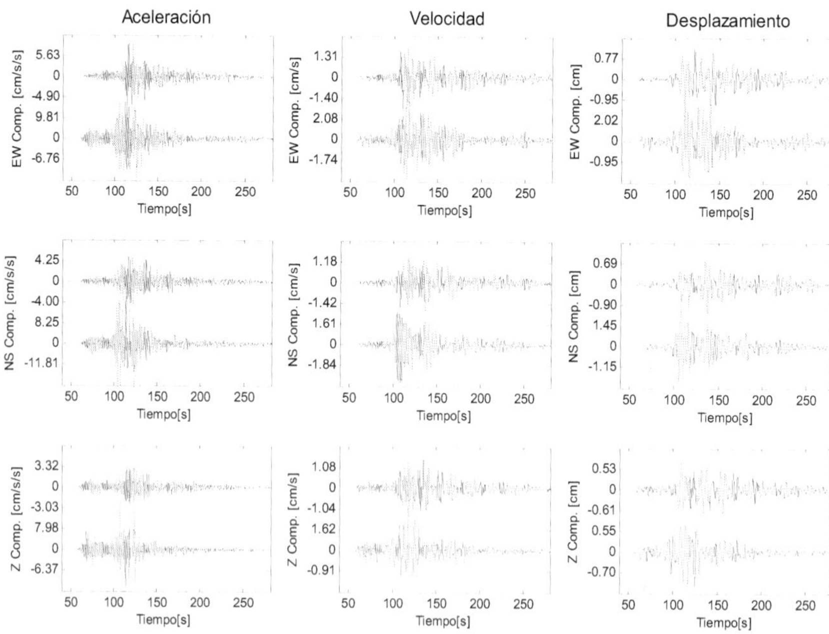

Fig. 1. Comparison between synthetics and observed records. Shown in Blue, synthetics records simulated using empirical Green's function method. Shown in red, observed records. Columns from left to right: acceleration, velocity, and displacement. Rows from top to bottom: EW, NS, and Z components.

10 major cities of Colima state where the earthquake spread are important factors to model the seismic source. To generate the model applied to the empirical Green's function method (EGFM) the Tecomán earthquake (Mw = 7.5) was used as target event and, the November 19, 2006 (Mw = 5.5) earthquake was used as element event.

In this investigation data from 4 broad band and 1 acceleration sensor of regional stations were used. These five stations provided good azimuthal coverage of Tecomán earthquake. The process of modeling the target event was done in 4 stages, each one involving one, two three and four SMGA for each stage respectively. The observed waveforms were adjusted gradually by the synthetics waveforms and the residual values progressively decreased in each stage from 1 to 3 SMGA's. The model with 4 SMGA showed an increased value in residual and poor adjustment. Thus the best fitting was obtained by modeling the target event with 3 SMGA. This model presents the best fit in terms of the lowest residual. In addition to the above, the model keeps a close resemblance with the dislocation model found by Yagi *et al.* (2004).

At the recording site of MANZ, the authors made a spectral analysis to compare weak and strong ground motions in order to identify if some energy is biased concentrated at certain frequencies in the interval of 1-10 Hz. The synthetics waveforms and Fourier spectrums shown in figures 1 and 2 show a good fitting with the observed ones in the five stations. The latter roughly corresponds to the dislocation model found by Yagi *et al.* (2004).

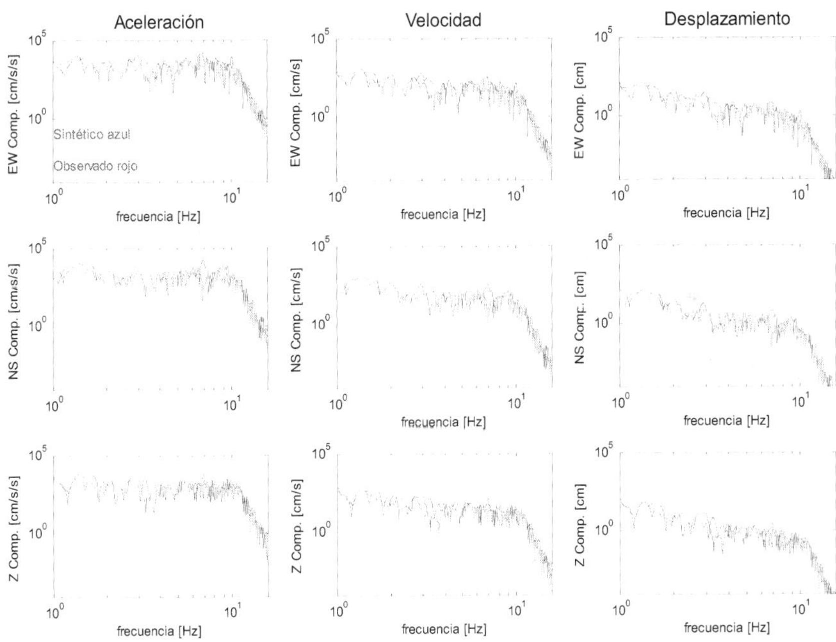

Fig. 2. Comparison between synthetics and observed Fourier spectra. Blue color, synthetics Fourier spectra simulated using EGFM. Red color, observed fourier spectra. Columns from left to right: acceleration, velocity, and displacement. Rows from top to bottom: EW, NS, and Z components.

As part of our investigation we quantified individual and average characteristics of SMGA, rupture area, dislocation time and rise time. These values are related against the seismic moment of these events (Mo = 1.6 x 10^{20} Nm); the results obtained were compared with relationships proposed by Somerville *et al.* (2002) for subduction earthquakes (table 1). These relationships involved the seismic moment (Mo), the fault area, dislocation time, and the rise time. In addition to the above the seismic moment and the following characteristics of SMGA: total area of the SMGA were considered in the comparison, area of bigger SMGA, radius of bigger SMGA, and distance from hypocenter to nearest SMGA.

	A: This study	B. Somerville *et al.* (2002)	C. Ratio (=A/B)
Rupture area	5.95E+03	8.01E+03	7.43E-01
Dislocation time	3.00E+01	4.80E+00	6.25E+00
Rise time	4.00E-01	2.38E+00	1.68E-01
Total area of SMGA	7.40E+01	2.00E+03	3.70E-02
Area of largest SMGA	3.55E+01	1.30E+03	2.73E-02
Radio of largest SMGA	3.36E+00	2.20E+01	1.53E-01
Hypocentral distance of nearest SMGA	1.37E+01	2.10E+01	6.53E-01

Table 1. Comparison (column C) between the relationship proposed by Somerville *et al.* (2002) for subduction earthquakes (column B) and the results of this study (column A).

The results of these comparisons show that the relationships between Mo and rupture area, Mo and hypocentral distance to the nearest asperity, adjust moderately well, which is not the case for the rest of the relationships described above. As commented before, studies conducted by others investigators to simulate Mexican earthquakes show the same poor adjust with Somerville *et al.* (2002) relations. The next doubt arises: Somerville *et al.* (2002) relations are appropriate to be applied in the simulation of strong ground motion for Mexico subduction zone?. To answer the last question we conducted a study to estimate new relationships using only data from Mexican subduction earthquakes.

3. Source scaling relationship of Mexican subduction earthquakes for the prediction of strong ground motions

In this study authors use fault slip models from Mexican subduction zone to investigate the source scaling relationships, and made a global compilation of source parameters to examine their relationships with Seismic Moment.

In the past, for several years many studies have been carried out to investigate source scaling of earthquakes whose objective is to understand the self similarity. This is an important topic in the develop of source scaling relationships. These relationships provide a way to understand the rupture mechanism, also provide deterministic parameters in the prediction of strong ground motions. The heterogeneities in the slip and stress drop distributions controls the generation of source ground motion, proving then that are important to characterize the heterogeneities of past earthquakes in constructing a source model for reliable prediction of strong ground motions.

A quantitative criteria for the rupture area estimation and asperity area from large subduction earthquakes was proposed by Somerville *et al.* (2002). In their criteria, the asperity is defined as a rectangular area whose slip is 1.5 or more times larger than the average slip over the fault. In their study, Somerville *et al.* (2002) compiled the slip models of ten large subduction earthquakes, and obtained empirical scaling relationships between inner and outer source parameters as well as seismic moment. In other hand, Miyake *et al.* (2003) used broadband strong ground motion simulations. In that study they found that the strong motion generation area, which is defined as a high slip velocity or a high stress drop area on the source fault, coincides spatially with the location of asperity or the large slip area of the heterogeneous kinematic slip models.

In our study, we construct the scaling relationship of the source parameters, and compiled slips models of 9 large earthquakes in Mexican subduction zone from kinematic slip models developed by several investigators. Our objective was to make a comparison between this new relationships that use only data from Mexican subduction earthquakes versus Somerville *et al.* (2002) who uses data from large subduction earthquakes word wide. The earthquakes and source parameters analyzed in our study are listed in table 2 and are shown in figure 3.

No	Eartquake	Date (dd/mm/yy)	Depth* (km)	Mw*	Mo « (dina-cm)	Area « (km²)	Strike; Dip; Rake * (°)	Reference
1	Michoacán	19/09/1985	17.00	8.01	1.10 E + 28	25020	300; 14; 72	Mendoza and Hartzell (1989)
2	Manzanillo	09/10/1995	15.00	7.96	1.15 E + 28	17000	309; 14; 92	Mendoza and Hartzell (1999)
3	Michoacán	11/01/1997	40.00	7.10	6.06 E + 26	1312.5	292; 18; -106	Santoyo *et al.* (2005)
4	Petatlán	14/03/1979	15.00	7.39	1.72 E + 27	14400	293; 14; 90	Mendoza and Hartzell (1997)
5	Playa Azul	25/10/1981	15.00	7.25	7.00 E + 26	2700	300; 14; 90	Mendoza *et al.* (1993)
6	San Marcos	25/04/1989	15.00	6.90	2.39 E + 26	2520	276; 10; 66	Zuñiga *et al.* (1993)
7	Tecomán	22/01/2003	20.00	7.50	2.05 E + 27	5950	300; 22; 93	Yagi *et al.* (2004)
8	Zihuatanejo	21/09/1985	20.00	7.42	2.49 E + 27	3500	300; 14; 100	Mendoza *et al.* (1993)
9	Oaxaca	30/09/1999	40.00	7.47	1.72 E + 27	3712.5	295; 50; -82	Castro R and Euclides Ruiz (2005)

Table 2. Source parameters of nine large subduction earthquakes used in this study.

A rectangular geometry of the fault was utilized for the finite fault. The rupture area and asperity area were extracted following the procedure proposed by Somerville *et al.* (1999). The definition of asperity used in this study follows the one proposed by Somerville *et al.* (1999). An asperity is defined to enclose fault elements whose slip is at least 1.5 times larger than the average slip over the fault and is subdivided if any row or column has an average slip less than 1.5 times the average slip. The asperity is then trimmed until all of the edges have an average slip equal or larger than 1.25 times the slip averaged over the entire rupture area. The discretization of the fault into fault element, place limits on the size of the smallest asperity. In view of this discretization that requires an asperity to have a minimum of two elements if the slip of each is 2 or more times the average slip; a minimum of 4 elements if the slip of each is 1.5 or more times the average slip, or the slip of one is 2 or more or the slip of two others is 1.5 or more times the average slip.

The 9 earthquakes have a total of 20 asperities, with earthquakes composed from 1 to 5 asperities. Total number of asperities of the 9 the earthquakes was 20 with an average of 2.22 asperities. For irregular shape asperities, the procedure of Somerville *et al.* (1999) can generate contrasting solutions depending upon whether we start with row-wise or column-wise operations. In this case we mainly make subjective selection of the best solution based on our knowledge of the earthquake. This subjective selection generally works well and produces a reasonable value for the combined area of asperities.

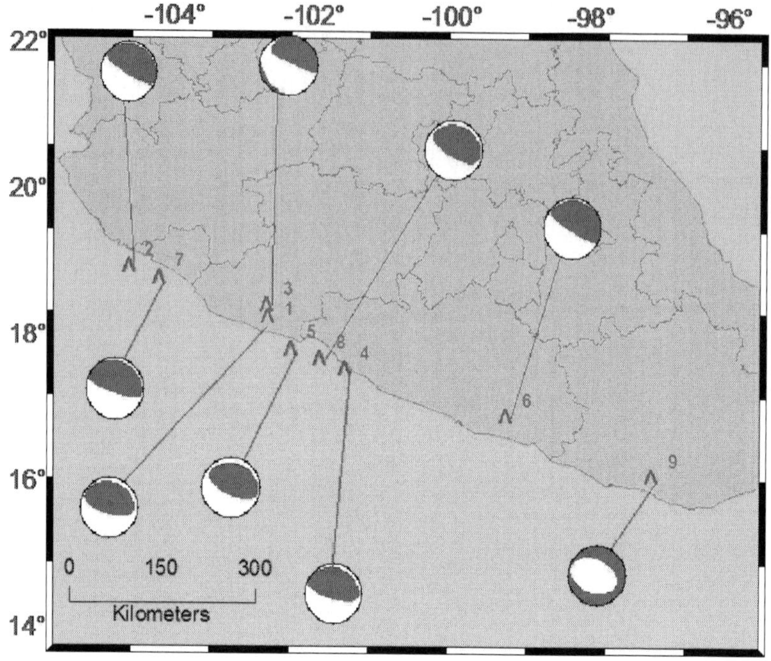

Fig. 3. Distribution of earthquakes used in this study and corresponding focal mechanism.

We carried out regression analyses of the obtained fault parameters for the large subduction Mexican earthquakes listed in table 2 with moment magnitudes range from Mw 6.9 to 8.1. We compared fault parameters with those for large subduction world - wide earthquakes of Mw 7.1 to 8.1 summarized by Somerville *et al.* (2002).

Table 3 show the relationship between seismic moment and each seismic source parameter obtained in this study. These scaling relationships are important for establishing general rules for developing source models for simulating strong ground motions. For each parameter it is first shown the unconstrained equation, followed by the constrained equation to be self similar. The self similar model is convenient to use, and in many instances its use can be justified because provides a reasonable good description of nature Somerville *et al.* (1999).

In this study of large subduction Mexican earthquakes, we find that the scaling of fault parameters with seismic moment fit reasonably well by a self similar model. For the case of the relation of average slip versus seismic moment the unconstrained relation suggest a nonself-similar scaling, for this case the physical interpretation suggest the absence of scale factor in the average slip of Mexican subduction earthquakes.

Rupture area vs. seismic moment.	A=3.69 x $10^{-15}M_o^{2/3}$
Average slip vs. seismic moment.	D=6.60 x $10^{-8}M_o^{1/3}$
Combined area of asperities vs. seismic moment.	A_2=6.56 x $10^{-16}M_o^{2/3}$
Area of largest asperity vs. seismic moment.	A_l=4.96 x $10^{-16}M_o^{2/3}$
Average number of asperities	2.2
Area of fault cover by asperities	0.194
Average slip contrast	2.42
Hipocentral distance to center of closets asperities vs. seismic moment.	R_A=8.84 x $10^{-9}M_o^{1/3}$
Hipocentral distance to center of largest asperities vs. seismic moment.	R_a=1.19 x $10^{-8}M_o^{1/3}$
Slip duration vs. seismic moment.	A_l=2.11 x $10^{-9}M_o^{1/3}$

Table 3. Scaling relations of slip models is assuming self similarity.

3.1 Average slip versus seismic moment

The relationship between average slip D and seismic moment determined without constraining the slope is:

$$D = 6.35 \times 10^{-2} M_o^{0.1138}$$

Constraining the slope to be 1/3, the relation is:

$$D = 6.60 \times 10^{-8} M_o^{1/3}$$

As show in figure 4 comparisons of constrained equations indicates that the estimated average slip of large Mexican subduction earthquakes is larger than constrained equation provided by Somerville *et al.* (2002).

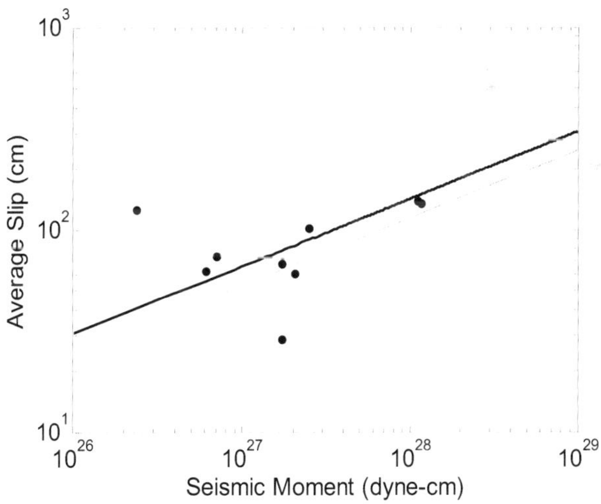

Fig. 4. Relationship between average slip and seismic moment. Dots represent individual events, solid line is the result of this study and dashed line represents the results obtained by Somerville *et al.* (1999).

3.2 Rupture area versus seismic moment

The relationship between rupture area A and seismic moment determined without constraining the slope is:

$$A = 8.96 \times 10^{-15} M_o^{0.6525}$$

Constraining the slope to be 2/3, the relation is:

$$A = 3.69 \times 10^{-15} M_o^{2/3}$$

As show in figure 5, comparison of constrained equation indicates that the estimated rupture area of large Mexican subduction earthquakes is shorter than constrained equation provided by Somerville *et al.* (2002).

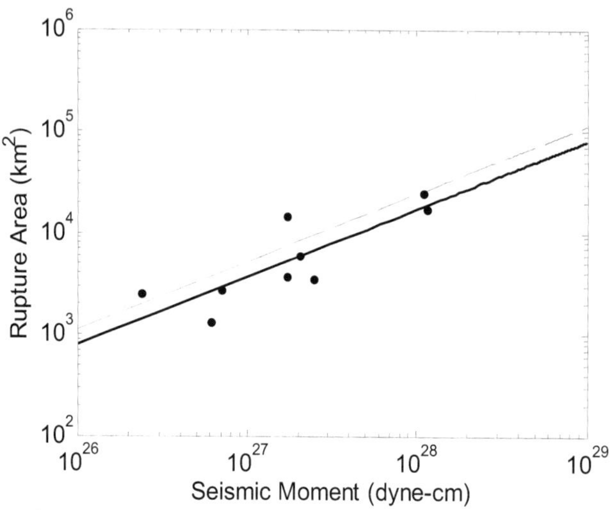

Fig. 5. Relationship between rupture area and seismic moment. Dots represent individual events, solid line is the result of this study and dashed line represents the results obtained by Somerville *et al.* (1999).

3.3 Combined area of asperities versus seismic moment

The relationship combined area of asperities A_2 and seismic moment determined without constraining the slope is:

$$A_2 = 6.21 \times 10^{-18} M_o^{0.7409}$$

Constraining the slope to be 2/3, the relation is:

$$A_2 = 6.56 \times 10^{-16} M_o^{2/3}$$

As show in figure 6, comparison of constrained equation indicates that the estimated combined area of asperities of large Mexican subduction earthquakes is shorter than constrained equation provided by Somerville *et al.* (2002).

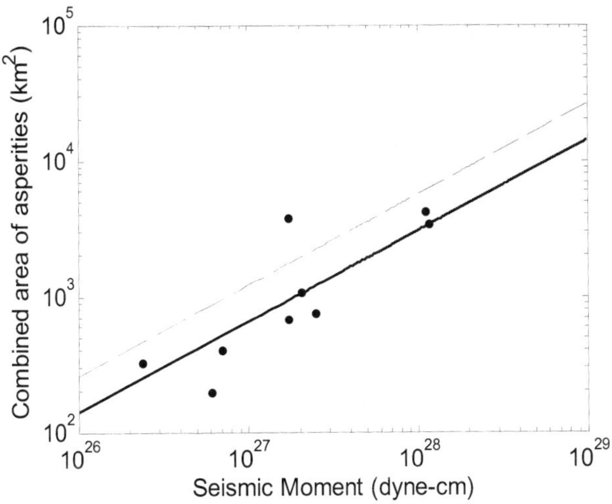

Fig. 6. Relationship between combined area of asperities and seismic moment. Dots represent individual events, solid line is the result of this study and dashed line represents the results obtained by Somerville *et al.* (1999).

3.4 Area of largest asperity versus seismic moment

The relationship between the area of largest asperity A_l and seismic moment determined without constraining the slope is:

$$A_l = 2.11 \times 10^{-15} M_o^{0.6436}$$

Constraining the slope to be 2/3, the relation is:

$$A_l = 4.96 \times 10^{-16} M_o^{2/3}$$

As show in figure 7, comparison of constrained equation indicates that the estimated area of largest asperity of large Mexican subduction earthquakes is shorter than constrained equation provided by Somerville *et al.* (2002).

3.5 Hipocentral distance to center of closets asperity versus seismic moment

The relationship between the hipocentral distance to the closes asperity R_A and seismic moment determined without constraining the slope is:

$$R_A = 2.61 \times 10^{-6} M_o^{0.2427}$$

Constraining the slope to be 1/3, the relation is:

$$R_A = 8.84 \times 10^{-9} M_o^{1/3}$$

As show in figure 8, comparison of constrained equation indicates that the estimated hipocentral distance to closest asperity of large Mexican subduction earthquakes is shorter than constrained equation provided by Somerville *et al.* (2002).

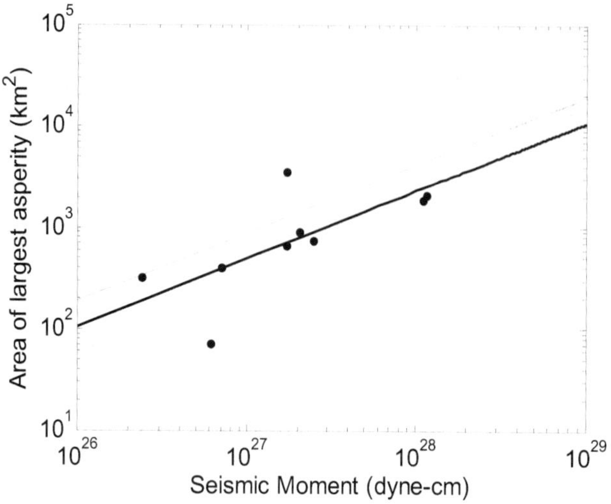

Fig. 7. Relationship between area of largest asperity and seismic moment. Dots represent individual events, solid line is the result of this study and dashed line represents the results obtained by Somerville *et al.* (1999).

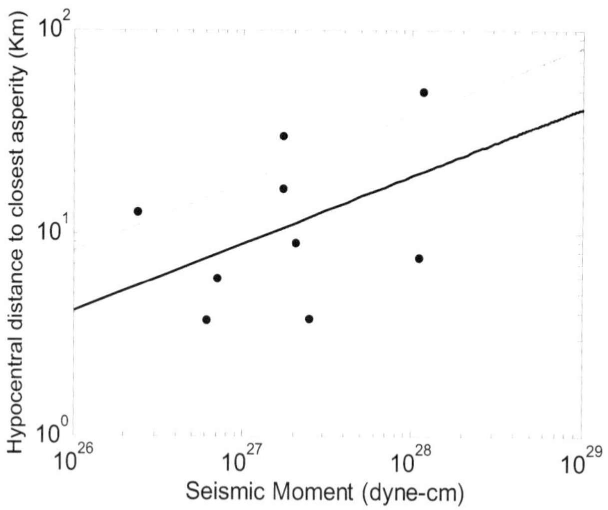

Fig. 8. Relationship between distance from de hypocenter to the center of the closest asperity and seismic moment. Dots represent individual events, solid line is the result of this study and dotted line represents the results obtained by Somerville *et al.* (1999).

The comparison between the parameters estimated both relations compared here, shows that the relationships proposed in this study provides lower estimation than relations proposed by Somerville *et al.* (2002), particularly in the cases of the total rupture area and the combined area of asperities. Curiously only in the case of average slip the relationship proposed in this study provides larger estimation than the relationship proposed by Somerville *et al.* (2002). The relationship between seismic moment versus total rupture area and seismic moment versus average slip for large Mexican subduction earthquakes might explain the result obtained by Ramirez-Gaytán *et al.* (2010), Garduño (2006) and Aguirre Gonzalez in the simulation of April 25, 1989 San Marcos Mexico earthquake. If seismic moment is the result of the product of area, average slip, and rock shear modulus, then a decrease in the area of seismic source and combined area of asperities implies a necessary increase of average slip in order to keep a similar seismic moment. However in this case (the relationship of average slip versus seismic moment) the unconstrained relationship show a no self-similar scaling, this suggest the absence of common or constant scale factor in the average slip of Mexican subduction earthquakes.

4. References

Castro R, Euclides Ruiz-Cruz. (2005) "Stochastic Modeling of the September 1999 Mw 7.5 Earthquake, Oaxaca, Mexico", *Bull. Seismol. Soc. Am.*, 96,6, 2259-2271.

Garduño, N. (2006). "Procesos de la fuente del sismo del 15 de Julio de 1996 usando el método de la función de Green empírica y algoritmos genéticos", *Thesis, Facultad de Ingeniería., Universidad Nacional Autónoma de México,* pp. 111.

Irikura, K. (1986). Prediction of strong accelerations motions using empirical Green's function, 7th *Japan Earthq. Eng. Symp.*, 151–156.

Mendoza C, and S. Hartezll. (1989). "Slip distribution of the 19 September 1985 Michoacan, Mexico eartquaque: near source and teleseismic constrains", *Bull. Seismol.Soc. Am.*, 79, 655-669.

Mendoza C. (1993) "Coseismic slip of two large Mexican earthquakes from teleseismic body waveforms: Implications for asperity interaction in the Michoacan plate boundary segment". *J. Geophys. Res.*, 98, 8197–8210.

Mendoza C, and Hartzell S. (1997)." Finite fault analysis of the 1979 March 14, Petatlan, Mexico earthquake using teleseismic P waveforms", *Geophys. J. Inter.*, 121, 675-683.

Mendoza C, Hartzell S. (1999). "Fault Slip Distribution of the 1995 Colima-Jalisco, Mexico, Earthquake", *Bull. Seismol. Soc. Am.*, 89, 1338-1344.

Miyake, H., T. Iwata, and K. Irikura (2003). Source characterization for broadband ground motion simulation: Kinematic heterogeneous source model and strong motion generation area, *Bull. Seismol. Soc. Am.*, 93, 2531–2545

Ramírez-Gaytán, A., J. Aguirre, and C. Huerta (2011). Tecomán Earthquake: Physical Implications of Seismic Source Modeling, Applying the Empirical Greens Function Method, and Evidence of non-Linear Behavior of Ground. *ISET J. Earthq. Tech.*, 47, 1-23.

Santoyo M, Shri K, Takeshi M. (2005) "Source process and stress change associated with the 11 January, 1997 (Mw=7.1) Michoacán, Mexico, inslab earthquake", *Geofis. Int.*, 44, 4, 317-330.

Somerville, P., K. Irikura, R. Graves, S. Sawada, D. Wald, N. Abrahamson, Y. Iwasaki, T. Kagawa, N. Smith, and A. Kowada (1999). Characterizing crustal earthquake slip models for the prediction of strong motion, *Seism. Res. Lett.*, 70, 59-80.

Somerville, P., T. Sato, I.Toru, N. Collins, K. Dan, and F. Hiroyuki (2002). Characterizing subduction earthquake slip models for the prediction of strong motion, *Proc. of the 11th Symp. of Earthq. Engrng.*, 1, 163-166, (in Japanese).

Yagi, Y., T. Mikumo, J. Pacheco, and G. Reyes (2004). Source rupture process of Tecomán, Colima, México earthquake of January 22, 2003, determined by joint inversion of teleseismic body-wave and near-source data, *Bull. Seismol. Soc. Am.*, 94, 1795–1807.

Zúñiga F, Gutierrez C, Nava E, Lermo J, Rodriguez M, Coyoli R. (1993). "Aftershocks of the San Marcos, earthquake, of april 25, 1989 and its implications for the potential of the Acapulco-San Marcos region". *Pure Appl. Geophys.*, special issue on Subduction Zone Earthquakes, 140, 287-30.

Part 2

Site Characterization

5

Evaluation of Linear and Nonlinear Site Effects for the M_W 6.3, 2009 L'Aquila Earthquake

C. Nunziata[1], M.R. Costanzo[1], F. Vaccari[2] and G.F. Panza[2,3]
[1]Dipartimento di Scienze della Terra, Univ. Napoli Federico II,
[2]Dipartimento di Geoscienze, Univ. Trieste,
[3]The Abdus Salam International Centre for Theoretical Physics, ESP-SAND Group, Trieste,
Italy

1. Introduction

An effective strategy for the seismic risk mitigation needs the use of advanced seismological methodologies for a realistic estimate of the seismic hazard and, consequently, to reduce earthquake damage through a preventive evaluation of vulnerability and actions for structure safety. Prediction of earthquakes and their related effects (expressed in terms of ground shaking) can be performed either by a probabilistic approach or by using modelling tools based, on one hand, on the theoretical knowledge of the physics of the seismic source and of wave propagation and, on the other hand, on the rich database of geological, tectonic, historical information already available. Strong earthquakes are very rare phenomena and it is therefore statistically very difficult to assemble a representative database of recorded strong motion signals that could be analyzed to define ground motion parameters suitable for seismic hazard estimations. That is, the probabilistic estimation of the seismic hazard is a very gross approximation, and often a severe underestimation, of reality.

A realistic and reliable estimate of the expected ground motion can be performed by using the Neo-Deterministic Seismic Hazard Analysis (NDSHA), an innovative modelling technique that takes into account source, propagation and local site effects (for a recent review see Panza et al., 2011). This is done using basic principles of physics about wave generation and propagation in complex media, and does not require to resort to convolutive approaches, that have been proven to be quite unreliable, mainly when dealing with complex geological structures, the most interesting from the practical point of view.

The NDSHA approach has been used, among others, in the framework of the UNESCO-IUGS-IGCP project 414 "Seismic Ground Motion in Large Urban Areas", to evaluate ground motion of a group of Large Urban Areas and Megacities in the world representative of a broad spectrum of seismic hazard severity (Panza et al., 2004).

A M_W 6.3 earthquake struck on 6 April 2009, at 01.32 GMT, the Abruzzo region (central Italy). The L'Aquila town, located few km northeast to the main shock epicentre, and several villages located nearby, suffered heavy damages and the casualties were about 300. The damage level generally corresponded to intensity ≤ VIII MCS, with few maximum values ≥

IX MCS generally associated to construction vulnerability and, in some cases, to site amplification effects (Fig. 1).
In the past, destructive earthquakes originated in the L'Aquila basin such as the 1349, I=IX–X; the 1461, L'Aquila, I=X and the 1703, I=X (CPTI working group, 2004). Seismic hazard maps based on geological fault slip-rate data show that strong events (intensities ~IX) can hit L'Aquila with short recurrence time of approximately 250±50 years (Roberts et al., 2004). Boncio et al. (2004) estimated a maximum expected earthquake magnitude of 6.1–6.4 for the L'Aquila fault segment in Paganica, and stronger events are expected for other segments of the same fault system or other neighboring faults for their impressive post-glacial fault scarps (Papanikolaou et al., 2010 and references therein). Based on trenching investigations, Galli et al. (2002) support that the Campo Imperatore fault, only 20 km away from L'Aquila, can give a Magnitude 7.0 earthquake. Moreover, the 1703 (M_W ~6.7) earthquake produced surface ruptures >10 km and a maximum vertical displacement of 1m in the neighbouring Arischia fault (Blumetti, 1995). These ruptures are almost one order of magnitude larger than the ruptures produced by the 6 April L'Aquila earthquake (0.1-0.3 m), implying that the surrounding faults have the capacity to generate significantly stronger events.
For seismic hazard assessment, in order to prevent damage from even more energetic and dangerous earthquakes at L'Aquila, it is necessary to compute realistic seismograms.
Aim of this paper is to compute the seismic ground motion at L'Aquila for the 6 April 2009 earthquake by the NDSHA approach and evaluate nonlinear effects with equivalent-linear approach, by assuming literature variations of shear modulus and damping with strain.

2. Main shock recordings

The M_W 6.3 earthquake of 6 April 2009, hereafter called main shock, had a pure normal faulting mechanism, dipping at about 45° to the SW, with location at about 9 km of depth. Few days later, the aftershock activity migrated from the south-east of L'Aquila towards the NE at Arischia and Campotosto. The distribution of the aftershocks defined a complex, 40 km long and 10-12 km wide, NW trending extensional structure (Fig. 1).
The epicentral area of the main shock corresponds to the upper and middle Aterno river valley which is characterised by the high variability of the geologic and geomorphologic patterns. The valley is superimposed on a Quaternary lacustrine basin of tectonic origin (Fig. 2). The depth of the Quaternary deposits is variable, from about 60 m in the upper Aterno river valley to more than 200 m in the middle Aterno river valley. The 2009 seismic sequence was recorded by accelerometers of Rete Accelerometrica Nazionale (RAN) network, managed by the Italian Protezione Civile, some of which located at L'Aquila (AQK station) or in the NW of it (Fig. 2). They are equipped with three-component sensors set to 1 or 2 g full-scale, coupled with 24-bit digitizers. The permanent station AQU is operating since 1988 as part of the Mediterranean Network (MedNet), managed by the Italian Istituto Nazionale di Geofisica e Vulcanologia (INGV). AQU is equipped with a very broadband Streckeisen STS-1 sensor and Quanterra digitizer.
After the 2009 seismic sequence, geological and new interpretation of gravity studies have been performed at L'Aquila by Protezione Civile (2009) to reconstruct the shallow 200-300 m of subsoil. Based on these data, the stations turn out to be set on different geological

conditions. Stations AQG, AQA, AQV and AQM are located in the so called Coppito plain, a small graben filled of alluvial deposits. The stations AQM and AQG are installed on maiolica (whitish limestones) formation, whereas AQA and AQV are set on alluvial deposits with a thickness of 36 m and 46.5 m, respectively. The stations AQU and AQK are set on breccias (so called megabreccias), 80 m and 60 m thick, respectively, lying on alluvial deposits with thickness of 150 m and 250 m, respectively.

V_S models of the superficial 30 m of Aterno alluvial soils have been defined in the Coppito area (located in Fig. 2) (Costanzo et al., 2011) from the nonlinear inversion with Hedgehog method (Panza et al., 2007 and references therein) of the group velocity dispersion curves of the fundamental mode extracted with the FTAN method (e.g. Levshin et al., 1989; Nunziata, 2010). The models are characterized by an average velocity of 190 m/s.

Recordings of the main shock put in evidence the variability of the geological setting, with the maximum peak ground acceleration (PGA) of 0.6 g recorded at the AQV station (Fig. 3), and with high frequency content. Moreover, the recorded vertical accelerations, in some cases, are comparable or higher than the horizontal accelerations. Despite the high accelerations, low displacements (4-13 cm) are observed at periods higher than 1 s, with the exception of the AQK station, that shows displacements of 10 cm and 25 cm, in the vertical and horizontal components, respectively (Akinci et al., 2010). The high frequency content and the very short duration (2-5 s) of the ground motion, together with the building eigenperiods of 0.2-0.5 s, allowed fortunately to reduce the number of the collapses.

Spectral H/V ratios of the horizontal and vertical components evidence a strong variability, which prevents their use to recognize the resonance frequency of the station sites. The only exception is the station AQK (Fig. 4) that confirms the resonance frequency of 0.6 Hz obtained in the area by previous studies (De Luca et al., 2005). The crucial point is the definition of representative V_S profiles in the uppermost hundred metres, significant for the

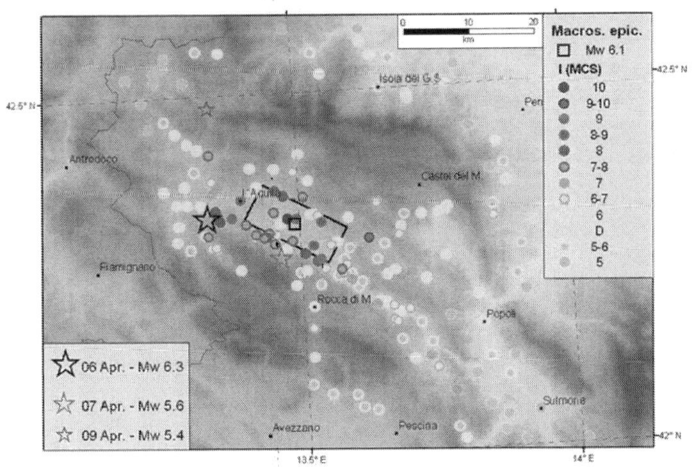

Fig. 1. Map of the effects of the 6 April 2009 earthquake and relative seismogenic box (from http://www.mi.ingv.it/eq/090406/quest.html)

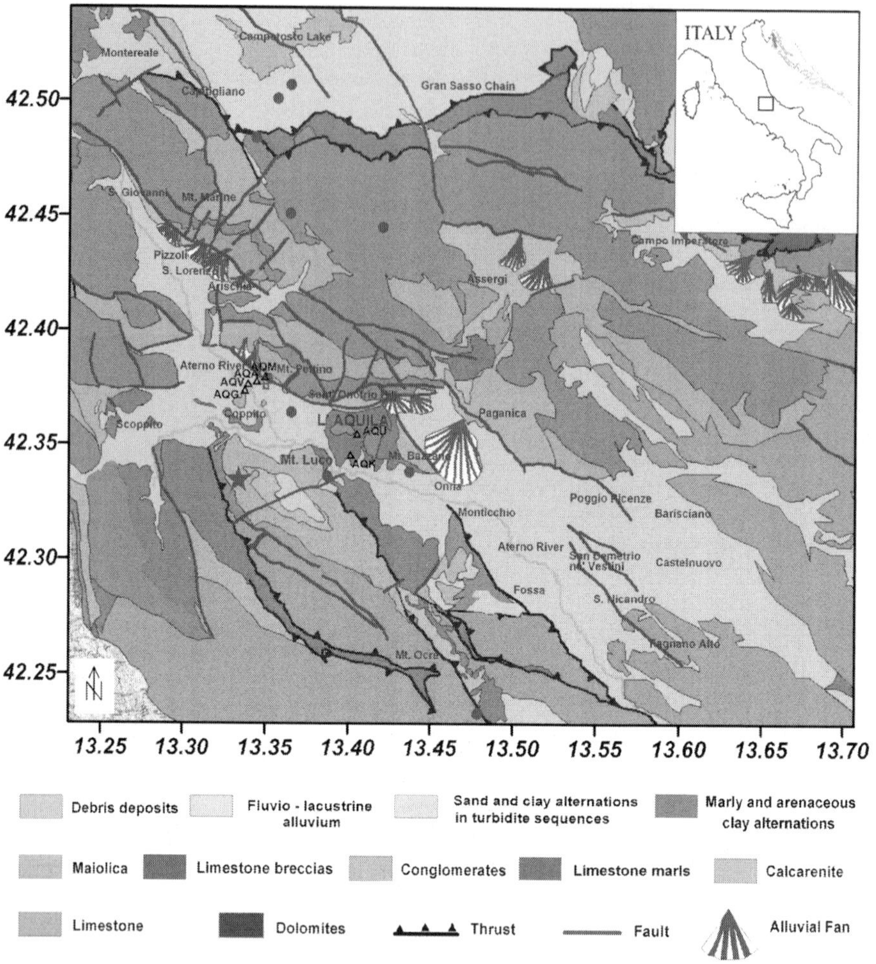

Fig. 2. Geological map of L'Aquila (modified from Vezzani and Ghisetti, 1998) with location of the RAN stations (black triangles), the main shock epicentre (blue star), the events in April 2009 (blue dots) and the site of the active seismic experiment (red square).

evaluation of site amplification effects. A good example is shown in Fig. 5 for the AQV station (located in Fig. 2) about the comparison of the frequency of the maximum peak of the H/V spectral ratio, relative to the main shock, with the 1D spectral amplifications, computed with SHAKE program (Schnabel et al., 1972), by assuming two different V_S data sets. The V_S profiles are relative to: 1. FTAN-Hedgehog analysis of surface measurements (Costanzo et al., 2011); 2. cross-hole measurements beneath the AQV station. Once more, such comparison evidences how the cross-hole (and down-hole) point-like measurements, even though quite precise, may not be representative of the average seismic path (e.g. Nunziata et al., 2004; Nunziata, 2007).

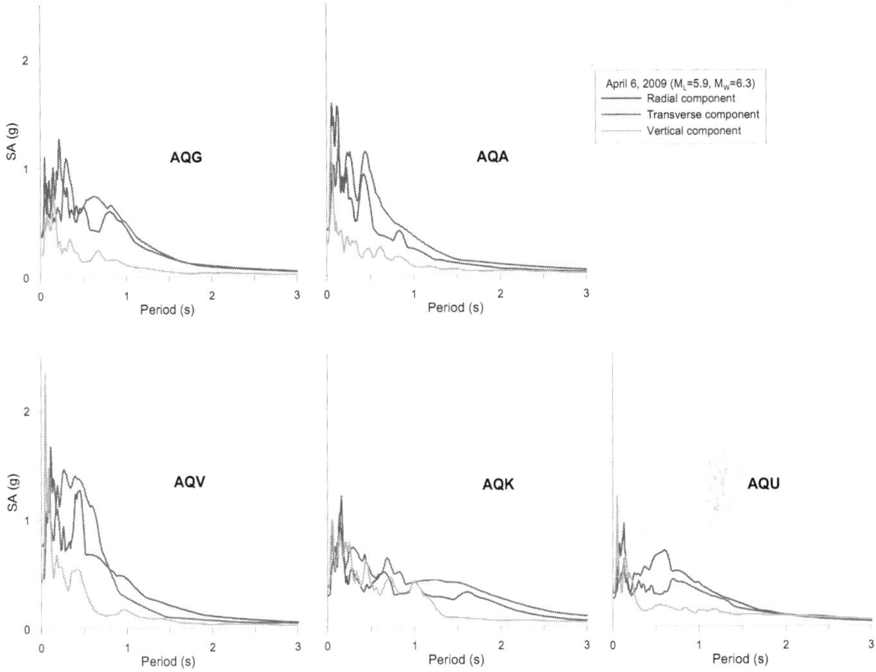

Fig. 3. Recorded acceleration response spectra of the main shock at the stations in the Aterno valley (located in Fig. 2)

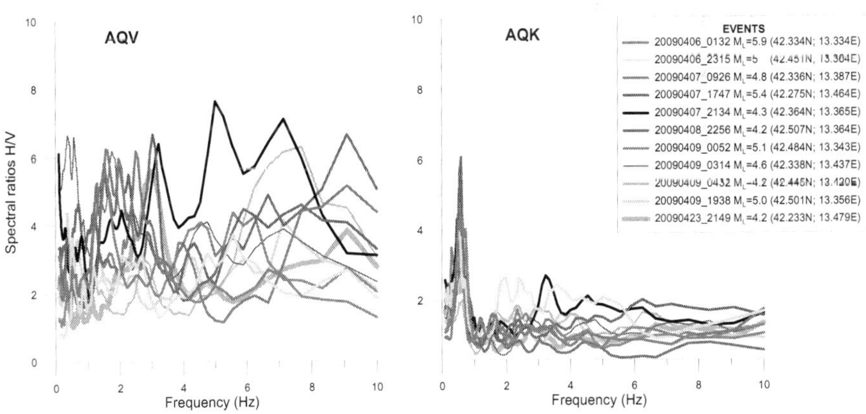

Fig. 4. Spectral ratios H/V relative to events recorded in April 2009 at AQV and AQK stations (located in Fig. 2).

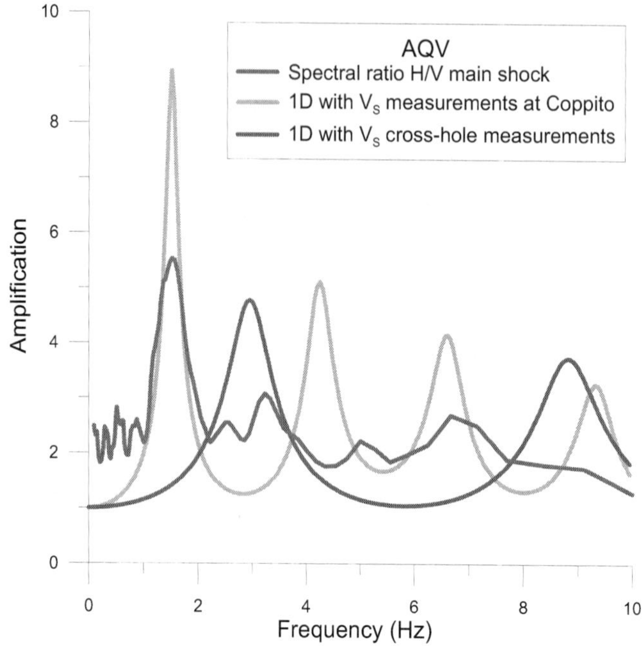

Fig. 5. Comparison between the resonance frequency estimated from the spectral ratio H/V of the main shock recorded at the AQV station and the 1D amplifications (SHAKE program) computed by assuming the V_S velocity profiles vs. depth measured by cross-hole measurements and obtained by FTAN-Hedgehog methods in the same lithotypes at Coppito (located in Fig. 2).

3. Ground motion modelling

Simulations of the 2009 L'Aquila earthquake have been performed with the NDSHA approach, using a hybrid method consisting of modal summation and finite difference methods (Panza et al., 2001 and references therein). This hybrid approach combines the advantages of both mode summation and finite difference technique. The path from the source up to the region containing the 2-D heterogeneities is represented by a 1-D layered anelastic structure. The resulting wavefield for both SH- and P-SV- waves is then used to define the boundary conditions to be applied to the 2-D anelastic region where the finite difference technique is used. Synthetic seismograms of the vertical, transverse and radial components of ground motion are computed at a predefined set of points at the surface (Fig. 6). Spectral amplifications are computed as response spectra ratios, RSR, i.e. the response spectra computed from the signals synthesized along the laterally varying section (2D) normalized by the response spectra computed from the corresponding signals, synthesized for the bedrock (1D). Two approximations have been considered to scale the seismogram to the desired scalar seismic moment: a scaled point-source approximation (Gusev, 1983 as reported in Aki, 1987), and an extended source approximation as proposed by Gusev and

Pavlov (2006) and Gusev (2011) and used by Gusev et al. (2008) for the modelling of Messina 1908 earthquake, where the rupturing process is modelled also in its random part.

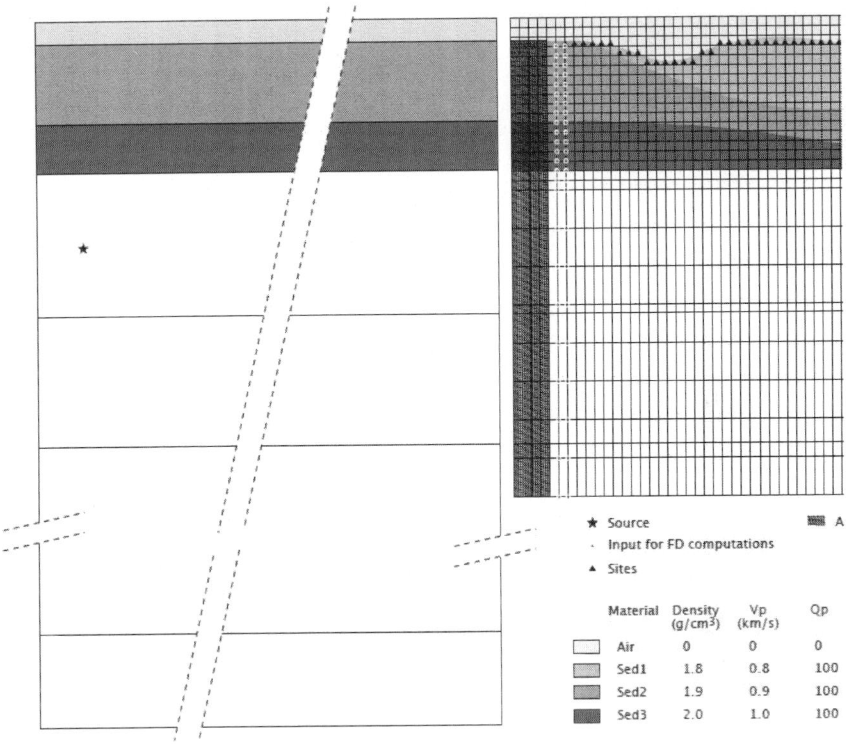

Fig. 6. Schematic representation of the hybrid method.

Modelling of the 6 April 2009 has been done along a SW-NE geological cross section at L'Aquila, close to the AQK station, and transverse to the Aterno river valley (Fig. 7a). Along the 2D model, the outcropping units are represented by megabreccias except in the Aterno river, where recent fluvial sediments are present. Megabreccias are on the top of lacustrine clays. Physical parameters of the lithotypes (Fig. 7b, Tab. 1) have been attributed on available data (De Luca et al., 2005). The geological subsoil beneath the AQV station is similar to that below the receiver R19. Such assumption is validated by the good agreement between synthetic and recorded accelerograms both at the AQV and AQK stations, despite the lack of detailed geological and seismic knowledge along the cross section (Fig. 8). Then, a parametric study of the expected ground motion has been done along the cross section by assuming extended and point sources.

Spectral amplifications of about 10 are computed in correspondence of the Aterno river alluvial sediments (Fig. 9) and for a wide frequency range (2-5 Hz), corresponding to resonance frequencies of several structure typologies. This result is very different from the maximum peaks of H/V spectral ratios at 0.6 Hz computed from strong and weak

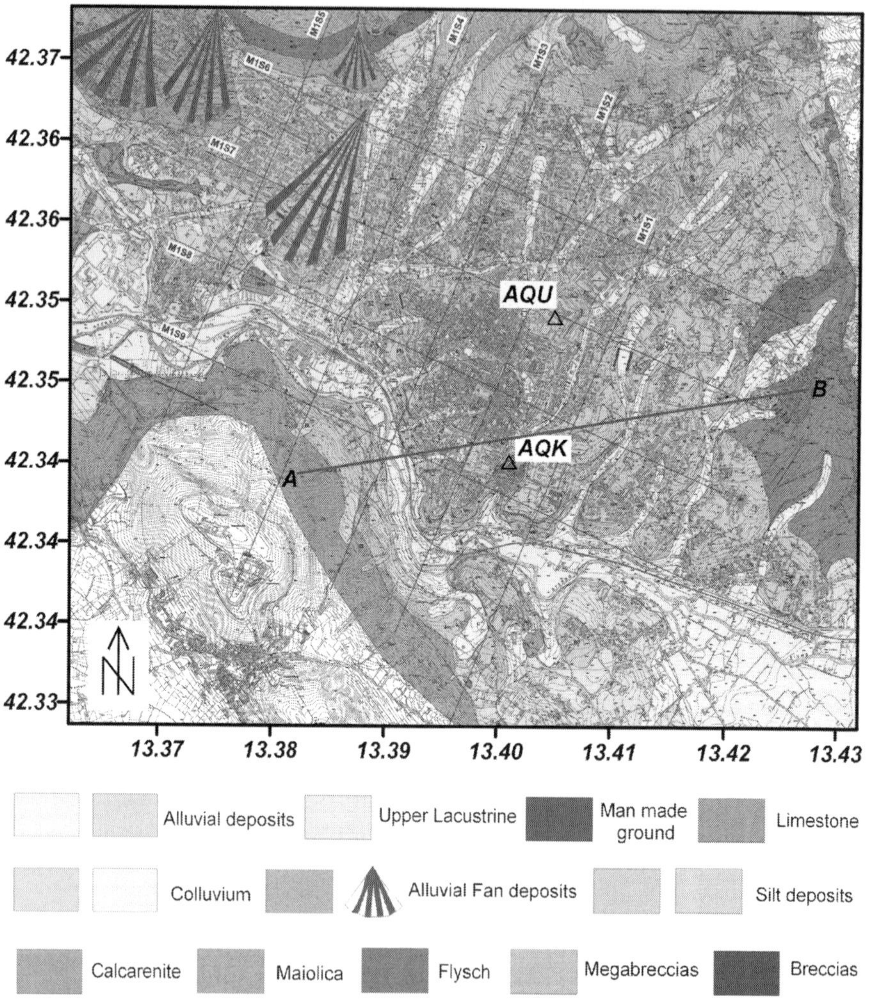
42.37
42.36
42.36
42.35
42.35
42.34
42.34
42.34
42.33
13.37
13.38
13.39
13.40
13.41
13.42
13.43
M1S1
M1S2
M1S3
M1S4
M1S5
M1S6
M1S7
M1S8
M1S9
AQU
AQK
A
B
N
Alluvial deposits
Upper Lacustrine
Man made ground
Limestone
Colluvium
Alluvial Fan deposits
Silt deposits
Calcarenite
Maiolica
Flysch
Megabreccias
Breccias

(a)

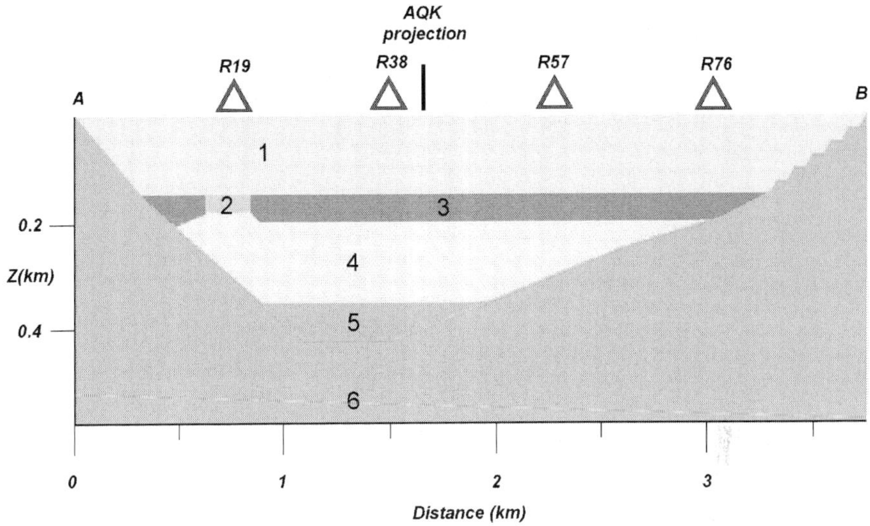

(b)

Fig. 7. (a) Location of the SW-NE cross section of the L'Aquila basin on the geological map 1:6000 (Protezione Civile, 2009); (b) computing cross section (from De Luca et al., 2005). Legend: 1. Air; 2. Aterno river recent deposits; 3. Megabreccias; 4. Upper lacustrine clays; 5. lower lacustrine clays; 6. Limestone. The physical parameters are reported in Tab. 1.

Name	Density (g/cm³)	V_P (km/s)	V_S (km/s)	Q_P	Q_S
Aterno Deposits	1.40	0.43	0.25	110	50
Megabreccias	2.00	1.56	0.90	220	100
Upper Lacustrine	1.80	0.86	0.50	220	100
Lower Lacustrine	1.80	1.12	0.65	220	100
Limestone	2.45	4.30	2.50	220	100

Table 1. Physical parameters of lithotypes

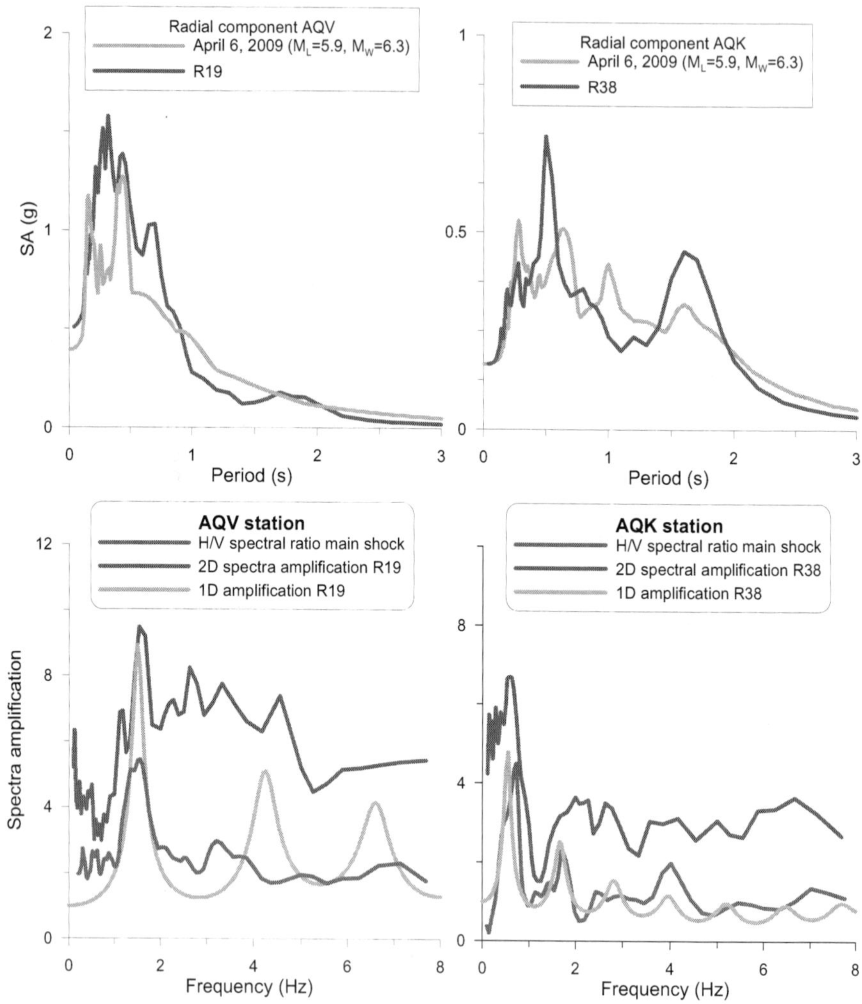

Fig. 8. Comparison at AQV and AQK stations of the: (top) recorded response spectra of the main shock with those computed at representative receivers R19 and R38, respectively (located in Fig. 7b); (bottom) recorded H/V spectral ratio with 2D and 1D (SHAKE program) spectral amplifications computed at receivers R19 and R38

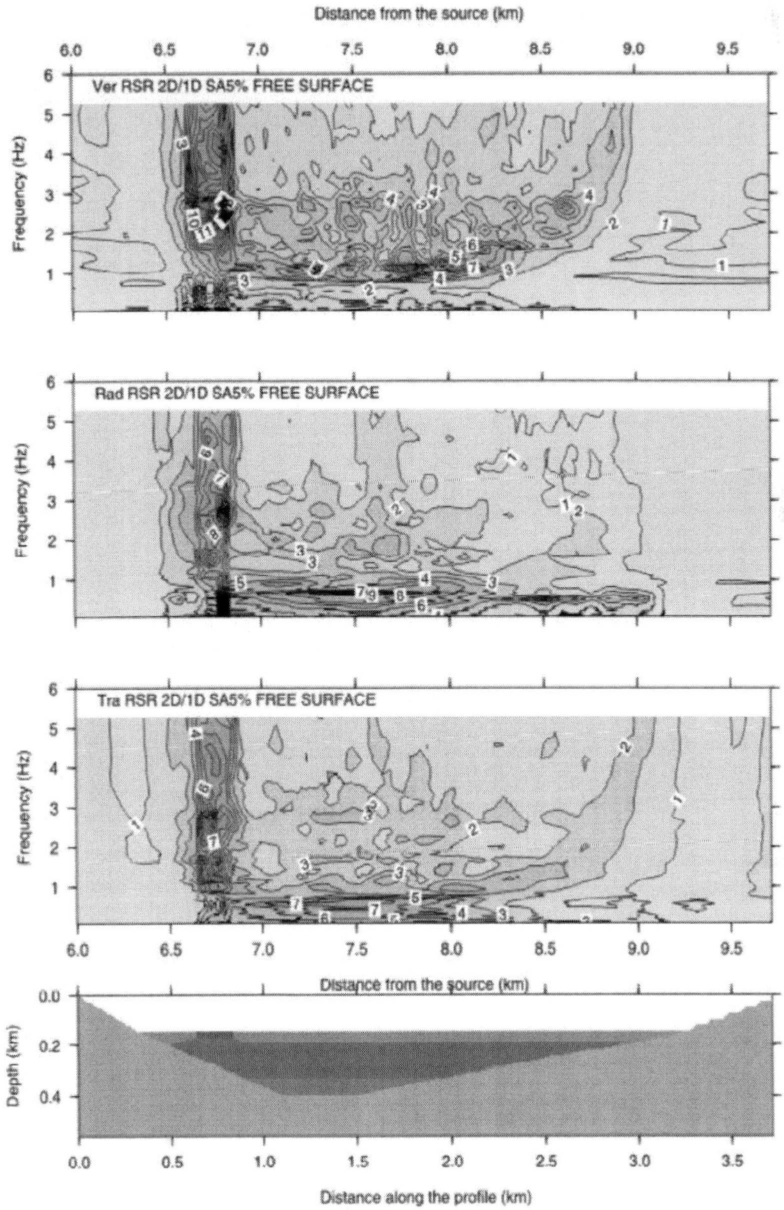

Fig. 9. Spectral amplifications (RSR 2D/1D) along the cross section of the L'Aquila basin. Response spectra are computed for 5% damping. From the top vertical, radial and transverse components of the computed ground motion.

earthquakes and noise along the same cross section (De Luca et al., 2005). The inadequacy of H/V spectral ratios is clearly seen even when they are computed from the same signals (synthetic) (Fig. 10). They do not show the strong amplifications where strong lateral heterogeneities are present. This result suggests great caution in the use of H/V ratios for seismic microzoning.

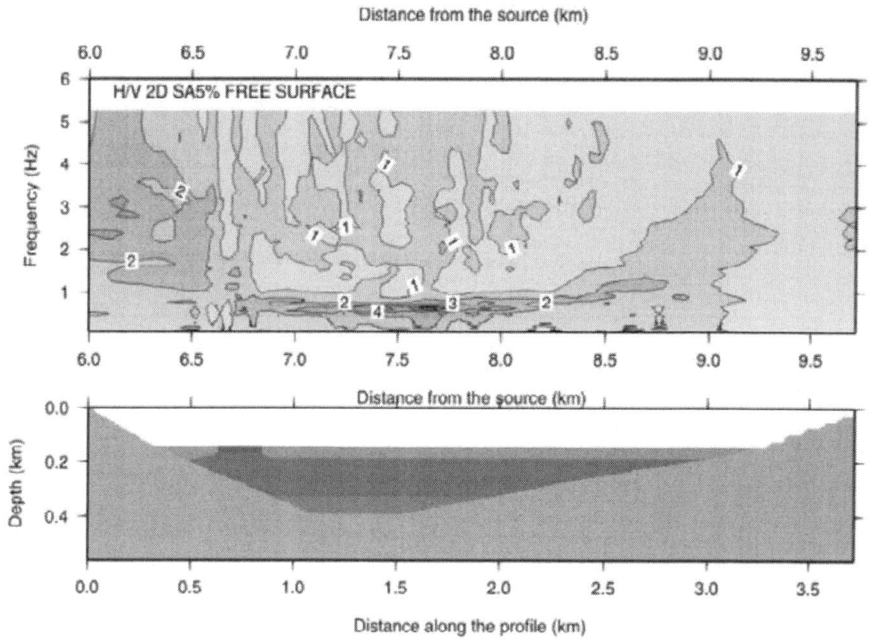

Fig. 10. Spectral amplifications along the cross section obtained from H/V spectral ratios computed on the same seismograms used for the 2D/1D amplifications shown in Fig. 9.

In November 2009, a map of buildings habitability at the historical center of L'Aquila, called red zone, has been published (L'Aquila Common Ordinance) (Fig. 11). It is evident that the majority of the buildings suffered serious damage (grade E). The cross section is representative of the historical center and, taking into account that generally 2-5 floor buildings are present, we can argue that spectral amplifications might have been responsible for damage, beside the near-field conditions.

Fig. 11. Map of building habitability (decreasing from F to A) at L'Aquila red zone (Common Ordinance, 2009).

4. Nonlinear effects

Nonlinear site effects, such as increase in damping and reduction in shear wave velocity as input strength increases, are commonly observed in the dynamic loading of soils using geotechnical models, especially at shear strains larger than 10^{-5} to 10^{-4}.

Typical test data were published by Idriss and Seed (1968) and have been extensively used in soil engineering since that time. The amplification function at a site is controlled by the wave velocity and damping in the soil layer hence nonlinear site effects can be expected in strong-motion seismology. Amplification function becoming amplitude dependent (difference between weak and strong motions) is an indication of nonlinearity (e.g. Beresnev and Wen, 1996).

Evidence of nonlinear behaviour has been reported beyond a threshold acceleration from 0.1 g to 0.2 g. Nonlinearity is considerable in cohesionless soil but may be negligible in stiff soils.

Explicit evidence of strong-motion deamplification, accompanied by shifts of resonant frequencies towards lower frequencies, is found in a number of events throughout the world. Field et al. (1997) reported that ground-motion amplification due to sediments for the main shock of the 1994 Northridge earthquake was up to a factor of two less than the amplification observed for its aftershocks. More recently Molchan et al. (2011) supplied a list of hot/cold spots, in the definition of Olsen (2000), sites identified in the Italian macroseismic data, which are related to local fault geometry rather than to soil conditions.

Direct seismological evidence of nonlinear site effects was reported using spectral ratio techniques for two-station pairs, including soil-to-rock and surface-to-borehole station pairs. An intriguing result was published by Archuleta *et al.* (1992). At the Garner Valley downhole array, one of the accelerometers was installed at the surface and one was located at a depth of 220 m in granite. Two earthquakes with nearly coincident hypocenters but different local magnitudes were recorded, and amplifications were computed from spectral ratios at 0 to 220 m. It resulted that weak-motion amplification was significantly larger than strong-motion amplification in a wide frequency band, from 3 Hz to 40 Hz and that the most pronounced resonance was slightly shifted toward lower frequencies.

Nonlinear soil responses have been also identified by the horizontal-to-vertical spectral ratio applied to weak and strong motion records (Wen et al., 2006).

An estimate of possible nonlinear site effects, when experimental data are not available, can be obtained through a parametric study of response spectra dependence on shear moduli variations with increasing shear strain. Strain-dependent soil properties can be incorporated in dynamic response analysis by using an equivalent-linear approach as proposed by Idriss and Seed (1968). The approach is based on the assumption that the nonlinear soil response can be simulated by a linear elastic model with damping, provided that its constants are assigned according to the average strain level achieved, which is typically taken to be 0.65 times the maximum strain. The equivalent-linear method was implemented in the program SHAKE (Schnabel *et al.*, 1972), which has become a common tool for the routine estimation of dynamic ground response in geotechnical applications.

4.1 Nonlinear effects at L'Aquila

Among the aftershocks recorded in April 2009, one event (7 April 2009) occurred along approximately the same azimuth, at a comparable epicentral distance of the main shock, but with smaller magnitude (M_L=4.8) (Fig. 2). Nonlinear site effects can be observed at the AQV station from the comparison of the H/V spectral ratios relative to the main shock and this last event (Fig. 12). A small reduction of the amplitude and a little shift of the main peak of the H/V ratio toward lower frequency can be observed at the AQV station, the only station set on alluvial soils recording the smaller size event too. Instead, such nonlinear effect is not observed at the AQK station, set on megabreccias.

Nonlinear site effects are estimated at the two receivers (R19 and R38 in Fig. 7b) representative of the L'Aquila basin, with (like AQV station) and without (like AQK station) Aterno river deposits. SHAKE program is used to compute seismograms on the top of the soil columns (limestone is assumed as bedrock), using as seismic input the 1D seismogram computed, at the specific receiver, with the modal summation technique, by assuming different sources, that is an extended source with unilateral rupture, an extended source with bilateral rupture, and a point source. In a further test the PGA of the point-source signal has been conventionally taken equal to 0.1 g.

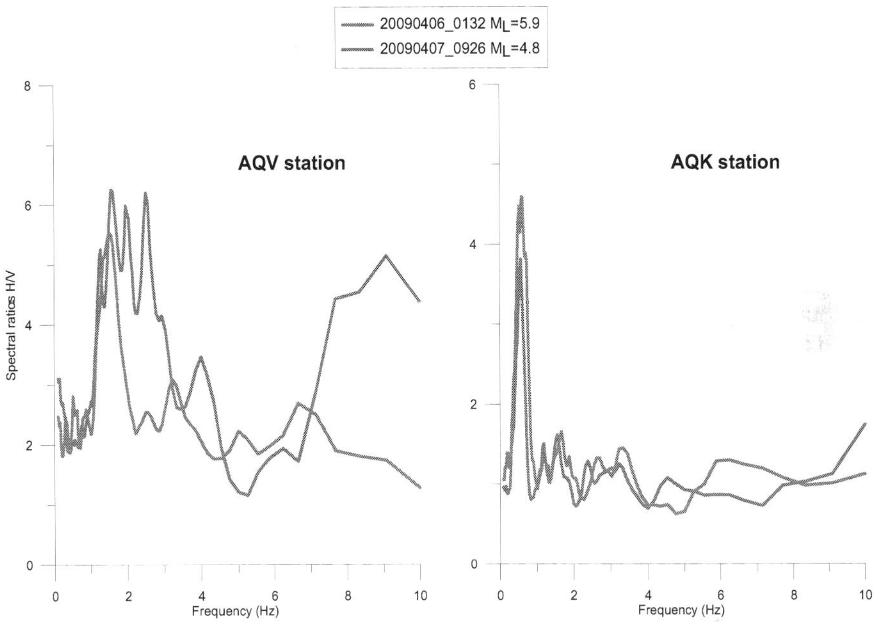

Fig. 12. H/V spectral ratios at AQV and AQK stations for two events with similar epicentral characteristics but different magnitude (location in Fig. 2).

Strain dependent shear modulus (G/Gmax) and damping (D/Dmin) of sand (pozzolana) (Guadagno et al., 1992), clay (CNR- GNDT, 2000) and peat (Vinale, 1988) are assigned, as plausible, to the Aterno river recent deposits, and of sand (pozzolana) to megabreccias and the lacustrine deposits (Fig. 13). Pozzolana sample is a cemented sand (hydrothermal hardening process) and can be considered as a stiff soil.

First of all, a remarkable difference between SHAKE (1D) and 2D (NDSHA) spectra exists, as expected, because of the lateral variations in the subsoil (Fig. 14). Response spectra computed with SHAKE by using the sand (pozzolana) soil nonlinear properties show nonlinear site effects (amplitude reduction and frequency shift to lower frequencies) more evident at receiver R19 and mostly for a point source (PGA=0.2 g). When considering PGA equal to 0.1 g and point source seismograms, such nonlinear effects turn out strongly reduced (Fig. 14 a-d).

A parametric study is performed at receiver R19 by assuming the seismic input corresponding to the point source and by attributing to the recent deposits of Aterno river the nonlinear soil properties of clay and peat (Fig. 15). A noticeable nonlinear effect appears

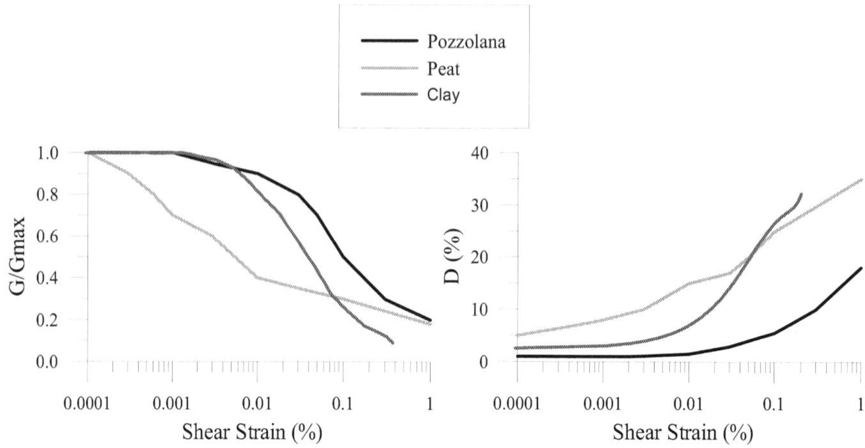

Fig. 13. Variation of shear modulus and damping with strain: pozzolana (Guadagno et al., 1992); clay (CNR-GNDT, 2000); peat (Vinale, 1988).

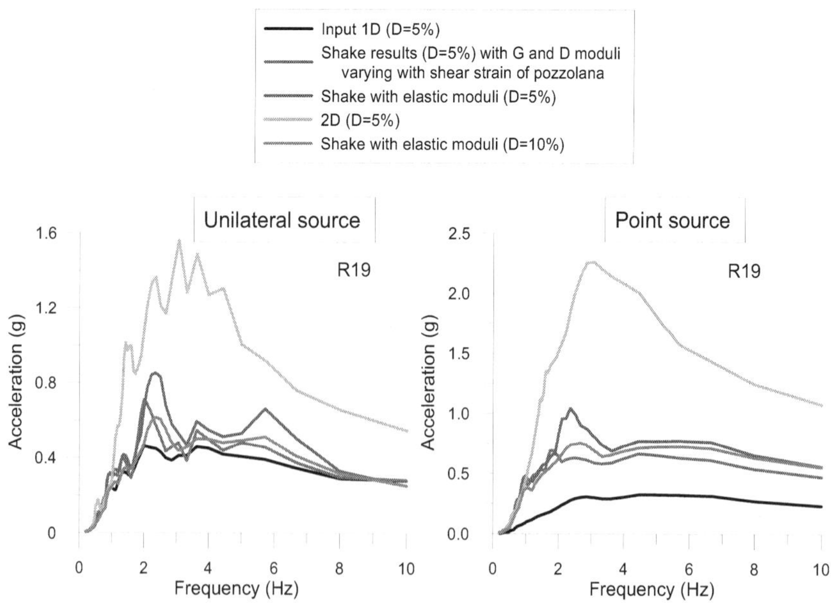

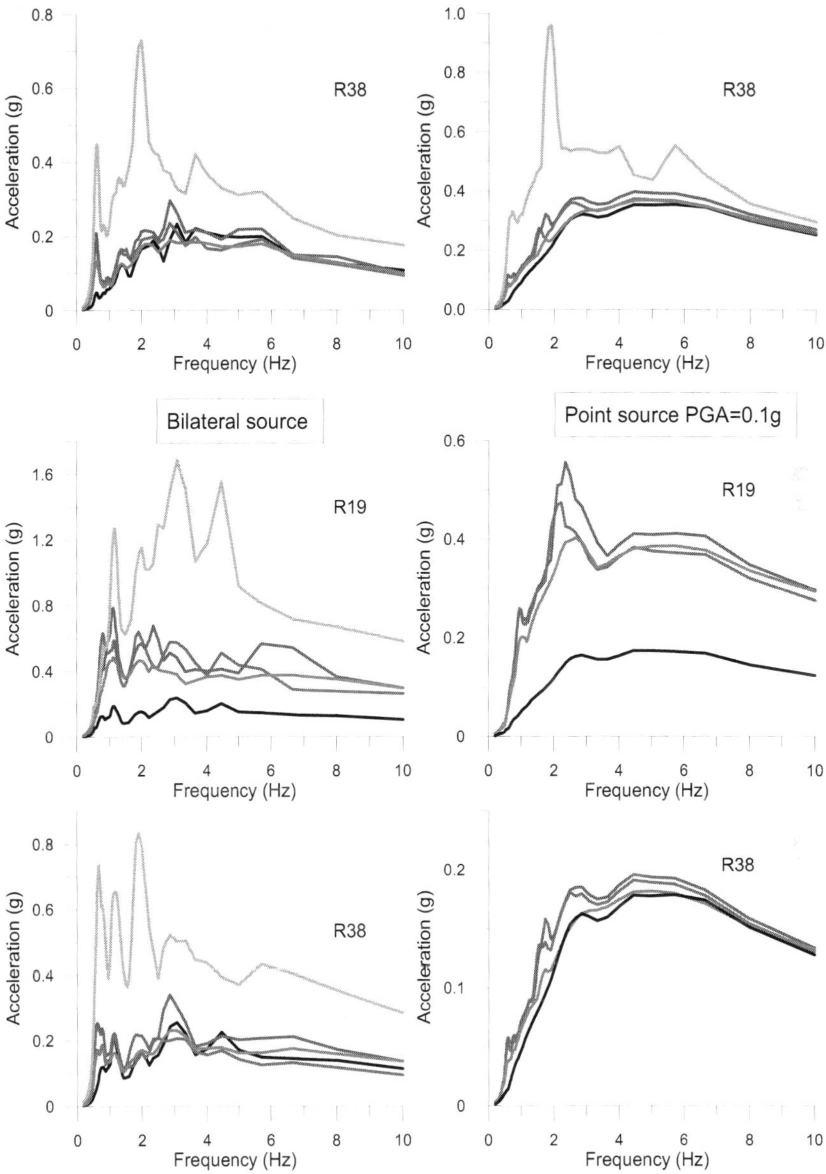

Fig. 14. Acceleration response spectra computed at R19 and R38 receivers with SHAKE program by using as input the seismograms computed, with mode summation technique, considering, from the left top: extended source with unilateral rupture; extended source with-bilateral rupture; point source. The fourth plot refers to the case in which the PGA of the point-source signal has been conventionally taken equal to 0.1 g.

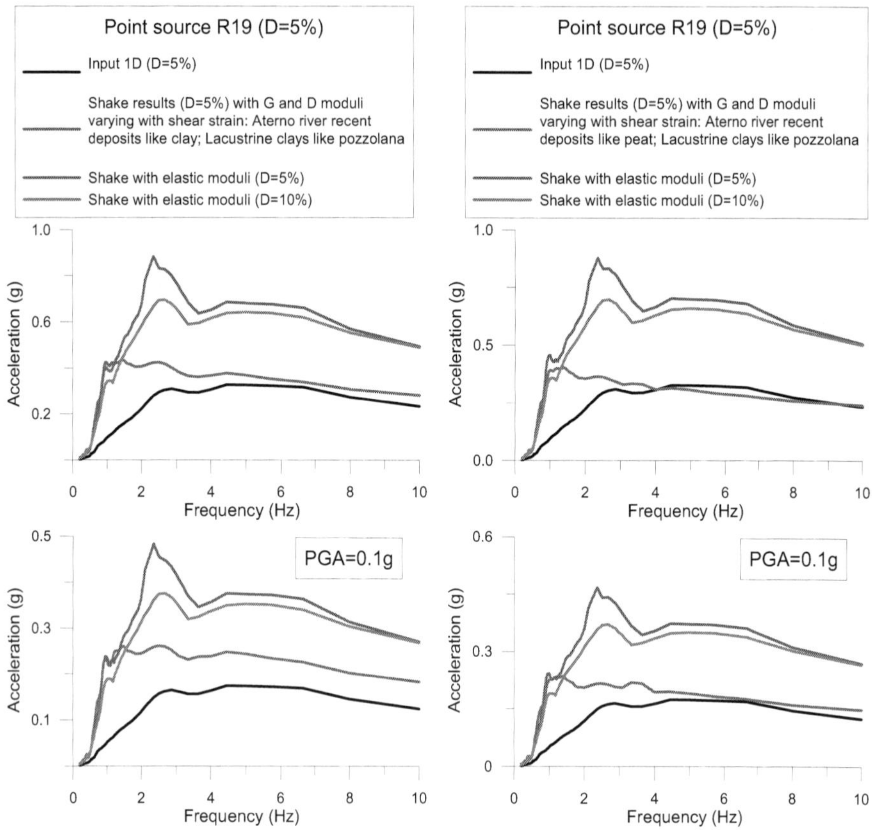

Fig. 15. Acceleration response spectra computed at R19 receiver with SHAKE program by using as input the seismograms computed with modal summation technique for the two point sources considered, by assigning to Aterno river deposits the strain-dependent dynamic parameters of clay and peat.

even in the point-source signal with PGA conventionally taken equal to 0.1 g. These results, though preliminary, as the real soil properties of L'Aquila basin are not known with sufficient detail, indicate that nonlinear effects may have locally affected the ground motion caused by the 6 April 2009 earthquake.

5. Conclusions

A realistic estimation of the ground motion at L'Aquila for the M_W 6.3 earthquake is obtained by the NDSHA approach, an innovative modelling technique that takes into account source, propagation and local site effects (Panza et al., 2001; 2011). The analysis of the H/V spectral ratios, both from recordings and computations, suggests that great caution must be paid in their use for reliable seismic microzoning. Another key point is the definition of V_S models representative of the seismic path, like those obtained from the nonlinear inversion of Rayleigh group velocities (Nunziata et al., 2004; Nunziata, 2007).

The simple parametric study of the variation of response spectra with the strain-dependent dynamic properties of soils, made using SHAKE, evidences that nonlinear effects may have affected the ground motion caused by the M_W 6.3 L'Aquila earthquake, as observed at the AQV station. Reduced spectral accelerations are due to (1) the amplitude dependent damping and (2) the shift of resonance frequencies towards lower frequencies due to the reduction of shear velocity. Nonlinearity under strong ground shaking depends on the physical properties of soils. Nonlinearity, that cannot be correctly simulated by the use of higher damping, may be considerable for soft clay and peat soils (PGA≥0.1 g) and negligible for stiffer materials for PGA ≥0.2g.

6. References

Aki, K. (1987). Strong motion seismology: Strong Ground Motion Seismology, NATO ASI Series C., D. Reidel Publishing Company, M.O. Erdik and M.N. Toksoz (Ed.), Dordrecht, Vol. 204, pp. 3-39.

Akinci, A.; Malagnini, L. & Sabetta, F. (2010). Characteristics of the strong ground motions from the 6 April 2009 L'Aquila earthquake, Italy. *Soil Dynamics and Earthquake Engineering,* Vol. 30, pp. 320–335.

Archuleta, R. J.; Scale, S. H.; Sangas, P. V.; Baker, L. M. & Swain, S. T. (1992). Garner Valley downhole array of accelerometers: instrumentation and preliminary data analysis. *Bull. Seism. Soc. Am.,* Vol. 82, pp. 1592-1621.

Beresnev, I. A. & Wen, K. L. (1996). Nonlinear ground response - a reality?. *Bull. Seism. Soc. Am.,*Vol. 86, pp. 1964-1978.

Blumetti, A. M. (1995). Neotectonic investigations and evidence of paleoseismicity in the epicentral area of the January–February 1703, Central Italy, earthquakes, in : Serva, L. and Slemmons, D. B.(Eds.), Perspectives in Paleoseismology. *Bull Assoc Eng Geol. Special Publication,* Vol. 6, pp. 83–100.

Boncio, P.; Lavecchia, G. & Pace, B. (2004). Defining a model of 3D seismogenic sources for seismic hazard assessment applications: The case of central Apennines (Italy). *J. Seismol.*, Vol. 8, pp. 407–423.

CNR-GNDT (2000). Cattaneo, M., Marcellini A. (Ed.), Terremoto dell'Umbria-Marche: Analisi della sismicità recente dell'Appennino umbro-marchigiano. *Microzonazione sismica di Nocera Umbra e Sellano,* 228 pp. + CD-ROM.

Costanzo M.R.; Gambale, V. & Nunziata, C. (2011). V_S of shallow crust structure in the L'Aquila basin (Italy), submitted to *Italian Journal of Geosciences.*

CPTI Working Group (2004). Catalogo Parametrico dei Terremoti Italiani, versione 2004 (CPTI04). INGV, Bologna. Available at: http://emidius.mi.ingv.it/CPTI04/.

De Luca, G.; Marcucci, S.; Milana, G. & Sanò, T. (2005). Evidence of Low-Frequency Amplification in the City of L'Aquila, Central Italy, through a Multidisciplinary Approach Including Strong- and Weak-Motion Data, Ambient Noise, and Numerical Modeling. *Bull. Seism. Soc. Am.,* Vol. 95, pp. 1469–1481.

EMERGEO Working Group (2010). Surface rupture of the 2009 L'Aquila earthquake (Italy). *Terra Nova,* Vol. 22, No. 1, pp. 43–51.

Field, E.H.; Johnson, P.A.; Beresnev, I.A. & Zengk, Y. (1997). Nonlinear ground-motion amplification by sediments during the 1994 Northridge earthquake. *Nature,* Vol. 390, pp. 599-602.

Galli, P.; Galadini, F.; Moro, M. & Giraudi, C. (2002). New paleoseismological data from the Gran Sasso d'Italia area (central Apennines). *Geophys. Res. Lett.*, Vol. 29, No. 7, Doi:10.1029/2001GL013292.

Guadagno, F. M.; Nunziata, C. & Rapolla, A. (1992). Dynamic parameters of volcaniclastic soils and rocks of Campi Flegrei (Naples, Italy). Volcanic seismology, IAVCEI Proceedings in volcanology, pp. 533-546, Springer-Verlag.

Gusev, A.A. (1983). Descriptive statistical model of earthquake source radiation and its application to an estimation of short period strong motion. *Geophysical Journal of the Royal Astronomical Society*, Vol. 74, pp. 787–800.

Gusev, A. A. & Pavlov, V. (2006). Wideband simulation of earthquake ground motion by a spectrum-matching, multiple-pulse technique. *First European Conf. on Earthquake Engineering and Seismology*, Geneva, Switzerland, Paper 408.

Gusev, A.A.; Pavlov, V.; Romanelli, F. & Panza, G.F. (2008). Low-frequency seismic ground motion at the pier positions of the planned Messina straits bridge for a realistic earthquake scenario. In: Seismic Engineering Conference Commemorating the 1908 Messina and Reggio Calabria Earthquake, Santini A. & Moraci N. (Ed.), AIP, 1020, pp. 362-369.

Gusev, A.A. (2011). Broadband Kinematic Stochastic Simulation of an Earthquake Source: a Refined Procedure for Application in Seismic Hazard Studies. *Pure Appl. Geophys.,* Vol. 168, pp. 155–200.

Idriss, I. M. & Seed, H. B. (1968.) Seismic response of horizontal soil layers, *J. Soil Mech. Found.*, Div. ASCE, Vol. 94, pp.1003-1031.

Levshin, A.L.;Yanovskaya, T.B.; Lander, A.V.; Bukchin, B.G.; Barmin, M.P.; Ratnikova, L.I. & Its, E.N. (1989). Seismic Surface Waves in a Laterally Inhomogeneous Earth, Keilis-Borok (Ed.), V.I., Kluwer, Norwell, Mass.

Molchan, G.; Kronrod, T. & Panza, G.F. (2011). Hot/Cold Spots in Italian Macroseismic Data. *Pageoph*, Vol. 168, pp. 739-752, DOI 10.1007/s00024-010-0111-3

Nunziata, C.; Natale, M.; & Panza, G.F. (2004). Seismic characterization of neapolitan soils. *Pageoph*, Vol. 161, No 5/6, pp. 1285-1300

Nunziata, C. (2007). A physically sound way of using noise measurements in seismic microzonation, applied to the urban area of Napoli. Engineering Geology, Vol. 93, No. 1-2, pp. 17-30.

Nunziata, C. (2010). Low shear-velocity zone in the Neapolitan-area crust between the Campi Flegrei and Vesuvio volcanic areas. *Terra Nova*, Vol. 22, pp. 208–217.

Olsen, K. B. (2000). Site amplification in the Los Angeles basin from three-dimensional modeling of ground motion. *Bull. Seismol. Soc. Am.*, Vol. 90, 6B, pp. 577–594.

Panza, G.F.; Romanelli, F. & Vaccari, F. (2001). Seismic wave propagation in laterally heterogeneus anelastic media: theory and applications to seismic zonation. *Advances in Geophysics*, vol. 43, pp. 1–95., Academic Press.

Panza, G. F; Paskaleva, I. & Nunziata, C. (Ed.) (2004). Seismic ground motion in large urban areas. *Pageoph* Topical Volume, Vol. 161, No 5/6, ISSN 0033-4553, Birkhauser.

Panza, G. F.; Peccerillo, A.; Aoudia, A. & Farina, B. (2007). Geophysical and petrological modelling of the structure and composition of the crust and upper mantle in complex of geodynamic setting: The Thyrrenian Sea and surrounding. Earth-Science Reviews, Vol. 80, pp. 1-46.

Panza, G.; Irikura, K.; Kouteva, M.; Peresan, A.; Wang, Z. & Saragoni, R. (Ed.) (2011). Advanced seismic hazard assessment, *Pageoph* Topical Volume, 752 pp., ISBN 978-3- 0348-0039-6 & ISBN: 978-3-0348-0091-4.

Papanikolaou, I.D.; Foumelis, M.; Parcharidis, I.; Lekkas, E.L. & Fountoulis, I.G. (2010). Deformation pattern of the 6 and 7 April 2009, Mw=6.3 and Mw=5.6 earthquakes in L'Aquila (Central Italy) revealed by ground and space based observations. *Nat. Hazard Earth Sys.*, Vol. 10, pp. 73–87.

Protezione Civile, 2009. Microzonazione Sismica dell'area aquilana (http://www.protezionecivile.it/jcms/it).

Roberts, G. P.; Cowie, P.; Papanikolaou, I. & Michetti, A. M. (2004). Fault scaling relationships, deformation rates and seismic hazards: An example from the Lazio-Abruzzo Apennines, central Italy. *J. Struct. Geol.*, Vol. 26, pp. 377–398.

Schnabel, P. B.; Lysmer, J. & Seed, H. B. (1972). SHAKE: a computer program for earthquake response analysis of horizontally layered sites. Report No. UCB/EERC 72/12, Earthquake Engineering Res. Center, Univ. California, Berkeley.

Vezzani, L. & Ghisetti, F. (1998). Carta Geologica dell' Abruzzo, Scala 1:100,000. S.EL.CA., Firenze.

Vinale, F. (1988). Caratterizzazione del sottosuolo di un'area campione di Napoli ai fini di una microzonazione sismica, *Rivista Italiana di Geotecnica,* Vol. 22, pp. 77-100.

Wen, K. L.; Chang, T. M.; Lin, C. M. & Chiang, H. J. (2006). Identification of nonlinear site response using the H/V spectral ratio method. *Terr. Atmos. Ocean. Sci.*, Vol. 17, pp. 533-546.

6

Active and Passive Experiments for S-Wave Velocity Measurements in Urban Areas

C. Nunziata, G. De Nisco and M.R. Costanzo
Dipartimento di Scienze della Terra,
Univ. Napoli Federico II
Italy

1. Introduction

One of the key parameters for the study of the effects of local site conditions is the S-wave velocity structure of unconsolidated sediments and the S-wave velocity contrast between bedrock and overlying sediments.

Detailed V_S profiles with depth can be measured with standard borehole logging and hole measurements, like down-hole and cross-hole. Such measurements are expensive, very local (point measurements) and may be not representative of large areas. Powerful methods for V_S measurements, that do not need drillings, are all based on the dispersion properties of Rayleigh wave phase and group velocities. Methods for phase velocity measurement of surface waves need recordings along dense arrays, with small geophone spacing, to avoid spatial aliasing, or, in case of 2 receivers, there is the problem of getting the right number of cycles and, hence, the analysis may lead to wrong values (Nunziata, 2005). Instead, the group velocity dispersion curve of the fundamental mode of Rayleigh waves can be extracted from the recorded signal at a single station by using the FTAN (Frequency Time Analysis) method (Nunziata, 2010 and references therein).

FTAN is appropriate to process surface wave data both for the identification and the separation of signals and for the measurement of signal characteristics other than phase and group velocities, like attenuation, polarization, amplitude and phase spectra. When not only the fundamental mode but also the higher modes are excited, FTAN method lets to estimate the gross Q values too. In fact, the comparison between synthetic seismograms computed with extreme Q values and experimental data is based on the relative amplitude of fundamental and higher modes (Nunziata et al., 1999). FTAN method is successfully employed both in seismological and engineering field (e.g. Nunziata et al., 2009; Nunziata, 2010). At urban sites, the impossible use of explosive sources or heavy masses blows, limits the penetration depth to the uppermost 20-30 m, depending upon the rock velocities. Recently, cross correlations of long time series of ambient seismic noise have been demonstrated to recover surface wave dispersion (Green function) over a broad range of distances, from a few hundred metres to several hundred kilometres (e.g. Weaver & Lobkis, 2001; Bensen et al., 2007; Nunziata et al., 2009). Detailed V_S profiles with depth are then obtained from the non linear inversion (Hedgehog method) of the average dispersion curve of the fundamental mode of Rayleigh group velocities.

Aim of this paper is to present examples of the FTAN and Hedgehog methods applied to both active and passive experiments, to obtain reliable V_S profiles to depths of 2 km in a complex urban area like Napoli, with high seismic and volcanic risk.

2. Methodologies

In the FTAN analysis, signal is passed through a set of narrow-band Gaussian filters with central frequencies varying in the frequency band of interest. The combination of all so filtered signals is a complex function $S(\omega_c, t)$ of two variables, filter central frequency ω_c and time, called frequency-time representation of a signal. A FTAN image is produced by displaying the logarithm of the square of the contour map of $|S(\omega_c, t)|$ amplitude and, for ω fixed, it represents the signal envelope at the output of the relevant filter. The group arrival time as a function of the central frequency of the Gaussian filter is determined from the peak of the envelope function. Typically period replaces filter central frequency and, being known the source-receiver distance, group velocity replaces time. One example of the FTAN analysis on a signal recorded with active seismic experiment at Napoli (Italy) is given in Figs. 1a-e.
An average dispersion curve of Rayleigh group velocities with errors is computed, from FTAN analysis made on a few (4-5 or more) signals, which can be inverted to determine V_S profiles versus depth. A non-linear inversion is made with the Hedgehog method (Panza et al., 2007 and references therein) that is an optimized Monte Carlo non-linear search of velocity-depth distributions. In the inversion, the unknown Earth model is replaced by a set of parameters (V_P, V_S, density and thickness) and the definition of the structure is reduced to the determination of the numerical values of these parameters. In the inversion V_S and thickness are variable parameters, while density is fixed and V_P is dependent on V_S through an assigned V_P/V_S ratio. In the inversion problem of V_S modeling, the parameter function is the dispersion curve of group velocities of Rayleigh fundamental mode.
Given the error of the experimental phase and/or group velocity data, it is possible to compute the resolution of the parameters, computing partial derivatives of the dispersion curve with respect to the parameters to be inverted (Panza, 1981). The parameterization for the inversion is defined so that the parameter steps are minima, subject to the condition $\sum_j \left(\partial V(T_i) / \partial P_j \right) \partial P_j = \sigma(T_i)$ where σ is the standard deviation of measurements, V is phase or group velocity, T_i is the i-th period, and P_j is the j-th parameter; in this way each parameter step represents a satisfactory measure of the uncertainty affecting each parameter. The theoretical phase and/or group velocities computed during the inversion with normal-mode summation are then compared with the corresponding experimental ones and the models are accepted as solutions if their difference, at each period, is less than the measurement errors and if the r.m.s. (root mean square) of the differences, at all periods considered, is less than a chosen quantity (usually 60–70% of the average of the measurement errors). All the solutions of the Hedgehog inversion differ by no more than ±1 step from each other. A good rule of thumb is that the number of solutions is comparable with the number of the inverted parameters.
From the set of solutions, we accept as a representative solution the one with the rms (root mean square) for phase and group velocities closest to the average rms for all the solutions, reducing in this way the projection of possible systematic errors into the structural model (Panza, 1981). Other selection criteria could be followed as described by Boyadzhiev et al. (2008). An example is shown in Fig. 2.

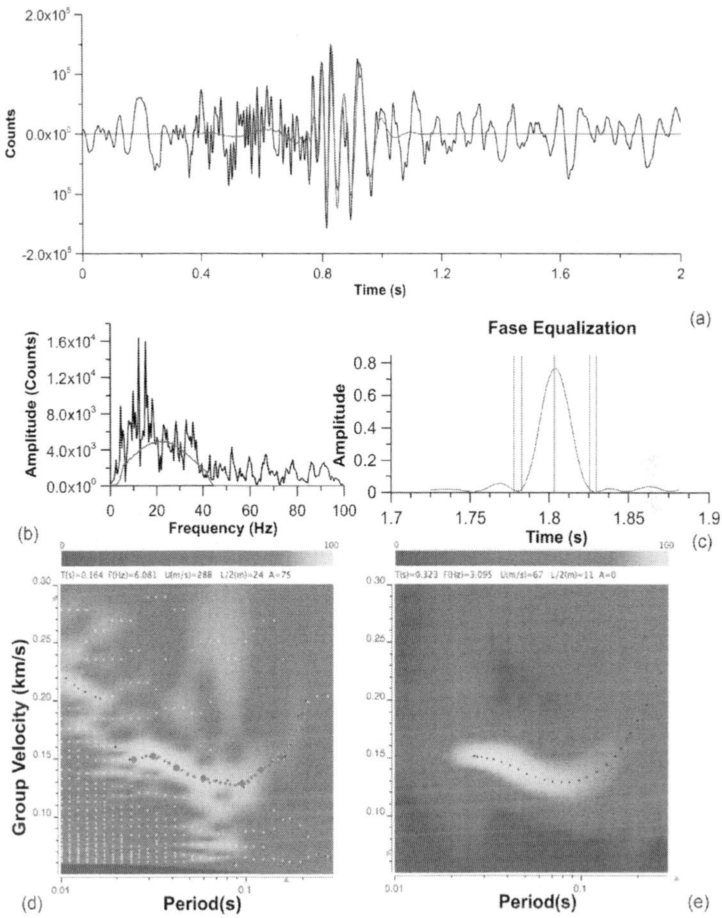

Fig. 1. Example of FTAN analysis on a signal of active experiment at Napoli (Scampia, located in Fig. 3) with 120m offset: (a) the raw waveform (black line); (b) Fourier spectrum amplitude of the signal; (c) a raw group velocity curve (green dots) is chosen by the analyst by picking maxima on the FTAN map. This raw group velocity dispersion curve is back Fourier transformed to get the dispersed signal. Phase-matched (anti-dispersion) filtering is performed on the chosen period-band to remove dispersion. (d) The anti-dispersed signal will collapse into a single narrow spike. Such operation has the only effect to alter the initial phase of the resulting signal, so it can be shifted to a convenient instant of time, for example, to the midpoint of the record. The collapsed waveform is then cut (vertical lines) from the surrounding time-series and re-dispersed to give the clean waveform. (e) The FTAN image of the cleaned waveform is computed and, using the same process applied to the raw waveform, the cleaned group velocity curve (blue dots) and fundamental mode waveform (red line in (a)) are obtained.

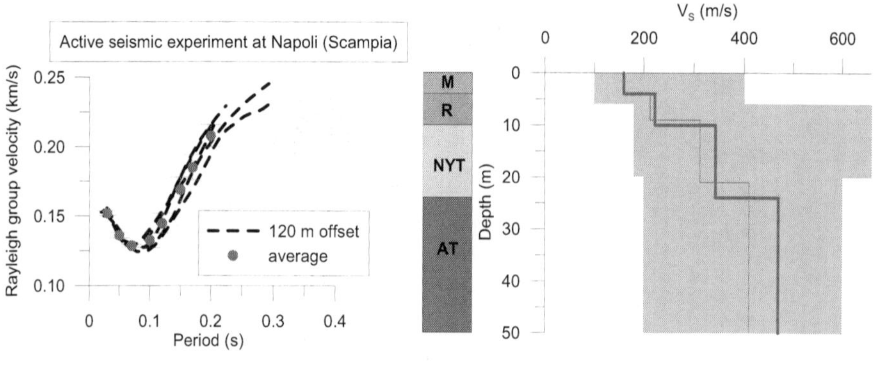

Fig. 2. (a): Dispersion curves of group velocity relative to receivers with 120m offsets at Napoli (Scampia, located in Fig. 3). (b): Shear wave velocity models: the grey area indicates the searched part of the parameter's space, while the accepted models are represented by the solid lines. The chosen solution has been selected as the closest to the known stratigraphy (red line). Legend: M man made ground; R recent pyroclastic products; NYT Neapolitan Yellow Tuff (soil facies); AT ancient tuff.

3. Active experiments

FTAN measurements have been successfully performed, at engineering scale, in italian urban areas with different soil and rock environments (e.g. Nunziata et al., 2004; Nunziata, 2005). A weight drop of 30 kg is used as source and one or more low frequency 4.5-1 Hz vertical geophones are used as receivers for offsets less or greater than 50 m, respectively. Only one receiver is requested, or, alternatively, a seismic refraction spreading, in order to evaluate an average group velocity dispersion curve from 4-5 receivers or 4-5 sources.
In the following some examples are reported at the neapolitan area to show the main advantages of FTAN method in complex geological settings of noisy urban areas.

3.1 FTAN measurements at Napoli

Several measurements have been performed at the urban area of Napoli for which, taking into account the stratigraphies, six geologically homogeneous zones have been recognized (Nunziata, 2004) (Fig. 3). The geological setting of Napoli is mainly characterized by pyroclastic materials, soil (pozzolana) and rock (tuff), produced by different eruptive centres at Campi Flegrei and Somma-Vesuvio volcanoes. The most widespread lithotype is the Neapolitan Yellow Tuff (NYT, 15 ka) which constitutes the skeleton of the historical urban area. FTAN-Hedgehog V_S models represent average values over distances of 50-100m and are more suitable than down-hole (DH) and cross-hole (CH) point measurements for seismic response analysis. Beside this, the good agreement of FTAN-Hedgehog with CH measurements (Nunziata et al., 2004), has allowed to enrich database and to acquire experience enough to select, for each zone, some V_S models for the evaluation of the spectral amplification (Nunziata, 2004). In fact, because of the lack of recordings of strong ground motion at Napoli, the only way to estimate site effects is to compute them. Starting from the

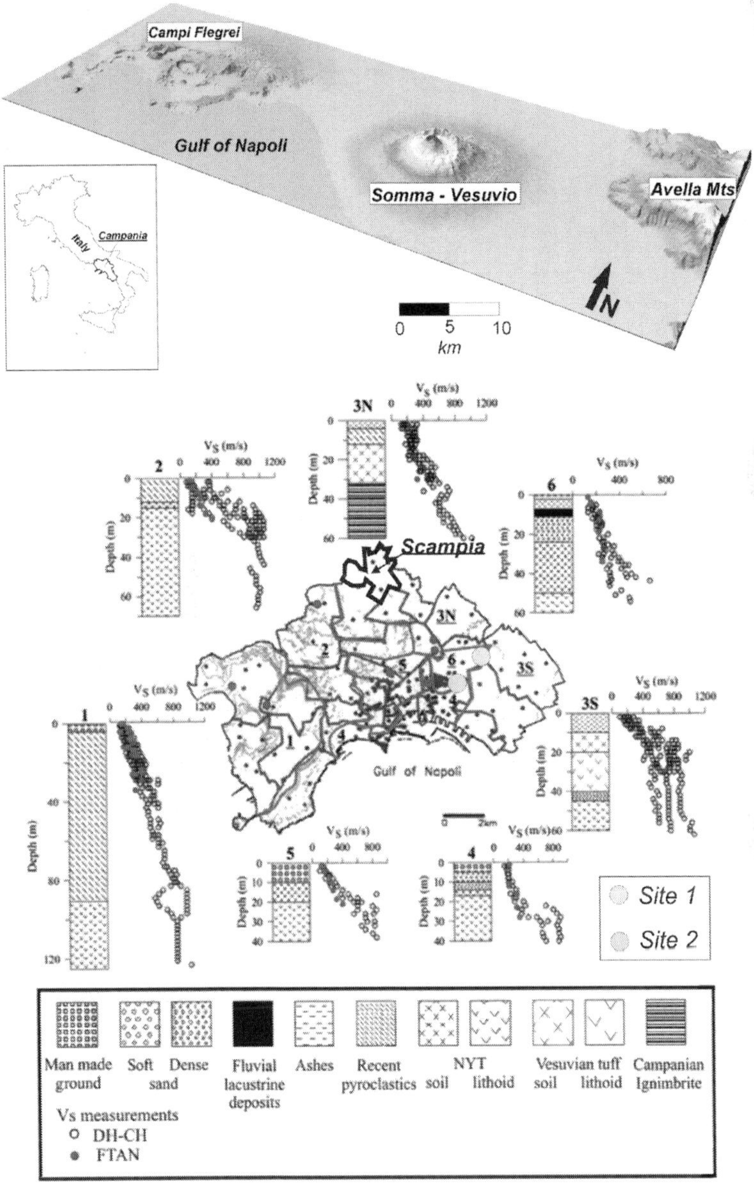

Fig. 3. Urban map of Napoli showing the quarter limits and the representative stratigraphic column for each of the six geological zones (bold underlined numbers), together with drillings (black dots), FTAN measurement sites (blue dots) and all available V_S measurements (modified after Nunziata et al., 2009).

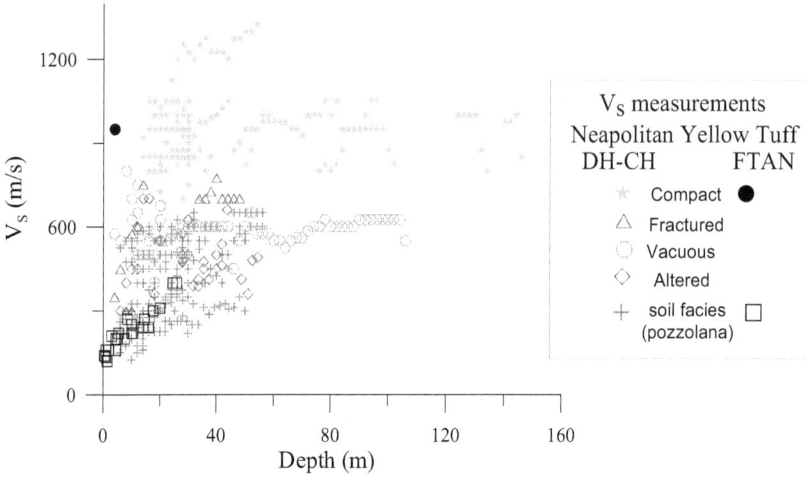

Fig. 4. V_S velocities of Neapolitan Yellow Tuff, both soil and lithoid facies (modified after Nunziata et al., 2004).

good fit between the first strong event recorded close to Napoli (about 20km far), that is the 1980 earthquake (Ms=6.9), and seismograms computed with mode summation technique, the detailed geological and seismic information of the neapolitan subsoil were used to compute quite realistic ground motion at Napoli for the 1980 earthquake (Nunziata, 2004) and the 1688 scenario earthquake (Nunziata et al., 2011). The propagation of the waves from the source to the complex laterally varying structure is computed with the mode summation technique, and in the laterally heterogeneous structure, it is computed with the finite difference method (Panza et al., 2001 and references therein). Site amplification effects were estimated in terms of spectral amplification, defined as the response spectrum at a site in the 2-D structural model, normalized to the response spectrum computed for the 1-D average reference model. Average and maximum response spectra and spectral amplifications were computed for all V_S models at each zone and proposed for zoning purposes. The need of doing robust V_S measurements vs. depth in volcanic settings is strictly dependent on their wide ranges of variations. They are the consequence of the profound differences in the physical properties and textural conditions that can be found even in the same formation. An additional important factor responsible for the observed scatter in the V_S values is the different hardening degree, due to the diagenetic process. As an example, V_S measurements relative to NYT, both in soil and lithoid facies, are shown in Figure 4.

3.2 FTAN measurements at Portici (Somma-Vesuvio)

Somma-Vesuvio is a very densely populated area, and accurate V_S measurements are requested for the volcanic and seismic hazards. FTAN measurements performed at the Royal Palace of Portici, famous town for the highest population density in Italy, have given detailed V_S models for the shallower 40 m (Fig. 5). These results are important for seismic zoning and give precious volcanological information in terms of thickness of the erupted products. In fact, taking into account the stratigraphy of a close drilling, it resulted that a Vesuvio shallow lava (medieval or 1631 eruption) is characterized by V_S of 500 m/s, which is higher than that of Somma lava at 30 m of depth, and that pyroclastic deposits have low V_S velocities around 200-300 m/s.

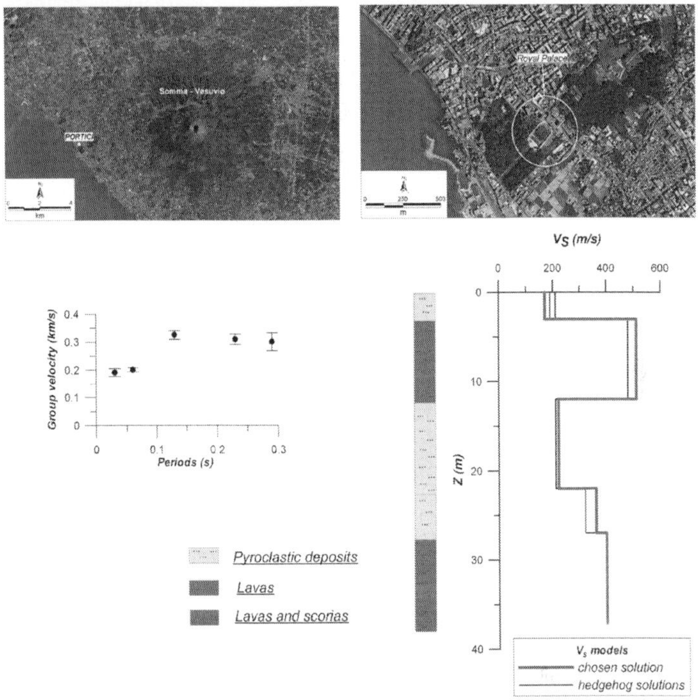

Fig. 5. Results of nonlinear inversion (Hedgehog method) of the average dispersion curve of Rayleigh group velocities, with error bars, together with a close drilling stratigraphy.

4. Passive experiments

Passive methods are based on ambient vibrations or microtremors and can be used to infer V_S profiles vs. depth. One method is the H/V method of Nakamura (1989), defined as the ratio between the mean of the Fourier spectra of the horizontal components and the spectrum of the vertical component, which has proven to be a convenient technique to estimate the fundamental frequency of soft deposits (e.g. Lermo & Chavez-Garcia, 1994). Another method is the Noise Cross-correlation Function (NCF) based on the cross correlation of simultaneous noise recordings at two sites, which allows to recover surface wave dispersion (Green function) over the site distance.

4.1 Single station

If V_S models representative of average geological structures are available, the measured main peak of the average H/V spectral ratios is in agreement with the ellipticity computed, from the models, of the fundamental mode Rayleigh wave (Nunziata, 2007). The ellipticity at each frequency is defined as the ratio between the horizontal and vertical displacement

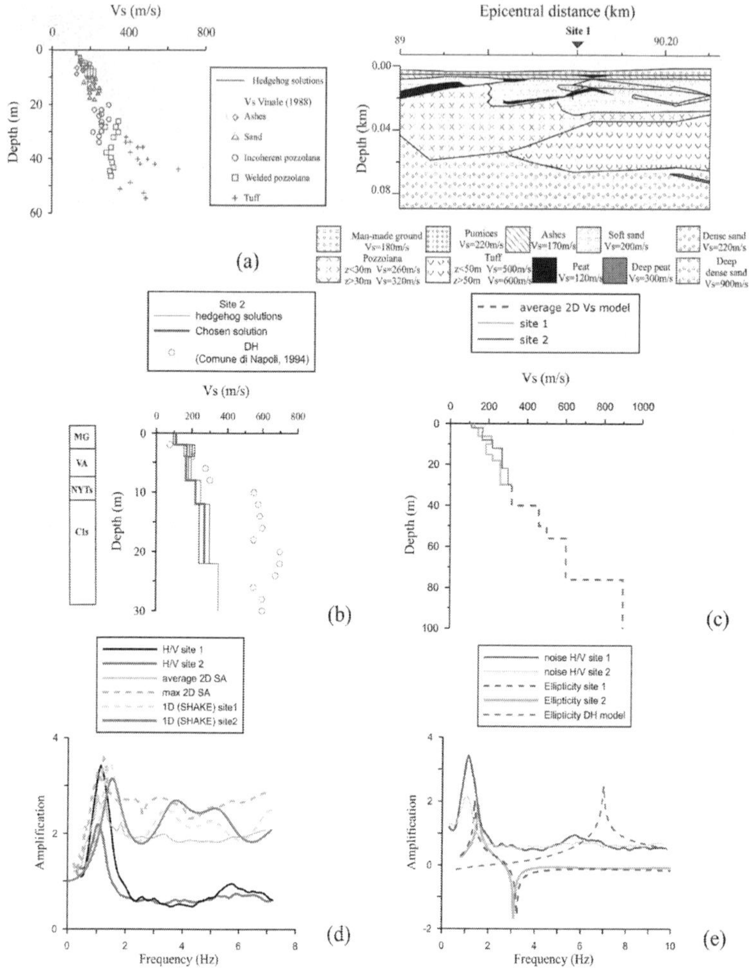

Fig. 6. Interpretation of noise H/V ratio at Napoli (zone 6 in Fig. 3). (a) Geological cross section representative of site 1 and shear velocities of the soils assigned based on Hedgehog solutions and DH and CH measurements. (b) Comparison between DH V_S measurements (circles) and the velocity range defined from the Hedgehog non-linear inversion of FTAN measurements at site 2. Legend: MG=Man made Ground; VA=Vesuvius Ashes; NYTs=Neapolitan Yellow Tuff (soil facies); CIs=Campanian Ignimbrite (soil facies). (c) V_S models (chosen Hedgehog solutions) at sites 1 and 2. At depths greater than 30m, V_S have been attributed on average DH-CH measurements (Vinale, 1988) as shown in (a). (d) Noise H/V ratios, synthetic 2D (hybrid method along the cross section in (a)) and 1D (SHAKE program) spectral amplifications at sites 1 and 2. 1D amplification computed for the DH velocity profile nearby site 2 is also shown. (e) Noise H/V ratios and logarithm of ellipticity absolute values computed for the V_S profiles shown in (c) and the DH velocity profile in (b).

eigenfunctions in the P-SV case, at the free surface. The ellipticity of the fundamental mode of Rayleigh waves computed for the V_S velocity models obtained by FTAN-Hedgehog methods, integrated at greater depths, down to the seismic bedrock, both by geological information and eventual down- and cross-hole measurements, has been compared with the ambient noise H/V ratio (Nunziata, 2007). The interpretation of noise measurements, carried out with Kinemetrics Quanterra Q330 station and a 3 component Episensor broadband sensor, has been also done through the comparison with the computed SH wave spectral amplifications, both 2D (Nunziata, 2004) and simplified 1D with SHAKE program (Schnabel et al., 1972).

As an example, we show the interpretation of noise measurements performed at two sites, 2km apart, at Poggioreale quarter, zone 6 (Figs. 6 a-e). Site 1 is located at the Centro Direzionale area with many skyscrapers built after 1980 earthquake and very detailed geological and geophysical information. The area was a marsh recently drained both for urban development and for the reduction of water supply. The subsoil is mainly formed by man-made ground, alluvial soils (ashes, sands, peat), loose and slightly cemented pozzolanas, NYT tuff and marine sands (Fig. 6a). Taking into account several DH and CH measurements, a good agreement has resulted with FTAN measurements and it has been possible to attribute V_S velocities at depths greater than 30m along a cross section through noise measurement site 1 (Nunziata, 2004 and references therein).

Instead, at site 2, strong discrepancy resulted between the V_S profiles vs depth obtained with FTAN-Hedgehog methods and the Down-hole measurements (Comune di Napoli, 1994) in a nearby drilling, despite the very good agreement between the representative Hedgehog solutions at the two sites since peat is characterized by similar velocities of host pyroclastics (Fig. 6b). At greater depth, average velocities increasing from 320 m/s, typical of tuff soils, up to 900 m/s, at 76m of depth, have been assigned on the basis of DH and CH measurements in the nearby Centro Direzionale (Fig. 6c). Hence, the seismic cross section through site 1 can be considered representative of site 2 as well. Two-dimensional spectral amplifications (5% damping) computed along the cross section are characterized by main peaks at about 1 Hz for the transverse component. H/V noise measurements show a good agreement with the average and maximum 2D spectral amplifications, while 1D amplifications have the first main peak at frequencies a little bit higher than H/V noise (Fig. 6d). Instead, the 1D amplification computed for the V_S profile measured in the drilling close to site 2 has a very high frequency content with the main peak at about 7 Hz. Agreement is also observed between H/V noise and ellipticity computed for the V_S models attributed at sites 1 and 2 as shown in Fig. 6c (Fig. 6e), whereas ellipticity computed for DH velocity profile has a peak at very high frequency (about 7 Hz).

4.2 Two stations

Cross correlations of long time series of ambient seismic noise have been demonstrated to recover surface wave dispersion (Green function) over a broad range of distances, from a few hundred meters to several hundred Kilometers (e.g. De Nisco & Nunziata, 2011 and references therein). This approach is based on earlier theoretical studies after Weaver and Lobkis (2001), who demonstrated that the cross correlation of a diffuse wavefield is related to local transient response. A diffuse field Φ in a finite body, in the point x at the time t, may be expressed as:

$$\Phi(x,t) = \Re \sum_{n=1}^{\infty} a_n u_n(x) \exp\{i\omega_n t\} \tag{1}$$

Where a_n are the complex modal amplitudes and u_n the real orthogonal mode shapes and $\Re$ indicates the real part of a complex quantity.
The cross correlation of the diffuse field at points x and y is:

$$\langle \Phi(x,t)\Phi(y,t+\tau)\rangle = \frac{1}{2}\Re\sum_{n=1}^{\infty} F(\omega_n)\, u_n(x)\, u_n(y) \exp\{-i\omega_n\tau\} \tag{2}$$

If F is almost constant, i.e. $F(\omega) \simeq$const, it turns out that the cross correlation is equivalent to the time derivative of the Green's function of the medium between x and y:

$$G_{xy}(\tau) = \sum_{n=1}^{\infty} u_n(x)\, u_n(y)\, \frac{\sin\omega_n\tau}{\omega_n} \qquad (\text{for } \tau > 0, \text{ otherwise } 0) \tag{3}$$

The basic idea of the method is that a time-average cross correlation of a random, isotropic wavefield computed between a pair of receivers will result in a waveform that differs only by an amplitude factor from the Green function between the receivers. Ambient seismic noise can be considered as a random and isotropic wavefield both because the distribution of the ambient sources responsible for the noise randomizes when averaged over long times and because of scattering from heterogeneities that occur within the Earth. Several researchers have used the noise cross correlation instead of the time derivative (e.g. Campillo & Paul, 2003; Shapiro & Campillo, 2004). This assumption is acceptable from the ambient noise recorded on broadband seismic stations, typically with relatively small bandwidth, being the difference between the cross correlation and its derivative a phase shift.
The assertion that the impulse response (Green's function) can be retrieved from cross correlation of the diffused fields (noise) in two receiving points is based on the time-reversal symmetry of the Green's function (Derode et al., 2003). Thus, for a perfectly homogeneous distribution of sources surrounding the two points, fully immersed in the scattering medium, the exact impulse response can be recovered from either the causal (t>0) or the anticausal (t<0) part of the cross correlation stacked for all sources, that is the cross correlation is symmetric. However, considerable asymmetry in amplitude and spectral content is typically observed, which indicates differences in both the source process and distance to the source in the directions radially away from the stations (Larose et al., 2004; Bensen et al., 2007). Yet, in the seismic experiments described by Campillo and Paul (2003) and Sabra et al. (2005), the Green's function was reconstructed from one-sided cross correlation because of a preferential direction. Many authors, using the spatial reciprocity of the Green's functions (De Nisco & Nunziata, 2011 and references therein), average positive and negative parts of NCF and impose symmetry. In most cases, this procedure enhances the signal to noise ratio (SNR) and effectively mixes the signals coming from opposite directions, which helps to homogenize the source distribution. If the signature of the Green function is recognized in only one part of the NCF or if there is a large time shift between the positive and negative part, seismic noise could have a preferred orientation source, and it is not possible to design the geometry of the array so that correct wave velocity can be computed. If the possible preferred orientation source is identified and the array is installed perpendicular to it, the Green function is obviously seen only in one part of NCF.
Experiments of noise cross correlation have been successfully conducted at Napoli over distances ranging from 50 m to about 4 km (Fig. 7). Two broadband Kinemetrics Quanterra

Q330 stations equipped with 3-component Episensor broadband FBA (Force Balance Accelerometer) sensors have been employed.

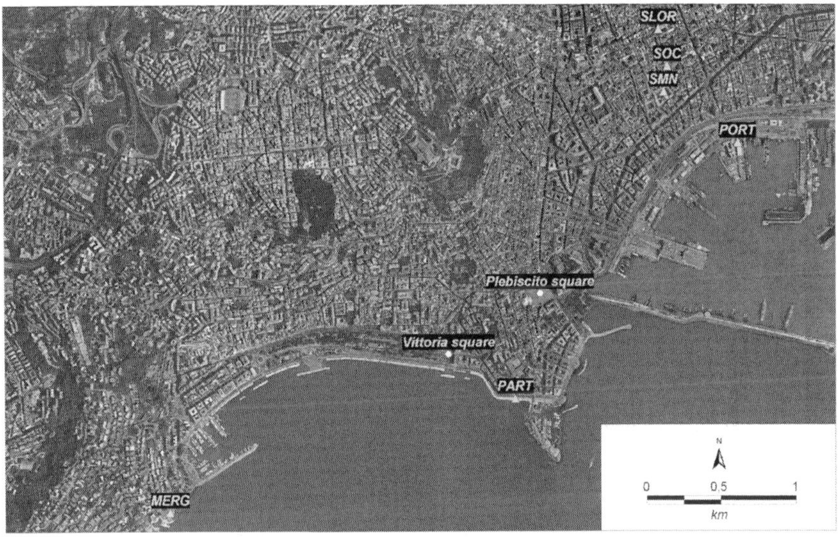

Fig. 7. Location of the seismic stations employed in the cross correlation experiments at the urban area of Napoli.

4.2.1 Signal analysis

The analysis of noise cross correlation consists of the steps that are illustrated in Figure 8 and refer to measurements at 4 km receiver spacing (path MERG-PORT located in Fig. 7). A one-bit normalization is applied to the vertical components of the recorded noise (Fig. 8A), which retains only the sign of the raw signal by replacing all positive amplitudes with a 1 and all negative amplitudes with a −1, in order to increase the signal-to-noise ratio. After average removal, power spectra are evaluated to discern the frequency band of interest (Fig. 8B). Signals are iteratively band-pass filtered (Butterworth filters) to enhance the dispersed wave trains in the cross correlation (Fig. 8C). The cross correlations are computed and then stacked with the Seismic Analysis Code (SAC) (Goldstein et al., 2003). One-sided cross correlations have generally resulted indicating differences in both the source process and distance to the source in the directions radially away from the stations (Larose et al., 2004; Bensen et al., 2007), or the presence of a preferential direction (Campillo & Paul, 2003; Sabra et al., 2005). Then FTAN analysis is performed to extract the fundamental mode of Rayleigh waves (Figs. 8 D-E).

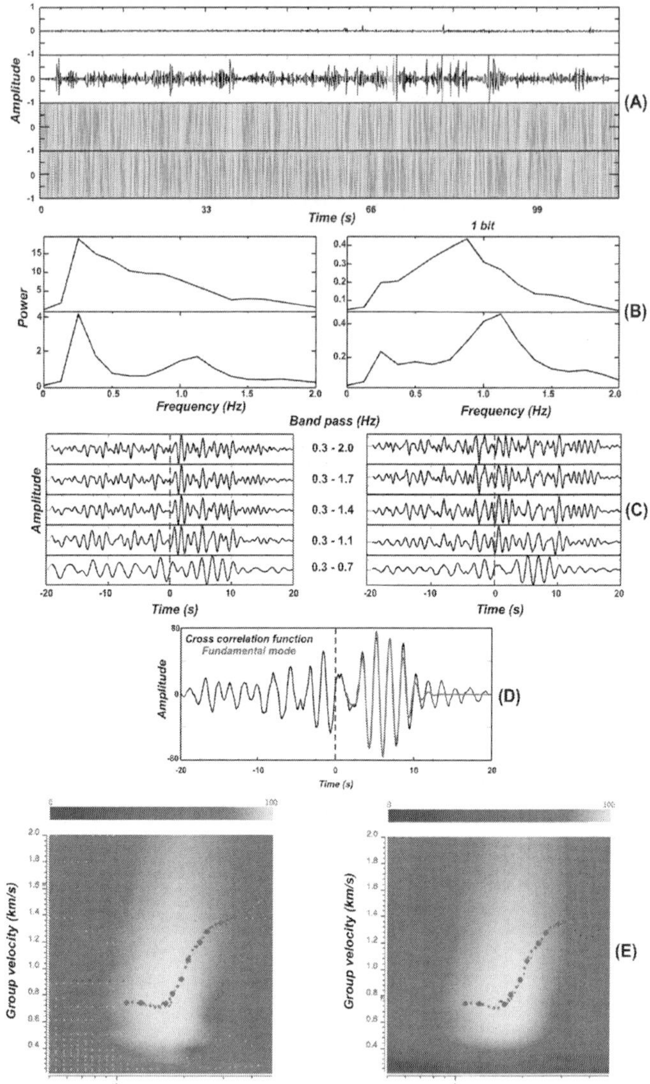

Fig. 8. Example of NCF analysis: (A) noise recordings with 1 Hz vertical geophones 4 km apart (1 and 2 receivers) at Mergellina (MERG) and harbour (PORT) (located in Fig. 7) and 1-bit normalized signals; (B) power spectra of noise recordings (on the left) and 1-bit normalized signals (on the right); (C) cross correlation of signals shown in (A) and band-pass filtered; (D) Cross correlation of 1-Bit 0.3-0.7 Hz band-pass filtered signals and the fundamental mode Green function extracted with FTAN method (red line). (E) Raw and cleaned FTAN maps of the cross correlation [bottom of (C)].

Three experiments have been conducted at the historical centre of Napoli (SMN, SOC and SLOR in Fig. 7) being one (SMN), permanently installed and continuously recording at the rock foundations of the SS. Marcellino and Festo monumental complex (www.geosism.unina.it). The other Q330 station was mobile and recorded ambient noise, for at least 1 hour, using a 100-Hz sampling frequency, at S. Lorenzo Maggiore basilica (SLOR) and the Sociology Faculty of University (SOC). The SMN-SOC and SMN-SLOR distances are 260 m and 400 m, respectively.

The retrieved V_S models (shown along the SMN-SLOR in Fig. 9) clearly show at 20-35 m below ground level a V_S increase to about 600 m/s. According to stratigraphies and velocity ranges for NYT tuff (Fig. 4), it can be argued that such V_S values are consistent with the presence of altered NYT tuff, typically found on the top of compact tuff and called "cappellaccio". Compact NYT tuff (V_S of about 800 m/s) is found 10-15 m deeper. The average thickness of the NYT tuff formation is 50m. At greater depth, a further increment of V_S is observed, to 970 m/s at 70-150 m of depth. This V_S distribution versus depth is very consistent with the stratigraphy of a deep borehole (400 m) drilled in the Plebiscito square, in front of the Royal Palace, and close to the investigated area. In fact, in the shallower 160 m, layers of NYT tuff, 80 m thick, and Whitish tuff, 80 m thick, were found. Then an important stratigraphic result can be deduced: the thickening of the NYT tuff layer moving towards west, that is in the direction of the eruptive centre (Nunziata et al., 2009).

Another experiment has been performed at the Partenope street (PART in Fig. 7), with highly chaotic traffic, by noise recordings with a 24 bit Geometrics StrataVisor seismograph and 1 Hz vertical geophones (Geospace GS-1), along a spreading with geophone distance of 180 m. The V_S distribution vs. depth is very consistent with the stratigraphy of a deep borehole at Vittoria square (Fig. 7), close to the investigated area (Fig. 10). It turns out that the investigated area is characterized by fractured and compact tuffs below a shallow layer of man made ground material, laid to construct and protect the street from sea actions.

Very recently, a further experiment has been performed over a distance of 4 km in order to define the thickness of the tuff cover. Two 24 bit digital tromographs with a wide frequency range (0.1-200 Hz) have simultaneously recorded noise for 1 hour at Mergellina (MERG) and harbour (PORT) (for their location see Fig. 7). The results are very important as, for the first time, structures have been defined below Napoli at 2 km of depth (Fig. 11). The obtained V_S models are in agreement with V_S velocities obtained by Nunziata (2010) along the Vesuvio-Campi Flegrei path, crossing Napoli and its gulf. The agreement with the data relative to Mofete (Campi Flegrei) drillings, that is V_S computed from V_P sonic log measurements (AGIP, 1987) by assuming a reasonable V_P/V_S ratio of 1.8, and ultrasonic measurements on saturated specimens (Zamora et al., 1994) is quite impressive and the following interpretative structural model can be formulated.

The first 0.5 km consists of tuffs while tuffs and tuffites are present at 0.5-1.2 km depth; tuffs and tuffites with lava interbedding, probably thermometamorphic, might be present at depths of 1.2-2 km. At these depths, both the agreement with ultrasonic velocity measured on a conglomerate sample (Bernard & Zamora, 2003) and the stratigraphy at Plebiscito square (Fig. 9), suggest that the presence of highly fractured sedimentary rocks cannot be escluded. The compact sedimentary horizon, with a V_S of 3.6-3.7 km/s, has been found below the Neapolitan area at about 3 km of depth (Nunziata, 2010).

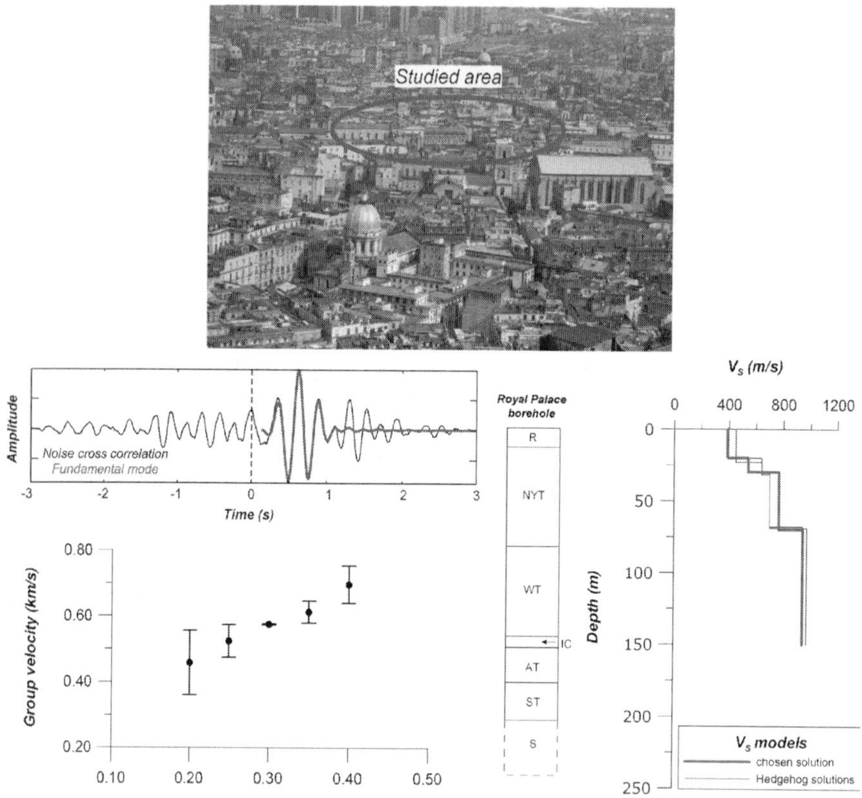

Fig. 9. The V_S solutions and the chosen solution (red line) for SMN-SLOR path obtained from the non-linear inversion with Hedgehog method of the average dispersion curve of the fundamental mode, with error bars, are compared with the stratigraphy of the deep drilling at Plebiscito square (located in Fig. 6). Legend: R=man made ground material and recent pyroclastic deposits; NYT=Neapolitan Yellow Tuff; WT=Whitish tuff; IC= Campanian Ignimbrite; AT= Ancient Tuffs; ST= Tuffs and sedimentary rocks; S= Sedimentary rocks. The cross correlation of 1-Bit 3-6 Hz band-pass filtered signals and the fundamental mode Green function extracted with FTAN method (red line) are also shown.

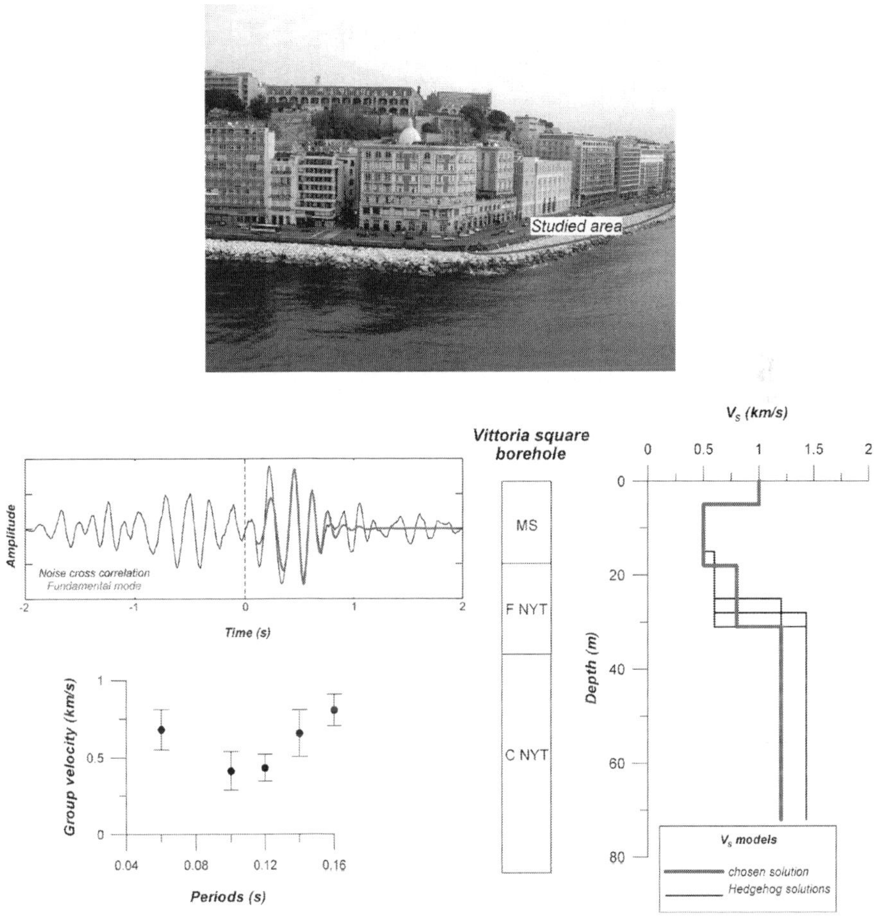

Fig. 10. The V_S solutions and the chosen solution (red line) for the spreading at Partenope st. with geophone distance of 180 m. The statigraphy is relative to Vittoria square borehole (located in Fig. 7). The photo of the investigated area, the cross correlation of 1-Bit 6-25 Hz band-pass filtered signals, the fundamental mode Green function extracted with FTAN method (red line), and average dispersion curve of the fundamental mode, with error bars, are also shown. Legend: MS = Marine sands; F NYT = Fractured Neapolitan Yellow Tuff; C NYT = Compact Neapolitan Yellow Tuff.

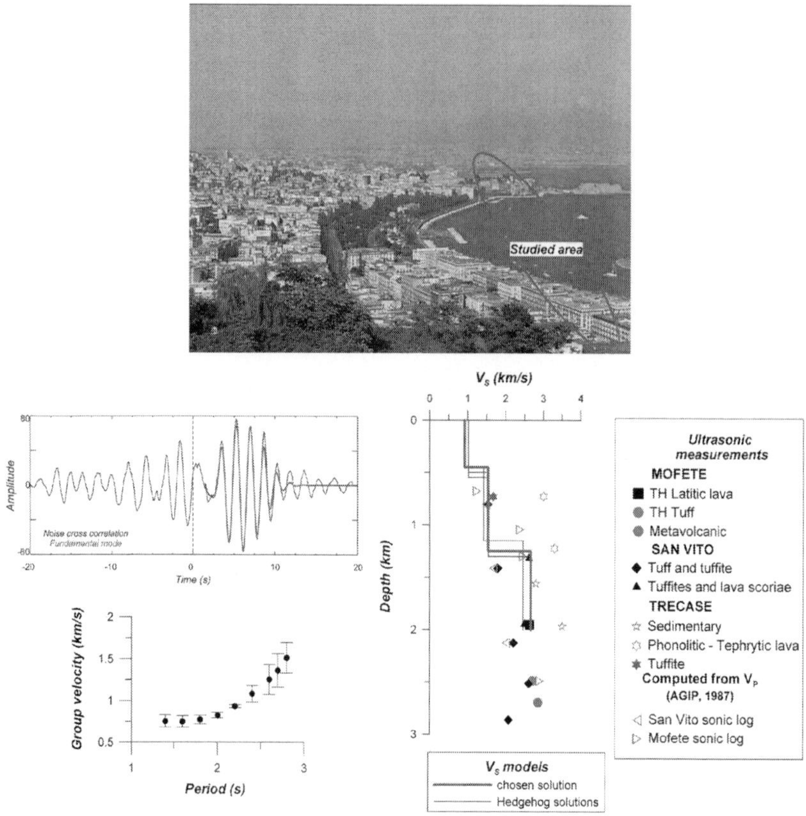

Fig. 11. The V_S solutions and the chosen solution (red line) for the MERG-PORT path (Fig. 7) are shown together with the average dispersion curve of the fundamental mode, with error bars, and comparison between 1-Bit 0.3-0.7 Hz band-pass filtered signals cross correlation and the fundamental mode Green function extracted with FTAN method (red line). Ultrasonic measurements on specimens from Campi Flegrei (Zamora et al., 1994) and Vesuvio (Trecase drilling) (Bernard & Zamora, 2003) are reported together with V_S values computed from V_P sonic log measurements (AGIP, 1987).

5. Conclusions

The results obtained in Napoli metropolitan area, with the non linear inversion of Rayleigh wave group velocity dispersion curve of the fundamental mode extracted with FTAN method from both active seismic surveys and noise cross correlation, show that the procedure is a powerful and reliable instrument to get V_S profiles versus depth in urban areas. The proposed methodology is low cost, as one (active experiments) or two (passive experiments) receivers are requested on ground surface and is particularly suitable for urban areas as doesn't require spreadings. The depth of penetration is mainly controlled by the distance and the soil velocities.

A new intriguing prospective is open by noise cross correlation as deep structures can be investigated in absence of specific energy sources. This is promising in terms of both seismic zoning and, even more, of volcanological zoning as stratigraphies can be reconstructed whereas they would be blind because of unaccessible outcrops in densely urbanized areas. Stratigraphic correlations can be assessed and used for volcanic hazard scenarios.

6. Acknowledgements

We wish to thank dr. P. Lazzaro for help with FTAN and Hedgehog analysis of active seismic signals at Scampia.

7. References

AGIP, 1987. Geologia e geofisica del sistema geotermico dei Campi Flegrei. Internal Report, pp. 17.

Bensen, G.D.; Ritzwoller, M.H.; Barmin, M.P.; Levshin, A.L.; Lin, F.; Moschetti, M.P.; Shapiro, N.M. & Yang, Y. (2007). Processing seismic ambient noise data to obtain reliable broad-band surface wave dispersion measurements. *Geophysical Journal International* No.169, pp. 1239-1260.

Bernard, M. L. & Zamora, M. (2003). Mechanical properties of volcanic rocks and their relations to transport properties, In the CD of: *The internal structure of Mt. Vesuvius*, P. Capuano, P. Gasparini, A. Zollo, J. Virieux, R. Casale & M. Yeroyanni (Eds)., Liguori Editore.

Boyadzhiev, G.; Brandmayr, E.; Pinat, T. & Panza, G.F. (2008). Optimization for non-linear inverse problems. *Rendiconti Lincei*, Vol.19, No.2, pp. 209.

Campillo, M. & Paul, A., (2003). Long-Range Correlations in the Diffuse Seismic Coda. *Science*, Vol. 299, No. 5606, pp. 547-549.

Comune di Napoli (1994). Indagini geologiche per l'adeguamento del P.R.G. alla legge regionale 07.01.1983 n. 9 in difesa del territorio dal rischio sismico.

De Nisco G. & Nunziata, C., (2011). V_S profiles from noise cross correlation at local and small scale. *Pure and Applied Geophysics*, Vol.168, No.3-4, pp. 509-520.

Derode, A., Larose, E., Tanter, M., de Rosny, J., Tourim, A., Campillo, & M., Fink, M. (2003). Recovering the Green's function from field-field correlations in an open scattering medium. *Journal of the Acoustical Society of America*, Vol.113, pp. 2973–2976.

Goldstein, P.; D. Dodge; M. Firpo & L. Minner, 2003. SAC2000: Signal processing and analysis tools for seismologists and engineers, In: *International Handbook of Earthquake and Engineering Seismology*, W.H.K. Lee; H. Kanamori; P.C. Jennings; & C. Kisslinger (Eds.), Academic Press, London, United Kingdom.

Larose, E.; Derode, A. ; Campillo, M. & Fink, M. (2004). Imaging from one-bit correlations of wideband diffuse wavefields. *Journal of Applied Physics*, Vol.95, pp. 8393–8399.

Lermo, J. & Chávez-García, F. J. (1994). Are microtremors useful in site response evaluation?. *Bulletin of the Seismological Society of America*, Vol.84, No.5, pp. 1350-1364.

Keilis-Borok V. I. (1989). *Seismic Surface Waves in a Laterally Inhomogeneous Earth*, Springer.

Nakamura, Y. (1989). A method for dynamic characteristics estimation of subsurface using microtremor on the ground surface, *QR of RTRI*, Vol.30, pp. 25-33.

Nunziata, C.; Costa G.; Natale, M.; Vuan, A. & Panza, G.F. (1999). Shear wave velocities and attenuation from Rayleigh waves, In: *Pre-failure Deformation Characteristics of Geomaterials*, 365-370, Balkema.

Nunziata, C. (2004). Seismic ground motion in Napoli for the 1980 Irpinia earthquake. *Pure and Applied Geophysics*, Vol.161, No.5/6, pp. 1239-1264.

Nunziata, C.; Natale, M. & Panza, G.F. (2004). Seismic characterization of neapolitan soils. *Pure and Applied Geophysics*, Vol.161, No.5-6, pp. 1285-1300.

Nunziata, C. (2005). FTAN method for detailed shallow V_S profiles. Geologia *Journal of technical & environmental geology*, No.3, pp. 25-43, Available from www.consiglionazionalegeologi.it.

Nunziata, C. (2007). A physically sound way of using noise measurements in seismic microzonation, applied to the urban area of Napoli. *Engineering Geology*, Vol.93, No. 1-2, pp. 17-30.

Nunziata, C.; De Nisco, G. & Panza, G.F. (2009). S-waves profiles from noise cross correlation at small scale. *Engineering Geology*, 105, No.3-4, pp. 161-170.

Nunziata, C. (2010). Low shear-velocity zone in the Neapolitan-area crust between the Campi Flegrei and Vesuvio volcanic areas. *Terra Nova*, Vol.22, No.3, pp. 208-217.

Nunziata C.; Sacco C. & Panza G.F. 2011. Modeling of ground motion at Napoli for the 1688 scenario earthquake. *Pure and Applied Geophysics*, Vol.168, No.3-4, pp. 495-508.

Panza G.F. (1981). The resolving power of seismic surface wave with respect to crust and upper mantle structural models, In: *The solution of the inverse problem in Geophysical Interpretation*, R. Cassinis (Ed.), 39-77, Plenum Press, New York, USA.

Panza, G.F.; Romanelli, F. & Vaccari, F. (2001). Seismic wave propagation in laterally heterogeneous anelastic media: theory and applications to the seismic zonation. *Advances in Geophysics*, Vol.43, pp. 1-95.

Panza, G. F.; Peccerillo, A.; Aoudia; A. & Farina, B. 2007. Geophysical and petrological modelling of the structure and composition of the crust and upper mantle in complex of geodynamic setting: The Thyrrenian Sea and surrounding. *Earth-Science Reviews*, Vol.80, pp. 1-46.

Sabra, K. G.; Gerstoft, P.; Roux, P. & Kuperman, W.A. (2005). Extracting time-domain Green's function estimates from ambient seismic noise. *Geophysical Research Letters*, Vol.32, No.2 L03310.

Schnabel B.; Lysmer J. & Seed H., 1972. *Shake: a computer program for earthquake response analysis of horizontally layered sites*, E.E.R.C. 72/12, Earthquake Engineering Research Center, University of California, Berkeley, USA.

Shapiro, N.M & Campillo, M. (2004). Emergence of broadband Rayleigh waves from correlations of the ambient seismic noise. *Geophysical Research Letters* Vol.31, L07614.

Vinale, F. (1988). Caratterizzazione del sottosuolo di un'area campione di Napoli ai fini di una microzonazione sismica. *Rivista Italiana di Geotecnica*, Vol.22, pp. 77-100.

Zamora, M.; Sartoris, G. & Chelini, W. (1994). Laboratory measurements of ultrasonic wave velocities in rocks from the Campi Flegrei volcanic system and their relation to other field data. *Journal of Geophysical Research*, Vol.99 pp. 13553–13561.

Weaver, R. L. & Lobkis, O. I. (2001). On the emergence of the Green's function in the correlations of a diffuse field*Journal of the Acoustical Society of America*, Vol.110, pp. 3011-3017.

7

Site-Specific Seismic Analyses Procedures for Framed Buildings for Scenario Earthquakes Including the Effect of Depth of Soil Stratum

P. Kamatchi[1], G.V. Ramana[2], A.K. Nagpal[2] and Nagesh R. Iyer[3]
[1]Senior Scientist, CSIR-Structural Engineering Research Centre, Chennai
[2]Professor, Indian Institute of technology, Delhi
[3]Director, CSIR-Structural Engineering Research Centre, Chennai
India

1. Introduction

The importance of the effect of sediments above bedrock in modifying the strong ground motion has been long recognized (Boore, 2004; Boore & Joyner, 1997; Idriss & Seed, 1970; Seed & Idriss, 1969; Lam et al., 2001; Govindarajulu et al., 2004; Tezcan et al., 2002; Bakir et al., 2005; Kamatchi et al., 2007) in literature. The nature of soil that changes the amplitude and frequency content has a major influence on damaging effects of earthquake. To account for these effects, most of the seismic codes, for example the Indian code (IS 1893 (Part 1) 2002) has defined response spectra for three types of soil viz., hard soil, medium soil and soft soil. As an improvement over this approach, amplification factors based on empirical and theoretical data (Borcherdt, 1994) have been introduced in International Building Codes (IBC, 2009; ASCE 7 2005) for site classes A to E for the short period range and long period range based on the average shear wave velocity of top 30 m soil stratum. For site class F (soft soil) it has been recommended that site-specific analysis need to be carried out.

However, Sun et al. (2005) showed that the site coefficients specified in IBC 2000 (IBC 2000) are not valid for Korean Peninsula due to the large difference in the depth of bedrock and the soil stiffness profile. Further, building codes are highly simplified tools and do not adequately represent any single earthquake event from a probable source for the site under consideration. Recently, it is being suggested (Heuze et al., 2004) that in addition to use of seismic code provisions, site-specific analysis which includes generation of strong ground motion at bedrock level and propagating it through soil layers (Heuze et al., 2004; Mammo 2005; Balendra et al., 2002) and arriving at the design ground motions and response spectra at surface should also be carried out.

In this chapter the procedure to carry out site-specific seismic analysis of framed buildings is illustrated with examples for Delhi city. Rock outcrop motions are generated for the scenario earthquakes of magnitude, M_w = 7.5, M_w = 8.0 and M_w = 8.5. Three actual soil sites of Delhi have been modeled and the free field surface motions and the response spectra are obtained. It has been observed that the PGA amplifications and the response spectra of the three sites are quite different for the earthquakes considered. The same has reflected in considerable difference in base shear and displacement response of reinforced concrete (RC) framed buildings when assumed to be located on the chosen three sites.

Major revisions are taking place in seismic codes towards performance based design (ATC 40 1996; FEMA 273&274, 1996). Response spectra which forms the basis for the demand curve in performance based design is of interest to many structural engineers. Having understood the importance of site-specific analysis, in this chapter, procedure for carrying out performance evaluation of buildings for site-specific earthquakes accounting for different depths of soil stratum is illustrated with an example.

Further, addressing to the complexity involved in carrying out site-specific seismic analysis, as an alternative, artificial neural network (ANN) based methodology is proposed in this chapter. The steps involved in the development of ANN models for site-specific seismic analysis viz., the process of identification of input and output parameters by carrying out sensitivity analysis, identification of suitable architecture and training of ANN, checking the performance of ANN and validating the performance of ANN with examples are illustrated in this chapter for Delhi city.

2. Steps involved in site-specific seismic analysis for scenario earthquake

Major steps involved in site-specific seismic analysis are identification of scenario earthquake, generation of strong ground motion for a rock site, propagation of ground motion through soil stratum and to arrive at the surface level response spectrum.

2.1 Identification of scenario earthquake

When this study was carried out (Kamatchi, 2008), not much of information was available on earthquake catalogue for Delhi capital city of India. Hence the scenario earthquakes have been chosen deterministically as per the information available from literature. Seismologists (Bilham et al., 1998; Singh et al., 2002) have reported that three major thrust planes viz., Main Central Thrust (MCT), Main Boundary Thrust (MBT) and Main Frontal Thrust (MFT) exist in Himalayas due to the relative movement of Indian plate by 5 cm/year with respect to Eurasian plate. Khattri (1999) has estimated the probability of occurrence of a great earthquake of moment magnitude 8.5 from the large unbroken segment called central seismic gap (Fig. 1) between MBT and MCT in the next 100 years to be 0.59. Delhi is situated at a distance of roughly 200 km from MBT and 300 km from MCT. Hence in the present study scenario earthquakes for Delhi city are assumed to be originated from central seismic gap of Himalayan region.

2.2 Generation of strong ground motion

Generation of artificial strong motions using stochastic models by identifying major fault zones and propagating seismic waves generated at these potential sources to the sites of interest is well accepted in literature (Boore 1983&2003; Beresnev & Atkinson 2002). In this process, path effects and anelastic attenuation effects predicted by the empirical and theoretical models (Beresnev & Atkinson 2002) are used. For source representation, point source models (Boore & Atkinson 1987) or finite source models (Hartzell, 1978) are widely used.

Stochastic simulation procedure for ground motion generation based on seismological models using point source model has been proposed by Boore (1983&2003). In this procedure the band limited Gaussian white noise is windowed and filtered in the time domain and transformed into frequency domain. The Fourier amplitude spectrum is scaled to the mean squared absolute spectra and multiplied by a Fourier amplitude spectrum obtained from source path effects. Then, the spectrum is transformed back to time domain and the time history is obtained.

From the analysis of recorded ground motions, it has been reported (Beresnev & Atkinson 1997) that point source models are not capable of reproducing the characteristic features of large earthquakes (M_w > 6) viz., long duration and radiation of less energy at low to intermediate frequencies (0.2-2 Hz). Simulation of strong ground motion from finite fault model has been developed by Beresnev and Atkinson (1997&1998). In this model, the fault rupture plane is represented with an array of sub-faults and the radiation from each sub-fault is modeled as a point source similar to Boore's model (Boore, 1983). According to finite source model, the fault rupture initiates at the hypocenter and spreads uniformly along the fault plane radially outward with a constant rupture velocity triggering radiation from sub-faults in succession. The improvements to finite source model viz., extended finite source model (Motazedian & Atkinson, 2005) implementing the effects of radiated energy on sub-fault size and dynamic corner frequency, are reported in literature when this chapter is written, however finite sources model (Beresnev & Atkinson 1997&1998) is used in the studies reported in this chapter. The details of the Fourier amplitude spectrum adopted in the present study and the assumptions made are illustrated subsequently.

The Fourier amplitude spectrum $A(\omega)$ of the point source of one element (sub-fault) is defined (Boore, 1983; Boore & Atkinson 1987; Brune, 1970) as

$$A(\omega) = \omega^2 S(\omega) P(\omega) G(R) A_n(\omega) \qquad (1)$$

Where, ω is the angular frequency, $S(\omega)$ is the source function, $P(\omega)$ is the filter function for high frequency attenuation, $G(R)$ is the geometric attenuation function, $A_n(\omega)$ is anelastic whole path attenuation function. $S(\omega)$, $P(\omega)$, $G(R)$ and $A_n(\omega)$ are further defined below:

2.2.1 Source function, S(ω)

The shape and amplitude of the theoretical source spectrum (ω²-model, Aki (1967) is given by,

$$S(w) = \frac{PFR^{\theta\phi} m_o}{4\pi\rho\beta^3 R} \frac{1}{\left[1+(\omega/\omega_c)^2\right]} \qquad (2)$$

where, P is the partition factor to represent one horizontal component, F is the free surface amplification factor, $R^{\theta\phi}$ is the spectral average for radiation pattern, m_o is the seismic moment of a sub-fault, ω_c is the corner frequency, ρ is the density in the vicinity of the source in g/cm³, β is the shear wave velocity in km/sec and R is the epicentral distance in km. In the simulation of ground motion for Delhi region in the present study, the values of different parameters are adopted (Singh et al., 2002) as $P=1/\sqrt{2.0}$; F=2.0; $R^{\theta\phi}$ = 0.55; ρ=2.85 gm/cc and β = 3.6 km/sec.

The moment magnitude (M_w) which defines the size of earthquake is related (Hanks & Kanamori 1979) to seismic moment (M_o) of the earthquake by,

$$M_w = 0.67 \log M_o - 10.7 \tag{3}$$

The rupture area (A) and sub-fault length (Δl) corresponding to a moment magnitude of earthquake can be calculated from empirical equations (Beresnev & Atkinson, 1998) as follows,

$$\log A = M_w - 4.0 \tag{4}$$

$$\log \Delta l = -2 + 0.4 M_w \tag{5}$$

For a sub-fault of equal dimensions ($\Delta w = \Delta l$, Δw, Δl being the dimensions of the sub-fault) the seismic moment of a sub-fault, m_o is given by

$$m_o = \Delta\sigma \Delta l^3 \tag{6}$$

where $\Delta\sigma$ is the stress parameter (Beresnev & Atkinson, 1998). The number of sub sources (N_{sub}) to be summed to reach the target seismic moment (M_o) is given by

$$N_{sub} = \frac{M_o}{m_o} \tag{7}$$

The corner frequency ω_c governs the acceleration amplitude and controls the frequency content of the earthquake at source is given by,

$$\omega_c = \frac{2 y_r z_s \beta}{\Delta l} \tag{8}$$

where, y_r is the constant representing the ratio of rupture velocity to shear wave velocity of source which is set to a value of 0.8 by Beresnev and Atkinson (1997), z_s is the parameter indicating maximum rate of slip also known as strength factor. The value of z_s may vary from 0.5 to 2.0 and in the present study a value of 1.4 (Singh et al., 2002) has been adopted for the simulation of earthquake motions for Delhi region.

2.2.2 Filter function for high frequency attenuation, P(ω)

In order to account for the high frequency attenuation by the near-surface weathered layer either a fourth order Butterworth filter with cut off frequency $\omega_m = 2\pi f_{max}$ or a spectral decay parameter kappa (κ) is widely used in stochastic models. In the present study, Butter worth filter function $P(\omega)$ with cutoff frequency f_{max} = 15 Hz (Singh et al., 2002) has been adopted.

$$P(\omega) = \left[1 + (\omega / \omega_m)^8\right]^{-1/2} \tag{9}$$

2.2.3 Geometric attenuation factor, G(R)

Geometric attenuation accounts for the decay and type of seismic waves. According to Singh et al. (2002) and Herrmann and Kijko (1983) for a distance of twice the crustal thickness the body waves dominate (direct seismic shear waves) and after that surface waves dominate (reflected L_g waves). Depending on the earth's crust thickness tri-linear or bilinear relationships are used for the calculation of G. The thickness of crust near Delhi has been reported to be 45 - 50 km (Tewari & Kumar, 2003) and bilinear relationship (Eq. 10) is adopted in the present study.

$$G(R) = \frac{1}{R} \qquad R<100 \text{ km} \tag{10a}$$

$$G(R) = \frac{1}{\sqrt{100R}} \qquad R>100 \text{ km} \tag{10b}$$

2.2.4 Anelastic whole path attenuation factor, $A_n(\omega)$

The anelastic whole path attenuation factor $A_n(\omega)$ which represents wave energy loss due to radiation damping of rock is accounted by this factor $A_n(\omega)$,

$$A_n(\omega) = e^{-\omega R / 2Q\beta} \tag{11}$$

where Q is the quality factor. The Q factor depends on the wave transmission quality of rock. For Himalayan arc region Q factor has been estimated by Singh et al. (2002) from the available earthquake records as follows.

$$Q(f) = 508 f^{0.48} \tag{12}$$

where, f is the frequency in Hz.

2.2.5 Simulation of time history

The Fourier amplitude spectrum derived from the section above gives the frequency content of the earthquake ground motion. This frequency information is combined with random phase angles in a stochastic process to generate artificial ground motion (Boore, 1983) for each sub-fault. Simulations from each sub-fault are lagged and summed to get the time history of earthquake.

Duration of the sub-fault time window, T_w is represented as the sum of its source duration, T_s and distance dependant duration, T_d (Beresnev & Atkinson, 1997; Boore 2003).

$$T_w = T_s + T_d \tag{13}$$

In references Beresnev and Atkinson (1997) and Boore (2003), T_s is taken as proportional to inverse of the corner frequency ($1/f_c$) and T_d is taken as $0.05R$.

Finite fault simulation program (FINSIM) has been widely used for the generation of ground motions of large size earthquakes (Atkinson & Beresnev, 2002; Beresnev & Atkinson 1998; Roumelioti & Beresnev 2003; Singh et al., 2002) and hence has been adopted in the present study.

The seismological parameters (Table 1) used in the generation of rock outcrop motions for Delhi region have been broadly adopted from Singh et al. (2002). In order to minimize the noise due to random fault rupture in the simulation, 15 ground motions have been generated for each earthquake magnitude. One of the simulations of the time histories generated for rock outcrop (Ridge observatory) in the present study have been compared (Fig. 2), with one simulation obtained from Singh (2005), for each of the magnitudes 7.5, 8.0, 8.5.

Parameters	M_w = 7.5	M_w = 8.0	M_w = 8.5
Fault orientation	Strike 300° Dip 7°	Strike 300° Dip 7°	Strike 300° Dip 7°
Fault dimension along strike and dip (km)	56 by 56	125 by 80	240 by 80
Depth of focus (km)	11	16	16
Stress parameter (bars)	50	50	50
No. of sub-faults	5x5	8x5	16x5
No. of sub-sources summed	28	57	339
Duration Model	$1/f_c$ +0.05R	$1/f_c$ +0.05R	$1/f_c$ +0.05R
Quality factor	$508f^{0.48}$	$508f^{0.48}$	$508f^{0.48}$
Windowing function	Saragoni-Hart	Saragoni-Hart	Saragoni-Hart
f_{max} (Hz)	15	15	15
Crustal shear wave velocity (km/sec)	3.6	3.6	3.6
Crustal density (kN/m^3)	2.8	2.8	2.8
Radiation strength factor	1.4	1.4	1.4

Table 1. Seismological parameters for strong motion generation

3. Propagation of strong ground motion through soil layer using one dimensional equivalent linear analysis

One dimensional equivalent linear vertical wave propagation analysis is the widely used numerical procedure for modeling soil amplification problem (Idriss, 1990; Yoshida et al., 2002). In one dimensional wave propagation analysis, soil deposit is assumed to be having number of horizontal layers with different shear modulus (G), damping (ξ) and unit weight (ρ). In the linear analysis, G and ξ are assumed to be constant in each layer. Since the soil will be subjected to nonlinear strain (Yoshida et al., 2002) even under small earthquake excitation equivalent linear analysis is preferred over linear analysis and the equivalent linear analysis program SHAKE (Ordonez, 2000, Schnabel et al., 1972) is used in the present study. Equivalent linear modulus reduction (G/G_{max}) and damping ratio (ζ) curves generated from laboratory test results are adopted from Vucetic and Dobry (1991) depending on the plasticity index of different soil layers. Since SHAKE is a total stress analysis program (Schnabel et al., 1972) depth of water table has not been considered in the analysis.

3.1 Typical soil strata for Delhi region

Three actual soil sites designated as site 1, site 2 and site 3 located in Delhi as shown in Fig. 3 are chosen in the present study. The layer wise soil characteristics (medium type) and the depth to the base of the layer from the surface is given in Tables 2 to 4. The shear wave velocity, V_s measurements are not available for the sites chosen. However the variations of N values with depth are available from the geotechnical data as given in Tables 2 to 4. The variation of shear wave velocity along the depth in the present study is obtained by using the correlations suggested for Delhi region by Rao and Ramana (2004) as given in eq. 14.

$$V_s = 79\, N^{0.43} \quad \text{(sand)} \tag{14a}$$

$$V_s = 86\, N^{0.42} \quad \text{(silty sand/sandy silt)} \tag{14b}$$

Layer No.	Description	Thickness (m)	Depth to the bottom of each layer (m)	SPT (N values)	Shear wave velocity v_s (m/sec)	Plasticity Index (%)	Total unit weight (kN/m³)
1	Sandy silt of low plasticity	3.5	3.5	13	252.55	7	16.3
2		1.5	5	17	282.67		
3		1.5	6.5	20	302.64		
4		1.5	8	23	320.94		
5	Sandy silt of low Plasticity	1.5	9.5	28	348.58	7	16.9
6		1.5	11	32	368.69		
7		1.5	12.5	35	382.83		
8		1.5	14	37	391.87	6	18.1
9		1.5	15.5	42	413.30		18.5
10		1.5	17	47	433.29		18.5
11	Rock	-		-	1500	-	24.0

Table 2. Geotechnical profile at Site 1

Layer No.	Description	Thickness (m)	Depth to the bottom of each layer (m)	SPT (N values)	Shear wave velocity v_s (m/sec)	Plasticity Index (%)	Total unit weight (kN/m³)
1	Clayey silt of low plasticity	1.5	1.5	9	216.41	11	16.9
2		1.5	3.0	9	216.41	15	17.4
3	Sandy silt	1.5	4.5	12	229.97	Non Plastic	17.4
4	Fine sand	1.5	6.0	12	229.97		17.2
5		1.5	7.5	12	229.97		17.1
6		1.5	9.0	13	238.03		17.1
7		1.5	10.5	15	253.13		17.1
8		1.5	12.0	19	280.21		17.1
9		1.5	13.5	20	286.46		17.7
10		1.5	15.0	21	292.54		17.7
11		1.5	16.5	26	320.68		17.7
12	Sandy silt of low plasticity	1.5	18.0	31	363.81		17.7
13		1.5	19.5	41	409.14		17.7
14		1.5	21.0	41	409.14	6	19.8
15		1.5	22.5	41	409.14		19.8
16	Rock	-		-	1500	-	24.0

Table 3. Geotechnical profile at Site 2

Layer No.	Description	Thickness (m)	Depth to the bottom of each layer (m)	SPT (N values)	Shear wave velocity v_s (m/sec)	Plasticity Index (%)	Total unit weight (kN/m³)
1	Sandy silt	3.5	3.5	5	169.07	Non Plastic	16.3
2		1.5	5.0	6	182.52		16.3
3		1.5	6.5	7	194.73		16.3
4		1.5	8.0	9	216.41		17.1
5		1.5	9.5	11	235.44		17.1
6		1.5	11.0	14	260.54		17.1
7		1.5	12.5	13	252.55		17.4
8		1.5	14.0	27	343.30		17.4
9	Clayey silt	1.5	15.5	36	387.39	15	17.7
10		1.5	17.0	32	368.69	15	17.7
11		1.5	18.5	13	252.55	15	17.7
12		1.5	20.0	28	348.58	15	17.7
13	Sandy silt	1.5	21.5	45	425.45	Non Plastic	18.1
14		1.5	23.0	28	348.58		18.1
15		1.5	24.5	42	413.30		18.1
16		1.5	26.0	44	421.45		18.5
17		1.5	27.5	47	433.29		18.5
18		1.5	29.0	More than 50	444.70		18.5
19		1.5	30.5		444.70		19.8
20		1.5	32.0		444.70		19.8
21		1.5	33.5		444.70		19.8
22		1.5	35.0		444.70		19.8
23		1.5	36.5		444.70		19.8
24		1.5	38.0		444.70		19.8
25	Rock	-			1500	-	24.0

Table 4. Geotechnical profile at Site 3

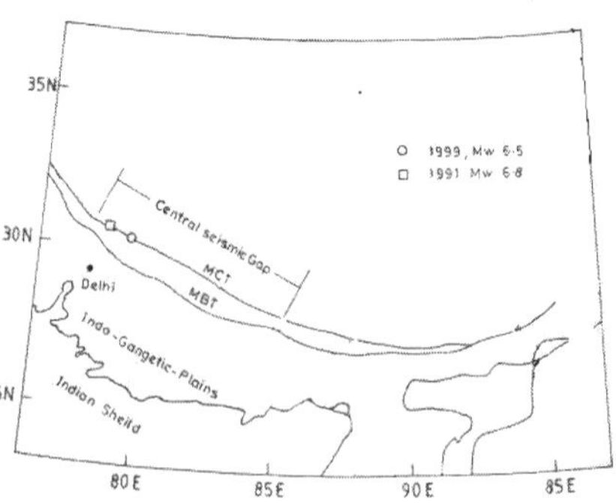

Fig. 1. Central seismic gap of Himalayan region

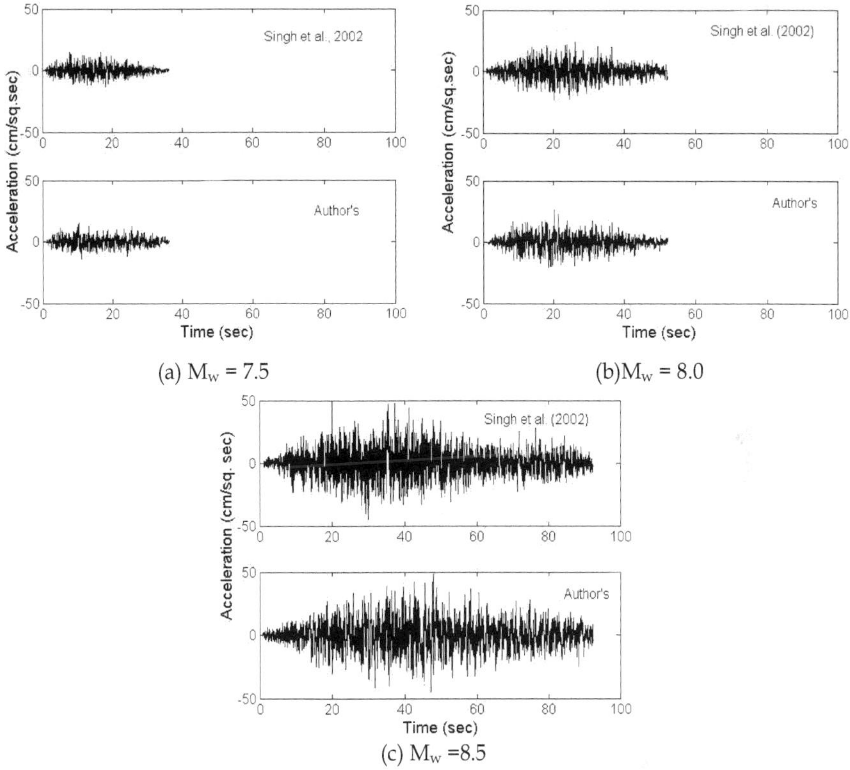

Fig. 2. Comparison of artificial ground motions generated for a rock site at Delhi for earthquakes from central seismic gap with similar generations from literature

Fig. 3. Location of three soil sites of Delhi region

4. Response of three sites for the scenario earthquakes

The amplification ratios and response spectra are the engineering outputs required to calculate the response of buildings for site-specific analysis. Shear wave velocity of bedrock (quartzite) at Delhi is reported to be in the range of 1000 m/s to 2000 m/s (Parvez et al., 2004). The rock outcrop motions simulated as per the procedure given in section 2.2 are considered as bedrock motions for soil sites and the same are propagated through the soil strata of the three sites and the free field motions are obtained. As a typical case the bedrock level and free field motions at the top of three sites for one simulation of earthquake (M_W = 7.5) for the three magnitudes are shown in Fig. 4.

The variations of average amplification ratios of 15 earthquake simulations for the three sites are obtained. As a typical case, variations of average amplification ratios for M_W = 8.5 earthquake is shown in Fig. 5 for the three earthquake magnitudes. It can be seen from the results that the peak amplification ratios as well as the frequencies at which the peak amplification ratios occur are quite different for the three sites. The fundamental time periods of the three sites for different earthquakes are given in Table 5. Difference in site period for larger magnitude earthquake is due to nonlinear response of soil sites for higher magnitudes.

M_W	**Site 1**	**Site 2**	**Site 3**
7.5	0.18	0.31	0.38
8.0	0.19	0.31	0.38
8.5	0.19	0.35	0.42

Table 5. Fundamental time periods of the three sites in seconds

The average ratios of PGA of free field motion to the PGA of bedrock motions for the three sites are shown in Table 6. Also shown in Table 6 are the average peak ground accelerations (PGA) for the 15 simulations of bed rock motions and free field motions for site 1, site 2 and site 3. It can be observed that the PGA amplifications of the three sites are different for the three magnitudes of earthquake. Response spectra (5% damping) for the 15 simulations of free field motions and their average, on top of three sites have been obtained for all the three earthquake magnitudes. Typically for M_w = 7.5, these have been shown for site 2 in Fig. 6. Further, the comparison of average response spectra for the three sites for the earthquake magnitudes M_w = 7.5, M_w = 8.0, M_w = 8.5 are shown in Fig. 7. From the comparisons it can be inferred that the shapes of the response spectra vary quite significantly for the three sites under the same earthquake.

M_W	**Average PGA (cm/sec^2)**				**Average PGA amplification**		
	Bedrock motions	**Free field motions**					
		Site 1	**Site 2**	**Site 3**	**Site 1**	**Site 2**	**Site 3**
7.5	15.74	31.73	51.73	39.60	2.02	3.29	2.52
8.0	23.36	53.48	53.10	60.10	2.29	2.27	2.57
8.5	46.47	100.82	100.01	113.32	2.17	2.15	2.44

Table 6. Average PGA of bedrock motion, free field motions and average PGA amplification of the three sites

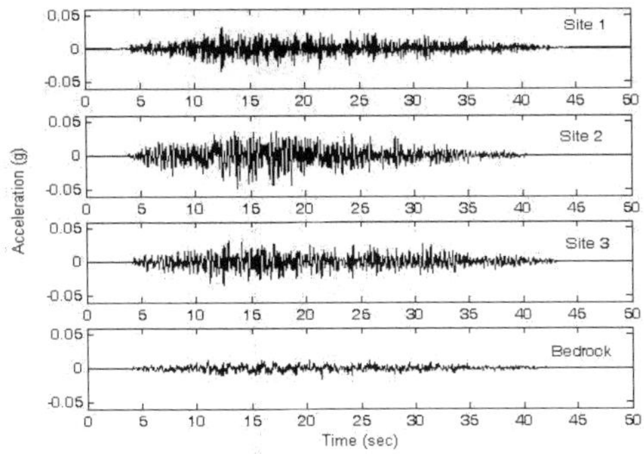

Fig. 4. Bedrock level and free field motions at the top of three sites for one simulation of earthquake, M_w =7.5

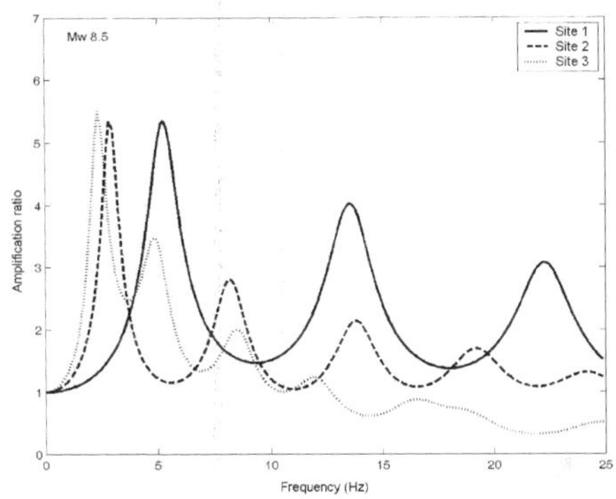

Fig. 5. Variations of average amplification ratios for the three sites, M_w = 8.5

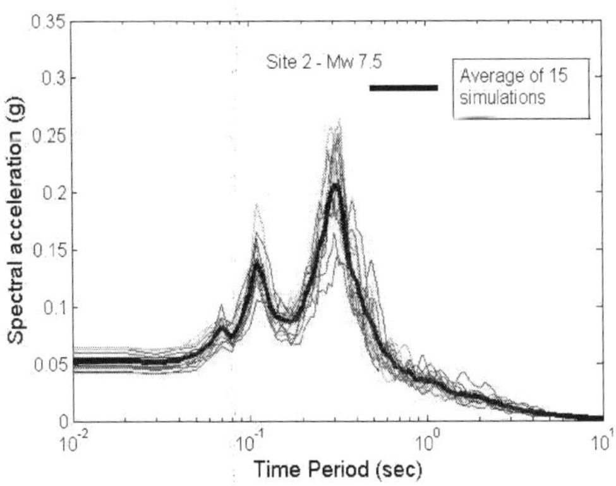

Fig. 6. Response spectra (5% damping) for the 15 simulations of free field motions and their average, M_w =7.5; Site 2

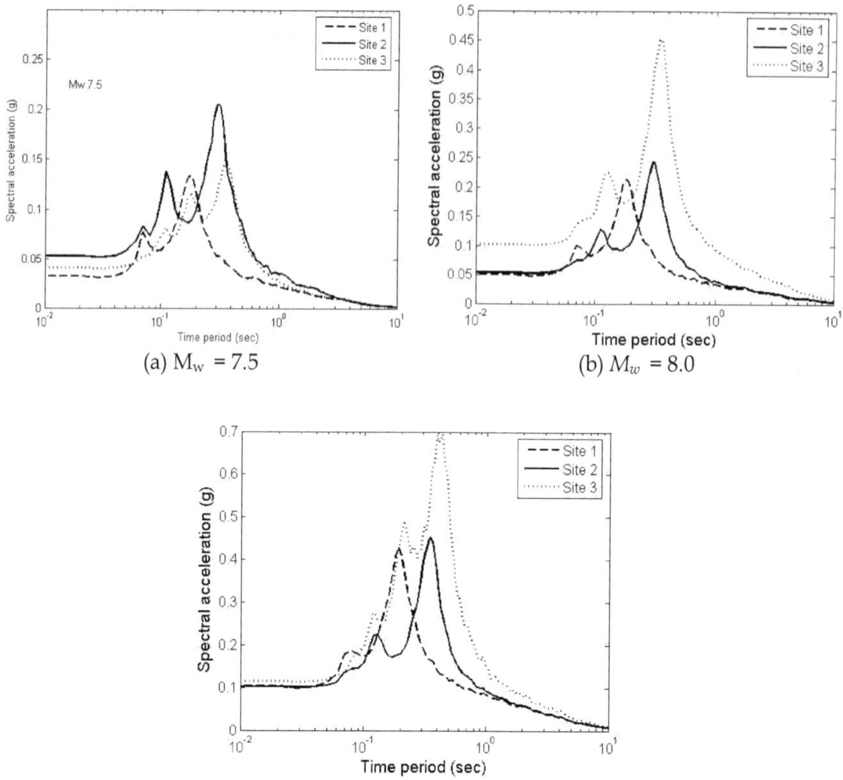

(a) M_W = 7.5

(b) M_w = 8.0

(c) M_w = 8.5

Fig. 7. Comparison of average response spectra for the three sites

5. Response of buildings on the three sites

It is clearly seen from the comparison of response spectra, that buildings situated on the three sites will be subjected to different force levels during the same earthquake. In order to demonstrate the site-specific response of buildings, a three storey building and a fifteen storey building designated as B1 and B2 with plan details as shown in Fig. 8 are chosen for the present study. The buildings are assumed to be situated on the three soil sites (Fig. 3) chosen at Delhi. The earthquake is applied in y direction. Both the buildings are assumed to be having frames as stiffening elements with uniform beam and column sections along the height of the building. The beams are assumed to be axially rigid and have infinite flexural rigidity. All the columns are square and are assumed to be axially rigid. The structural properties of buildings are given in Table 7. The storey shears have been obtained by response spectrum method as per IS 1893(Part 1)-2002, (2002). In the evaluation of storey shears response reduction factor has been taken equal to one.

Building	Moment of inertia of square columns 10^{-4} (m^4)		Storey height (m)	No. of stories	Mass of all the floors except top floor (kN-sec²/m)	Mass of the top floor (kN-sec²/m)
B1	11.1 (Frames 1,8)	21.0 (Frames 2,3,4,5,6,7)	3.0	3	410	205
B2	108.0 (Frames 1,4)	108.0 (Frames 2,3)	3.5	15	500	300

Table 7. Structural properties of the buildings

The comparison of storey shears for buildings B1 and B2 for site-specific earthquakes and storey shears obtained by considering the three sites as medium soil sites(MS) for design basis earthquake(DBE) as per seismic code IS 1893 (Part 1)-2002, (2002) for Delhi are shown in Fig. 9. For both the buildings storey shears obtained as per IS 1893 (Part 1)-2002, (2002) are different from the storey shears from site-specific analysis. Comparison of displacement responses for B1 and B2 are given in Fig. 10. It may be noted that, larger variation in base shear and displacement response for different sites for B1 is observed due to the proximity of fundamental time period (0.3 sec) of B1 to the site periods (Table 5).

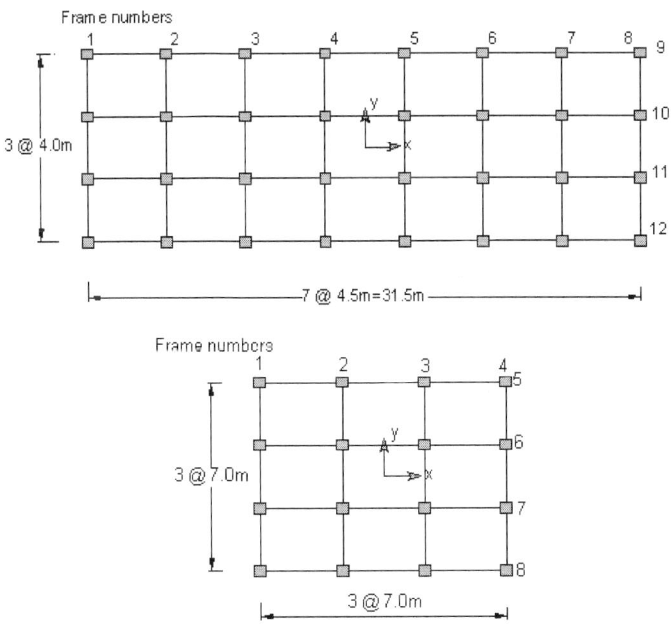

Fig. 8. Plan of buildings B1, B2

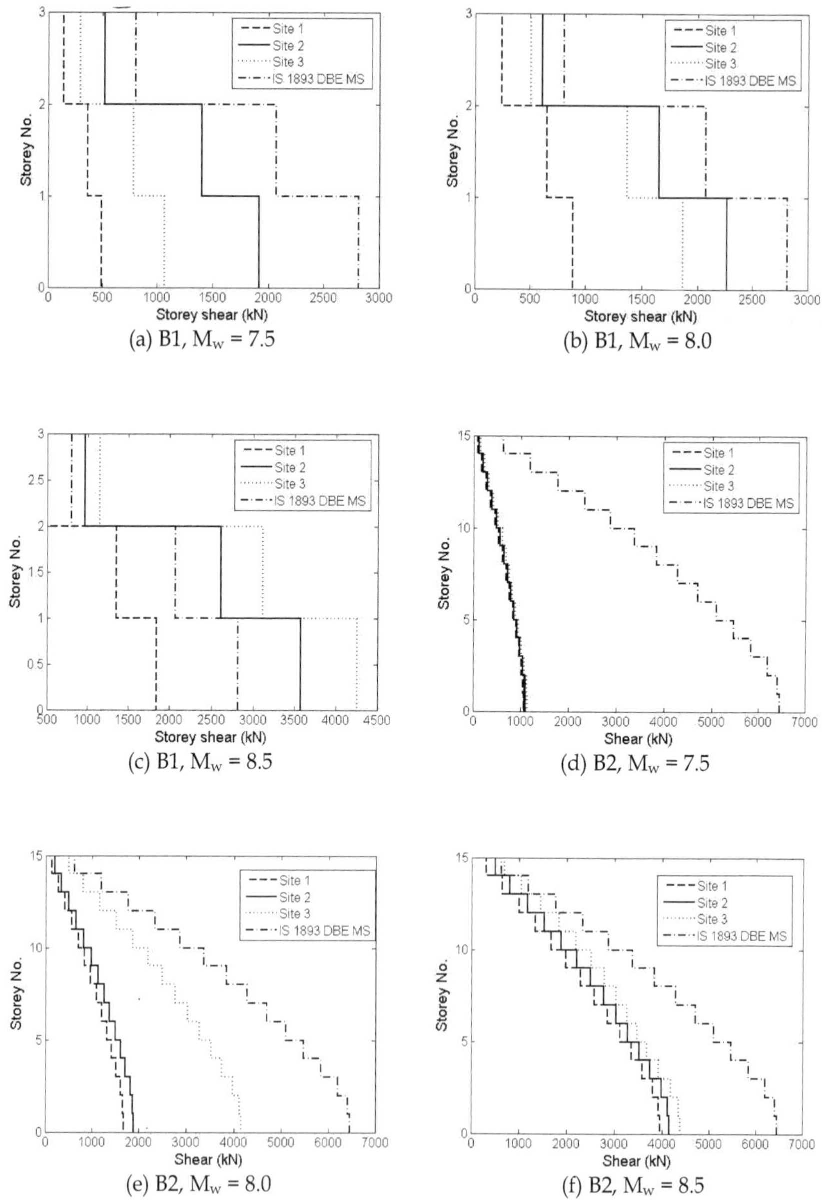

Fig. 9. Comparison of elastic storey shear of B1 and B2 situated on three sites at Delhi

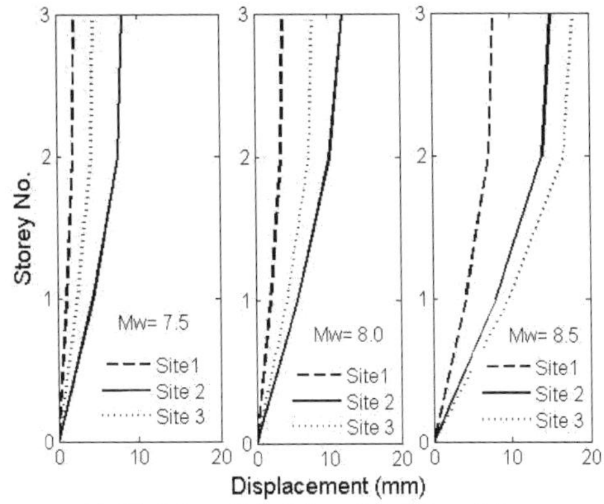

(a) B1, $M_w = 7.5$, $M_w = 8.0$, $M_w = 8.5$

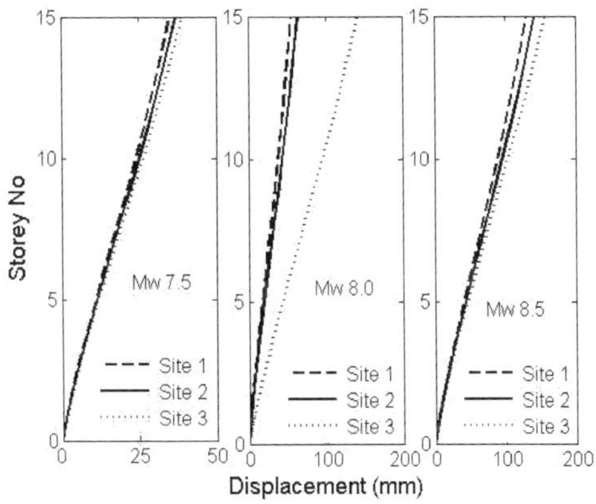

(b) B2, $M_w = 7.5$, $M_w = 8.0$, $M_w = 8.5$

Fig. 10. Comparison of elastic displacement response of B1 and B2 situated on the three sites at Delhi

6. Performance evaluation of RC framed building for site-specific earthquake

Inelastic response of buildings plays vital role in earthquake resistant design and performance based procedures aim to evaluate the inelastic response of building. After establishing the procedure and studying the elastic response of framed building for site-specific scenario earthquake, in this section the procedure to determine the inelastic response of building for site-specific earthquake is illustrated in performance based environment.

Two important elements of seismic performance evaluation of buildings are demand spectrum and capacity spectrum. Demand spectrum is the representation of the severity of the ground motion while capacity spectrum depicts the ability of the structure to withstand forces of specific nature. While carrying out performance evaluation for site-specific earthquake, code based response spectrum needs to be replaced with site-specific spectrum and the same will be considered as demand spectrum. Capacity spectrum method (ATC 40, 1996; FEMA 273&274, 1996) has provisions to modify a demand spectrum to account for lengthening of the period or increase in the damping of the structure.

The average response spectra for the three sites are represented in Acceleration Displacement Response Spectra (ADRS) format in Fig. 11. It is seen that for the same time period of the building the spectral acceleration and spectral displacement are different for the three sites. This clearly indicates that the same building will be subjected to different levels of damage due to the difference in elastic or inelastic displacement experienced by the building for the same earthquake when situated on different sites for Delhi region.

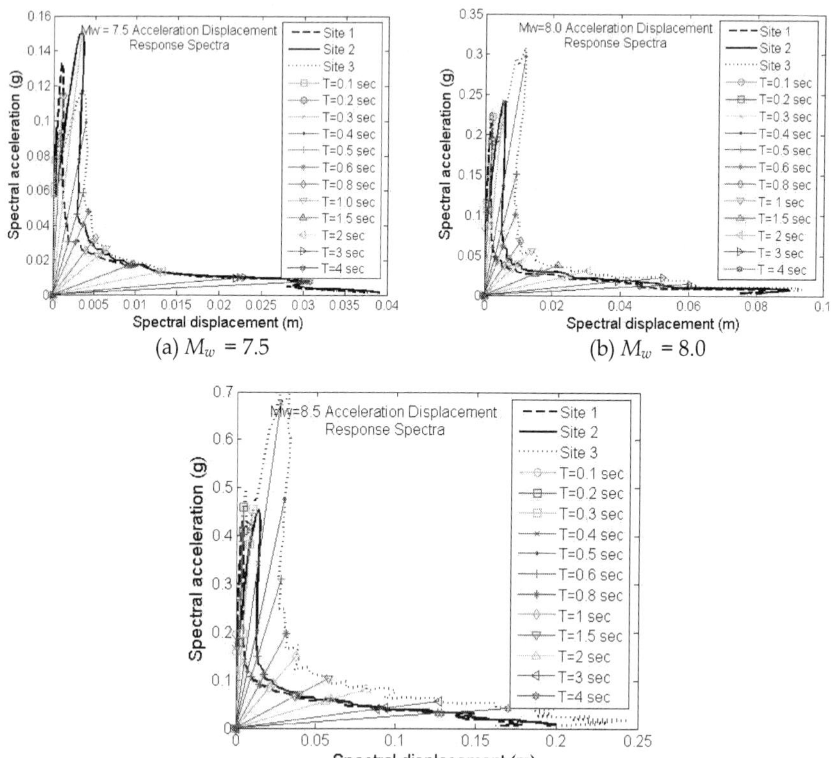

(a) M_w = 7.5 (b) M_w = 8.0 (c) M_w = 8.5

Fig. 11. Comparison of response spectra for three sites in ADRS format

6.1 Generation of depth dependant site-specific demand spectrum

For carrying out site-specific performance evaluation of buildings for Delhi city, a multi-storey building is assumed to be located on different depths of soil strata at Delhi. Here, instead of considering only individual sites, in the absence of complete borelog details for greater depths, representative soil sites of different depths are considered. The thickness of alluvium above the bedrock at Delhi varies significantly and according to a report by Central Ground Water Board (CGWB) (2002), variation is from less than 50 m to more than 300 m (Fig. 12). In the present study 8 representative soil strata defined by depths 10m, 20m, 30m, 50m, 75m, 100m, 150m and 200m have been chosen. Shear wave velocity, modulus reduction curve and damping curve are the other important properties that influence the modification of ground motion through soil layer. For shear wave velocity, regression relations (Eq. 15) have been suggested by Satyam (2006) based on seismic refraction and MASW tests. Based on the measured values, Delhi has been divided into three regions (Fig. 13) viz., (i) south and south central Delhi, (ii) north and north western Delhi and (iii) trans Yamuna Delhi. Separate shear wave velocity models viz., V_{s1}, V_{s2}, V_{s3} has been proposed for each region as given in equation 15.

$$V_{s1} = 281\, D_s^{\,0.08} \text{ (South and South Central Delhi)} \tag{15a}$$

$$V_{s2} = 217\, D_s^{\,0.13} \text{ (West and North Western Delhi)} \tag{15b}$$

$$V_{s3} = 140\, D_s^{\,0.24} \text{ (Trans Yamuna, Delhi)} \tag{15c}$$

where D_s is the depth of soil stratum below the ground level in m.
However, the model V_{s3} only has been considered for the performance evaluation of building in the present study. For dynamic characteristics, from the large number of borelog data available for Delhi region, it is observed that the Plasticity Index (PI) of soils at Delhi region varies from 0% to 15%. Modulus reduction curves and damping curves for Delhi soil corresponding to PI=0% (Non plastic), PI=15% (low plasticity) soil have been adopted as explained in section 3 from Vucetic and Dobry (1991). For rock, modulus reduction curves and damping curves have been chosen from Schnabel and Seed (1972).

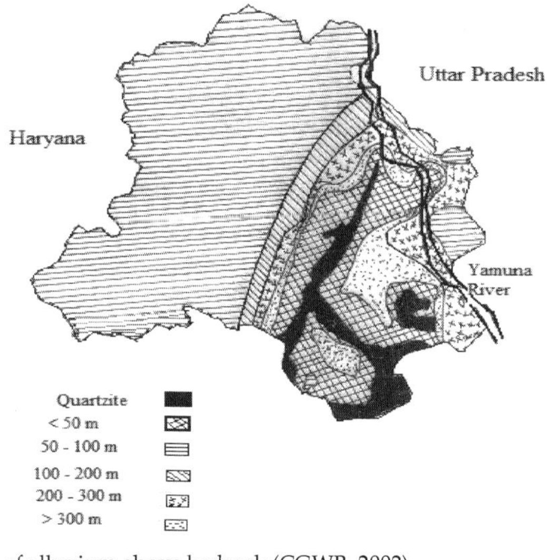

Fig. 12. Thickness of alluvium above bedrock (CGWB, 2002)

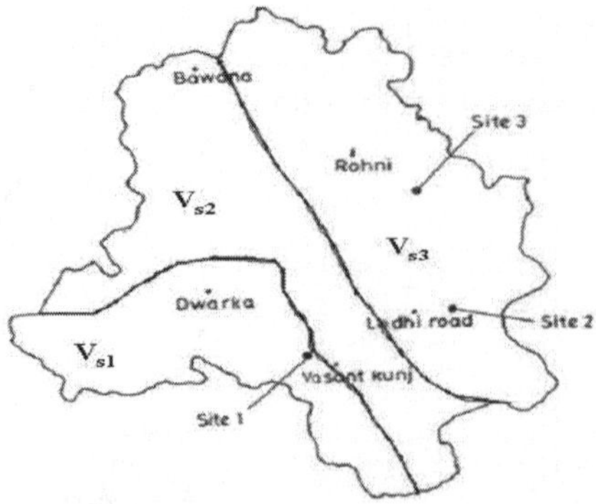

Fig. 13. Three zones of Delhi based on surface geology

Artificial ground motions are generated for rock outcrop site for earthquake magnitude M_w = 8.5 with the parameters given in Table 1, and further propagated through different depths of soil stratum. Ground motions are obtained at the top of representative soil sites by conducting equivalent linear one dimensional wave propagation analysis as explained earlier, using the program SHAKE2000 (Ordonez, 2000). The time periods of the 8 different soil stratum depths considered in the present study are given in Table 8. Using the surface ground motions, the average Depth Dependent Response Spectra (DDRS) of 15 random simulations of the ground motions at soil surface have been obtained corresponding to 2%, 5%, 10%, 15%, 20% and 25% damping.

Sl. No.	**Depth of soil stratum (m)**	**Time period (sec)**
1.	10	0.19
2.	20	0.34
3.	30	0.47
4.	50	0.67
5.	75	0.9
6.	100	1.2
7.	150	1.6
8.	200	2.0

Table 8. Time periods of soil strata

After carrying out the site-specific analysis the response spectra of a site is available in the standard spectral acceleration (S_a) versus time period (T) format and the same is converted to Acceleration-Displacement Response Spectra (ADRS) format using eq. 16.

$$S_{di} = \frac{T_i^{\ 2}}{4\pi^2} S_{ai} g \tag{16}$$

where:
S_{di} = spectral displacement ordinate in m
S_{ai} = spectral acceleration ordinate in units of g

T_i = time period of the building in secs
g = acceleration due to gravity in m/s^2
i = i^{th} point of the spectra

6.2 Generation of capacity spectrum from capacity curve through nonlinear static analysis

Plan of an eight storey building chosen (designated as B3) for the present study is shown in Fig. 14. Overall length and width of the building are 11.4m and 10.9m, respectively. Height of the building is 23.6m. Cross section and reinforcement details of the beams and columns are modeled as given in the construction drawings of the building. The total lumped mass due to dead and participating live loads of the building for the bottom six stories is equal to 179.2 tons while the lumped mass for seventh and eighth stories is equal to 90.1 tons and 17.9 tons, respectively. Building is modeled using SAP2000 computer program (2004), with default PMM hinge properties for column and default M3 properties for beam. Displacement controlled nonlinear static pushover analysis has been carried out for the 3D building model and the capacity curve of the building is obtained. Further, the capacity curve is transformed to capacity spectrum using Eqns. 17 and 18.

$$S_{aj} = \frac{V_j / W}{\alpha_1} \tag{17}$$

$$S_{dj} = \frac{\Delta_{roof}}{PF_1 \cdot \varphi_{1,roof}} \tag{18}$$

where:
V_j = base shear at the jth point of the capacity curve
W = weight of the building as sum of dead load and percentage live load
α_1 = modal mass coefficient for the first natural mode
Δ_{roof} = roof displacement
PF_1 = modal participation factor for the first natural mode
$\phi_{1,roof}$ = amplitude at roof level in first natural mode

6.3 Determination of performance point

In the present study, site-specific demand spectrum for 2%, 5%, 10%, 15%, 20% and 25% damping are obtained for the building. According to ATC 40 (1996), effective damping (β_{eff}) of the building during earthquake excitation is combination of viscous damping that is inherent in the building (about 5%) and hysteretic damping (β_o) (that is related to the area inside the hysteretic loops formed when the earthquake force is plotted against the structural displacement). In view of this, it is required to modify the demand spectrum to account for the effective damping of the structure. An iterative method as suggested by authors (Kamatchi et al., 2010[a]) is used to determine the performance point, wherein, demand spectrum has to be updated in each iterative cycle till convergence is achieved. In the process, effective damping is obtained as per the procedure suggested in ATC 40 (1996).

6.3.1 Performance points for the chosen building

The capacity and demand curves for the eight different depths of soil stratum are obtained for building and shown in Fig. 15. For the soil stratum depths of 10m, 20m, 30m, 50m and 200m, the 5% demand curve intersect the capacity curve in the elastic response region. For the other depths (75m, 100m and 150m), the intersection points are found to lie in inelastic response region. For these three depths, spectral reduction factors are applied to 5% demand spectra and the performance points are obtained as per ATC 40 (1996). This is carried out through number of trials as shown in Fig. 16. The trial performance points are arrived by using effective damping (β_{eff}), Spectral reduction factor for acceleration predominant region (SR_A) and velocity predominant region (SR_v) corresponding to soil stratum depths of 75m, 100m and 150m. The base shear (V_b) and roof displacements(Δ_{inel}) corresponding to the final performance points of the building for 75m, 100m and 150m depths of soil stratum are compared with corresponding values obtained for DBE earthquake and Medium soil site conditions as per IS 1893(Part 1)-2002 (2002) in Table 9.

Depth of soil stratum (m)	V_b (kN)			Δ_{inel} (m)		
	Site-specific	DBE	% difference	Site-specific	DBE	% difference
75	1235.91	928.56	33.10	0.071	0.049	44.90
100	1381.7	928.56	48.80	0.093	0.049	89.80
150	1338.05	928.56	44.10	0.077	0.049	57.14

Table 9. Comparison of V_b and Δ_{inel} for reduced spectra with DBE

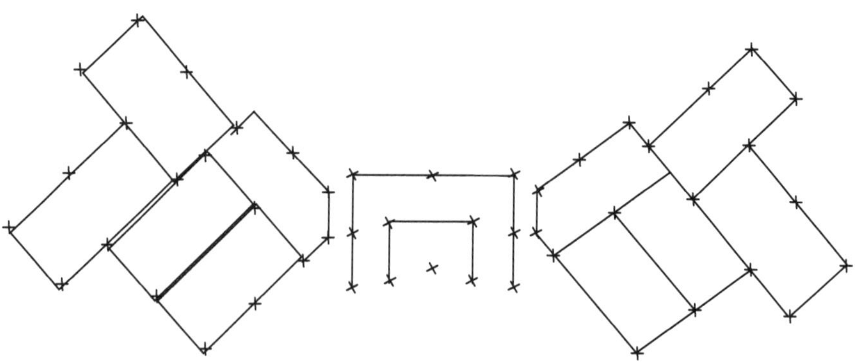

Fig. 14. Plan of the building, B3

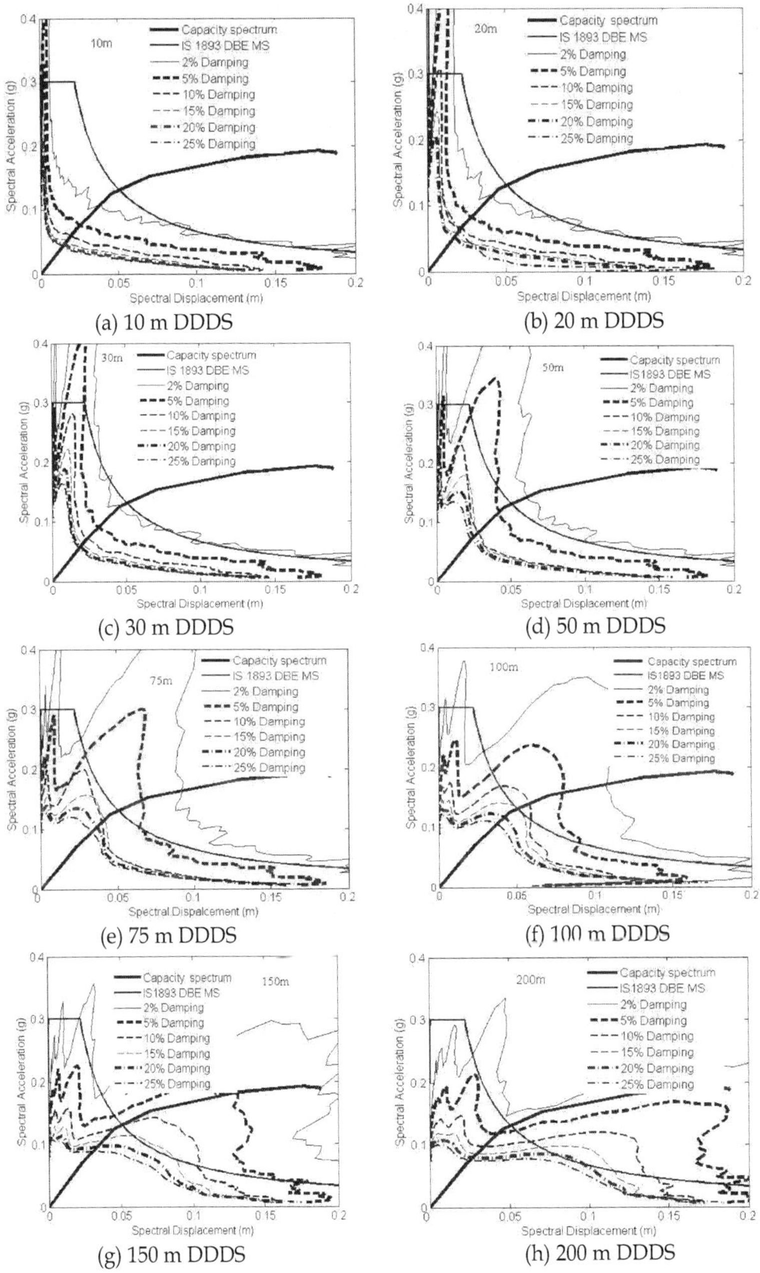

Fig. 15. Capacity and Demand spectra for different damping ratios

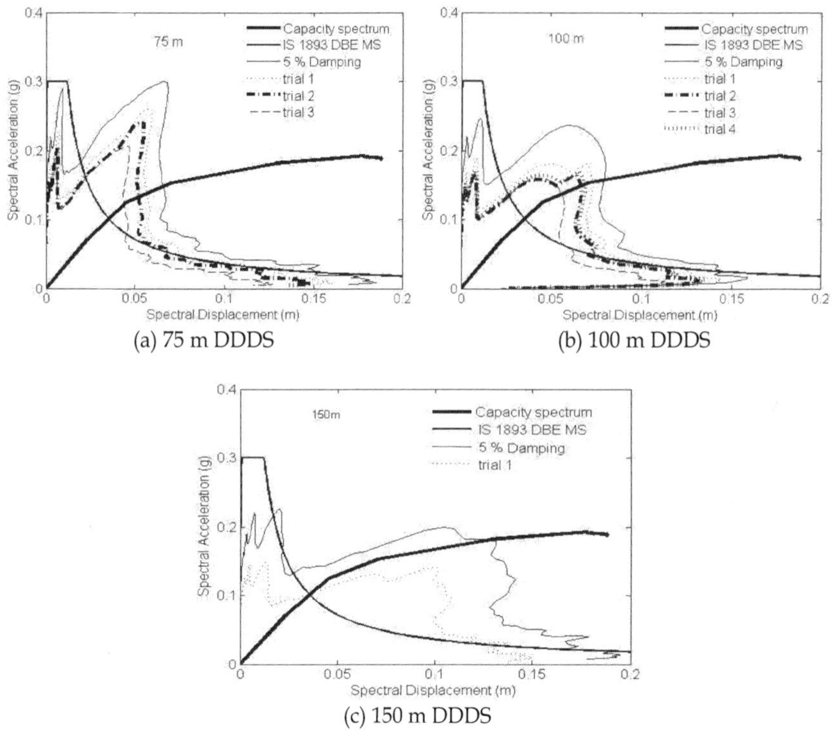

Fig. 16. Performance points using ATC 40 (1996) procedure for different soil stratum depths, DDDS and IS 1893(Part 1)-2002 medium soil demand spectra

7. Development of artificial neural networks for site-specific seismic analysis of buildings

In the earlier sections, procedure for carrying out site-specific analysis for scenario earthquakes is demonstrated with examples of RC framed buildings. It is accepted in literature, site-specific seismic analysis is mandatory for sites of specific soil (F type) (IBC 2009) and for earthquake-resistant design of important and critical structures. It is being insisted in this chapter that, even for other structures, and for medium soil, it is preferred to carry out a detailed site-specific analysis to arrive at the design force levels. Site-specific analysis requires considerable modelling and computational effort w.r.t. representing the generation as well as propagation of strong ground motion and it's effect on the structures. As an alternative, ANN models are generated in this chapter for the prediction of site-specific spectral accelerations.

Standard feed forward back propagation neural network algorithm with one hidden layer has been adopted to implement this. The main advantage of the neural network model lies in calculation of the realistic site-specific design base shear values without the need to generate strong ground motions and complex modelling of the soil profile.

7.1 Input and output parameters for ANN models

Initially the following parameters are identified as probable input parameters and the average spectral acceleration (S_a/g) is identified as the output parameter for sensitivity analysis.

Moment magnitude of the earthquake (M_w),
Shear wave velocity of rock half-space (V_{sr}),
Average plasticity index of the soil stratum (PI),
Average soil density of the soil stratum (γ),
Depth of the soil stratum above the bedrock (h),
Shear wave velocity model (V_{sm}),
Damping ratio of SDOF oscillator (ξ) and
Time period of the SDOF oscillator (T_b)

Based on the results of the sensitivity studies (Kamatchi, 2008; Kamatchi et al., 2010[b]) for the development of neural network models five input parameters viz., M_w, h, ξ, T_b, V_{sm} and one output parameter S_a/g are chosen.

7.2 Choosing the configuration of the neural networks

Multilayer neural network with neurons in all the layers and fully connected in a feed forward manner has been chosen for the present implementation. Sigmoid function (with output in the range of 0 to 1) is used for activation and the back propagation learning algorithm is used for training. The feed forward back propagation algorithm has been used successfully for many civil engineering applications and is considered as one of the most efficient algorithms for engineering applications (Adeli, 2001). One hidden layer has been chosen for the network and the number of neurons in the hidden layer is decided in the learning process by trial and error.

7.3 Training of neural network

As the number of data sets is large, for the sake of convenience in handling the data during training, it has been decided to have two neural networks. The soil stratum depths up to 75 m have been taken in the first network designated as NET1 with rest of the parameters assuming all the values in their respective ranges. Similarly, the second network designated as NET2 includes soil stratum depths up to 200 m and has number of patterns as that for NET1, i.e., 49815. The data has been randomly partitioned (Reich & Barai, 1999), with two third of the data sets being used for training and the remaining are used for testing. The number of sampling points for the input parameters and the number of data sets are shown in Table 10.

Input parameter					Number of data sets		
M_w	V_{sm}	ξ	h(m)	T_b (sec)	Total	Training	Testing
3	3	3	15	123	49815	35000	14815

Table 10. Number of sampling points for the input parameters and number of data sets for NET1 and NET2

Normalisation of input and output parameters is carried out and the factors are available in Table 11. Input parameters are normalised in linear sense while the output parameter (S_a/g) is normalised in logarithmic scale to accommodate the exceptionally large variation in its value. Further, a bias has been added to satisfy the mathematical validity. The training is carried out using the Stuttgart Neural Network Simulator (SNNS, 1999). For training, the number of neurons in the hidden layer has been varied and several trials have been carried out. Architecture of the network (5-14-1) with 5 neurons in the input layer, 14 neurons in the hidden layer and 1 neuron in the output layer has been found to predict the results with good accuracy for the network NET 1 and 5-15-1 architecture is found to predict good results for NET 2. The standard back propagation algorithm with learning rate (η) equal to 0.9 has been used for training. The Table 12 gives the mean square error at this stage. The percentage root mean square error of 14,815 test patterns is found to be 3.479% and 3.658% for the two networks NET1 and NET2, respectively. Using weights and biases of the trained neural networks, a simple program has been developed to arrive at the spectral acceleration values for everyday use in a design office.

Network	Normalization factors for parameters					
	Input					Output
	M_w	V_{sm}	ξ	h(m)	T_b(sec)	Log (S_a/g)
NET1	1/10	1/4	7	1/80	1/5	(+3.5)/4
NET2	1/10	1/4	4	1/210	1/5	(+3.0)/3

Table 11. Normalization factors used in training

Network	Number of neurons in layers			MSE		Epochs
	Input	Hidden	Output	Training	Testing	
NET1	5	14	1	0.00024	0.00027	44,000
NET2	5	15	1	0.00031	0.00033	65,000

Table 12. Configuration of networks, mean square errors and number of epochs

7.4 Checking the performance of the neural networks

Coefficient of correlation (R_c) (Kerh & Ting, 2005) often used to represent the degree of accuracy of a prediction by ANN is given by:

$$R_c = \frac{\sum_{i=1}^{n}(x_i - \bar{x})(y_i - \bar{y})}{\sqrt{\sum_{i=1}^{n}(x_i - \bar{x})^2 \sum_{i=1}^{n}(y_i - \bar{y})^2}} \tag{19}$$

where, x_i and $\bar{x}$ are targeted and its averaged values, y_i and $\bar{y}$ are estimated and its averaged values and n denotes the number of data in the analysis. Square of the coefficient of correlation of the networks NET1 and NET2 for both the training and test patterns are found to be more than 0.9 which guarantees better performance of network for any new input.

7.5 Estimation of base shear for framed buildings using ANN models

A six storey building designated as B4 (Fig. 17) and fifteen storey building, B2 with details as given in section 5 (Fig.8, Table 7), are considered for the validation of ANN models. Both the buildings are assumed to be situated in Delhi and analyzed for earthquakes having moment magnitude of 7.8 and 8.2 and originate from central seismic gap of Himalayan region. The building B4 is assumed to be at a location with shear wave velocity model V_{s2} and on a soil stratum of depth 52 m above the bedrock. The building B2 is assumed to be in a location with shear wave velocity model V_{s3} and on a soil stratum of depth 105 m above the bedrock. The first three time periods of buildings for B4, B2 are evaluated as T_{b4} =0.679sec, 0.23sec, 0.14sec and T_{b2} = 1.62sec, 0.491sec, 0.359sec respectively. The damping ratios for B4 and B2 are assumed to be 0.03 and 0.06 respectively for all the modes. The input values and the corresponding output S_a/g values predicted by neural network along with those obtained by a detailed SSA are given in Table 13. It may be noted that parameters of these validation examples have not been used in training or testing of the neural network.

Network	Building	Input values for network					S_a/g	
		M_w	V_{sm}	ξ	h(m)	T_b(sec)	ANN	SSA
NET1	B4	7.8	2	0.03	52	0.68	0.1449	0.1462
		7.8	2	0.03	52	0.23	0.1786	0.1163
		7.8	2	0.03	52	0.14	0.1291	0.0943
NET2	B2	8.2	3	0.08	105	1.63	0.0753	0.0753
		8.2	3	0.08	105	0.49	0.1123	0.1257
		8.2	3	0.08	105	0.36	0.1282	0.1324

Table 13. Input parameters and the output from the neural network

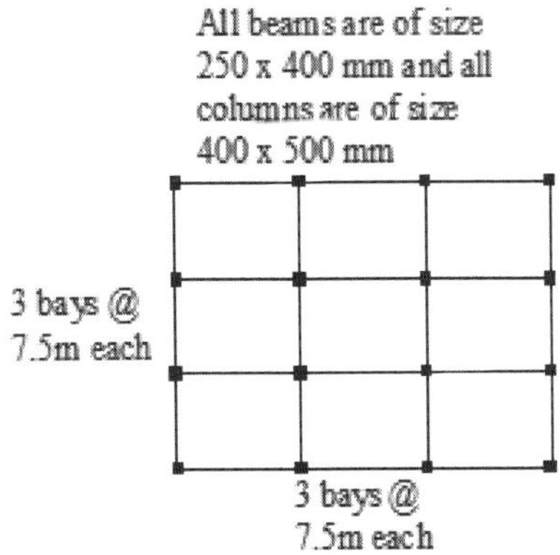

Storey height 3m

Fig. 17. Plan of building, B4

Storey level shears obtained for buildings B4 and B2 by Site-specific seismic analysis and those obtained using predicted S_a/g values from ANN are presented in Table 14 for comparison. The root Mean Square percentage Errors (MSE) in the predicted storey shears of buildings B4 & B2 are 1.43%, 4.3% respectively. The maximum percentage errors in the predicted storey shears for B4 & B2 are 2.99% and 4.98%, respectively. This has to be seen in the background that the time involved for the generation of strong motion for a given magnitude of earthquake, ground response analyses through soil layer and the computation of surface level spectral acceleration involves huge effort and enormous time and computational cost. The neural network based on the proposed methodology is, on the other hand, capable of predicting the spectral acceleration values at a fraction of these resources.

Network	Building	Storey No.	Storey shear (kN)		
			SSA	ANN	% error
NET1	B4	1	1317.9	1323.7	-0.44
		2	1232.4	1242.0	-0.78
		3	1081.3	1090.0	-0.81
		4	876.7	877.4	-0.09
		5	622.5	613.9	1.38
		6	317.9	308. 4	2.99
NET2	B2	1	1533.9	1609.5	-4.93
		2	1524.3	1599.5	-4.93
		3	1477. 3	1550.5	-4.96
		4	1395.0	1464.4	-4.98
		5	1306.	1370.8	-4.97
		6	1210.7	1270.2	-4.92
		7	1109.5	1163.0	-4.83
		8	1002.4	1049.4	-4.69
		9	889.6	929.7	-4.50
		10	770.9	804. 0	-4.28
		11	646.5	672.5	-4.02
		12	516.9	536. 1	-3.70
		13	383.1	395.6	-3.28
		14	247.3	253.5	-2.49
		15	115.5	115.0	0.38

Table 14. Comparison of storey shear

8. Summary

In this chapter, the importance and procedures to carry out site-specific seismic analysis of framed buildings for long distance large magnitude earthquakes including the effect of depth of soil stratum are illustrated for Delhi capital city of India, with scenario earthquakes from central seismic gap of Himalayan region as an example. Steps involved in carrying out site-specific seismic analysis for scenario earthquakes are discussed. Rock outcrop motions are generated for Delhi for the scenario earthquakes of magnitude, M_w = 7.5, M_w = 8.0 and M_w = 8.5. Three actual soil sites (medium soil type) are modeled and the free field surface motions and the response spectra are obtained. It has been observed that the PGA amplifications and the response spectra of the three sites are quite different for the earthquakes considered. It is clear from response of two RC framed building that the performance of the buildings will be different when situated on three different soil sites. From the studies made, it can be concluded that, it may be necessary to perform the site-specific analyses of buildings at sites having medium types of soil as well.

Having established the importance of carrying out site-specific seismic analysis for moderate sites and RC framed buildings, the procedure to carry out seismic performance evaluation of existing building for site-specific earthquake is demonstrated for an eight storey building assumed to be situated on different depths of soil stratum for Delhi region. Rock outcrop motions generated for earthquake of M_w = 8.5 are propagated through different depths of representative soil stratum and depth dependant demand spectrum are obtained. The effect of depth dependant response spectrum on the performance of building is studied for eight different depths of soil stratum above bedrock. From the studies made, it is clear that considering the design spectra suggested by seismic codes and only the top 30 m soil stratum to include the effects of soil amplification may not ensure safe seismic performance of a building. It is further seen that the site-specific earthquake and the depth of soil stratum have significant influence on the performance of the building both in terms of inelastic displacement as well as inelastic base shear.

Procedure to develop ANN models to rapidly estimate the site-specific spectral acceleration of structures is illustrated with Delhi as an example. From the results of sensitivity studies conducted earlier for Delhi city, moment magnitude of the earthquake, M_w, depth of the soil stratum above the bedrock, h, shear wave velocity model, V_{sm}, damping ratio, ξ and time period, T_b, of the SDOF oscillator are identified as governing input parameters for predicting the output of spectral acceleration, Sa/g.

Excellent performance of the trained neural networks has been demonstrated by the calculated values of coefficient of correlation. Trained neural networks are validated by using borelog data of an actual soil site. In addition, the neural networks have been validated for two different buildings assumed to be located in Delhi city. Performance of ANN models developed are checked and validated with number of examples. Validation results for two different buildings included in this chapter indicates that the root mean square percentage error is within tolerable values for both the buildings analysed in the present study.

The procedures suggested in this chapter is suitable for carrying out site-specific seismic analysis for framed buildings for any region for which information about seismic hazard in terms of scenario earthquake and the relevant geological and geotechnical details of the region are made available.

9. Acknowledgement

This chapter is published with kind permission of the Director, CSIR-Structural Engineering Research Centre. The first author sincerely acknowledges the support extended by Dr. J. Rajasankar, Senior Principal Scientist, Dr. K. Balaji Rao, Senior Principal Scientist, Dr. S. Arunachalam, Chief Scientist, Dr. N. Lakshmanan, Former Director, CSIR-Structural Engineering Research Centre in preparation of this chapter.

10. References

Adeli, H. (2001). Neural networks in civil engineering: 1989-2000. *Computer Aided Civil and Infrastructure Engineering*, 16(2), pp. 126-42.

Aki, K. (1967). Scaling law of seismic spectrum. *J. Geophys. Res.*, 72, pp. 1217-1231.

American Society of Civil Engineers ASCE 7. (2005) Minimum Design Loads for Building and other Structures, Virginia.

Applied Technology Council, ATC-40. (1996). Seismic evaluation and retrofit of concrete buildings, 1-2, California.

Atkinson, G. M., and Beresnev, I. A. (2002). Ground motions at Memphis and St.Louis from M7.5-8.0 earthquakes in the New Madrid seismic zone. *Bull. Seism. Soc. Am.*, 92(3), pp.. 1015-1024.

Bakir, B.S., Yilmaz, M. T., Yakut, A., and Gulkan, P. (2005). Re-examination of damage distribution in Adapazari: geotechnical considerations. *Eng. Struct.*, 27(7), pp. 1002-1013.

Balendra, T., Lam, N.T.K., Wilson, J.L., and Kong, K.H. (2002). Analysis of long-distance earthquake tremors and base shear demand for buildings in Singapore. *Eng. Struct.;* 24(1), pp. 99-108.

Beresnev I.A, Atkinson G.M. (1997). Modeling finite - fault radiation from the ωn spectrum. *Bull. Seism. Soc. Am.*, 87(1), pp. 67-84.

Beresnev, I.A., Atkinson, G.M. (1998). FINSIM - a FORTRAN program for simulating stochastic acceleration time histories from finite faults. *Seism. Res. Lr.*, 69(1), pp. 27-32.

Beresnev, I.A., Atkinson, G.M. (2002). Source parameters of earthquakes in eastern and western north America based on finite fault modeling. *Bull. Seism. Soc. Am.*, 92(2), pp. 695-710.

Bilham, R., Blume, F., Bendick, R., and Gaur, V.K. (1998). The geodetic constrains on the translation and deformation of India: Implication for future great Himalayan earthquakes. *Cur. Sci.*, 74(3), pp. 213-229.

Boore, D. M. (1983). Stochastic simulation of high frequency ground motions based on seismological models of the radiated spectra. *Bull. Seism. Soc. Am.*, 73(6), pp. 1865-1894.

Boore, D. M. (2003). Simulation of ground motion using the stochastic method. *Pure App. Geophy.*, 160(3-4), pp. 635-676.

Boore, D.M. (2004). Can site response be predicted. *J. Earth. Eng.*, 8(1), pp. 1-41.

Boore, D.M., Atkinson, G.M. (1987). Stochastic prediction of ground motion and spectral response at hard-rock sites in Eastern North America. *Bull. Seism. Soc. Am.*, 77(2), pp. 440-467.

Boore, D.M., Joyner, W.B. (1997). Site amplifications for generic rock sites. *Bull. Seism. Soc. Am.*, 87(2), pp. 327-341.

Borcherdt, R. D. (1994). Estimates of site-dependent response spectra for design methodology and justification. *Earth. Spect.*, 10(4), pp. 617-653.

Brune, J. (1970). Tectonic stress and the spectra of seismic shear waves from earthquakes. *J. Geophys. Res.*, 75(26), pp. 4997–5009.

CGWB. (2002), Delhi environmental status report. Pollution Monitoring and Technical Corporate Division, Central Ground Water Board. New Delhi, India.

FEMA 273, NEHRP guidelines for the seismic rehabilitation of buildings;

FEMA 274, Commentary, Washington (DC): Federal Emergency Management Agency, 1996.

Govindarajulu, L., Ramana, G.V., HanumanthaRao, C., and Sitharam, T.G. (2004). "Site-specific ground response analysis." *Curr. Sci.*, 87(10), 1354-1362.

Hanks, T. C., and Kanamori, H. (1979). "A moment magnitude scale." *J. Geophys. Res.*, 84(BS), 2348–2350.

Hartzell, S. H. (1978). "Earthquake aftershocks as Green's functions." *Geophy. Res. Lett.*, 5(1), 1-4.

Herrmann, R.B., and Kijko, A. (1983). "Modeling some empirical vertical component Lg relations." *Bull. Seism. Soc. Am,*. 73(1), 157-171.

Heuze, F., Archuleta, R., Bonilla, F., Day, S., Doroudian, M., Elgamal, A., Gonzales, S., Hoehler, M., Lai, T., Lavallee, D., Lawrence, B., Liu, P.C., Martin, A., Matesic, L., Minster, B., Mellors, R., Oglesby, D., Park, S.,. Riemer, M., Steidl, J., Vernon, F., Vucetic, M., Wagoner, J., and Yang, Z. (2004). "Estimating site-specific strong earthquake motions", Soil Dy. Earth. Eng., 24(3), 199–223.

IBC. (2000). "International Building Code." International Code Council.

IBC. (2009). "International Building Code." International Code Council.

Idriss, I. M. (1990). "Response of soft soil sites during earthquakes." *Proc. Symp. to Honor Prof. H. B. Seed,* Berkeley, CA, 273-289.

Idriss, I. M., and Seed, H. B. (1970). "Seismic response of soil deposits." *J. Soil Mech. Foun. Div.,* Proc. ASCE, Mar, 631-638.

IS 1893: (Part 1): (2002). "Criteria for earthquake resistant design of structures - Part1: General provisions and buildings." *Bureau of Indian Standards,* New Delhi.

Kamatchi P. (2008). "Neural network models for site-specific seismic analysis of buildings." Ph. D. thesis, Department of civil engineering, Indian Institute of Technology, Delhi.

Kamatchi, P. Rajasankar, J. Nagesh R. Iyer, Lakshmanan, N. Ramana, G.V. and Nagpal, A.K. (2010[a]). "Effect of depth of soil stratum on performance of buildings for site-specific earthquakes" Soil Dy. Earth. Eng., 30, 647-661.

Kamatchi, P. Rajasankar, J. Ramana, G.V. and Nagpal, A.K. (2010[b]). "A neural network based methodology to predict site-specific spectral acceleration values" Earth. Eng. Eng. Vib., 9, 459-472.

Kamatchi, P., Ramana, G.V., Nagpal, A.K. and Lakshmanan, N. (2007). "Site-specific response of framed buildings at Delhi for long distance large magnitude earthquake" Jnl. Struct. Eng. 34(3), 227-232.

Kerh, T. & Ting, S.B. (2005), Neural network estimation of ground peak acceleration at stations along Taiwan high-speed rail system, Eng. App. Art. Int., 18(7), 857-866.

Khattri, K. N. (1999). "An evaluation of earthquakes hazard and risk in northern India." *Him. Geo.,* 20, 1-46.

Lam, N.T.K., Wilson, J.L., and Chandler, A.M. (2001). "Seismic displacement response spectrum estimated from the frame analogy soil amplification model." *Eng. Struct.,* 23(11), 1437-1452.

Mammo T. Site-specific ground motion simulation and seismic response analysis at the proposed bridge sites within the city of Addis Ababa, Ethiopia. Eng. Geo. (2005); 79(3-4):127-150.

Motazedian, D. and Atkinson, G. M. (2005). "Stochastic finite-fault modeling based on dynamic corner frequency." , Bull. Seism. Soc. Am. 1997; 95(3):995-1010.

Ordonez, G.A. (2000). "SHAKE 2000: A computer program for the I-D analysis of geotechnical earthquake engineering problems."

Parvez, I. A., Vaccari, F. & Panza, G. F. (2004), Site-specific microzonation study in Delhi metropolitan city 2-D modeling of SH and P-SV waves, Pure and Applied Geophysics, 161(5-6), 1165-84.

Rao, H. Ch., and Ramana, G. V. (2004). "Correlation between shear wave velocity and N value for Yamuna sand of Delhi." *Proc. Int. Conf. Geotech. Eng.*, UAE, 262-268.

Reich, Y. & Barai, S. V. (1999), Evaluating machine learning models for engineering problems, *Artificial Intelligence in Engineering,* 13(3), 257–72.

Roumelioti, Z., and Beresnev, I.A. (2003). "Stochastic finite-fault modeling of ground motions from the 1999 Chi-Chi, Taiwan, earthquake: application to rock and soil sites with implications for nonlinear site response." *Bull. Seism. Soc. Am.,* 93(4), 1691–1702.

SAP2000: Linear and Nonlinear Static and Dynamic Analysis and Design of 3 Dimensional structures", Version 9.0, Computers and Structures Inc., California, USA, 2004.

Satyam, N. D. (2006). "Seismic microzonation studies for Delhi region." *Ph. D. thesis* submitted to Indian Institute of Technology, Delhi.

Schnabel, P. B., Lysmer, J., and Seed, H. B. (1972). "SHAKE, a computer program for earthquake response analysis of horizontally layered sites." *Report No. EERC 72 - 12,* Earthquake Engineering Research Center, University of California, Berkeley, California.

Schnabel, P.B. Seed, H.B. Accelerations in Rock for Earthquakes in the Western United States. Report No. EERC 72-2, University of California, Berkeley, 1972.

Seed, B.H., and Idriss, I.M. (1969). "Influence of soil conditions on ground motions during earthquakes." *J. Soil Mech. Found. Div.,* Proc. ASCE, 95(1), 99-137.

Singh S.K, Mohanty W. K, Bansal B.K, Roonwal G.S. (2002). "Ground motion in Delhi from future large/great earthquakes in the central seismic gap of the Himalayan arc." Bull. Seism. Soc. Am.; 92(2):555-569.

Singh, S.K. (2005). Personal communication.

SNNS. (1999), Stuttgart Neural Network Simulator User Manual. Version 4.2, Sttutgart, Germany.

Sun, C.-G., Kim, D.-S., and Chung, C.-K. (2005). "Geologic site conditions and site coefficients for estimating earthquake ground motions in the inland areas of Korea." *Eng. Geo.,* 81 (4), 446-469.

Tewari, H.C., and Kumar, P. (2003). "Deep seismic sounding studies and its tectonic implications." *J. Vir. Expl.,* 12, 30-54.

Tezcan, S.S., Kaya, I. E., Bal, E., and Ozdemir, Z. (2002). "Seismic amplification at Avcilar, Istanbul." Eng. Struct., 24(5), 661-667.

Vucetic, M., and Dobry, R. (1991). "Effect of Soil Plasticity on Cyclic Response." *J. Geotech. Eng.,* ASCE, 117(1), 89-107.

Yoshida, N., Kobayashi, S., Suetomi, I., and Miura, K. (2002). "Equivalent Linear method considering frequency dependent characteristics of stiffness and damping." *Soil Dyn. Earth. Eng.,* 22(3), 205-222.

8

Microtremor HVSR Study of Site Effects in Bursa City (Northern Marmara Region, Turkey)

Elcin Gok[1] and Orhan Polat[2]
[1]*Dokuz Eylul University, Earthquake Research and Implementation Center, Izmir*
[2]*Dokuz Eylul University, Engineering Faculty, Department of Geophysics, Izmir*
Turkey

1. Introduction

Local site effects are one of the most important aspects in the assessment of seismic hazard. Local site response can be investigated by empirical and theoretical methods. Theoretical methods allow a detail analysis of the parameters considered in the evaluation; however, they require information of the geological structure (Dravinski et al., 1996). Empirical methods are based on seismic records on sites with different geological condition from which relative amplitudes and dominant periods may be determined directly. This approach requires of a large number of earthquakes. In regions with low seismicity, it would be necessary to wait for a long time to obtain a complete data set. For this reason, the use of ambient seismic noise is becoming popular as an alternative (Bard, 1998).

Recording and analyzing ambient noise is simple. A few minutes of microtremor data are usually sufficient. Microtremors are present continuously in time and space. A single three-component station is the only instrument required. Routine spectral techniques can be easily applied to estimate the dominant frequency of vibration of the sedimantary structure. These frequencies of vibration are closely related to the physical features of the site under study, i.e., layer thicknesses, densities and wave velocities. Estimates of these frequencies are useful to constrain the physical properties at a given site.

The Nakamura technique (Nakamura, 1989), based on the horizontal to vertical spectral ratio (HVSR), has been commonly used to estimate the site effects. Later it has been extended to both weak motions (Ohmachi et al., 1991; Field & Jacob, 1993, 1995); and strong motions (Lermo & Chavez-Garcia, 1994; Theodulidis & Bard, 1995; Suzuki et al., 1995). Lermo & Chavez-Garcia (1993) applied this technique to estimate the empirical transfer function from the intense S-wave part of a small sample of earthquake records obtained in three cities of Mexico. Their results showed that the HVSR can estimate the dominant frequency at a site based on earthquake data.

Suzuki et al. (1995), using both microtremor and strong motion data in Hokkaido, Japan, showed that the dominant frequency obtained from HVSR was in good agreement with the predominant frequency estimated from the thickness of an alluvial layer. Lermo & Chavez-Garcia (1993) compared transfer functions computed using the Haskell method agreement with the HVSR. Lermo & Chavez-Garcia (1994) verified that the underlying assumptions of Nakamura's technique are consistent with the propagation of Rayleigh waves.

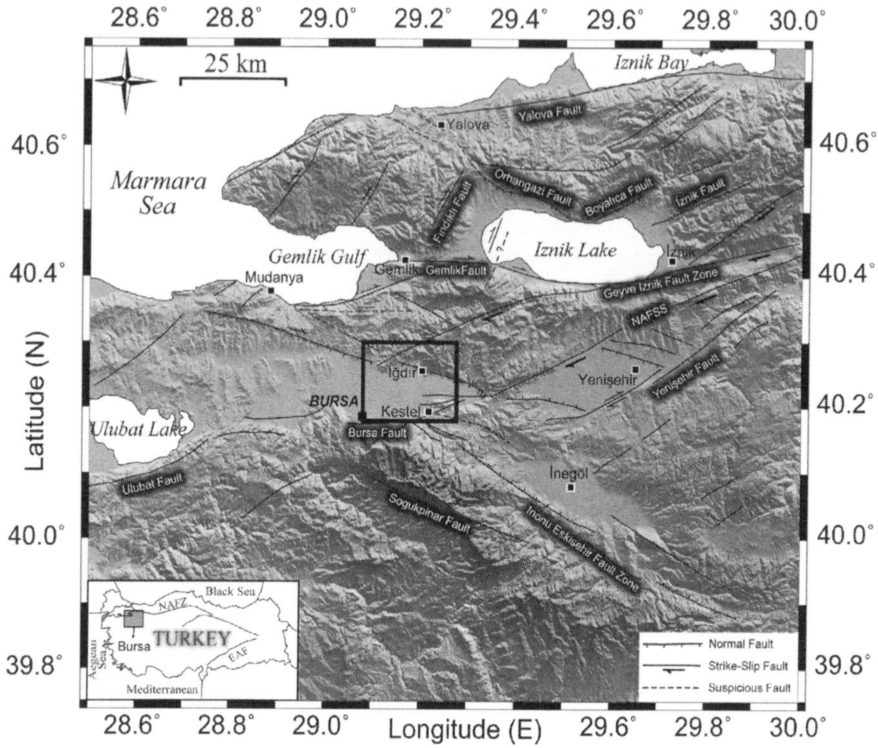

Fig. 1. Map of Bursa. The box indicates the study area. NAFZ: North Anatolian Fault Zone, NAFSS: Southern strand of the North Anatolian Fault Zone, EAF: East Anatolian Fault Zone.

2. Tectonic and geological setting

The region of study is surrounded by many active faults; Gemlik Fault (GF), Geyve-Iznik Fault Zone (GIFZ), Yenişehir Fault, Bursa Fault (BF), Inonu-Eskisehir Fault Zone (IEFZ). The main lithological units in the vicinity of Bursa are Quaternary alluvial deposits and Neogene basement rocks. The thickness of the Quaternary deposits is larger than 300m where those are as Neogene units vary from 50 to 200m. in Bursa basin (Imbach 1997; Topal et al., 2003). South of Bursa, Paleozoic and methamorfic units are present. The simplified geological map of the study area, modified from MTA (General Directory of Mineral Research and Exploration), is shown in Figure 2.

Bursa city is located in the southern Marmara Region, characterised by significant historical and instrumental seismicity (Figure 1). Two strong earthquakes, with maximum intensities X and IX EMS-98, occurred in 1855. Seismicity is related with the activity of southern branch of the NAFZ.

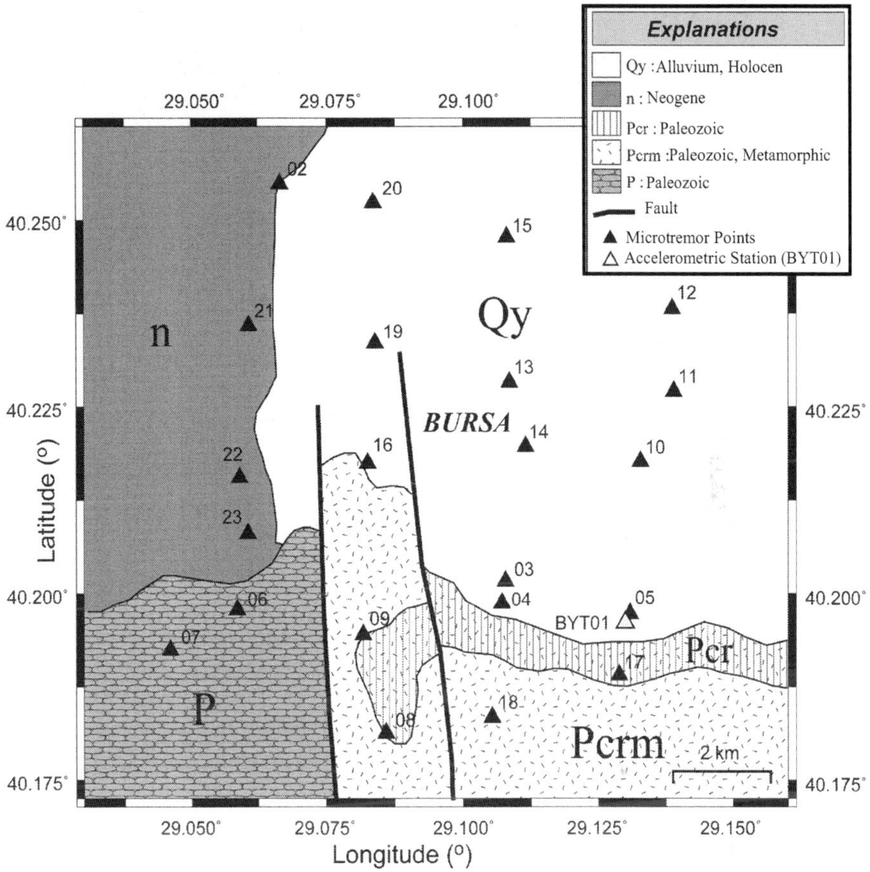

Fig. 2. Simplified geological map of Bursa region. Black triangles indicate points of microtremor measurements; open triangle shows the location of BYT01 station (Modified after MTA, General Directory of Mineral Research and Expolaration).

3. Method

The microtremor HVSR method is generally used for microzonation and site responses studies. It considers that the amplification produced by a surface layer can be estimated from the ratio between the horizontal and vertical spectral amplitudes. This method is known as the Nakamura's technique.

The method supposes that microtremors are composed of Rayleigh waves which propagate in a surface layer over a half-space (Dravinski et al., 1996; Lermo & Chavez-Garcia, 1994). The motion at the interface between the layer and the half-space is not affected by the source effect. Moreover, the horizontal and vertical motions at the interface have similar amplitude due to the ellipticity of the Rayleigh waves.

Horizontal to Vertical spectral ratio is related to the ellipticity of Rayleigh waves which is frequency dependent (Bard, 1998; Bonnefoy-Claudet et al., 2006). HVSR showes a sharp peak at the fundamental frequency of the sediments, if there is a high impedance contrast between the sediments and the bottom bedrock. Criticism of the HVSR method was often related to the fact that there is no common practice for data acquisition and processing (Mucciarelli & Gallipoli, 2001). Attempts to provide standards were only made recently (SESAME, 2004). It is widely accepted today that the frequency of the peak of HVSR showes the fundamental frequency of the sediments. Its amplitude depends mainly on the impedance contrast with the bedrock and cannot be used as site amplification. Comparisons with results of standard spectral ratio method have also shown that the HVSR peak amplitude sometimes underestimates the actual site amplification. (Bard, 1998; Gosar & Martinec, 2009)

4. Microtremor measurements and analyses

4.1 Instruments and data

A single seismic station was used for the microtremor measurements. It was composed of a three-component seismometer with GPS time, the passing band of this system in DC to 100 Hz. Our sampling was 100 sps, reducing the frequency to the band below 50 Hz. We recorded data at 22 different points. Record duration was set to 30 minutes. The mean distance between recording sites is approximately 2 km. The sensors were buried in the ground at each site.

4.2 HVSR analyses

Microtremor measurements were made at 22 sites (Figure 2). Their locations were selected to avoid the influence of trees, sources of monochromatic noise, rivers, and strong topographic features. HVSR analysis was performed following SESAME (2004). Recorded time series were visually inspected to identify possible inaccurate measurements and transient pulses. Each record was split in windows between 15 to 30 s long %5 overlapping windows for which amplitude spectra in a range 0.5–20 Hz were computed using a cosine taper with 10% smoothing and Konno & Ohmachi smoothing with a constant of 40 (Konno, & Ohmachi, 1998). HVSR was then computed as the average of both horizontal component spectra divided by the vertical spectrum for each window. After produced HVSR dominant frequency and maximum amplification were determined. Figures 3 and 4 show an example of the results.

The smallest dominant frequency values (≤2 Hz) were obtained in the northern part of the basin, covered by the thick Neogene and Quaternary sediments (points 19, 21, 22 in Figure 3 and 13, 14, 12 in Figure 4). Frequencies in the range 2 to 4 Hz were observed on Paleozoic sediments of moderate thickness (points 08, 09, 17 in Figure 3). Dominant frequencies larger than 5 Hz was obtained on Paleozoic and metamorphic rocks (06, 07 in Figure 3 and 04, 05 in Figure 4). These values are characteristic for most of the Bursa area.

In some cases the microtremor measurements were unable to provide an estimate of dominant frequency (Figure 4). The possible reasons are: wide peak, two or more peaks in a spectrum, flat spectral ratio and very small amplitude of the peak.

Figure 4a shows an example of wide peak that can not be associated to a resonant frequency. Probably due to the several impedance contrasts at various depths, HVSR sometimes resulted in two or more peaks with similar amplitudes. In Figure 4b, the two peaks are well separated in frequency, so it can be the boundary between soft sediments and rock is related to the peak at 1.3 Hz. The second peak at 5 Hz may be related to Paleozoic rocks. However, in the case shown in Figure 4c, there are two peaks of the same amplitude at 1 Hz and 13

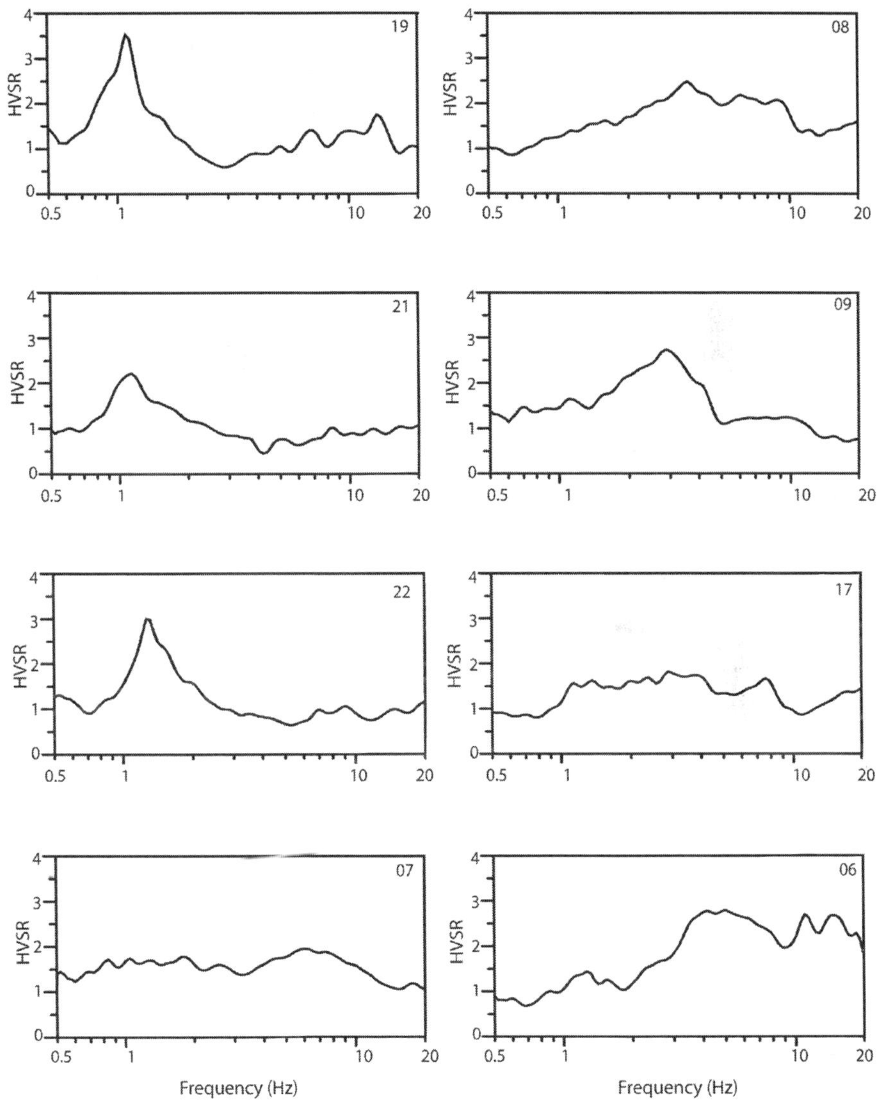

Fig. 3. Examples of HVSR for the measurements points (06, 07, 08, 09, 17, 19, 21 and 22)

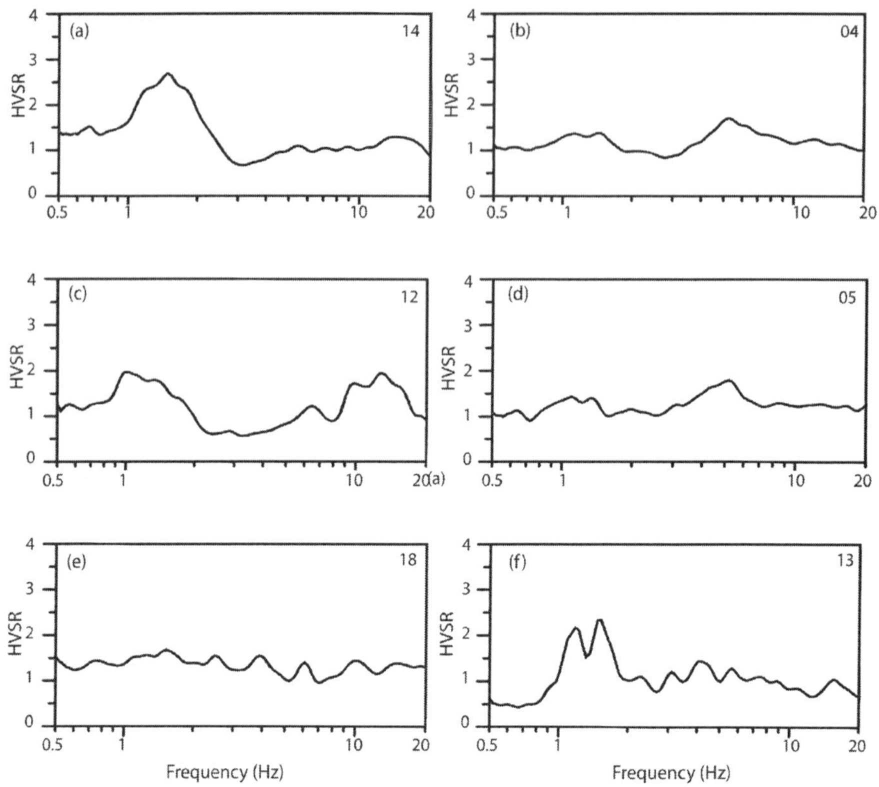

Fig. 4. Some examples of microtremor measurements for which determination of the dominant frequency may be problematic (a) wide peak, (b) two peaks, (c) artificial source of noise, (d) artificial noise which frequency can be determined, (e) almost flat spectral ratio, (f) group of peaks.

Hz. In such cases, we were unable to identify which one corresponds to the most significant geological boundary. Another example (Figure 4d) shows two different peaks at the 1.2 Hz frequency and 5 Hz. Artificial noise is seen on the first peak but the real peak of HVSR is at a higher frequency (5 Hz). In some cases, we compared the dominant peak frequency with that from neighbouring measurements with more clear peaks. For some measurements, we obtained almost flat spectral ratios (Figure 4e) with maximum amplitudes smaller than 1.5 Hz. We found no clear peak for this point but it may be correlated with Paleozoic rocks. In Figure 4f, two peaks are observed around 1 Hz. The shape of this HVSR curve indicates that the peak is at a similar frequency, but since it is contaminated with artificial noise, it cannot be accurately identified.The amplitudes of the peaks of HVSR are mostly in the range 1-2 Hz in Figure 5. Only in a few cases they are larger than 5 Hz.

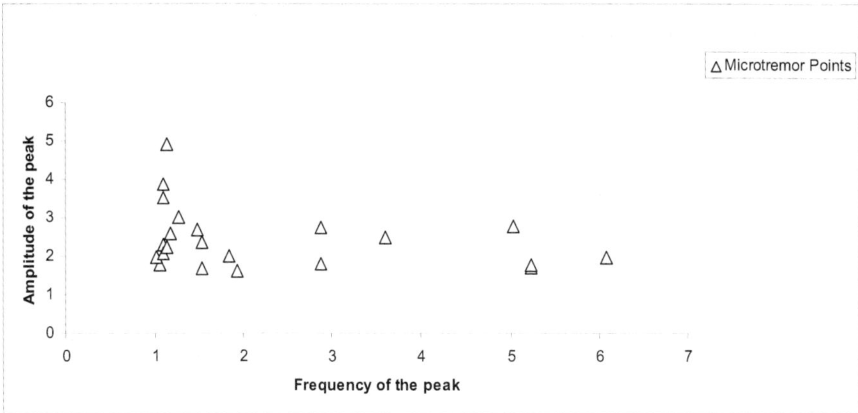

Fig. 5. Amplitude vs. frequency graph of HVSR peaks

4.3 Time-dependent HVSR

The common procedure to compute the HVSR spectral ratio relies on average amplitude spectra of the three components of motion. Some researchers such as Almendros et al., (2004) have suggested that this aproach may lead to errors. Perturbations of the wavefield may occur during the recording period and be recorded together with the microtremor data. Usually, these transients are easily identified in the spectra, and the analysis can be performed using only on data windows free of perturbations in order to obtain reliable results. In these cases, artificial peaks appear in the HVSR (Figure 4). These peaks affect the spectral ratio and produce inaccurate results. Because of this problem, time-dependent HVSR has also been used to estimate spectral ratios. This approach consists of compiling HVSR to successive data windows along the traces. This procedure creates several HVSR functions that can be represented a two-dimensional contour plots versus frequency and time. This plot, that is called ratiogram, represents the evolution of the HVSR in the same way that a spectrogram represents the evolution of the spectrum versus frequency and time. (Almendros et al., 2004)

In this study, we selected a window of 25 s and slided it at intervals of 5 s along the traces. This length is suitable for the numerical fast Fourier transform (FFT) algorithm for frequencies larger than 0.5 Hz. For each window we calculated the amplitude spectra of the three components using an FFT algorithm, and smoothed it using a cosine window. Frequency-dependent window lengths have also been used keeping a constant number of cycles (Kind et al., 2005). We computed the HVSR separately for all time intervals and plotted them. An example is as a fuction of time shown in Figure 6.

Three component microtremor data was shown in Figure 6a. Using the standard technique, average HVSR are computed from individual windows (Figure 6b). We observed the presence of a dominant peak at about 1.2 Hz and we can conclude that the site produces amplification for this frequency. Figure 6c shows the time-dependent HVSR which is stationary, at least during particular time periods. An average HVSR could be obtained by stacking the HVSRs.

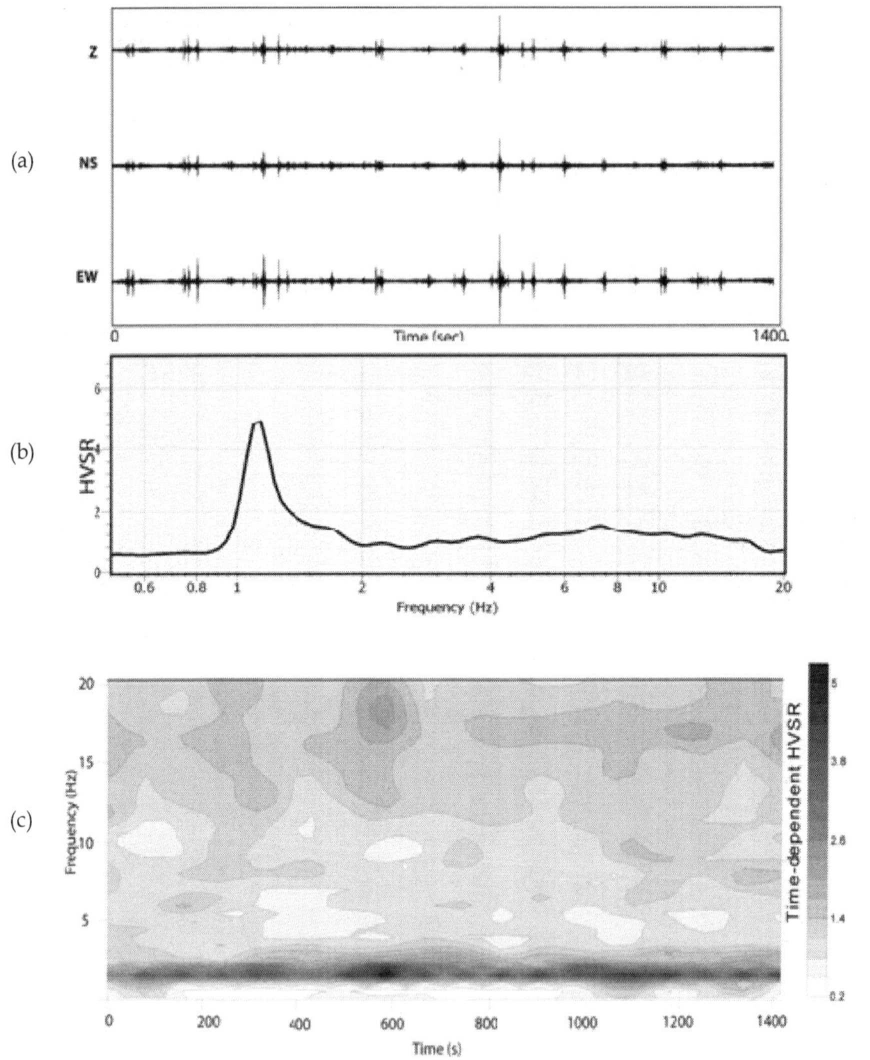

Fig. 6. Example of the application of the time dependent HVSR method (a) three-component microtremor data, (b) average HVSR using standard procedure, (c) ratiogram representing the HVSR as a function of frequency and time.

5. HVSR results using earthquake data

Lermo and Chavez-Garcia (1993) presented that the Nakamura (1989) technique could be applied to the S-wave part of the earthquakes, and the HVSR ratios provided amplitude of the soil deposits. We applied the HVSR ratios of the S-wave window for the recorded

ground motions to sites in the BYT01 for site effect estimation. The Fourier spectrum of ground motion for each event was obtained using the HVSR method.
Earthquake records from an accelerographic station (Figure 2) deployed in the city have been obtained. We used them to compare the results obtained from microtremor survey. A location of the station is given in Table 1.

Station Coordinates	Altitude (m)	Recorder Type	Recorder Serial Nr.
40.18240N 29.12960E	193	Etna	5035

Table 1. Coordinates of Station BYT01

This station has recorded four shallow earthquakes (depths smaller than 19 km.) with magnitudes (Md) between 3.6 and 5.2. Locations of the events are given in Table 2. Spectral ratios have been computed using the HVSR technique (Figure 7). We have used events for which the signal to noise level is larger than 3 in the frequency range 0.5-20 Hz. The selected window has duration of 15 second beginning 2-3 sec before S-wave arrival. The analysis included a cosine taper before Fourier transform and smoothing with a factor of 40 using the window by Konno & Ohmachi, 1998.

Earthquake Date and Time (GMT)	**Earthquake Coordinates**	**Depth**	**Magnitude (Md)**
20/10/2006	40.2519N-27.9792E	16.7	5.2
24/10/2006	40.4221N-28.9937E	7.9	5.2
25/10/2006	40.3698N-29.0059E	10.7	3.6
19/12/2006	40.3400N-28.3200E	18.5	4.2

Table 2. Recorded earthquakes in BYT01 (location parameters were taken from AFAD-ERD).

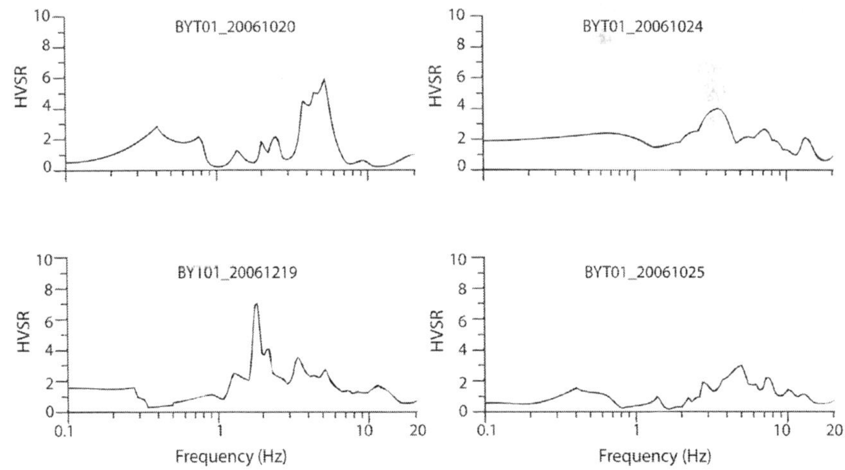

Fig. 7. The HVSR results of four earthquakes.

A dominant frequency around 5 Hz is observed for events 20061020 and 20061025 in the HVSR results (Figure 7). The BYT01 station is very close to the microtremor point Nr. 05

which shows on Figure 4b. We found the dominant frequency at 5 Hz on point Nr. 05 and these two earthquakes are related to the result of microtremor point and they show similar results. In Figure 7, events 20061024 and 20061219 show different dominant frequency, between 2 Hz and 4 Hz.

6. Conclusion and discussion

The 22 values of dominant frequency and maximum relative amplification (HVSR) were used to draw the contours shown in Figure 8. The contours of dominant frequency values coincide with surficial geology (Figure 8a), the maximum amplification values vary between 1 and 5 (Figure 8b).

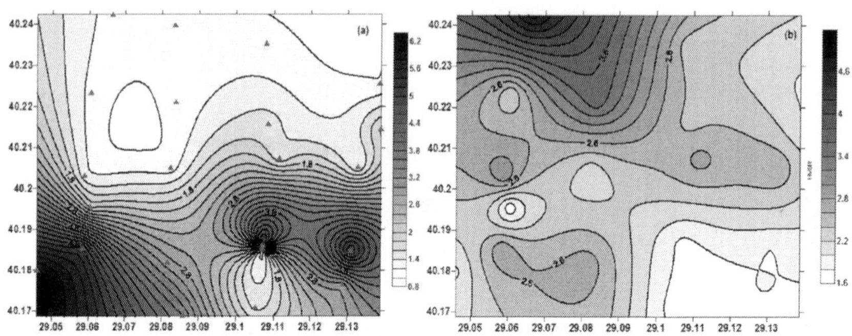

Fig. 8. (a) Dominant frequency and (b) HVSR map of the study area.

Figure 9 shows examples of ratiograms obtained at two different sites. In each case, the top pannel shows the three-components of ambient noise: The bottom pannel shows the calculated ratiogram and the right pannel shows the average HVSR. The gray scale on the right represents the values of the time-dependent HVSR in both ratiograms. In the first case (Figure 9a), the average HVSR does not show a dominant frequency. The flat response is seat with an amplification level approximately equal to one. In the second case (Figure 9b); a clear dominant frequency of 1.2 Hz appears throughout the duration of the records. Ratiograms like these have been calculated for the entire data set.

In general, the smaller values of dominant frequency show that (1-2 Hz) correlate with alluvium and Neogene sediments. Peaks at larger frequencies are correlated to Paleozoic and metamorphic rocks. Our measurements show that there are transient zones between different geologic structures (alluvium and Paleozoic rocks).

The map of fundamental soil frequency derived from free-field microtremor measurements should be confirmed by independent information from boreholes, geophysical investigations or earthquake recordings in the future, since the interpretation of microtremors is restricted to identifying the resonance frequency and gives no information on the amplification of seismic ground motion. The HVSR provides an estimate of the bandwidth over which the ground motion is amplified. This is especially important for any microzonation.

Three-component microtremor measurements were conducted at 22 sites in the northern section of the Bursa city, where the different geological structures in the study area outcrop. The fundamental frequencies of the sediments show a range between of 0.5 and 20 Hz. The lower frequencies (below 2 Hz) correspond to the Holocene and neogene deposits overlain

by alluvium, forming a small basin.The higher frequencies correspond to Paleozoic and metamorphic rocks. However, variations over short distances are large.
In addition to microtremor data, earthquake records were also used to compute HVSR. The HVSR analysis of four earthquakes and microtremor at 22 points gives similar results: Dominant frequencies and spectral ratios correlate well with geological structures in the Bursa city.

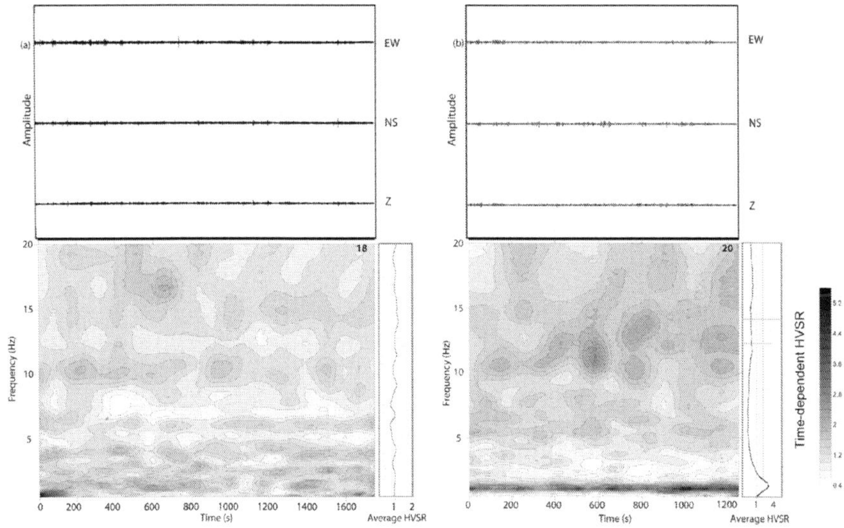

Fig. 9. Two examples of ratiograms and average HVSRs obtained from microtremors recorded at stations a) 18 and b) 20.

Microtremor measurements at 22 points and analysis of contribute valuable preliminary microzonation and site response information. However a more complete study of city-scale earthquake hazard, it is still necessary. More microtremor points and events are necessary to understand site response in Bursa City.

7. Acknowledgements

This research is a part of PhD study of Elcin Gok. We thank to the Earthquake Research Department of Presidency of Disaster and Emergency Management Directorate (AFAD) belongs to the Prime Ministry of Turkey for their data support. We also appreciate to Zafer Akcig, Senol Ozyalin and Zulfikar Erhan for facilitating our works in the field. This work was supported by BAP of Dokuz Eylul University (Project Nr. 2006.KB.FEN.007).

8. References

Almendros, J., Luzon, F., & Posadas, A. (2004). Microtremors analysis at Teide Volcano (Canary Islands, Spain): assessment of natural frequencies of vibration using time-dependent horizontal-to-vertical spectral ratios, *Pure and Applied Geophysics, 161*, 1579–1596.

Bard, P. Y. (1998). Microtremor measurements: A tool for site effect estimation? In Okada & Sasatani (Eds), *Proc. 2nd Intl. Symp. on The Effects of Surface Geology on Seismic Motion*, pages 1251-1279. Balkema.

Bonnefoy-Claudet, S., Cornou, C., Bard, P. Y., Cotton, F., Moczo, P., Kristek, J., & Fa¨h, D. (2006). H/V ratio: A tool for site effects evaluation. Results from 1-D noise simulations. *Geophysical Journal International 167*, 827–837.

Dravinski, M., Ding, G., & Wen, K. L. (1996). Analysis of Spectral Ratios for Estimating Ground Motion in Deep Basins. *Bulletin Seismological Society of America 86*, 646–654.

Field, E.H., & Jacob, K., (1995). A comparison and test of various site response estimation techniques, including three that are non reference-site dependent. *Bulletin Seismological Society of America 85*, 1127–1143.

Field E, & Jacob K. (1993). The theoretical response of sedimentary layers to ambient seismic noise. *Geophysical Research Letters, 20* (24), 2925-2928.

Gosar, A. & Martinec, M. (2009) Microtremor HVSR study of site effects in the Ilirska Bristica town area (S. Slovenia), *Journal of Earthquake Engineering 13*, 50–67.

Imbach, T., (1997). Geology of Mount Uludag with emphasis on the genesis of the Bursa, Northwest Anatolia, Turkey. In: Shindler, C. & Pfister, M. (Eds.), *In Active Tectonics of Northwestern Anatolia-The Marmara Poly-Project*, Hochschulverlag AG an der ETH Zurich-Swiss, p. 239-266.

Kind, F., D. Fah, & D. Giardini (2005). Array measurements of S-wave velocities from ambient vibrations. *Geophysical Journal International, 160*, 114–126.

Konno, K. & Ohmachi, T. (1998). Ground-motion Characteristics Estimated from Spectral Ratio between Horizontal and Vertical Components of Microtremor. *Bulletin Seismological Society of America, 88*, 228–241.

Lermo, J., & Chavez-Garcia, F. J. (1993). Site effect evaluation using spectral ratios with only one station *Bulletin Seismological Society of America, 83*, 1574–1594.

Lermo, J., & Chavez-Garcia, F. J. (1994). Are microtremors useful in the site response evaluation? *Bulletin Seismological Society of America, 84*, 1350–1364.

Mucciarelli, M., & Gallipoli, M. R. (2001). A critical review of 10 years of microtremor HVSR technique. *Bolletino di Geofisica Teorica ed Applicata 42*, 255–266.

Nakamura, Y. (1989). A method for dynamic characteristics estimation of subsurface using microtremor on the ground surface. *Quarterly Report of the Japanese Railway Technical Research Institute (RTRI), 30(1)*, 25–33.

Ohmachi, T., Nakamura, Y., & Toshinawa, T. (1991). Ground motion characteristics in the San Francisco Bay area detected by microtremor measurments. In S. Prakash (Ed.), *Proceedings of the Second International Conference on Recent Advances in Geotechnical Earthquake Engeneering and Soil Dynamics*, March 11–15, St. Louis, Missouri, Univ. of Missouri-rolla, p. 1643– 1648.

SESAME, (2004*). Guidelines for the implementation of the H/V spectral ratio technique on ambient vibrations: measurements, processing and interpretation*, 62 pp.,http://sesame-fp5.obs.ujfgrenoble.fr/Delivrables/Del-D23

Suzuki, T., Adachi, Y., & Tanaka, M. (1995). Application of Microtremor Measurements to the Estimation of Earthquake Ground Motions in Kushiro City during the Kushiro-Oki Earthquake of 15 January 1993. *Earthquake Eng. Struct. Dyn. 24*, 595–613.

Theodulidis, N. P. & Bard, P. Y. (1995). Horizontal to Vertical Spectral Ratio and Geological Conditions: an Analysis of Strong Motion Data from Greece and Taiwan (SMART-1). *Soil Dyn. Earthquake Eng. 14*, 177–197.

Topal, T., Doyuran, V., Karahanoglu, N., Toprak, V., Suzen, M.L, & Yesilnacar, E., (2003). Microzonation for earthquake hazard: Yenisehir settlement, Bursa, Turkey. *Engineering Geology, 70*, 93-108.

Revisions to Code Provisions for Site Effects and Soil-Structure Interaction in Mexico

J. Avilés[1] and L.E. Pérez-Rocha[2]
[1]*Instituto Mexicano de Tecnología del Agua*
[2]*Instituto de Investigaciones Eléctricas*
México

1. Introduction

The seismic hazard in Mexico has been re-evaluated recently and now we can estimate the maximum acceleration on rock for any given site in the country. This is the starting point for constructing site-specific earthquake design spectra that explicitly include the effects of soil dynamic amplification (site effects). The effects of soil-structure interaction (SSI) can be accounted for in two stages: first in the elastic design spectrum, considering the enlargement of the period and the increase in damping; and then in the strength reduction factor, taking into consideration the global ductility reduction.

Site effects refer to the dynamic amplification of ground motion due to the local geotechnical characteristics of the subsoil. These effects are reflected in the seismic design coefficients specified by building codes in terms of site- and structure-response amplification factors. The SSI effects, on the other hand, refer to the modification of the foundation motion with respect to the free-field ground motion due to the flexibility of the supporting soil. In building codes, however, these effects are generally accounted for modifying the dynamic properties (natural period and damping ratio) of the original structure and evaluating the response of the modified structure to the prescribed free-field motion specified by a design spectrum.

A new approach to determine earthquake design spectra including site and SSI effects has been incorporated in the CFE Seismic Design Code (MDOC), a model design code in Mexico (Tena-Colunga et al., 2009). The previous version of the MDOC was published by the Federal Electricity Commission in 1993, so an in-depth review was mandatory in order to update the code for 2008. In the absence of state seismic codes, the MDOC is legally used in the entire nation for earthquake design of different structure types. The MDOC provides expressions to determine earthquake design spectra at any location in the country, which makes unnecessary the use of conventional zoning maps. These spectra, which have realistic shape and size for elastic response, are then reduced by two separate factors to account for the nonlinear structural behavior and overstrength.

2. Site effects

It is well recognized that seismic hazard varies significantly throughout the country and that it cannot be described in detail by means of regional spectra for different types of soil.

Thus, it is necessary to construct site-specific earthquake design spectra, which depend mainly on the proximity from the place to the tectonic sources and on the local subsoil conditions. In the seismic behavior of structures, several response factors associated with the source, the wave's path, the site and the structure itself are involved. In order to simplify the problem, a design earthquake motion at the bedrock is specified by the MDOC, in such a way that the effects of source and wave's path are considered implicitly. In this way, it only remains to account for the site and SSI effects on the structural response. To do this, the simplified reference model shown in fig. 1 is used. This is formed by a modal oscillator placed on a rigid foundation that is embedded into an equivalent stratum with elastic bedrock.

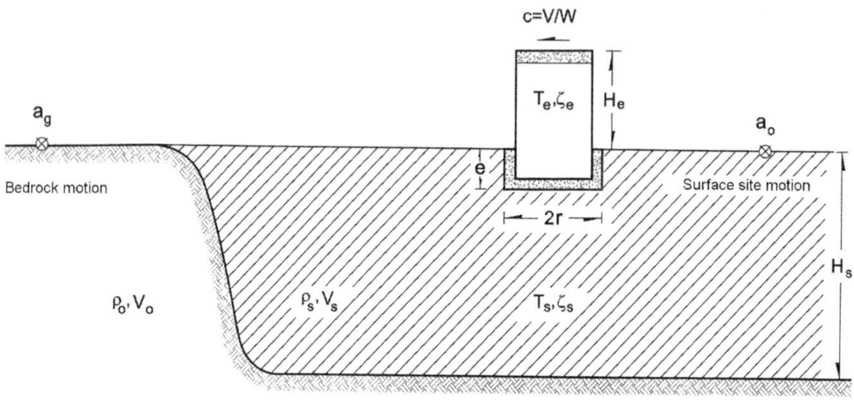

Fig. 1. Simplified reference model to account for site effects and SSI.

For the analyses presented here, a soft soil site (UAPP) located in the city of Puebla with dominant period T_s=1.25 s, soil/bedrock impedance ratio $p_s=\rho_s V_s/\rho_o V_o$=0.2, Poisson's ratio ν_s=0.4 and hysteretic damping ratio ζ_s=0.05 has been considered. The value of the site period corresponds to a shear-wave velocity V_s=80 m/s and stratum thickness H_s=25 m.

Based on the considered model, a new approach to specify earthquake design spectra for arbitrary locations in Mexico has been developed. These spectra realistically represent the levels of strength and displacement demands that would take place in single elastic structures during the design earthquake motion. It is evident that the multi-degree-of-freedom effects in real buildings are not accounted for.

2.1 Acceleration design spectrum

In the MDOC, the seismic hazard was re-evaluated with the use of both deterministic and probabilistic approaches, using spectral attenuation relations developed specifically for the different seismic sources affecting Mexico. The map of fig. 2 shows the nationwide distribution of peak rock acceleration, a_g, for design of standard occupancy structures. This map was produced with a computer program developed for this purpose. The approach proposed to construct elastic design spectra is based on the value of this ground-motion parameter. Next, site- and structure-response factors are developed to account for the peak dynamic amplification of soil and structure responses, respectively. The nonlinear soil behavior is considered with two additional factors, one for the site period shift and other for the site response reduction, using soil properties (shear modulus and damping ratio) consistent with the shear strain.

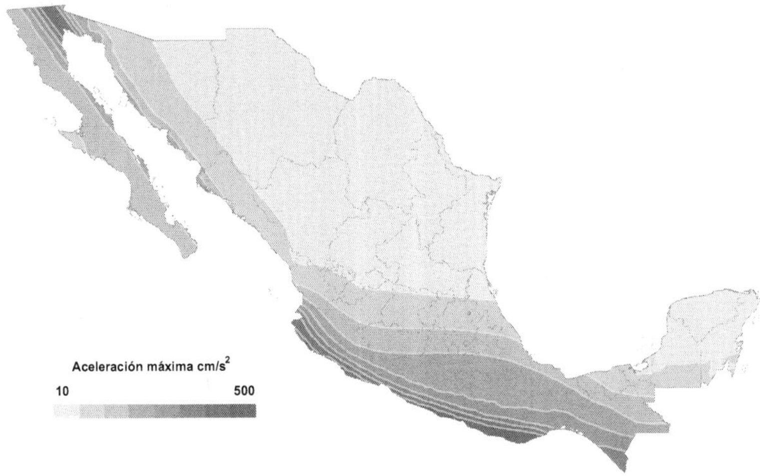

Fig. 2. Distribution of peak rock acceleration in Mexico for design of standard occupancy structures.

With these general ideas in mind, the following steps have to be taken to construct site-specific earthquake design spectra:

1. Compute the distance factor as $F_d=a_g/500$, which is equal to unity near the subduction seismic source. This parameter expresses not only the seismic-wave attenuation with distance, but also the filtering of the high-frequency components of the earthquake excitation.
2. From geotechnical information of the site soil profile, compute the dominant soil period as follows:

$$T_s = 4\sqrt{\left(\sum_{n=1}^{N}\frac{h_n}{G_n}\right)\left(\sum_{n-1}^{N}\rho_n h_n (w_n^2 + w_n w_{n-1} + w_{n\ 1}^2)\right)} \tag{1}$$

where G_n and ρ_n are the shear modulus and mass density of the nth layer of thickness h_n; w_o=0 at the bedrock and

$$w_n = \frac{\sum_{i=1}^{n} h_i/G_i}{\sum_{i=1}^{N} h_i/G_i}; \quad n = 1,2,...,N \tag{2}$$

is a static approximation for the fundamental mode of vibration. With T_s known, the effective shear-wave velocity $V_s=4H_s/T_s$ is computed over the depth $H_s=\Sigma h_n$ This novel procedure is found to give more accurate results than those obtained by using the average shear-wave velocity of the surficial soils, which ignores the layer sequence in the soil profile.

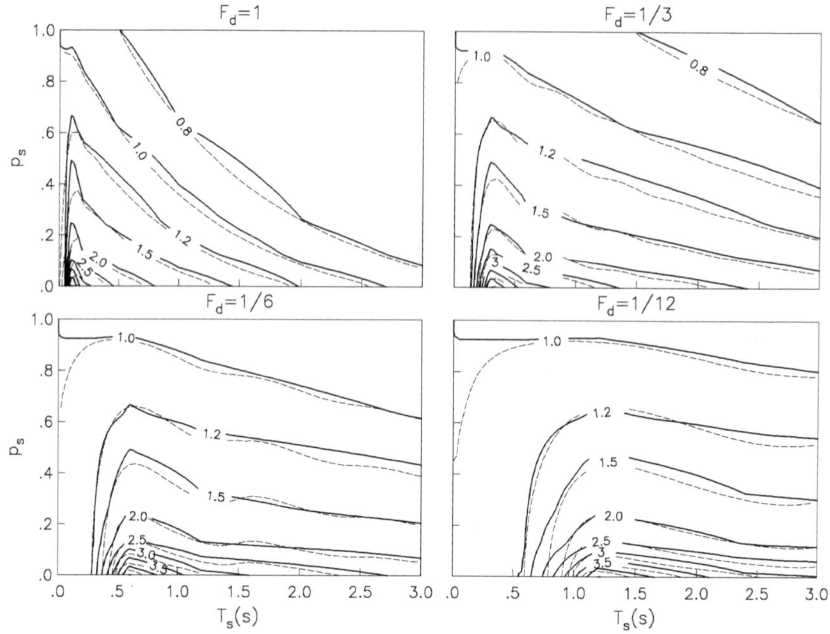

Fig. 3. Contours of F_s derived from site response analysis (dashed line) and by linear interpolation of data in table 1 (solid line).

3. Assuming linear soil behavior, the site-response amplification factor F_s=a_d/a_g is obtained. The values for this factor are based on site response analysis, using the input power spectrum of the rock excitation (Park, 1995) and through application of the random vibration theory (Boore & Joyner, 1984) to predict peak responses. The theoretical results are shown in fig. 3 and the discrete values specified by the MDOC are tabulated in table 1 as a function of $T_s' = T_s\sqrt{F_d}$ and the impedance ratio p_s between soil and bedrock.

T_s' / p_s	0.00	0.05	0.10	0.20	0.50	1.00	2.00	3.00
1.000	1.00	1.00	1.00	1.00	1.00	1.00	1.00	1.00
0.625	1.00	1.08	1.23	1.12	1.00	1.00	1.00	1.00
0.250	1.00	1.18	1.98	1.60	1.40	1.12	1.00	1.00
0.125	1.00	1.20	2.64	2.01	1.69	1.32	1.00	1.00
0.000	1.00	1.22	4.51	3.17	2.38	1.75	1.19	1.00

Table 1. Values of the site amplification factor F_s.

4. Depending on the level of shaking, soil/rock impedance ratio and soil type, the following factors are used to account for the nonlinear soil behavior:

$$F_n^d = \begin{cases} 1-(1-\hat{F}_n^d)\dfrac{T_s}{1.5}, & \text{if } T_s \le 1.5\text{ s} \\ \hat{F}_n^d, & \text{if } T_s > 1.5\text{ s} \end{cases} \quad (3)$$

p_s \ F_d	0.00	0.10	0.20	0.30	0.40	0.50	0.75	1.00
1.000	1.00	0.97	0.93	0.90	0.86	0.83	0.75	0.71
0.625	1.00	0.95	0.91	0.89	0.85	0.82	0.71	0.68
0.250	1.00	0.93	0.87	0.82	0.77	0.73	0.63	0.56
0.125	1.00	0.92	0.84	0.75	0.67	0.64	0.58	0.53
0.000	1.00	0.90	0.78	0.66	0.58	0.54	0.53	0.50

Table 2. Values of the nonlinear factor $\hat{F}_n^d$ for sands and gravels.

p_s \ F_d	0.00	0.10	0.20	0.30	0.40	0.50	0.75	1.00
1.000	1.00	0.98	0.95	0.91	0.87	0.85	0.79	0.75
0.625	1.00	0.97	0.94	0.93	0.90	0.88	0.81	0.79
0.250	1.00	0.96	0.93	0.91	0.87	0.85	0.77	0.74
0.125	1.00	0.93	0.85	0.76	0.70	0.67	0.61	0.56
0.000	1.00	0.82	0.63	0.46	0.36	0.32	0.31	0.28

Table 3. Values of the nonlinear factor $\hat{F}_n^d$ for clays and cohesive soils.

$$F_n^s = \begin{cases} 1-(1-\hat{F}_n^s)\dfrac{T_s}{1.5}, & \text{if } T_s \le 1.5\text{ s} \\ \hat{F}_n^s, & \text{if } T_s > 1.5\text{ s} \end{cases} \quad (4)$$

where the values of $\hat{F}_n^d$ and $\hat{F}_n^s$ are listed in tables 2-3 and 4-5, respectively. While F_n^d expresses the site response reduction due to an increase in damping, $1/F_n^s$ expresses the site period shift due to a decrease in stiffness. Note that these factors tend to unity for very short site period, corresponding to hard rock conditions.

p_s \ F_d	0.00	0.10	0.20	0.30	0.40	0.50	0.75	1.00
1.000	1.00	0.99	0.98	0.97	0.96	0.95	0.95	0.95
0.625	1.00	0.98	0.97	0.93	0.90	0.89	0.89	0.89
0.250	1.00	0.97	0.93	0.86	0.81	0.79	0.79	0.79
0.125	1.00	0.97	0.92	0.85	0.80	0.78	0.78	0.78
0.000	1.00	0.97	0.92	0.85	0.80	0.78	0.78	0.78

Table 4. Values of the nonlinear factor $\hat{F}_n^s$ for sands and gravels.

F_d \ p_s	0.00	0.10	0.20	0.30	0.40	0.50	0.75	1.00
1.000	1.00	1.00	1.00	1.00	1.00	1.00	1.00	1.00
0.625	1.00	1.00	1.00	0.99	0.99	0.99	0.99	0.99
0.250	1.00	0.99	0.98	0.96	0.94	0.93	0.93	0.93
0.125	1.00	0.98	0.95	0.90	0.86	0.84	0.84	0.84
0.000	1.00	0.95	0.88	0.77	0.69	0.67	0.66	0.66

Table 5. Values of the nonlinear factor $\hat{F}_n^s$ for clays and cohesive soils.

5. The peak soil acceleration is obtained from the peak rock acceleration multiplied by the site and nonlinear factors, as follows:

$$a_o = F_n^d F_s a_g \tag{5}$$

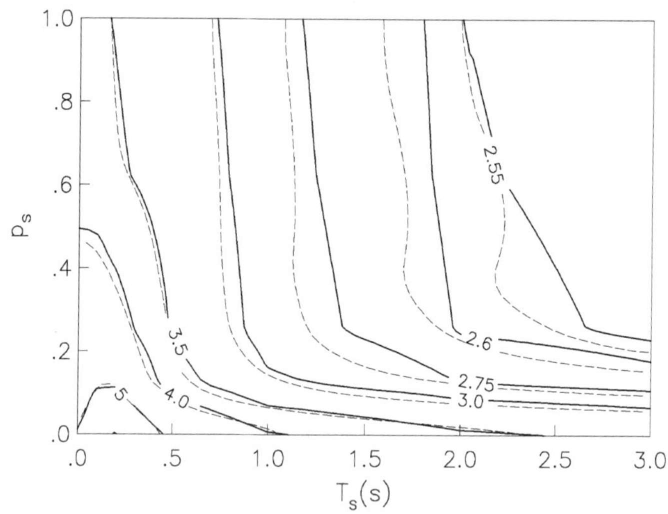

Fig. 4. Contours of F_r derived from site-structure response analyses (dashed line) and by linear interpolation of data in table 6 (solid line).

6. The seismic coefficient that defines the plateau height of the design spectrum is given by

$$c = F_r a_o \tag{6}$$

where F_r is the structure-response amplification factor. The values for this factor are based on the random vibration analysis of a single oscillator subjected to a base excitation passed through the site soil profile. The theoretical results are shown in fig. 4 and the discrete values specified by the MDOC are listed in table 6 as a function of T_s and p_s. As the distance factor has little influence on these results, it has been ignored.

T_s / p_s	0.00	0.05	0.10	0.20	0.50	1.00	2.00	3.00
1.000	2.50	2.50	2.50	2.50	2.50	2.50	2.50	2.50
0.625	2.50	3.80	3.74	3.57	3.26	2.81	2.56	2.51
0.250	2.50	4.36	4.41	4.27	3.45	2.85	2.59	2.53
0.125	2.50	4.74	4.91	4.90	3.70	3.06	2.75	2.65
0.000	2.50	5.27	5.66	6.02	4.81	4.05	3.58	3.40

Table 6. Values of the structural amplification factor F_r.

7. The lower and upper periods of the flat part of the design spectrum are given by

$$T_a = 0.35\frac{T_s}{F_n^s} \geq 0.1\text{ s} \tag{7}$$

$$T_b = 1.2\frac{T_s}{F_n^s} \geq 0.6\text{ s} \tag{8}$$

These expressions are intended to cover not only the peak structural response at the first soil period, but also that at the second one ($\approx T_s/3$). The upper period is taken 20% greater than the site period to account for differences between the computed and actual values of T_s.

8. In terms of the natural vibration period T_e and viscous damping ratio ζ_e, the acceleration design spectrum has the following basic representation:

$$\frac{Sa}{g} = \begin{cases} a_o + (\beta c - a_o)\dfrac{T_e}{T_a}, & \text{if } T_e < T_a \\ \beta c, & \text{if } T_a \leq T_e < T_b \\ \beta c\dfrac{T_b}{T_e}, & \text{if } T_b \leq T_e < T_c \\ \beta c\dfrac{T_b}{T_c}p_c\left(\dfrac{T_c}{T_e}\right)^2, & \text{if } T_e \geq T_c \end{cases} \tag{9}$$

where $p_c = k + (1-k)(T_c/T_e)^2$, with $k = 2.3 - 1.6T_s \geq 0.2$, and

$$T_c = \begin{cases} 2\text{ s}, & \text{if } T_b < 2\text{ s} \\ T_b, & \text{if } T_b \geq 2\text{ s} \end{cases} \tag{10}$$

$$\beta = \begin{cases} \beta_o, & \text{if } T_e < T_c \\ 1 + (\beta_o - 1)\dfrac{T_c}{T_e}, & \text{if } T_e \geq T_c \end{cases} \tag{11}$$

where $\beta_o = (0.05/\zeta_e)^{0.45}$. In the specification of the design spectrum, a nominal damping value of 5% is considered. To account for the supplemental damping due to SSI or mechanical damping devices, the reduction factor β has been introduced. This tends to unity for long-period ordinates, which are independent of the damping value.

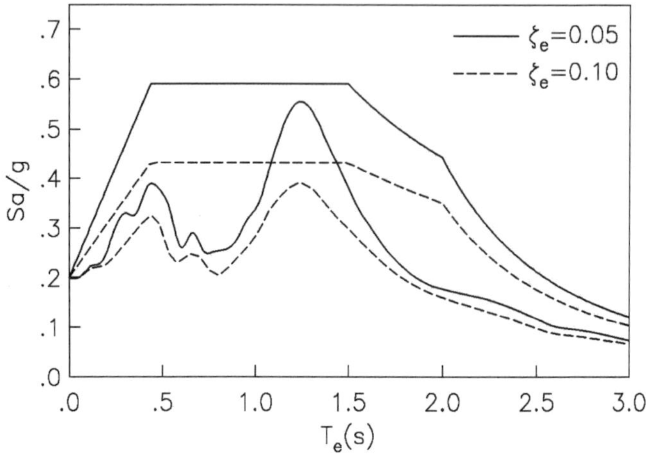

Fig. 5. Acceleration, response and design spectra for site UAPP considering two values of damping.

Following the procedure described above, site-dependent elastic design spectra can be constructed, the shape and size of which are based on the knowledge of peak rock acceleration, site-source distance, dominant soil period and soil/bedrock impedance ratio. For site UAPP, the ensuing spectra for 5 and 10% of damping are shown in fig. 5, along with the corresponding response spectra for the 15 June 1999 Tehuacán earthquake recorded at this site and scaled to the peak rock acceleration specified by the MDOC, without any change in the frequency content and duration characteristics. This normal faulting earthquake of magnitude Mw=7.0 occurred inland 125 *km* from the city of Puebla. Here, it is used as the input control motion at the ground surface.

2.2 Displacement design spectrum

The spectral shapes for $T_e<T_c$ have been in use for many years in Mexican building codes. For $T_e>T_c$, however, a new descending branch is proposed in order to have a better description of the displacement design spectrum *Sd*. Specifically, the limit of this spectrum for very long period must tend to the peak ground displacement D_{max}. In view of the relationship between spectral displacement and acceleration,

$$Sd = \frac{T_e^2}{4\pi^2} Sa \tag{12}$$

this long-period limit can only be achieved if the acceleration design spectrum decays at least as fast as T_e^2. For a slower decay, the displacement design spectrum tends incorrectly to infinity as the structure period increases. As can be seen in fig. 6, the observed spectral displacements at site UAPP are well represented by the code spectral displacements. It is interesting to note that values of Sd larger than D_{max} can occur for natural vibration periods nearby the site period.

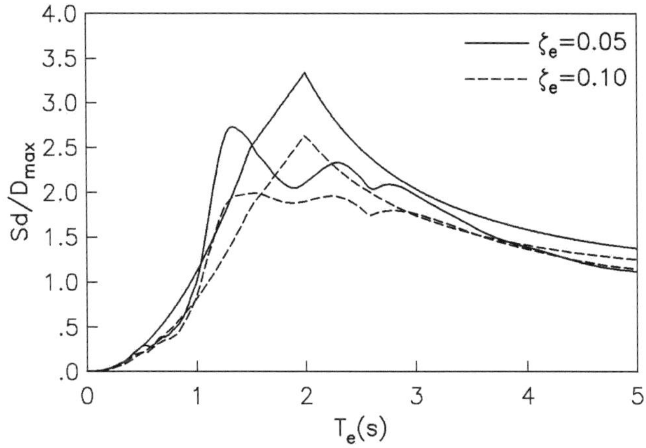

Fig. 6. Displacement, response and design spectra for site UAPP considering two values of damping.

When $k<1$, the peak spectral displacement occurs at $T_e=T_c$ and is given by

$$Sd_{max} = \frac{g}{4\pi^2}\beta_o c T_b T_c \tag{13}$$

If $k\geq 1$, the peak spectral displacement occurs at $T_e=\infty$ and converges to the peak ground displacement given by

$$D_{max} = \frac{g}{4\pi^2} k c T_b T_c \tag{14}$$

From eqns. 13 and 14, it can be found that

$$k = \frac{D_{max}}{Sd_{max}/\beta_o} \tag{15}$$

Notice that parameter k has a physical meaning. It represents the ratio of peak ground displacement to peak spectral displacement for 5% of damping. The code values for this parameter cover a wide variety of site conditions, from hard rock (k=1.5) to very soft soils (k=0.2).

3. Soil-structure interaction

The design approach used in most current codes to take the SSI effects into account has not changed over the years: a replacement oscillator represented by the effective period and damping of the system. The most extensive efforts in this direction were made by Veletsos (1977) and his coworkers. Indeed, their studies form the basis of the SSI provisions currently in use in the US building codes. Although this approach does not account for the ductile capacity of the structure, it has been implemented in many codes in the world for the convenience of using standard fixed-base spectra in combination with the effective period and damping of the system. Nevertheless, seismic regulations that allow reductions in the design base shear by ductility and SSI separately should be taken with caution. This deficiency has been recognized in the last revision to the SSI procedures in the NEHRP design provisions (Stewart et al., 2003). In the MDOC, the SSI effects are expressed by a shift in the fundamental period T_e, an increase in the damping ratio ζ_e and a reduction in the ductility factor Q_e, as a function of the foundation flexibility H_eT_s/H_sT_e. If a design spectrum is specified for a given site, then the earthquake loads and displacements can be computed by entering with the effective period $\tilde{T}_e$, damping $\tilde{\zeta}_e$ and ductility $\tilde{Q}_e$, just as though the structure were fixed at the base.

3.1 Effective period and damping

For elastic conditions, the system's period and damping are defined as the natural period and damping ratio of a replacement oscillator whose resonant harmonic response is equal to that of the SSI system. Introducing some permissible simplifications, the following expressions can be obtained (Avilés & Pérez-Rocha, 1996):

$$\tilde{T}_e = (T_e^2 + T_h^2 + T_r^2)^{1/2} \tag{16}$$

$$\tilde{\zeta}_e = \zeta_e \frac{T_e^3}{\tilde{T}_e^3} + \frac{\zeta_h}{1+2\zeta_h^2} \frac{T_h^2}{\tilde{T}_e^2} + \frac{\zeta_r}{1+2\zeta_r^2} \frac{T_r^2}{\tilde{T}_e^2} \tag{17}$$

where $T_h = 2\pi(M_e/K_h)^{1/2}$ and $T_r = 2\pi(M_e(H_e+e)^2/K_r)^{1/2}$ are the natural periods associated with the rigid-body translation and rocking of the structure with mass M_e, whereas $\zeta_h = \pi C_h/\tilde{T}_e K_h$ and $\zeta_r = \pi C_r/\tilde{T}_e K_r$ are the damping ratios of the soil for the translational and rocking modes of the foundation. The terms K_h, K_r and C_h, C_r are the frequency-dependent springs and dampers by which the soil is replaced for the two vibration modes of the foundation. The springs account for the stiffness and inertia of the soil, whereas the dampers for the energy dissipation by hysteretic behavior and wave radiation in the soil.

The SSI effects on the period and damping are shown in fig. 7 for high-rise (H_e/r=5) and low-rise (H_e/r=2) structures with embedded foundation (e/r=1) in a soil deposit (H_s/r=5). The system period increases with respect to the fixed-base period as the foundation flexibility increases, especially for the high-rise structure. While the system damping increases for the low-rise structure, it may be smaller than the fixed-base damping for the high-rise structure. The damping reduction due to an increased structural response is particularly important for tall buildings, which are more effectively excited by rocking of the foundation. In the MDOC, the value of $\tilde{\zeta}_e$ cannot be taken less than 0.05, the nominal damping value implicit in the design spectrum.

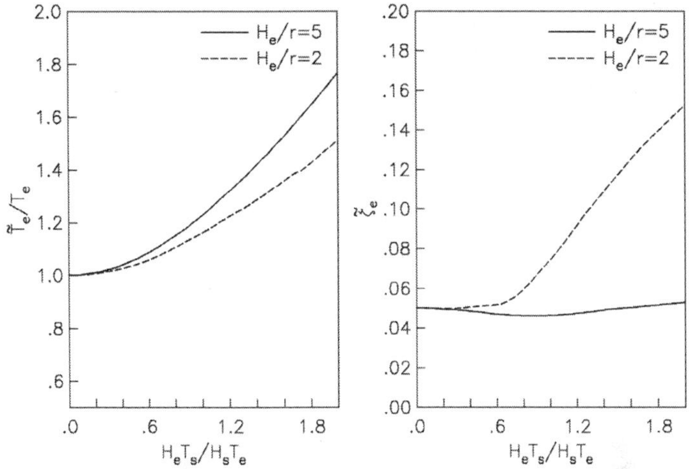

Fig. 7. Effect of SSI on the fundamental period and damping ratio of high- and low-rise structures on flexible foundation.

3.2 Effective ductility

To take the nonlinear structural behavior into account, an equivalent ductility factor is needed to be defined. By equating the maximum plastic deformation of an elastoplastic replacement oscillator with that developed in the SSI system under monotonic loading, the system ductility is found as (Avilés & Pérez-Rocha, 2003)

$$\tilde{Q}_e = 1 + (Q_e - 1)\frac{T_e^2}{\tilde{T}_e^2} \tag{18}$$

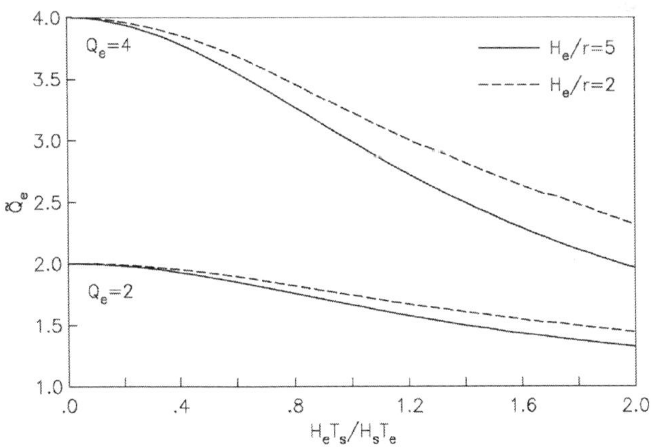

Fig. 8. Effect of SSI on the ductility factor of high- and low-rise structures on flexible foundation.

As shown in fig. 8, the global ductility of the system $\tilde{Q}_e$ reduces with respect to the allowable ductility of the structure Q_e as the foundation flexibility increases. The influence of the structure slenderness is relatively less important. Although the foundation flexibility reduces the ductility factor, the capacity of structural ductility remains unchanged. This apparent paradox stems from the fact that the response of the replacement oscillator includes not only the displacement of the structure itself, but a rigid-body motion of the foundation as well. It is the presence of this motion that reduces the ductility factor.

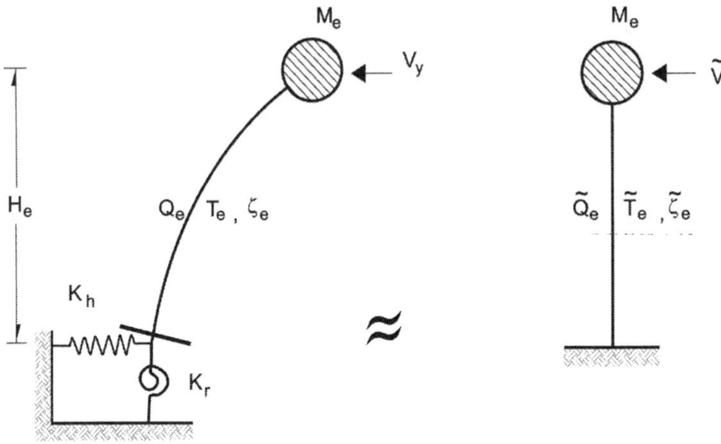

Fig. 9. Analogy between the SSI system and a replacement oscillator.

From the analogy between the SSI system and a replacement oscillator excited by the same base motion, see fig. 9, it is found that their yield resistance and peak displacement are interrelated by

$$V_y = \tilde{V}_y \tag{19}$$

$$Sd = \frac{T_e^2}{\tilde{T}_e^2}\frac{Q_e}{\tilde{Q}_e}\tilde{S}d \tag{20}$$

The difference between the relative inelastic displacement Sd and the total inelastic displacement $\tilde{S}d$ is due to the contribution by the translation and rocking of the foundation. Furthermore, the elastic displacement developed in the replacement oscillator results from the flexibilities of both the structure and foundation.

For a specific case with H_eT_s/H_sT_e=1.33, fig. 10 shows strength spectra obtained with this approach using the input control motion. Base-shear coefficients with ($\tilde{C}_y = \tilde{V}_y / M_e g$) and without ($C_y = V_y / M_e g$) SSI are plotted against the fixed-base period. For Q_e=1, the strength spectrum with SSI shifts toward shorter periods and is a bit less amplified than the strength spectrum without SSI. For Q_e=4, the resonant peaks associated with the first and second vibration modes of the soil tend to disappear.

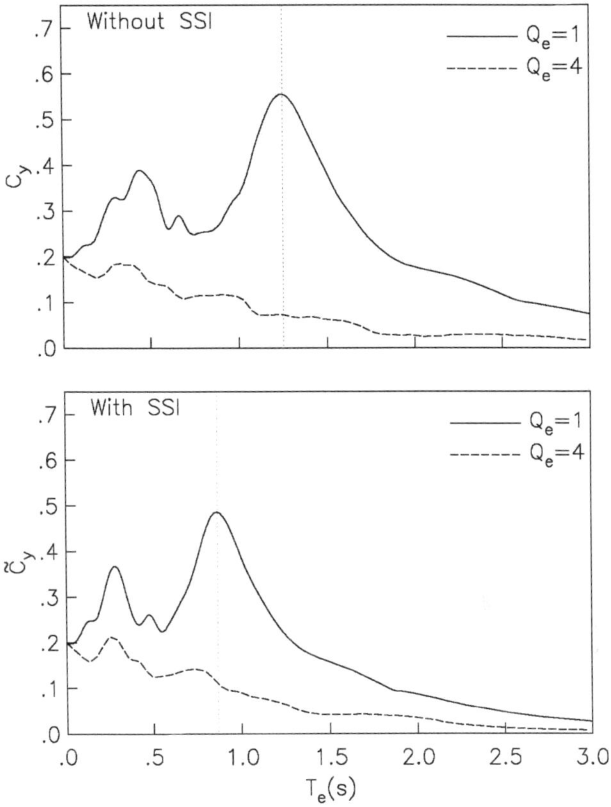

Fig. 10. Strength spectra with and without SSI for elastic and inelastic behavior.

3.3 Strength-reduction factor

For code-designed structures, it is common practice to make use of a strength-reduction factor for estimating inelastic design spectra by reducing elastic design spectra. For a given earthquake, this factor is defined as the ratio between the strength required to have elastic behavior and the strength required for the allowable ductility. The shape of this factor has been extensively studied for the fixed-base condition, using recorded motions and theoretical considerations. In particular, Ordaz and Pérez-Rocha (1998) observed that it depends on the ratio of the elastic displacement spectrum to the peak ground displacement as follows:

$$Q'_e = 1 + (Q_e - 1)\left(\frac{Sd(T_e, \zeta_e)}{D_{max}}\right)^{1/2} \tag{21}$$

It is apparent that period and damping dependency of Q'_e is implicit in *Sd*. A simplified version of eqn. 21 implemented in the MDOC is the following:

$$Q'_e = \begin{cases} 1 + (Q_e - 1)\sqrt{\dfrac{\beta}{k}}\dfrac{T_e}{T_b}, & \text{if } T_e \le T_b \\ 1 + (Q_e - 1)\sqrt{\dfrac{\beta p_b}{k}}, & \text{if } T_e \ge T_b \end{cases} \tag{22}$$

where $p_b = k + (1-k)(T_b/T_e)^2$. In developing eqn. 22, the following considerations were made: For simplicity, it was decided to have a linear variation between $Q'_e(0) = 1$ and $Q'_e(T_b) = Q'_{max}$, with $Q'_{max} = 1 + (Q_e - 1)\sqrt{Sd_{max}/D_{max}}$ being the maximum value can be reached. The shape of $Q'_e(T_e > T_b)$ results from replacing the corresponding displacement spectrum in eqn. 21. For very long period, β=1 and p_b=k and hence Q'_e tends to Q_e, as dictated by theory.

The fixed-base reduction rule given by eqn. 21 is more general than others reported in the literature, because its period and damping dependence is properly controlled by the actual shape of the elastic displacement spectrum, and not by a smoothed shape obtained empirically. This rule may be readily implemented for flexible-base structures by merely replacing the terms $Q_e - 1$ by $(\tilde{Q}_e - 1)\tilde{T}_e^2/T_e^2$ (from eqn. 18) and Sd by $(T_e^2/\tilde{T}_e^2)\tilde{S}d$ (from eqn. 20), with which we have

$$\tilde{Q}'_e = \begin{cases} 1 + (\tilde{Q}_e - 1)\dfrac{\tilde{T}_e}{T_e}\sqrt{\dfrac{\tilde{\beta}}{k}}\dfrac{\tilde{T}_e}{T_b}, & \text{if } \tilde{T}_e \le T_b \\ 1 + (\tilde{Q}_e - 1)\dfrac{\tilde{T}_e}{T_e}\sqrt{\dfrac{\tilde{\beta}\tilde{p}_b}{k}}, & \text{if } \tilde{T}_e \ge T_b \end{cases} \tag{23}$$

The shapes of factors Q'_e and $\tilde{Q}'_e$ are displayed in fig. 11, along with the results given by the equal displacement rule (Veletsos & Newmark, 1960). Contrarily to what happens in many building codes, in this proposal the values of Q'_e can be larger than the ductile capacity Q_e when k<1 corresponding to soft soil sites. This behavior due to site effects is counteracted by SSI. The reason is that SSI tends to shift the structure period to the long-period spectral region, for which the equal displacement rule is applied. Although the representation is not perfect, the proposed reduction rule reproduces satisfactorily the general trends observed for the input control motion.

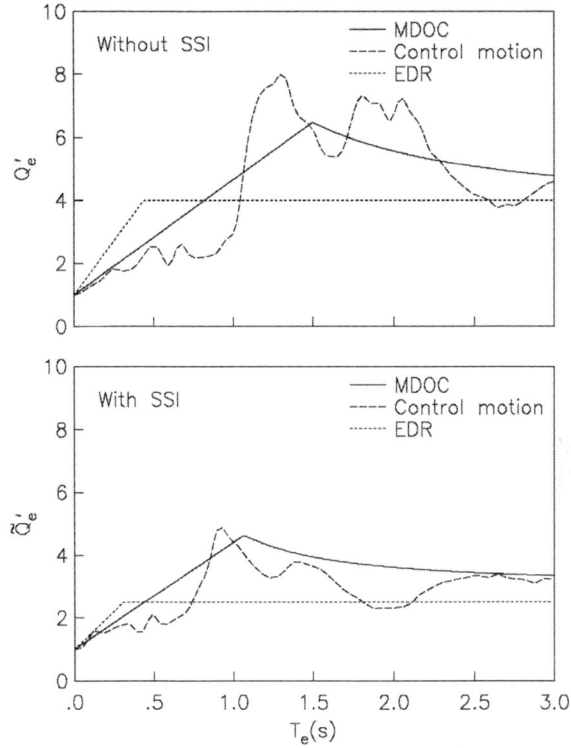

Fig. 11. Strength-reduction factors with and without SSI obtained from code provisions (MDOC and EDR) versus observations.

3.4 Design strength

In view of what has been discussed previously, the required base-shear coefficients with and without SSI can be computed as follows:

$$\tilde{C}_y = \frac{\tilde{S}a(\tilde{T}_e, \tilde{\zeta}_e)/g}{R\tilde{Q}'_e} \tag{24}$$

$$C_y = \frac{Sa(T_e, \zeta_e)/g}{RQ'_e} \tag{25}$$

The elastic acceleration spectra $\tilde{S}a$ and Sa are used to emphasize the fact that the former should be evaluated for $\tilde{T}_e$ and $\tilde{\zeta}_e$, and the latter for T_e and ζ_e. Notice that the overstrength reduction factor R is independent of SSI. Strength design spectra with and without SSI are exhibited in fig. 12, along with strength response spectra for the input control motion. It is clear that the latter spectra are safely covered by the former in the whole period range. Nevertheless, the conservatism inherent in smoothed design spectra overshadows some important changes by SSI, as those happening in the spectral region between the first and second soil periods.

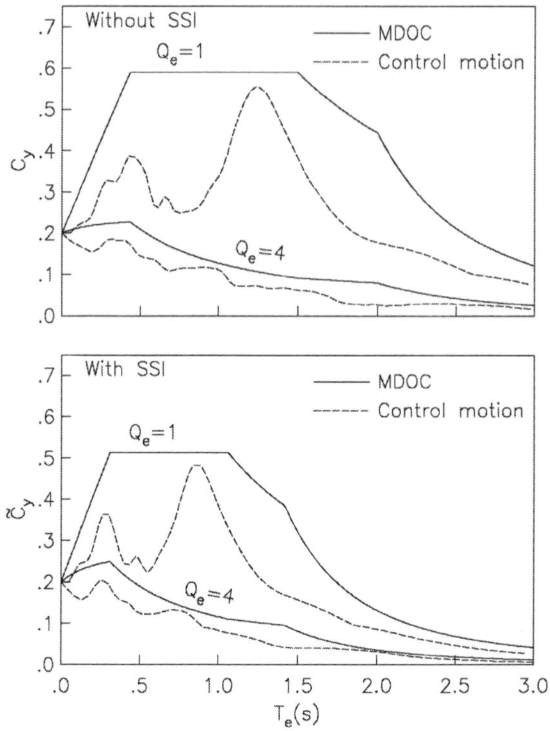

Fig. 12. Design and response strength spectra with and without SSI for elastic and inelastic behavior.

The use of the recommended SSI provisions will increase or decrease the design earthquake forces with respect to the fixed-base values, depending on the dynamic properties of the structure and soil and the characteristics of the earthquake excitation. The lateral displacements will undergo additional changes due to the contribution by the translation and rocking of the foundation. This latter motion may be particularly significant for tall buildings.

A convenient factor to account for modifications of the structural response due to SSI is given by the ratio $\tilde{V}_y/V_y = \tilde{C}_y/C_y$. The results shown in fig. 13 for this SSI factor, derived from the strength design spectra of fig. 12, illustrate the following points: The increments in the base shear are less important than the reductions. While the greater increments arise in nonlinear systems (Q_e=4), the greater reductions arise in linear systems (Q_e=1). The SSI factor can be used to modify the response quantities computed for the structure assumed to be fixed at the base. In the MDOC, the value of $\tilde{V}_y/V_y$ cannot be taken less than 0.7, nor greater than 1.2. It is seen, however, that the calculated reduction can be considerably larger than 30%.

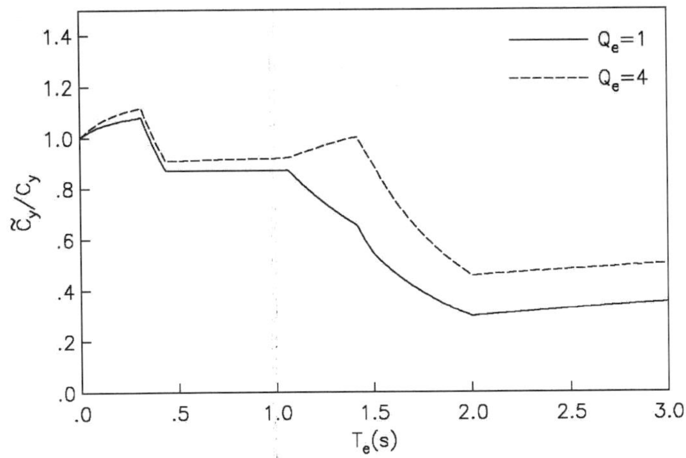

Fig. 13. Variation of the SSI factor for elastic and inelastic behavior.

Finally, the maximum displacement of the flexible-base structure relative to the ground can be determined as

$$\tilde{\delta}_{max} = \frac{\tilde{V}_y}{V_y}\left(\delta_{max} + \frac{V_y}{K_h} + (H_e + e)^2 \frac{V_y}{K_r}\right) \tag{26}$$

where $\delta_{max} = (V_y/K_e)Q_e$ is the maximum displacement of the fixed-base structure, with K_e being the lateral structural stiffness.

4. Conclusions

The site effects and SSI provisions described in this work have been incorporated in the 2008 MDOC seismic design code used in Mexico. A simplified model of the soil and structure that forms the basis of current design practice was investigated. A new approach for constructing site-specific earthquake design spectra was devised, which reflects some research advances made on site response and SSI. The approach is based on the peak rock acceleration determined with a computer program developed for this purpose. Improved site- and structure-response factors to account for the peak dynamic amplification of soil and structure responses were developed. The nonlinear soil behavior was considered with two additional factors, one for the site period shift and other for the site response reduction. These factors should be computed using soil properties consistent with the shear strain. To account for the nonlinear structural behavior, a known site-dependent strength reduction factor properly adjusted to include SSI was implemented. The SSI effects were expressed by a shift in the fundamental period and an increase in the damping ratio for the elastic condition, as well as a reduction in the nominal ductility factor. It was shown that the independent reduction of the design base shear by ductility and SSI is unsuitable, especially for very ductile structures. It is expected that with these improvements to code provisions, the earthquake response of code-designed structures will be assessed more accurately.

5. References

Avilés, J. & Pérez-Rocha, L.E. (1996). Evaluation of Interaction Effects on the System Period and the System Damping due to Foundation Embedment and Layer Depth. *Soil Dynamics and Earthquake Engineering*, Vol. 15, No. 1, pp. 11-27, ISSN 0267-7261

Avilés, J. & Pérez-Rocha, L.E. (2003). Soil-Structure Interaction in Yielding Systems. *Earthquake Engineering and Structural Dynamics*, Vol. 32, No. 11, pp. 1749-1771, ISSN 0098-8847

Avilés, J. & Pérez-Rocha, L.E. (2005). Design Concepts for Yielding Structures on Flexible Foundation. *Engineering Structures*, Vol. 27, No. 3, pp. 443-454, ISSN 0141-0296

Boore, D.M. & Joyner, W.B. (1984). A Note on the Use of Random Vibration Theory to Predict Peak Amplitudes of Transient Signals. *Bulletin of the Seismological Society of America*, Vol. 74, No. 5, pp. 2035-2039, ISSN 0037-1106

Ordaz, M. & Pérez-Rocha, L.E. (1998). Estimation of Strength-Reduction Factors for Elastoplastic Systems: A New Approach. *Earthquake Engineering and Structural Dynamics*, Vol. 27, No. 9, pp. 889-901, ISSN 0098-8847

Park, Y.J. (1995). New Conversion Method from Response Spectrum to PSD Functions. *Journal of Engineering Mechanics*, Vol. 121, No. 12, pp. 1391-1392, ISSN 0733-9399

Stewart, J.P.; Kim, S.; Bielak, J.; Dobry, R. & Power, M.S. (2003). Revisions to Soil-Structure Interaction Procedures in NEHRP Design Provisions. *Earthquake Spectra*, Vol. 19, No. 3, pp. 677-696, ISSN 8755-2930

Tena-Colunga, A.; Mena-Hernández, U.; Pérez-Rocha, L.E.; Avilés, J.; Ordaz, M. & Vilar, J.I. (2009). Updated Seismic Design Guidelines for Model Building Code of Mexico. *Earthquake Spectra*, Vol. 25, No. 4, pp. 869-898, ISSN 8755-2930

Veletsos, A.S. & Newmark, N.M. (1960). Effect of Inelastic Behavior on the Response of Simple Systems to Earthquake Motions. *Proceedings of the 2nd World Conference on Earthquake Engineering*, Tokyo and Kyoto, Japan

Veletsos, A.S. (1977). Dynamics of Structure-Foundation Systems, In: *Structural and Geotechnical Mechanics*, W.J. Hall, (Ed.), pp. 333-361, Prentice-Hall, Englewood Cliffs, USA

10

Influence of Nonlinearity of Soil Response on Characteristics of Ground Motion

Olga Pavlenko
Institute of Physics of the Earth, Russian Academy of Sciences
Russia

1. Introduction

Recent earthquakes, such as the 1985 Michoacan earthquake, the 1989 Loma Prieta earthquake, the 1994 Northridge earthquake, and the 1995 Kobe earthquake provided new experimental data on the soil behavior in strong ground motion, in particular, on the liquefaction phenomena, and discussions on the nonlinearity of soil behavior were induced (Lomnitz et al., 1995; Aguirre & Irikura, 1997; Field et al., 1997; O'Connell, 1999, etc.). Though nonlinear elastic properties of soils were studied in multiple laboratory experiments, and valuable laboratory experimental data are accumulated sometimes, this is not sufficient for understanding the soil behavior *in situ*, because soils often represent multi-component systems containing water, air, gases, etc., and strong ground motion can induce movement and redistribution of these components, i.e., changes in the properties of the soils. Experimental data on the soil behavior in strong motion *in situ* are still few, fragmental, and non-representative; and accumulation of these data is important for improving our understanding of soil behavior in strong motion.

In strong ground motion Hooke's law does not hold for subsurface soils, i.e., soils should be taken as nonlinear systems transforming incident seismic signals into movement on the surface. For studying nonlinear properties of systems, effective methods are developed in system analysis, so-called nonlinear system identification technique, based on the determination of higher-order impulse characteristics of the systems. An output of a nonlinear system is represented as the Volterra-Wiener series, i.e., a sum of the response of a linear system to the input signal and a number of nonlinear corrections, which are due to quadratic, cubic nonlinearity, and nonlinearities of higher (the 4-th, 5-th, etc.) orders. If we know the input and output of a nonlinear system, we can judge regarding the types and quantitative characteristics of the system nonlinearity (Marmarelis & Marmarelis, 1978). Nonlinear identification of soils in various geotechnical conditions seems to be promising, because it allows a better understanding of the behavior of soils and structures in strong ground motion. However, to apply methods of system analysis to studying nonlinear properties of soils, knowledge of stress-strain relations in the soil layers in strong motions is required. In this section, a method of estimation of nonlinear stress-strain relations in soils in strong ground motion is proposed based on vertical array data.

Numerous methods and programs developed for calculating the ground response in strong motion in various conditions do not allow estimation of stress-strain relations in soil layers in situ. Moreover, in cases of strong nonlinearity, there often remains some disagreement between the observed and simulated records. As is known, equivalent linear models (SHAKE, QUAD-4, FEADAM, LUSH, FLUSH) are not applicable for calculation of such complex phenomena as soil liquefaction. Programs DESRA (Lee & Finn, 1978), TARA and their modifications (Finn et al., 1986; Finn & Yogendrakumar, 1989) allow determination of the possible level of the pore pressure and the possibility of liquefaction, and they can be applied for the analysis of soil behavior after liquefaction. Changes in the pore pressure are related to the volumetric deformations in soils in drained conditions, and one-dimensional diffusion is included in the algorithm. Programs DYSAC2, DYNAFLOW, and SWANDYNE are considered to provide the best results (Arulanandan et al., 1995). Equations of motion of the liquid and solid phases are related to the equation of conservation of matter. Generation and dissipation of the pore pressure are connected with the deformation of the solid matrix due to the Biot equations (Biot, 1956). However, in every case simplifications and assumptions are applied, concerning the medium properties, as well as the mechanisms of the processes, therefore, any uncertainties and mistakes in modeling lead to an improper calculation of the soil movement.

At the same time, records of strong ground motion provided by seismic vertical arrays allow estimation of stress-strain relations in soil layers *in situ*. This chapter describes the method of estimation of stress-strain relations developed by Pavlenko & Irikura (2003) and its application to 1995 Kobe, 2000 Tottori, and 1999 Chi-Chi earthquakes. The method allows us to trace temporal changes in the stress-strain relations. Since the estimates are based only on real measurements, they are free of theoretical approximations and physical assumptions concerning mechanisms of processes arising in the medium in strong ground motion.

2. Estimation of nonlinear time-dependent soil behavior in strong ground motion based on vertical array data

Vertical array records of the 1995 Kobe earthquake were processed for three recording sites, Port-Island, SGK, and TKS. Distances to the closest point on the fault line are 2 km, 6 km, and 16 km, respectively. Figure 2.1 shows the locations of the sites, the major principal axes, and the epicenters of the main shock and aftershocks summarized by Disaster Prevention Research Institute of Kyoto University.

At Port Island, the vertical array contains four three-component accelerometers, installed at GL-0 m, GL-16 m, GL-32 m, and GL-83 m; the arrays at SGK and TKS sites consist of three three-component devices at GL-0 m, GL-24.9 m, and GL-97 m, and GL-0 m, GL-25 m, and GL-100 m, respectively (Fig. 2.2). We checked the directional drifts of the accelerometers by calculating the horizontal orbits of the long-period particle motions for the main shock and the aftershocks at different depths at the three sites. At Port Island, N19°W rotation at GL-83 m was detected and corrected; at SGK site, a reverse of NS component and N6°W rotation at GL-24.9 m and a reverse of NS component and N34°E rotation at GL-97 m were detected and corrected. At TKS site, N23°W rotation at GL-25 m and N9°E rotation at GL-100 m were detected and corrected. All these corrections agree with the conclusions of other authors.

The materials at the three sites are similar to one another: reclaimed soil, clays, sands, and gravel (Fig. 2.2). The profiling data at Port Island used for nonlinear simulation were taken from (Aguirre & Irikura, 1997). Shear wave velocity, shear modulus degradation, and maximum shear strain at SGK and TKS sites were taken from Soeda et al. (1999). These data were used to calculate the shear stress in failure τ_{max} and the low-strain shear modulus G_{max} in different layers, following Seed et al. (1984) and Sun et al. (1988). P-wave velocities, when exceeding ~1000 m/s, indicate saturation of soils with water. The underground water level lay at about 13 m at Port Island and ~3-5 m at SGK and TKS. The deepest layers were Pleistocene gravely soils and the upper layers consist of alternating Pleistocene gravel/clay layers, Holocene sand/clay layers and fill (Sato et al., 1996).

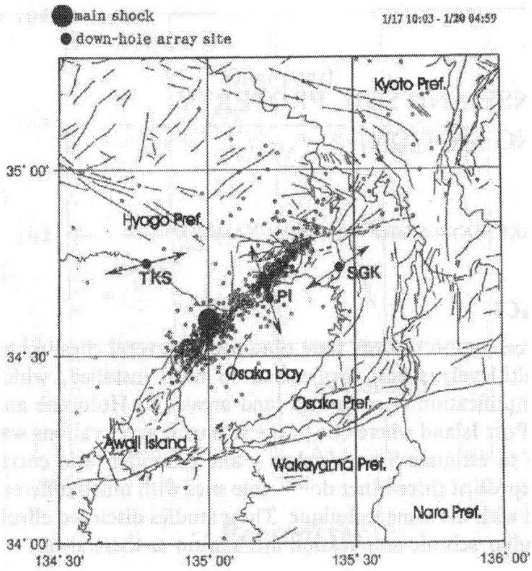

Fig. 2.1. Locations of vertical array sites Port Island (PI), SGK, and TKS around Osaka bay, the major principal axes, and the epicenters of the main shock and aftershocks (derived from Sato et al., 1996).

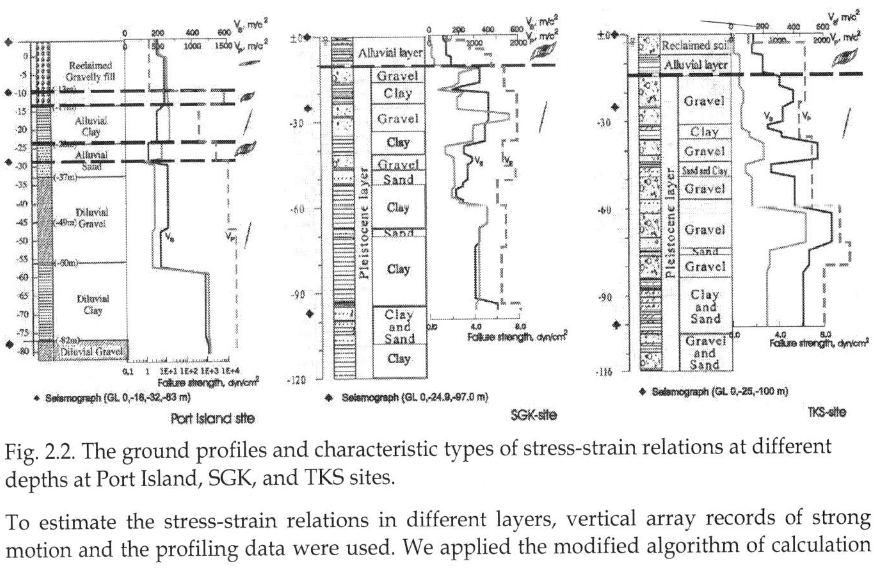

Fig. 2.2. The ground profiles and characteristic types of stress-strain relations at different depths at Port Island, SGK, and TKS sites.

To estimate the stress-strain relations in different layers, vertical array records of strong motion and the profiling data were used. We applied the modified algorithm of calculation of the propagation of a vertically incident shear wave in soil layers up to the surface, based on the lumped mass method (Joyner & Chen, 1975), and combined it with the method of "trial and error" to find the stress-strain relations showing the best agreement between the observed and simulated records. In our computations, stress and strain are normalized in the manner used by Hardin & Drnevich (1972): stress is normalized by multiplying by $1/\tau_{max}$, and strain is normalized by multiplying by G_{max}/τ_{max}. For calculations, the studied medium from the surface down to the location of the deepest device was divided into groups of layers, for which certain types of stress-strain relations were assumed (Fig. 2.2). Three main types of stress-strain curves were considered:

1. Those that are similar to laboratory experiments by Hardin & Drnevich (1972), to describe the behavior of dense soils at depths;
2. Those of "soft" type, similar to type (1), but with greater slope, being close to horizontal for large strains, for liquefied soils;
3. Those of "hard" type, declining to the stress axis at large strains, for water-saturated soils (terms "soft" and "hard" type stress-strain curves were introduced by Zvolinskii, 1982).

Sets of such curves were generated, and item-by-item examination was applied to find groups of curves showing the best-fit approximation to the observed data.

To account for temporal changes in the soil behavior, the records were divided into intervals of 1.5-seconds duration. Within each interval, the stress-strain relations were assumed to be stationary, and vary for different intervals.

Calculations (for horizontal components) were performed (i.e., the "best-fit" stress-strain relations in the layers were determined) successively, interval by interval. No discontinuities occur at the boundaries of the intervals, because, in the next interval, the whole cycle of loading (or unloading) was recalculated for the new curve from its beginning.

The results of the simulation with the "best-fit" stress-strain relations show a good agreement between the observed and simulated records at the three sites. The most intense and complex movements and liquefaction were observed at Port Island. The strongest horizontal accelerations were measured at the deepest point of 83 m, such as, ~500 Gal for the EW component and ~600 Gal for the NS component. According to the material structure and the results obtained in the previous analysis (Kawase et al., 1995), the medium was divided into five groups of layers, for which certain types of stress-strain relations were assumed (Fig. 2.2). Since liquefaction occurred in the upper layers, curves of type (2) were applied for surface soils above the level of the underground water (0–13 m). Soil layers at depths 13–18 m and 27–32.5 m are reclaimed gravely fill and alluvial sand saturated with water, therefore, curves of type (3) were used to describe the soil behavior in these layers (Fig. 2.2). Peaks, or sharp increases in the amplitudes of the observed records at depths of 16 m and 32 m confirm that curves of type (3) are relevant in this case. Stress-strain curves of type (1), which are considered to be typical for dense soils at depth, were assumed for alluvial clay layers at 18–27 m and for deep diluvial layers below 32.5 m. For successive 1.5-second time intervals, the groups of stress-strain relations were found, showing the best-fit approximation to the observed records. Figure 2.3 shows the observed and simulated records (a) and the obtained stress-strain relations at different depths (b) for ten successive time intervals.

The most noticeable changes in the soil behavior were observed in layers near the surface, in the upper 13 m: the stress-strain relations become more and more sloping with time, showing a substantial progressive reduction of the shear modulus and liquefaction. At

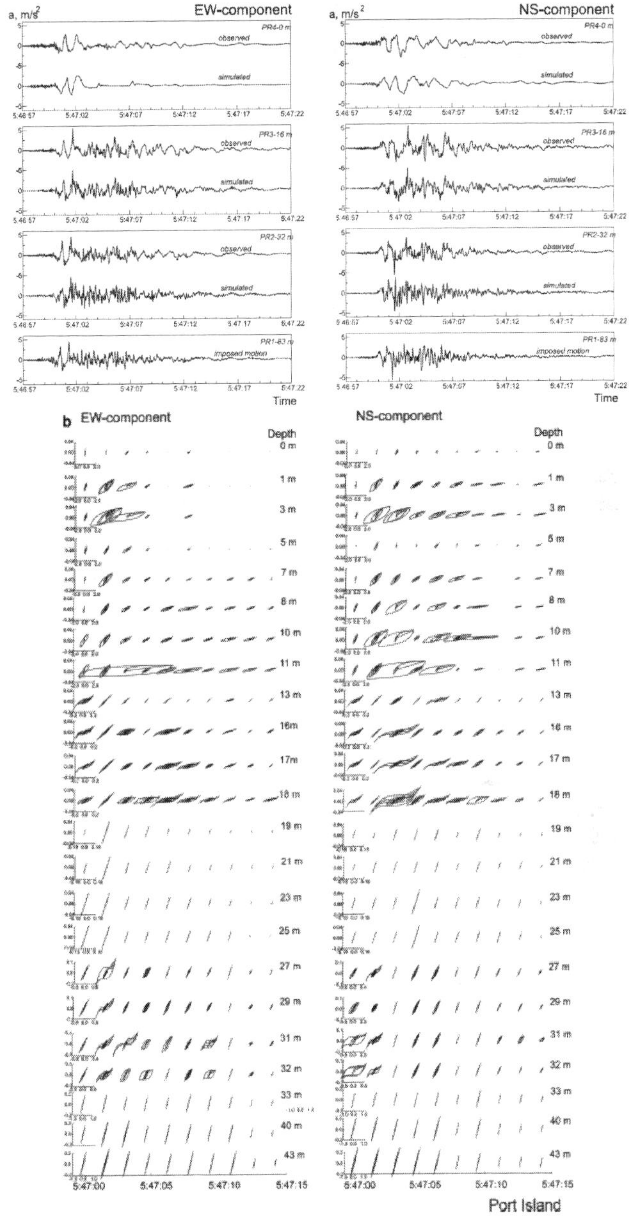

Fig. 2.3. The acceleration time history of the main shock in PI, observed and simulated (a), and the obtained stress-strain relations changing with time (b). The axes scales of the stress-strain relations are in relative units; the same for all time intervals at a given depth.

depths 13–18 m and 27–32.5 m, the obtained stress-strain curves show a slight reduction and the following recovery of the shear modulus. Below 32.5 m, no changes in the soil behavior are observed (Fig. 2.3b).

At SGK and TKS sites no liquefaction occurred, though sand boils observed after the quake around the TKS site indicate that liquefaction took place in the vicinities of this site. Maximum accelerations recorded at SGK site were also high (Fig. 2.4a), up to 650 Gal on the surface for the EW-component.

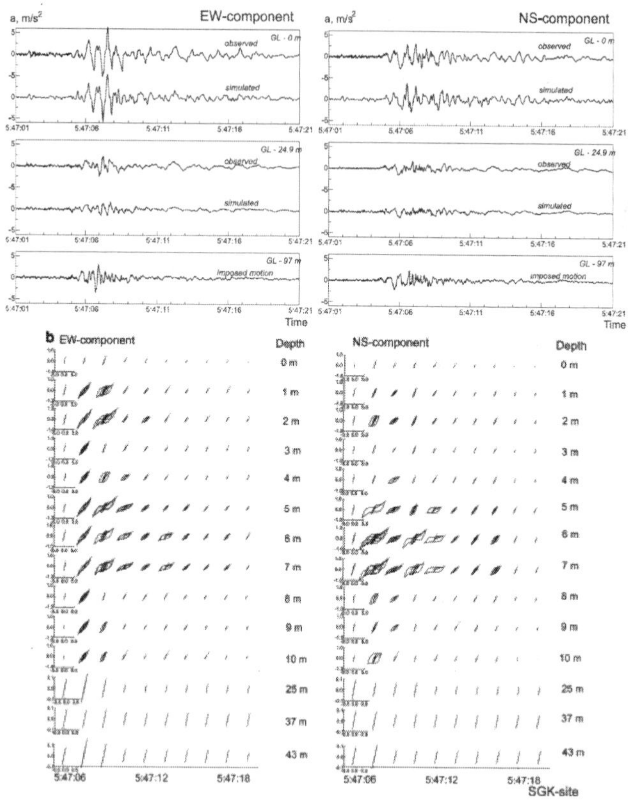

Fig. 2.4. The acceleration time history of the main shock in SGK site, observed and simulated (a), and the obtained stress-strain relations changing with time (b). The axes scales of the stress-strain relations are in relative units; the same for all time intervals at a given depth.

At this site, the soil behavior can be described by a relatively simple model: the soil profile is divided into two groups of layers, such as the near-surface alluvial layers (0–11 m), and the deeper layers (below 11 m), which are mostly gravel and clays. High values of P-wave velocities in the upper layers indicate the presence of water (Fig. 2.2), therefore, curves of type (3) are appropriate for the upper layers. Curves of type (1) were used for the deeper layers. Figure 2.4 represents the observed and simulated accelerograms (a) and the obtained stress-strain relations (b) for SGK site. The behavior of the layers below 11 m is stationary, whereas the stress-strain relations describing the behavior of the upper layers change with time: slopes of the curves decrease, then increase again, indicating reduction and recovery of the shear modulus in the upper layers.

At TKS site, maximum recorded accelerations were about 200 Gal, and the soil profile is represented by a water-saturated reclaimed fill and an alluvial layer in the upper part (0–14 m), and by gravel, clays, and sands in the deeper parts. Since the soil conditions are similar to that at SGK site, the choice of curves was also similar: curves of type (3) were selected for the upper 14 m, and curves of type (1) were chosen for the deeper parts. The most intense movements took place in the layers below the level of the underground water, at depths 4–7 m. At TKS site, the soil behavior in the upper layers changes with time similar to SGK site: we observe reduction and the following recovery of the shear modulus, whereas, the behavior of layers below 14 m is stationary.

The obtained stress-strain relations seem to represent a fairly good approximation to reality. On one hand, they show a good agreement between the observed and simulated data. On the other hand, they give a description of the process which is physically correct, i.e., the stress-strain relations obtained for Port Island show progressive liquefaction in the upper layers and a stable behavior in the deeper parts. Reduction and a following recovery of the shear modulus is obtained for the upper layers at SGK and TKS sites. Similarity of the stress-strain relations obtained for two horizontal components is an additional factor testifying to the validity of the solution. As shown above, the choice of the types of stress-strain relations in layers was physically justified. To check the possibility of different representations (i.e., vertical distributions of the stress-strain relations), calculations were performed several times for Port Island and SGK sites, with variations in the shapes of curves, however, every time similar distributions of the stress-strain relations in layers were obtained.

Numerical simulation of accelerograms of the Kobe earthquake at depths of locations of the recording devices has shown that, (1) at least within ~8-10 km from the fault plane, the nonlinearity in the soil response was substantially higher than that stipulated by conventional computer programs of the nonlinear ground response analysis, and (2) stress-strain relations of different types, depending on the composition of soil layers, their saturation with water, and depth, describe the behavior of the layers. In particular, the behavior of sandy water-saturated or wet subsurface soils is described by stress-strain relations of "hard" type, declining to the stress axis at large strains. In such soils, amplification of large-amplitude oscillations occurs, which is related to the "hard-type" nonlinearity of the soil response, like at SGK site during the 1995 Kobe earthquake.

3. Nonlinear behavior of soils revealed from the records of the 2000 Tottori (Japan) earthquake at stations of the Digital Strong Motion Network Kik-Net

During the Kobe earthquake, the nonlinear soil behavior was identified at sites located within ~16 km from the fault plane, and the content of nonlinear components in the soil response was estimated. It turned out to be rather high, up to ~60% of the whole intensity of the response, at ~2 km from the fault plane and about 10-15% of the intensity of the response at ~16 km from the fault plane (Pavlenko & Irikura, 2005). Such kind of analysis and estimations became possible because of the availability of vertical array records of the Kobe earthquake. Seismic vertical arrays usually contain two, three, or four three-component accelerometers, installed on the surface and at depths down to ~100 or ~200 m; one of the primary motivations for observations with borehole arrays is to understand nonlinear soil response.

The Tottori earthquake occurred in the Tottori Prefecture of Japan at approximately 1:30 p.m. on October 6, 2000. The earthquake mechanism was a strike-slip fault, and the fault rupture plane surface was about 30 km wide by 10 km deep, nearly vertical. This earthquake was recorded by stations of the Kiban-Kyoshin Digital Strong-Motion Seismograph Network (Kik-Net) at 220 sites located at epicentral distances of 7 to 626 km. It resembled the 1995 Kobe earthquake in its magnitude and focal depth and therefore represented a good opportunity to check the conclusions made based on records of the Kobe earthquake. Acceleration records of the 2000 Tottori earthquake, provided by the Kik-Net stations, show a clear evidence of the nonlinearity of soil response at sites located in near-fault zones, such as evident differences in shapes and spectra of records on the surface and at depth, emphasizing low-frequency oscillations on the surface. In this paragraph, we analyze accelerograms of the 2000 Tottori earthquake (Fig. 3.1) and reconstruct stresses and strains, induced in soil layers at sites, where the nonlinear soil behavior was identified. We estimate reduction of the shear moduli of the soil layers, caused by the strong motion, and discuss transformations of spectra of seismic waves in the soil layers. The Kik-Net stations contain two accelerometers, installed on the surface and at a depth of ~100 or ~200 m. Simultaneous records of the two devices allow us to simulate the behavior of soil layers from the surface down to the location of the deep device. For data processing, we chose stations located within ~80 km from the epicenter, where maximum recorded accelerations exceeded 100 Gal, assuming that at larger distances or at smaller accelerations the soil response is linear. The list of the studied Kik-Net stations with their epicentral distances, maximum recorded accelerations, and soil conditions is given in Table I.

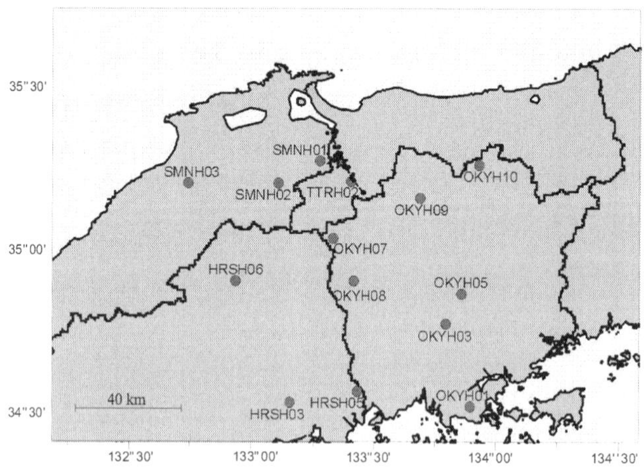

Fig. 3.1. Locations of the main shock of the 2000 Tottori earthquake and KiK-net stations in the vicinity of the epicenter (derived from the KiK-net website).

Nonlinear soil behavior was identified at five Kik-Net stations: TTRH02, SMNH01, HRSH06, SMNH03, and HRSH05 stations, located at 7 km, 8 km, 57 km, 57 km, and 80 km from the epicenter, respectively. At other stations, listed in Table I, either the profiling data were absent, or the thickness of soft subsurface soil layers was less than ~10 m and they were underlayed by dense rock, so that resonant phenomena predominated over nonlinear ones.

The profiling data, such as, the composition of the soil layers, P- and S- wave velocities, are provided by the Kik-Net website. At TTRH02 station, the upper 10.5 m represent sand and gravel (V_S~210 m/s); at SMNH01 station, the upper 12 m are fill soil with sand, gravel, silt, and cobble stone (V_S~290 m/s). At both stations below these layers down to ~100 m, granites, andesites, and basalts are identified with V_S gradually increasing from 340 m/s (at 10.5-20 m) to 790 m/s (at 42-100 m) at TTRH02 and from 550 m/s (at 12-22 m) to 2800 m/s (at 54-100 m) at SMNH01. At HRSH06 station, the upper 6 m represent weathered mudstone (V_S ~ 170 m/s), below which mudstones, conglomerates, and granites are found with V_S increasing from 270 m/s (at 6-19 m) to 1650 m/s (at 51-100 m). At SMNH03, the soil profile consists of fill soil, concrete, sand, and gravel with cobble stone (V_S ~ 240 m/s) in the upper 7 m and mudstones, sandstones, and porphyrites below with V_S increasing from 390 m/s (at 7-14 m) to 1300 m/s (at 51-100 m). At HRSH05, the upper 7.5 m of surface soil, clay, sand, and gravel (V_S ~ 280 m/s) are underplayed by more dense layers of sand and gravel with cobble stone, sandy clays, slates, and diabases with V_S increasing from 420 m/s (at 7.5-32.5 m) to 2390 m/s (at 80-100 m).

Site code	Site name	Max acceleration	Epicentral distance	Thickness of the soft layer (V_S< 300 m/s)
TTRH02	HINO	927.2 gal	7 km	11 m
SMNH01	HAKUTA	720.4 gal	8 km	10 m
SMNH02	NITA	564.0 gal	24 km	0 m
OKYH07	SHINGOU	179.7 gal	26 km	0 m
SMNH10	MIHONOSEKI	226.4 gal	31 km	no profiling data
OKYH09	YUBARA	283.8 gal	32 km	0 m
TTRH04	AKASAKI	218.1 gal	33 km	no profiling data
OKYH08	TETSUTA	238.5 gal	41 km	0 m
OKYH14	HOKUBOU	443.0 gal	45 km	no profiling data
SMNH12	YOSHIDA	258.6 gal	46 km	no profiling data
SMNH11	HIRATA	58.2 gal	52 km	no profiling data
OKYH10	KAMISAIBARA	280.7 gal	53 km	8 m
HRSH06	KUCHIWA	240.3 gal	57 km	18 m
SMNH03	SADA	154.8 gal	57 km	7 m
OKYH05	TAKEBE	149.0 gal	65 km	4 m
OKYH03	OKAYAMA	129.5 gal	69 km	7 m
OKYH11	SYOUOU	139.1 gal	74 km	9 m
SMNH05	HASUMI	121.5 gal	79 km	4 m
HRSH05	KANNABE	131.0 gal	80 km	7 m

Table 3.1. Maximum recorded accelerations, epicentral distances, and soil conditions at the Kik-Net stations located in near-fault zones of the 2000 Tottori earthquake (data derived from the Kik-Net website).

For calculations, the soil profiles were divided into two groups of layers, according to the profiling data. At all stations the groups of upper soft layers were distinguished (10.5 m at TTRH02, 12 m at SMNH01, 6 m at HRSH06, 7 m at SMNH03, and 7.5 m at HRSH05), and "hard-type" stress-strain relations were selected to describe the behavior of these layers. To describe the behavior of deeper layers at all the stations, we used stress-strain relations obtained in laboratory experiments by Hardin & Drnevich (1972). To account for temporal changes in the soil behavior, successive 1.5-second time intervals were analyzed; calculations were performed successively, interval by interval. Since the profiling data provided by the Kik-Net website contain only the composition of the soil and P- and S-wave velocities in the layers, we estimated other parameters used in calculations, such as shear stress in failure, τ_{max}, density, and damping in the soil layers. Also we defined more exactly S- wave velocity profiles at the closest to the fault plane stations TTRH02 and SMNH01. Density and damping were estimated based on the soil composition. To estimate other profiling parameters, we performed inversion using the genetic algorithm. For the inversion, we used twenty aftershocks of the Tottori earthquake recorded by the two stations during one month after the main shock. Intervals defining allowed values of S-wave velocities V_S in the soil layers (input data in the inversion problem) included V_S values provided by the Kik-Net website; limiting estimates of τ_{max} were based on empirical relationships, accounting for the composition of a soil layer, S- wave velocity and preexisting stress in the layer. For each model (combination of V_S and τ_{max} profiling values), we calculated the propagation of the aftershock waves in the soil layers; the behavior of the layers was described by the stress-strain relation obtained by Hardin & Drnevich (1972).

All the models were evaluated by the summary deviations of the simulated accelerograms from the recorded ones. The deviations were calculated as sums of the mean square "point-by-point" deviations and the differences of the mean intensities of the simulated and recorded accelerograms. The "best" model was produced after crossing of fifty initial models during about five hundred generations. Based on the obtained profiling data, stresses and strains induced in the soil layers at the five stations during the main shock of the earthquake were estimated. The obtained stresses and strains were used to trace changes in the shear moduli in the soil layers. Estimated stresses and strains at depths of 0 - 100 m or 0 - 200 m, changing with time during the strong motion for TTRH02 and SMNH01 stations are shown in (Figs 3.2 and 3.3).

A rather good agreement was obtained between the simulated and observed accelerograms for all the stations, though the agreement is worse than that achieved for SGK and TKS sites in simulating accelerograms of the 1995 Kobe earthquake. Evidently, the accuracy of simulation sufficiently depends on the available information on the parameters of the soil profiles. However, the applied method of simulation allows some correction of the soil parameters. Stress-strain relations used in calculations are defined in their normalized form in the manner proposed by Hardin & Drnevich (1972): stress is normalized by multiplying by $1/\tau_{max}$, and strain is normalized by multiplying by G_{max}/τ_{max}. During the calculations, stress-strain relations are selected, which satisfy the prescribed (probably, with some error) values of G_{max} and τ_{max} to simulate oscillations on the surface close to the observed ones. At the same time, the result, i.e., vertical distributions of stresses and strains in the soil layers, is expressed in absolute stress and strain units (Figs 3.2, 3.3); this decreases its dependence on the selected G_{max} and τ_{max}, and allows us to make conclusions about the soil behavior.

At the closest to the fault plane stations TTRH02 and SMNH01, the soil response was substantially nonlinear, as seen from Figs 3.2 and 3.3. At these stations, the behavior of the upper soil layers during the earthquake can not be described by a single stress-strain relation; an agreement between the observations and simulations can only be obtained if we describe the behavior of the upper layers by different stress-strain relations at different time intervals. Thus, strong ground motion changes rheological properties of the upper soil layers in near-fault zones.

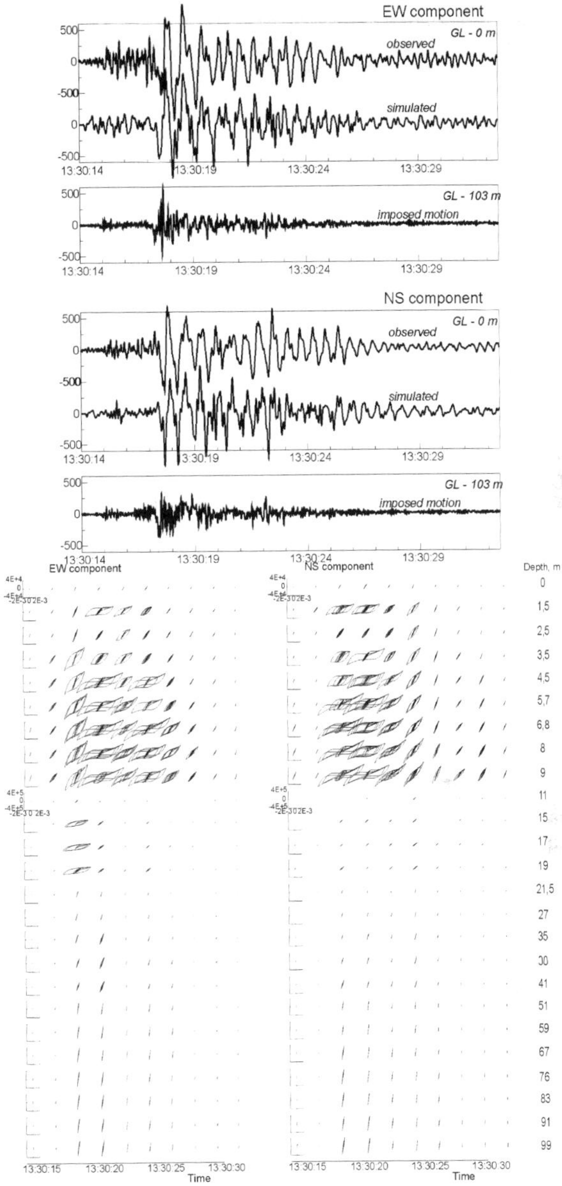

Fig. 3.2. The acceleration time histories of the main shock of the Tottori earthquake, observed and simulated, and estimated stress-strain relations in the soil layers, changing with time during the strong motion, at the TTRH02 station (stresses are given in kPa, strains - in strains, the same scales for depths 0 - 10 m and 11 - 100 m).

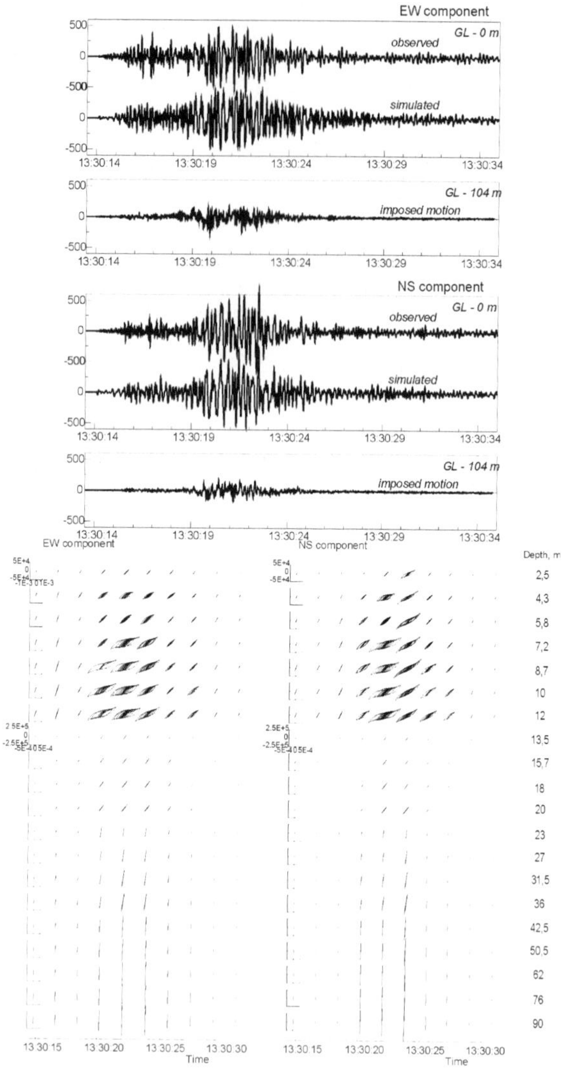

Fig. 3.3. Same as Figure 3.2 for the SMNH01 station (stresses are given in kPa, strains – in strains, the same scales for depths 0 – 12 m and 13 – 100 m).

Strains induced by the strong motion achieve 0.6% in the upper 10.5 m at TTRH02 station and 0.3% in the upper 12 m at SMNH01 station. At both stations resonant oscillations in the upper soil layers are observed; maximum stresses are as high as ~0.6 bars in the upper 10-12 m and increase with depth up to ~5 bars at 50-100 m, whereas strains decrease with depth. The shapes of the stress-strain curves in the upper soil layers reveal pore pressure development during the strong motion (Figs 3.2 and 3.3).

At stations HRSH06, SMNH03, and HRSH05, located at epicentral distances of 57 km and more, the behavior of the upper soil layers is described by the same stress-strain relation at all time intervals, rheological properties of the soil are not changed. During the 1995 Kobe earthquake, changes of the rheological properties of the upper soil layers were detected at Port Island and SGK sites, located within ~6 km from the fault plane. Thus, records of two earthquakes with magnitudes M_W ~6.7–6.8 and focal depths less than ~30 km allow a rough estimation of the area where rheological properties of the upper soil layers change, as ~7-8 km, or ~1/4 of the length of the fault. Stresses and strains in the soil layers at these stations are substantially lower than at TTRH02 and SMNH01: up to 0.12 bars and 0.05% in the upper 5-8 m; stresses increase up to ~0.5 bars at 50-100 m.

The estimated stresses and strains, changing with time during the strong motion, were used to trace changes in the shear moduli of the soil layers. As seen from Figs 3.2 - 3.3, these changes are observed in the upper 9-12 m of the soil profiles at the closest to the fault plane stations. The behavior of the deeper layers at these stations and the behavior of all the layers at remote stations HRSH06, SMNH03, and HRSH05 were stable. Shear moduli were calculated as slopes of stress-strain curves averaged over the groups of upper layers and over oscillations within each time interval during the strong motion. At TTRH02 and SMNH01 stations, the reduction of the shear moduli in the upper layers achieved ~60% of their initial values, at HRSH06 station the reduction of the shear moduli did not exceed 15 % of the initial value, though scattering does not allow accurate estimation, and at SMNH03 and HRSH05 stations it was negligible.

Though scattering of the obtained estimates of shear moduli reduction indicate some inaccuracy in our simulations, we can see from our results a total recovering of the shear modulus at TTRH02 station and its almost total recovering at SMNH01 station. Note that the recovering starts immediately following the decrease of the intensity of the strong motion, which testifies to a high permeability of the upper soil layers as composed of non-cohesive soils: sands, gravels, silts with cobble stones. Similar shear modulus behavior was observed at SGK and TKS sites during the 1995 Kobe earthquake (Pavlenko & Irikura, 2002).

Note a significant amplification of low-frequency oscillations on the surface at TTRH02 station, which is due to the "hard-type" nonlinearity of the soil response, as seen from Fig. 3.2. A similar tendency is seen in records of SMNH01 station. At other stations, the thickness of the soft soil layer is rather small, and resonance effects predominate in oscillations on the surface. As a result, transformations of spectra caused by the nonlinearity of the soil response are virtually insignificant.

As known, nonlinearity of the soil response induces changes in spectra of seismic waves propagating in the soil layers: the energy of the waves is redistributed over the spectral bands, because of mutual interactions of spectral components of the propagating waves. Low-frequency components are amplified, spectral peaks and spectral gaps smooth, and spectra of oscillations on the surface tend to take the form $E(f) \sim f^{-k}$. This spectral shape can be achieved in cases of strong nonlinearity (intense seismic waves and/or thick soft soil layers), whereas in cases of weak nonlinearity, we only see the tendencies of these spectral transformations. These tendencies are clearly seen in spectra of TTRH02 station. At other stations, however, resonance phenomena prevail, and spectral changes, induced by the soil nonlinearity, are inconspicuous.

4. Characteristics of soil response in near-fault zones during the 1999 Chi-Chi, Taiwan, earthquake

The Chi-Chi earthquake (Mw=7.6) that occurred in central Taiwan on September 21, 1999 was recorded by more than 400 strong motion devices island wide. The majority of near-fault records were obtained at soil sites, and at present they apparently represent the most complete database allowing a study of soil behavior during a strong earthquake in near-fault zones at various distances from the fault.

In (Pavlenko & Wen, 2008) we simulated acceleration time histories of the Chi-Chi earthquake at rock and soil sites and constructed models of the behavior of upper ~80 m of soil at 31 soil sites located within ~50 km from the fault. To construct models of soil behavior, we used a method similar to that developed for the estimation of soil behavior based on vertical array records (Pavlenko & Irikura, 2003; 2006). As input for soil layers, we used acceleration time histories of the Chi-Chi earthquake, simulated by stochastic finite-fault modeling with a slip distribution over the fault plane obtained by Chi et al. (2001). First, we simulated the acceleration time histories of the Chi-Chi earthquake at 18 rock sites, and comparing them with the observed ones, calibrated the calculation program, i.e., found input parameters for the stochastic simulation, such as, parameters of radiation of seismic waves from the source and parameters of their propagation: geometrical spreading, Q(f), kappa operator describing additional attenuation of the spectra, parameters defining the shape of the time window and some others. The earthquake source (119 km by 35 km) was represented as a set of 85 subfaults of 7 km by 7 km. At the second stage, we simulated acceleration time histories at soil sites, at the bottoms of soil layers. These signals were used as "inputs" to soil layers, i.e., prescribed motion at the base of soil columns (that is, we assumed infinity rigidity in the underlying medium). The technique and the constructed models of soil behavior at 31 soil sites are described in detail in (Pavlenko & Wen, 2008). The models represent vertical distributions of hysteretic relations of stresses and strains, induced in the upper tens of meters of soil and changing in time during the strong motion.

Models of soil behavior are constructed for 31 soil sites located at various distances from the fault plane within a wide range of azimuthal directions, and they allow us to obtain a general representation of soil behavior during the Chi-Chi earthquake. With these models, we can investigate some regularities in soil behavior during a strong earthquake at various distances from the fault plane.

In this paragraph, the constructed models of soil behavior are applied to estimate parameters, characterizing soil response during the Chi-Chi earthquake, such as, amplification of seismic waves by soil layers, stresses and strains induced by the strong motion in soil layers at different depths, reduction of shear moduli in soil layers due to strong motion, and nonlinear components of soil response. The distributions of these parameters around the fault plane are analyzed. Information on the soil sites, such as, the distances to the fault plane and site classification according to (Lee et al., 2001), as well as estimated parameters of soil response, are given in Figure 4.1 and in Table 4.1.

Maximum and average (averaged over the duration of the strong motion) stresses and strains induced by the strong motion in soil layers were estimated. To obtain these estimates, at each site, for each soil layer, maximum (during the strong motion) stresses and strains were found; to find their average estimates, absolute values of "limiting" (for loading and unloading cycles) stresses and strains corresponding to final points of loading and unloading on the hysteretic curves were averaged for each site and for each soil layer; then, averaging was performed over two horizontal components, NS and EW. Average values of stresses and strains were estimated for the upper 30 m of soil most representative from the viewpoint of engineering seismology.

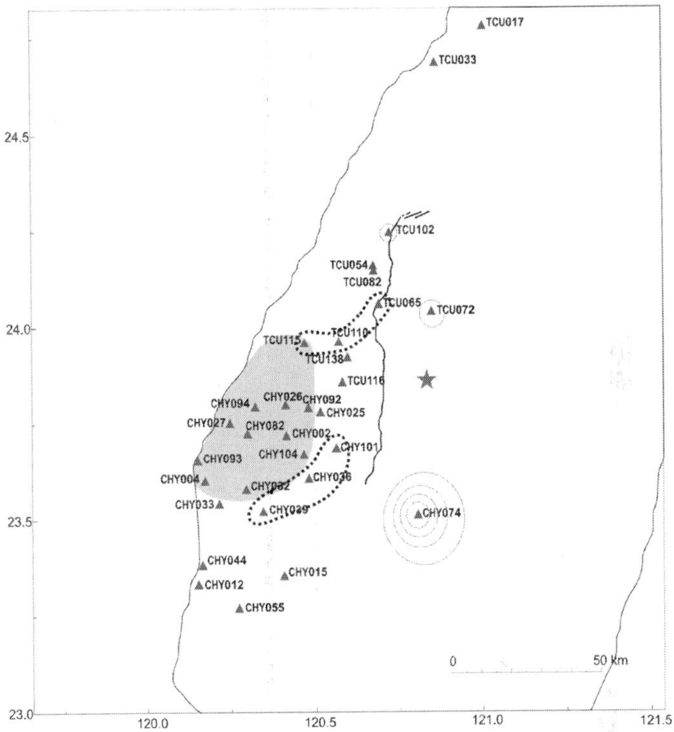

Fig. 4.1. Locations of the studied soil sites around the fault plane of the Chi-Chi earthquake. Dotted lines indicate areas showing signs of soil liquefaction; the area of basin-induced surface waves is marked by gray color.

Shear moduli reduction values were assessed for soil layers within the upper 30 m in the following way. For each site, average shear moduli were calculated as the ratios of stresses and strains averaged in time (over the strong motion duration), in depth (over the upper 30 m), and over two horizontal components; stresses and strains were taken as their absolute values corresponding to final points of loading and unloading cycles of the hysteretic curves. Shear moduli reduction was calculated as the difference between maximum values of shear moduli, calculated within small time intervals at the beginning parts of the strong motion records, and their average values, normalized by the average values and expressed in percent.

To estimate the contents of nonlinear components in soil response, methods of nonlinear system identification and the white-noise approach (Marmarelis & Marmarelis, 1978) were applied. Application of these methods to seismic data analysis is described in (Pavlenko, 2001; Pavlenko & Irikura, 2005).

Station Code	r, km	V_{s-30}, m/s	F_a	F_v	τ_{30}, Pa	γ_{30}, 10^{-5}	$\Delta G/G_{max-30}$, %	NI, %	Site Class (Lee *et al.*, 2001)
CHY101	1.9	260	1.5	2.0	1817	18.0	81	75	D
TCU110	2.4	207	0.7	2.9	1013	32.0	46	57	E
TCU116	3.4	483	1.0	1.8	852	0.3	32	16	E
TCU138	3.8	605	1.1	1.6	2140	0.54	10	24	D
TCU082	5.7	478	1.0	1.7	1162	0.3	15	15	D
CHY025	5.9	277	1.2	2.7	1487	11.5	41	62	E
TCU054	6.1	454	1.1	1.7	1494	0.8	21	11	D
TCU065	6.9	300	2.5	2.9	1677	9.3	63	54	D
CHY036	7.5	282	2.4	2.1	1995	9.6	73	61	D
TCU102	7.8	735	1.4	2.2	2325	3.4	51	56	D
CHY092	9.5	260	1.3	2.5	440	1.2	37	11	E
CHY104	9.5	225	2.2	2.9	1435	5.7	13	28	E
TCU115	11.4	190	1.4	3.1	654	17.0	34	50	E
CHY074	12.8	546	1.6	2.1	1824	3.5	44	25	C
TCU072	14.9	563	2.7	2.1	3481	6.2	37	48	D
CHY002	15.0	231	1.9	2.6	420	0.7	16	28	E
CHY026	16.1	226	1.2	2.0	530	8.0	22	33	E
TCU033	16.9	448	2.3	1.9	1642	0.5	12	22	D
CHY039	20.8	198	2.0	2.2	450	5.8	86	27	E
CHY094	24.9	227	1.6	2.7	1048	4.1	9	16	E
CHY032	25.6	202	3.1	3.2	780	2.9	9	33	E
CHY015	25.9	234	2.2	2.6	361	1.0	19	31	D
CHY082	26.6	209	2.2	3.0	211	0.8	13	25	E
TCU017	29.7	601	2.2	2.3	227	0.2	28	8	?
CHY027	32.1	221	3.9	3.4	645	4.9	13	24	E
CHY033	33.6	194	3.0	3.6	970	5.3	7	13	E
CHY004	38.2	279	2.8	3.0	1342	3.5	7	11	E
CHY093	41.2	199	2.8	3.4	1145	4.7	5	11	E
CHY055	41.5	240	3.7	3.0	1234	3.8	5	12	E
CHY044	43.0	193	3.0	2.9	576	2.2	5	30	E
CHY012	46.6	202	3.5	2.8	276	1.0	2	28	E

Table 4.1. Characteristics of soil response at 31 soil sites during the 1999 Chi-Chi earthquake

In this paragraph, the dependence of the described above parameters of soil response during the Chi-Chi earthquake on the distance from the fault plane is analyzed. All the studied soil sites are located in near-fault zones, and their distances from the fault plane are comparable with the fault length. Later on distance from the fault r designates the shortest distance of a site from the fault plane.

As seen from Table 4.1, the majority of sites (22 sites) possess soft soils in the upper 30 m: V_{s-30} is less than 300 m/s, whereas at other 9 sites possessing denser subsurface soils V_{s-30} varies within 448-735 m/s. So in many cases it was found reasonable to consider separately these two groups of sites, such as, sites with "softer" subsurface soils ($V_{s-30} \leq 300$ m/s) and sites with "harder" subsurface soils ($V_{s\text{-}30} \geq 450$ m/s). At the same time, the whole number of the studied sites is not large and virtually all of them are located in valleys to the west of the fault, so that further division of them into smaller groups considering the age and composition of soil layers is not reasonable.

4.1 Amplification of seismic waves in subsurface soils in near-fault zones during the Chi-Chi earthquake

The constructed models of soil behavior allowed a rather detailed study of amplification of seismic waves by subsurface soils during strong ground motion. Amplification factors were estimated for acceleration and velocity (Fa and Fv); the results are presented in Figure 4.2 and in Table 4.1.

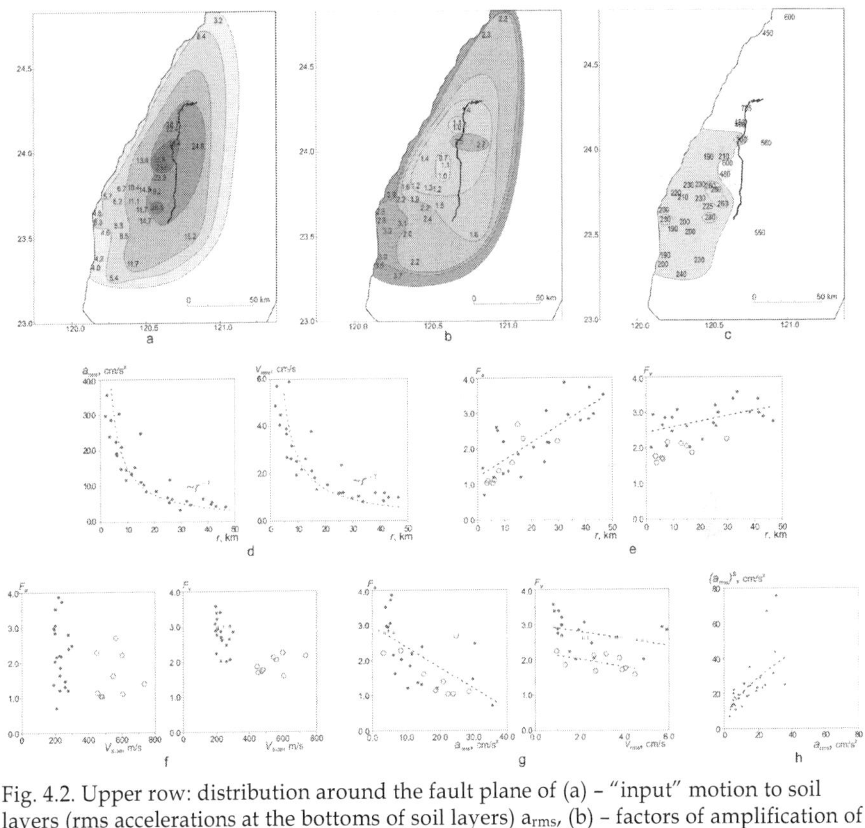

Fig. 4.2. Upper row: distribution around the fault plane of (a) – "input" motion to soil layers (rms accelerations at the bottoms of soil layers) a_{rms}, (b) – factors of amplification of seismic waves by soil layers for acceleration F_a, (c) – average S-wave velocities in the upper 30 m $V_{S\text{-}30}$. Lower row: (d) – "input" motion to soil layers a_{rms}, (e) – amplification factors for acceleration and velocity F_a and F_v as functions of the distance from the fault; (f), - the dependencies of the obtained amplification factors on the level of "input" motion to soil layers (f), on $V_{s\text{-}30}$ (g), and the dependency of rms accelerations on the surfaces $(a_{rms})_s$ on rms accelerations at the bottoms of soil layers a_{rms}.

The obtained results allow us to analyze the influence of various factors on the values of amplification of seismic waves by soil layers. Values of a_{rms} and v_{rms} decrease with increasing distance from the fault plane as $\sim r^{-1}$ (Fig. 4.2d), which corresponds to theoretical decrease of amplitudes due to geometrical spreading. Amplification factors Fa and Fv are greater than 1 and smaller than 4 for virtually all the sites (Fig. 4.2b, e-g; Table 4.1). As seen from the figures, amplification increases with distance (Fig. 4.2 d), i.e., with decreasing amplitudes of "inputs" to soil layers, a_{rms}, which is more clear for acceleration than for velocity. Linear functions can be used as approximating ones, however, scattering of the obtained estimates around linear regression lines is rather large, and it can be due to different factors.

As known, three main mechanisms of seismic wave transformation in subsurface soils are: (1) transition of seismic waves to upper soil layers with (usually) smaller values of $V_{s\text{-}30}$ and density, ρ, leading to amplification of seismic oscillations according to energy conservation law, (2) resonant phenomena the upper softer soil layers also leading to amplification, (3) nonlinearity of soil response, often leading to de-amplification of seismic oscillations. To evaluate the influence of different mechanisms on amplification of seismic waves in soil layers during the Chi-Chi earthquake, let's consider separately sites with "softer" and "harder" subsurface soils (in Fig. 4.2e-g they are shown by points and circles, respectively). As seen from the figure, amplification factors for acceleration, Fa, at "softer" and "harder" soil sites are close to each other, and can be approximated by the same function of the distance from the fault (or of the level of "input" motion), whereas, amplification factors for velocity, Fv, are substantially smaller at "harder" subsurface soils than at "softer" soils. This is clearly seen from Fig. 4.2f and agrees with seismological observations that "peak ground velocity and displacement show higher amplifications for soil sites than for rock sites (in our case, softer soil sites and harder soil sites), while peak ground acceleration is roughly independent of the site classification" (Aki & Irikura, 1991) (that is, does not show higher amplification at soft soil sites).

The obtained results are in a good agreement with the dependencies of amplification factors on the intensity of ground motion and on average S-wave velocities in upper 30 m of soil described in (Stewart et al., 2001). The authors discuss general regularities of reduction of amplification factors with increasing $V_{s\text{-}30}$ or reference motion amplitude, obtained by various researchers for various strong earthquakes. Preliminary data on the 1999 Chi-Chi earthquake, obtained by Stewart et al., are also in agreement with the results of this work (Fig. 4.2g).

The observed dependencies of amplification factors on the level of "input" motion to soil layers is evidently due to the influence of nonlinearity of soil response, which is higher at the closest to the fault sites. At rather large distances from the fault (~40-50 km), amplification factors Fa and Fv are close to each other (Fig. 4.2e), whereas at small distances, nonlinearity of soil response substantially decreases amplification for acceleration and, to a lesser extent, for velocity. This can be due to nonlinear transformations of seismic waves propagating in soil layers: their spectra tend to take the form $E(f) \sim f^{-n}$, when the high- and medium-frequency spectral components are decreased and the low-frequency components remain virtually at the same level. Obviously, the effect is stronger for accelerations than for velocities and displacements.

In (Pavlenko & Wen, 2008), it was concluded that at the majority of soil sites, soil behavior during the Chi-Chi earthquake was defined by resonant oscillations induced in soil layers (in the upper 40-60 m) during strong motion and by nonlinearity of soil response.

At near-fault sites, subsurface water-saturated sandy soils possess hard-type nonlinearity, where de-amplification of seismic oscillations due to nonlinear damping can be rather small, because: (1) stresses in soil layers rapidly increase at large strains (especially, in near-fault zones) due to pore-pressure built-up (and therefore, accelerations also increase); (2) nonlinear damping (which is proportional to areas within hysteretic curves) in can be rather small, as at TCU065 site, for example (Pavlenko & Wen, 2008). As a consequence, in near-fault zones in sandy water-saturated soils possessing hard-type nonlinearity, de-amplification of seismic oscillations due to nonlinearity often can not compensate their amplification stipulated by other two (linear) mechanisms; as a whole, we observe amplification of oscillations on the surface. This is clearly seen, for example, in vertical array records of the 1995 Kobe earthquake at SGK site: in the upper ~11 m of sandy soils possessing hard-type nonlinearity seismic oscillations were noticeably amplified, peak accelerations increased from 0.2-0.3 g at depths of ~25 m and 100 m up to ~0.7 g on the surface. At the same time at Port Island site, liquefied surface soils possessing soft-type nonlinearity substantially de-amplified seismic oscillations, especially their high-frequency components; peak accelerations at 83 m, 32 m, 16 m, and on the surface were almost similar, ~0.4-0.5 g (Pavlenko & Irikura, 2003).

Thus, amplification factors can be rather large even in cases of strong nonlinearity (at high levels of "input" motion, in near-fault zones) at sites possessing hard-type nonlinearity of subsurface soils, which is clearly seen in Fig. 4.2b: at TCU065 and TCU072 sites located very closely to the fault plane, amplification factors are rather high, Fa ~2.5-2.7.

Fig. 4.2h shows the obtained estimates of rms accelerations on the surfaces of soil, $(a_{rms})_S$, versus estimates of rms accelerations at the bottoms of soil layers, a_{rms}. This figure can be compared to the well-known findings of Idriss (1990), such as, plots peak ground acceleration (PHA) on rock versus PHA on soft soil (accounting for the fact that peak accelerations correlate well with rms accelerations, at least, in the case of the Chi-Chi earthquake). As a whole, the dependence shown in Fig. 4.2h agrees well with Idriss's data, but two points with increased $(a_{rms})_S$ values attract attention, which correspond to TCU065 and TCU072 sites, where subsurface soils possess strong hard-type nonlinear behavior. Our previous experience in studying soil behavior during strong earthquakes shows that we can almost always find areas in the closest vicinities of the fault plane, where soils possess such behavior.

Amplification of oscillations on the surface resulted from resonant phenomena in subsurface soils and hard-type soil nonlinearity are also observed at other near-fault sites, such as, CHY101 (Fa~1.5), CHY036 (Fa~2.4), TCU102 (Fa~1.4), CHY104 (Fa~2.2), TCU115 (Fa~1.4), CHY074 (Fa~1.6), CHY002 (Fa~1.9), etc.

"Hard" character of soil response nonlinearity virtually disappears at distances of more than ~12-15 km from the fault plane (Pavlenko & Wen, 2008), because of the decrease of "input" motion intensity. At rather large distances from the fault, amplification of seismic oscillations results from two described above linear mechanisms; whereas nonlinear effects decrease with distance and can not effectively deamplify seismic oscillations any more.

4.2 Stresses and strains induced in soil layers in near-fault zones during the Chi-Chi earthquake

Analyzing stresses and strains induced in soil layers by the Chi-Chi earthquake, it is also reasonable to distinguish sites with "softer" ($V_{s-30} \leq 300$ m/s) and "harder" ($V_{s-30} \geq 450$

m/s) soils. Fig. 4.3 represents estimated average (upper rows) and maximum (lower rows) stresses and strains induced in soil layers at the studied sites during the Chi-Chi earthquake.

Names of sites possessing "harder" subsurface soils are underlined. We can see from the figures that at the closest to the fault plane sites (within ~20 km from the fault), resonant phenomena are observed in the upper soft soil layers (down to ~40-60 m), i.e., trapping of seismic waves due to the impedance contrast between softer and neighboring denser layers. Stresses in soil layers increased with depth, whereas maximum strains usually occurred in the upper 15-40 m of soil (Fig. 4.3). At sites, where soft surface layers with Vs-30 ≤ 300 m/s were underlayed by denser layers, and impedance contrast was higher, resonant phenomena were more pronounced. Maximum strains were achieved at the closest to the fault plane sites, such as, TCU065, TCU110, TCU115, CHY101 (~0.6-0.8%), CHY025, and CHY036 (~0.4-0.6%). At other sites maximum strains in soil layers did not exceed ~0.1-0.4%. In (Pavlenko & Wen, 2008) it is concluded that liquefaction phenomena occurred at TCU065, TCU110, TCU115, CHY101, CHY036, and CHY039 sites.

The obtained estimates of average stresses and strains induced in the upper 30 m of soil during the strong motion, τ_{30} and γ_{30}, are given in Table 4.1, and the distributions of these parameters around the fault plane are shown in Fig. 4.4a, b. As seen from the figures, areas of softer subsurface soils with $V_{s\text{-}30} \leq 300$ m/s in bajada, to the west of the fault, correspond to zones of decreased stresses and increased strains. Average stresses and strains in the upper 30 m of soil, τ_{30} and γ_{30}, are shown as functions of the distance from the fault (Fig. 4.4d) and as functions of the level of "input" motion to soil layers, rms accelerations a_{rms} (Fig. 4.4f). At sites with "softer" subsurface soils (shown by points in the figures), average stresses and strains decrease with distance (in Fig. 4.4d, the corresponding approximations by functions ~ r^{-1} are shown), and they increase with the level of "input" motion, however scattering of estimates around approximating linear functions is rather large. At sites possessing "harder" subsurface soils, average stresses and strains (shown by circles) deviate from power functions and appreciably increase scattering. At these sites, average stresses are higher and average strains are lower than at sites with "softer" subsurface soils.

Average stresses and strains in the upper soil layers depend on the distance from the fault plane (the level "input" of motion), as well as on the mechanical parameters of a soil (V_S and densities). Stresses and strains naturally decrease with increasing distance from the fault, and, at the same time, stresses τ_{30} increase, and strains γ_{30} decrease with increasing $V_{s\text{-}30}$. Decreased stress values at some sites in the vicinity of the fault (for example, at CHY092 and CHY002) are obviously related to decreased $V_{s\text{-}30}$ values in the upper layers at these sites, whereas decreased strains values of at some other sites (for example, at TCU116, TUC138, TCU082, and TCU054) are due to relatively high $V_{s\text{-}30}$ in the upper soil layers at these sites.

Thus, according to the obtained estimates, at sites possessing "softer" soils, average strains decrease with distance from the fault more quickly than average stresses, probably because stresses in soil layers are defined to a larger extent by radiation from the earthquake source (level of "input" motion to soil layers) than by mechanical properties of soils, whereas strains more depend on soil properties, i.e., on resonant phenomena in upper soil layers.

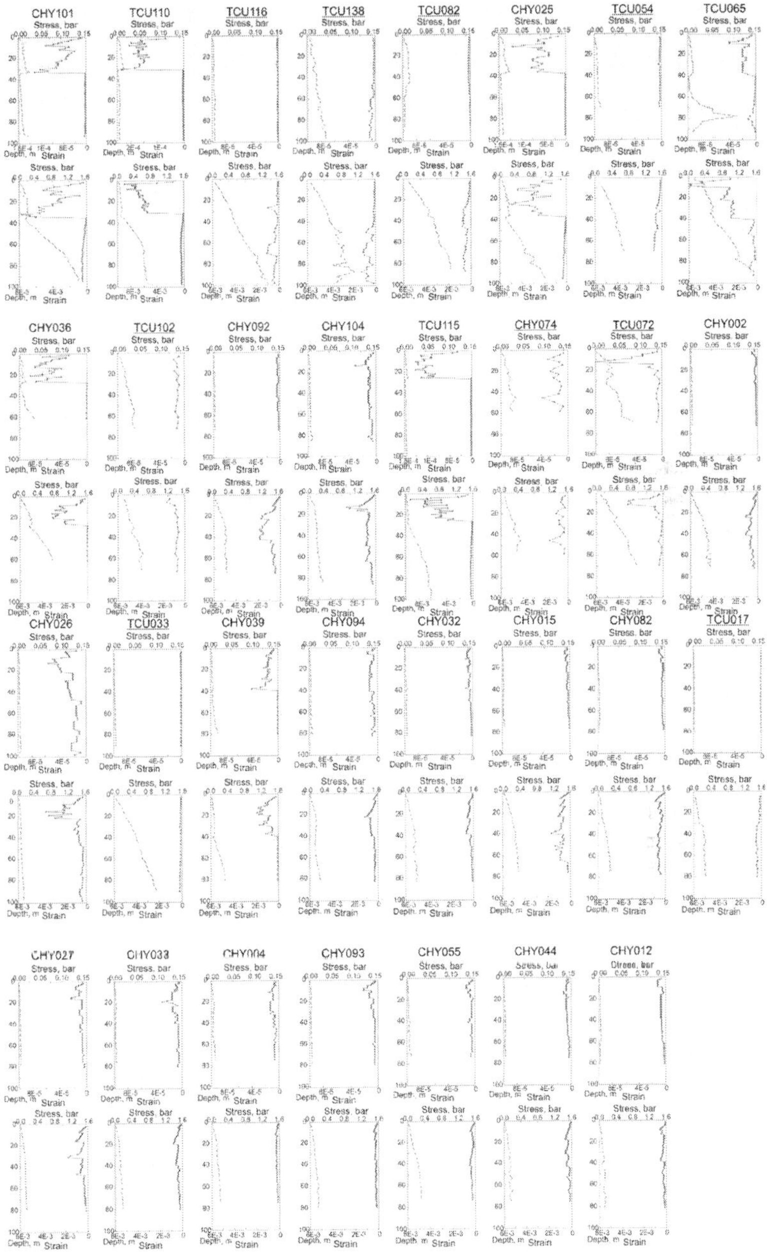

Fig. 4.3. Estimated average (upper rows) stresses (dash lines) and strains (solid lines), induced in soil layers during the Chi-Chi earthquake.

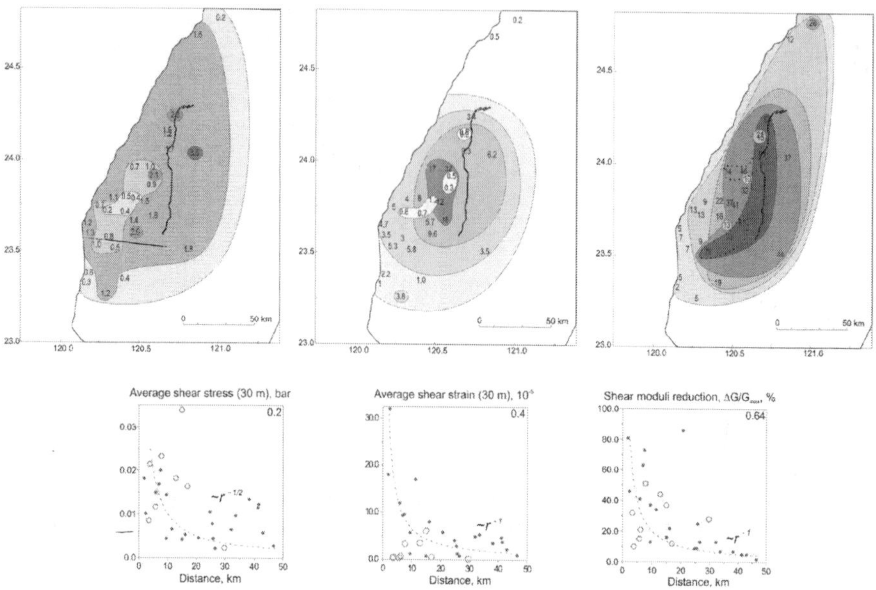

Fig. 4.4. Upper row: distribution around the fault plane of (a) – average stresses in the upper 30 m of soil τ_{30}, (b) – average strains in the upper 30 m of soil γ_{30}, (c) – estimates of shear moduli reduction in the upper 30 m of soil $\Delta G/G_{max\text{-}30}$ (dotted lines indicate areas showing signs of soil liquefaction). Lower row: (d) – average stresses τ_{30}, (e) – average strains γ_{30}, (f) – estimates of shear moduli reduction $\Delta G/G_{max\text{-}30}$ as functions of the distance from the fault.

4.3 Shear moduli reduction in soil layers in near-fault zones during the Chi-Chi earthquake

As concluded in (Pavlenko & Wen, 2008), at the closest to the fault plane sites soil behavior during the Chi-Chi earthquake was substantially nonlinear. Nonlinearity of soil response can be characterized by reduction of shear moduli in soil layers (Pavlenko & Irikura, 2005). Thus, the constructed models of soil behavior were used for estimating shear moduli reduction in the upper 30 m of soil, $\Delta G/G_{max\text{-}30}$. The obtained estimates are presented in Table 4.1; Fig. 4.4c, e, g show the distribution of $\Delta G/G_{max\text{-}30}$ estimates around the fault plane, and the dependencies of the estimates on the distance from the fault and on the level of "input" motion to soil layers, respectively.

The evaluated values of $\Delta G/G_{max\text{-}30}$ testify to high nonlinearity of soil response during the Chi-Chi earthquake: in the closest vicinity of the fault plane, at TCU110, CHY025, TCU065, TCU102, and CHY074 sites, shear moduli reduction in the upper 30 m achieved ~40-50% (Fig. 4.4c, e; Table 4.1). The highest $\Delta G/G_{max\text{-}30}$ values were obtained at CHY101, CHY036, and CHY039 sites, where the constructed models of soil behavior show substantial differences in the shapes of stress-strain curves at the beginning and at the end of the strong motion, indicating softening (probably, liquefaction) of surface layers. This area is marked by a dotted line in Figs 4.1 and 4.4 c.

High values of shear moduli reduction $\Delta G/G_{max-30}$ ~30-50% were obtained in the closest vicinity of the fault (within ~25 km from the surface rupture) at "soft" soil sites: TCU102, TCU110, TCU115, TCU072, CHY074, TCU116, CHY025, and CHY092 (Fig. 4.4c). At sites TCU138, TCU082, TCU054, and TCU033, possessing "harder" subsurface soils (V_{s-30} ~ 450-605 m/s), lower values of shear moduli reduction were obtained: $\Delta G/G_{max-30}$ ~10-20%. At sites CHY104, CHY002, CHY026, CHY094, and CHY032, shear moduli reduction can be underestimated because of the presence of basin-induced surface waves in records (Fig. 4.1), which noticeably increase amplitudes of oscillations on the surface and can not be simulated in our one-dimensional problem (Pavlenko & Wen, 2008).

With increasing distance from the fault, shear moduli reduction values decrease down to ~10% at 35-50 km and to ~5% at 50-70 km from the fault (Fig. 4.4c, e), where soil response becomes virtually linear.

In Fig. 4.4e, g shear moduli reduction values for "softer" and "harder" sites are shown by points and by circles, respectively. It is seen from the plots that shear moduli reduction estimates at "softer" sites decrease with distance approximately as ~r^{-1} and they grow in proportion to the level of "input" motion to soil layers. Points, noticeably deviating from the approximating functions, correspond to CHY039 and CHY036 sites with increased $\Delta G/G_{max-30}$, where the constructed models of soil behavior show liquefaction phenomena, and to CHY074 site with decreased $\Delta G/G_{max-30}$, possessing dense soils (Class C). As seen from the plots, $\Delta G/G_{max-30}$ estimates at "harder" sites substantially increase scattering of points around the approximating relationships (Fig. 4.4e, g).

Thus, reduction of shear moduli in the upper 30 m of soil at "softer" sites can be rather accurately described as a function, inversely proportional to the distance from the fault, ~r^{-1} and as a linear function of the level of "input" motion, a_{rms}. As mentioned above, strong ground motion during the Chi-Chi earthquake induced resonant amplification of seismic oscillations in the upper soft layers at many sites, and since soil conditions were similar at soil sites located mostly to the south and south-west of the fault (Fig. 4.1), average strains in the upper 30 m are proportional to the level of "input" motion to soil layers, and their decrease with distance from the fault can be approximately described as ~r^{-1}. Accordingly, shear moduli reduction values depend on distance and on the level of "input" motion in the same manner.

At the same time, at sites with "harder" subsurface soils, resonant phenomena were not observed in the upper 30 m of soil (according to our estimates, maximum strains at these sites correspond to depths of ~10 m and more (Fig. 4.3)), and estimates of shear moduli reduction depend not only on the distance from the fault (or on the level of "input" motion), but, to a greater extent than for "softer" soils, on the profiling data, which are very diverse at these sites, dispersed over a large area around the fault. Consequently, shear moduli reduction values can be quite different even at sites located at the same distances from the fault.

4.4 Nonlinear components of soil response in near-fault zones during the Chi-Chi earthquake

To estimate nonlinear components of soil response, models of soil behavior constructed in (Pavlenko & Wen, 2008) were tested by the Gaussian white noise. Linear and nonlinear components of soil response were distinguished, and their intensities were estimated in percent of the whole intensity of the response. The obtained estimates are given in Table 4.1.

Soft soils in the closest vicinity of the fault possess the highest nonlinearity of the response: according to our estimates, the whole nonlinear components of soil response achieve Nl ~ 60-80%, and nonlinear quadratic and cubic components are up to ~12-16% of the intensity of the response. At the same time, nonlinear residual components of soil response related to higher-order nonlinearities (the 4th order and higher) are also high, up to ~30-40% of the intensity of the response. Evidently, this indicates very high nonlinearity of soil response in near-fault zones during the Chi-Chi earthquake.

With increasing distance from the fault plane, nonlinear components of soil response decrease, and their dependence on distance can be described in the same manner as for other parameters, such as, approximately ~ r^{-1}. Accounting for sites possessing "harder" subsurface soils substantially increases scattering of points around the approximating functions. For sites with "softer" subsurface soils, the dependencies of the obtained estimates of nonlinear components on the level of "input" motion can be approximately described by linear functions. This is in agreement with our representations of soil nonlinearity: at soft soils, manifestations of nonlinearity increase with the level of "input" motion, i.e., with approaching to the fault.

At two closest to the fault plane sites possessing "harder" subsurface soils, such as, TCU072 and TCU102, nonlinearity of soil response was also high, evidently because of expressed resonant phenomena in the upper soft soil layers. At TCU072 site, resonant amplification of seismic oscillations occurred in the upper ~11 m of colluvium ($V_S \approx 250$ m/s), and at TCU102 site, resonant phenomena were related to the upper ~4-6 m of soft silty soils, which were softened during the strong motion. The behavior of these upper soft layers was substantially nonlinear and described by very sloping stress-strain relations (Pavlenko & Wen, 2008). Since surface soft layers were underlayed by hard breccia (at TCU072) and gravel (at TCU102), average S-wave velocities in the upper 30 m were higher than ~300 m/s at both sites, and the sites were assigned as possessing "harder" subsurface soils, however, nonlinear components at these sites satisfy regularities obtained for sites possessing "softer" subsurface soils. With increasing distance from the fault, nonlinearity of soil response decreases and at distances of ~40-50 km (~1/2 of the length of the fault plane), nonlinear components in soil response do not exceed 10-15%, according to our estimates.

5. Conclusions

Experimental data provided by recent large earthquakes, such as, the 1994 Northridge earthquake (M_W ~ 6.7), the 1995 Kobe (M_W ~ 6.8) and 2000 Tottori (M_W ~ 6.7) Japanese earthquakes, the 1999 Chi-Chi (M_W ~ 7.7) Taiwanese earthquake and others, have shown a clear evidence of the nonlinear behavior of subsurface soils in near-fault zones.

Acceleration records of the Kobe earthquake gave a bright illustration of the fact that in strong ground motion, maxima of energy of oscillations at soil sites shift to the lower-frequency domain. The fact was explained by the nonlinearity of the soil response: mutual interactions of spectral components of seismic waves propagating in soil layers lead to redistribution of energy of oscillations over the spectral band, and spectra of signals on the surface tend to take the form $E(f) \sim f^{-k}$.

Numerous surface records obtained during the 1999 Chi-Chi (Taiwan) earthquake possess similar features, indicating nonlinearity of the soil response in many places near the fault plane. Although these records can not be analyzed in the same manner as records of the Kobe earthquake because of the absence of borehole data, we can suppose that certain typical nonlinear distortions occur in seismic waves of similar intensities in subsurface soils.

Records of recent strong earthquakes, such as, the 1995 Kobe earthquake, the 2000 Tottori (Japan) earthquake, and the 1999 Chi-Chi (Taiwan) earthquake, were used to study the influence of nonlinearity of soil response on characteristics of ground motion.
In weak motion soil response can be considered as linear, whereas in strong motion it is nonlinear, and the degree of its nonlinearity depends on the intensity of oscillations. Nonlinearity of soil response leads to changes in spectra (which can be essential) and amplification of ground motion, as shown in the diagram:

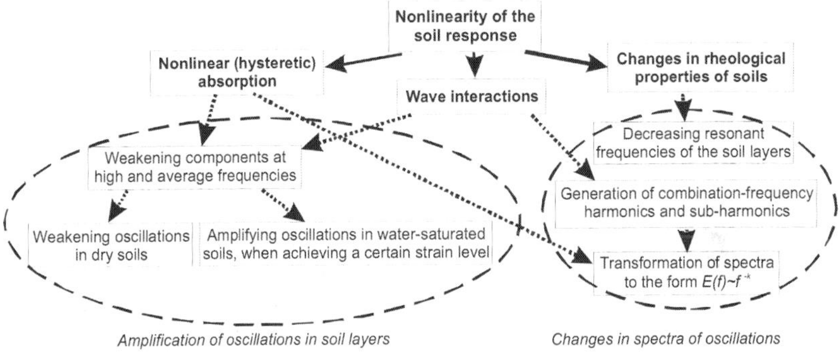

Fig. 5.1. Influence of nonlinearity of soil response on amplification and spectra of seismic waves propagating in soil layers.

In strong motion, nonlinear damping mechanisms are turning on, which lead to weakening of high- and average-frequency components. Low-frequency components are not weakened; they can be even amplified, because in cases when they were absent in input to soil layers they appear in the output, i.e., on the surface.
Changes in spectra of ground motion related to nonlinearity of soil response appear in shifting of resonant frequencies of soil layers to the low-frequency domain and in transformations of spectra of ground motion to the form $E(f) \sim f^{-k}$.
Amplification of seismic waves is decreased due to the nonlinearity of soil response (if compared with "linear" case) in dry soils (where the level of the underground water is lower than ~10 m) and, in lesser extent, in wet soils (where the level of the underground water is above ~10 m). In wet soils, nonlinear damping sometimes cannot compensate amplification of seismic waves resulting from (1) the transition of seismic waves to upper layers with smaller impedance values, and (2) resonant phenomena in the upper softer layers. As a whole, amplification can occur even in cases of high nonlinearity.
Thus, the degree of the nonlinearity of soil response depends on the composition of soil layers and on their saturation with water, as well as on the intensity and spectral composition of seismic waves, incident to soil layers. Since the composition and mechanical parameters of soils vary from place to place, any generalized dependencies of PGA (or oscillation intensity) on soils on magnitude-distance will surely possess large scatterings, which was noticed by many authors.
Therefore, the correct approach in seismic microzonation is calculation of space-dependent acceleration time histories of possible future strong earthquakes accounting for soil nonlinearity. The limitations of conventional programs for estimation of soil response are: underestimation of the degree of soil-response nonlinearity in near-fault zones, disregard of

the differences in nonlinear behavior of cohesive and non-cohesive soils (soft-type, hard-type, liquefaction), and disregard of changes in rheological properties of the upper soft layers in strong ground motion. Soil response should be estimated by methods of nonlinear analysis accounting for all points mentioned above.

According to the obtained estimates, based on records of the Kobe (1995), Tottori (2000), and Chi-Chi (1999) earthquakes, during crustal earthquakes with magnitudes M_W ~6.7-7.3, strong nonlinearity of the soil response (changes in rheological properties of the upper soil layers and shear modulus reduction of ~50-60% and more) are observed within an area up to ~¼ of the length of the fault plane. Within this area, noticeable manifestations of soil nonlinearity are observed in the upper 15-25-60 m of the soil profiles. At distances of ~ ½ of the length of the fault, nonlinearity is much weaker, even in soft subsurface soils. Stress-strain relations, suggested by Hardin & Drnevich (1972), adequately describe the behavior of soils at all depths in conditions of moderate dynamic loadings. In conditions of large loadings, in cases of strong nonlinearity, these stress-strain relations can be used to describe the behavior of dense soils at depths below some level, depending on the composition of the soil layers and their saturation with water, as well as on the magnitude and location of the earthquake; whereas the behavior of soft subsurface soils should be described by other, more "nonlinear" stress-strain relations. Such relations are found in this work.

Ignoring the features of soil behavior in strong ground motion leads to underestimation of maximum acceleration in near-fault zones, and mistaken estimates of spectra of oscillations on the surface at soil sites.

6. Acknowledgments

I thank Kansai Electric Power Company and Kobe City for acceleration records of the vertical arrays at SGK, TKS, and PI recording sites and I thank the website of the Kik-Net Digital Strong-Motion Seismograph Network for records of the 2000 Tottori earthquake and the profiling data. The profiling data at soil stations in Taiwan provided by NCREE are greatly appreciated. This study was partially supported by Grant-in-Aid for Scientific Research, No.11792026 and No.00099 (JSPS), from the Ministry of Education, Science, Sports, and Culture, Japan and by the National Science Council (Taiwan) under Grant No. NSC 95-2811-M-492-001.

7. References

Hardin, B. O. & Drnevich, V. P. (1972), Shear modulus and damping in soils: Design equations and curves, *Proc. Am. Soc. Civil Eng., J. Soil Mech. Found. Div.*, 98, 667-692.

Joyner, W. B. & Chen, T. F. (1975), Calculation of nonlinear ground response in earthquakes, *Bull. Seism. Soc. Am.*, 65, 1315-1336.

Marmarelis, P. Z. & Marmarelis, V. Z. (1978), *Analysis of Physiological Systems, The White-Noise Approach*, Plenum Press, New York and London.

Pavlenko, O. V. & Irikura, K. (2003), Estimation of nonlinear time-dependent soil behavior in strong ground motion based on vertical array data, *Pure Appl. Geophys*, 160, 2365–2379.

Pavlenko, O. V. & Wen, K.-L. (2008), Estimation of Nonlinear Soil Behavior during the 1999 Chi-Chi, Taiwan, Earthquake, *Pure Appl. Geophys.*, 165, 373-407.

Part 3

Seismic Hazard and Early Warning

11

Intraplate Seismicity and Seismic Hazard: The Gulf of Bothnia Area in Northern Europe Revisited

Päivi Mäntyniemi
Institute of Seismology, University of Helsinki
Finland

1. Introduction

Intraplate seismicity is a challenging subject to study, because earthquake observations accumulate slowly in plate interiors. It is known that great earthquakes can occur in these regions and that shallow earthquakes may break the surface, but examples are limited to a dozen or so. This has obvious implications to seismic hazard and risk. England & Jackson (2011) pointed out that the unanticipated earthquakes in continental interiors have claimed more human lives than earthquakes at plate boundaries in the past 120 years.

The present investigation deals with intraplate seismicity of the Fennoscandian (Baltic) shield in northern Europe. The seismicity of the region has been discussed by several authors over the years (e.g., Bungum et al., 1986; Gregersen et al., 1991). The investigation focuses on the Gulf of Bothnia bordered by Finland to the east and Sweden to the west (Fig. 1). The Gulf of Bothnia has long been recognized as one area of enhanced seismicity in the region; even the oldest seismicity maps based on written documentary records show how earthquakes occur on its coasts. It is a seismicity area in miniature: the current seismograph networks register relatively frequent micro-earthquake activity down to magnitude below 0, while the largest observed earthquakes had magnitude above 4. It is not clear whether magnitude 5 has been exceeded during the last three centuries, the time span of the available seismicity record, because the largest earthquakes occurred during the non-instrumental era and their magnitudes are affected by uncertainties. The first short-period seismographs suited for the registration of local earthquakes were installed in the study area in the latter half of the 1950s. All earthquake information prior to that time is defined as historical.

Emphasis is laid on the historical data in this study. Many of the largest earthquakes known occurred adjacent to the Gulf of Bothnia, but only those in 1883, 1888 and 1898 have been subjected to a more detailed analysis (Mäntyniemi, 2005, 2008). New macroseismic maps are presented for earthquakes of 27 November 1757, 14 July 1765, 13 October 1780, 26 May 1907, 31 December 1908, and 9 March 1909. Many previously unknown reports of these earthquakes were brought to light, when scanning the contemporary press. In addition, attention is paid to the location of the largest historical earthquakes. They are compared with recent instrumental records and reviewed against the seismo-tectonic setting. Evidence for larger earthquakes is analyzed using historical data. The potential for larger earthquakes in the area and implications for seismic hazard are discussed.

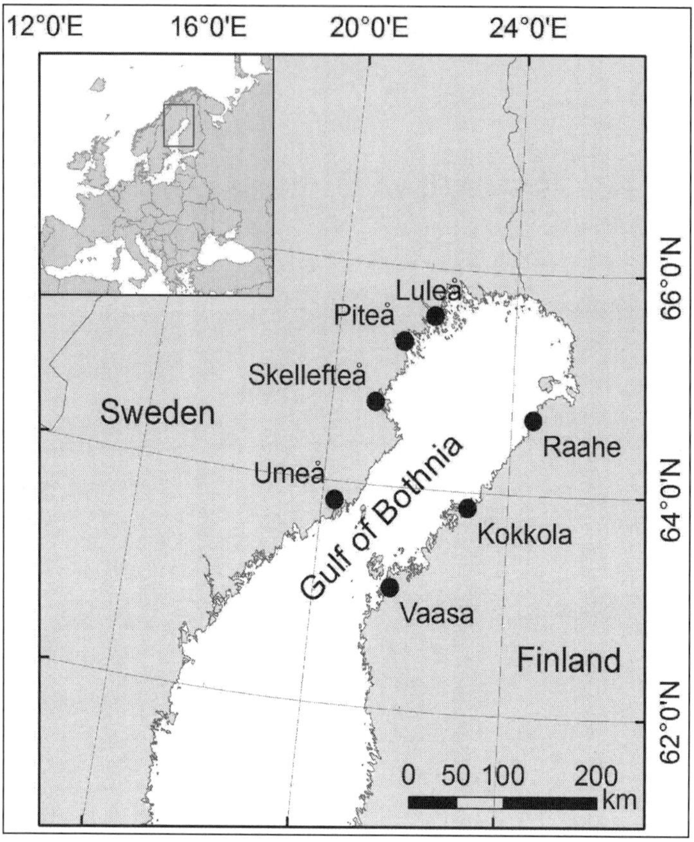

Fig. 1. Study area and place names mentioned in the text. Black dots denote towns. The light purple line in the north denotes the present-day state border between Sweden and Finland.

2. Seismicity features

Earthquake activity along plate boundaries is driven by tectonic forces. The study area undergoes postglacial uplift with a maximum uplift rate of about 10 mm/yr centered in the northern Gulf of Bothnia (e.g., Kakkuri, 1997). The uplift of land from the sea has been documented in Finland and Sweden for more than three centuries, so it appeared reasonable to explain the observed earthquake activity by land uplift, as many early geoscientists did. The advent of focal mechanism studies and stress observations provided arguments in favor of ridge-push from the mid Atlantic, because the NW-SE compression dominates in the region. Therefore the first-order stresses are generally attributed to forces at plate margins, the nearest plate margin being the mid-Atlantic ridge. The role of postglacial uplift in present seismicity remains an open question. Fjeldskaar et al. (2000) modeled the isostatic movements related to deglaciation and uplift. They concluded that, although the modeling fits the overall observations well, there are areas of misfit between them and the isostatic uplift model. The misfit was interpreted to reflect a tectonic component of the uplift. The Swedish east coast with the center northeast of the Gulf of Bothnia was one area of computed misfit. Also local features, such as sediment loading and topography, may affect the rupture of individual earthquakes, or different combinations of several factors.

2.1 Historical seismicity

Many towns were founded around the Gulf of Bothnia during the reign of king Gustavus II Adolphus of Sweden in 1611-1632. This increased the chances of documentation, and earthquakes felt on the shores of the Gulf of Bothnia are known since the 1700s. The oldest reports can be found in scientific essays and contemporary newspapers. They tend to be brief and sparse, so the historical method needs to be modified for observations stemming from the 1700s and most of the 1800s (Mäntyniemi et al., 2011). Local geologists started to use macroseismic questionnaires systematically in the 1880s.

The area of interest is quite challenging for macroseismic analyses, because it is crossed by sea, and also by a state border for more than two centuries. A crucial question in historical seismology is whether a paucity of observations in a given area in a given time interval results from an absence of earthquakes or earthquake reports. It is unclear how much earthquake reporting in the area was affected by the detachment of Finland from Sweden in 1809, when the Gulf of Bothnia became a border area. Interesting earthquake occurrences were reported in the latter half of the 1700s. After the separation, felt earthquake reports east of the Gulf of Bothnia ceased to be included in Swedish newspapers. Anyway, it is important to collect and display what is available, because not all available documentation on the effects of historical earthquakes has been used to date and macroseismic maps do not exist for every interesting earthquake in the area. Previously disregarded earthquake reports have been discovered in the contemporary press in particular. Below new macroseismic maps are presented for a selection of earthquakes felt on both coasts of the Gulf of Bothnia in the 1700s and 1900s. The maps consist of macroseismic data points (MDPs) that illustrate where effects of earthshaking were reported and indicate the strength of the effects, i.e. seismic intensity, when the available information is adequate.

2.1.1 Earthquakes in the latter half of the 1700s

An earthquake was felt at the bottom of the Gulf of Bothnia on 27 November 1757 between 6 and 7 am local time (Fig. 2). Reports about the felt effects have survived in local history and lore (cf. Sidenbladh, 1908; Renqvist, 1930). The use of oral accounts or reminiscences of earthquakes, not untypical of the study area, is discussed by Mäntyniemi et al. (2011). One macroseismic data point is uncertain. It may be related to another earthquake. However, the time of observation matches this earthquake except for the year (1752 instead of 1757). The time of day and time of year are considered the most reliable criteria of a genuine eyewitness report of an earthquake (Mäntyniemi et al., 2011).

Figure 3 shows the information available for the earthquake on the evening of 14 July 1765 (at about 9 pm local time). It is based on four letters from the surrounding provinces published in newspaper *Inrikes Tidningar* ("Domestic news") between 29 July and 23 September 1765. The intensity 1 (not felt) observation relies on a reminiscence published in the same newspaper on 6 November 1780. Only one observation is available for the earthquake on the morning of 13 October 1780, a situation sometimes encountered in historical seismology (Fig. 4). The assigned intensity is on the slightly damaging level, since fractured ovens were reported. The same town has sustained this kind of damages during other earthquakes as well (Mäntyniemi, 2007).

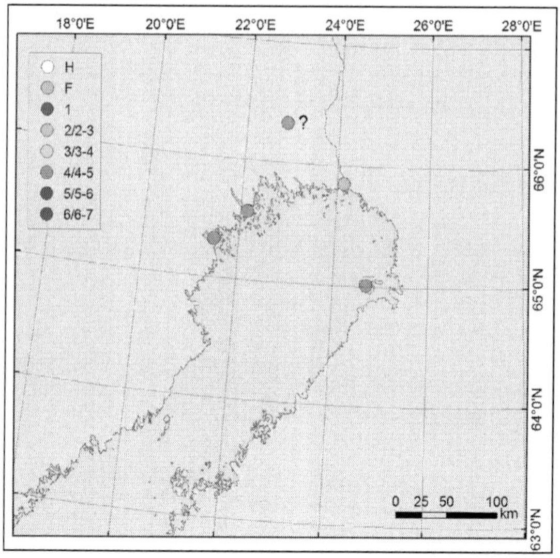

Fig. 2. A macroseismic map on the earthquake of 27 November 1757 between 6 and 7 am local time. Filled circles denote intensities given on the European Macroseismic Scale (Grünthal, 1998). The letter H stands for heard, F for felt. It is not certain whether the intensity point accompanied by a question mark belongs to this earthquake.

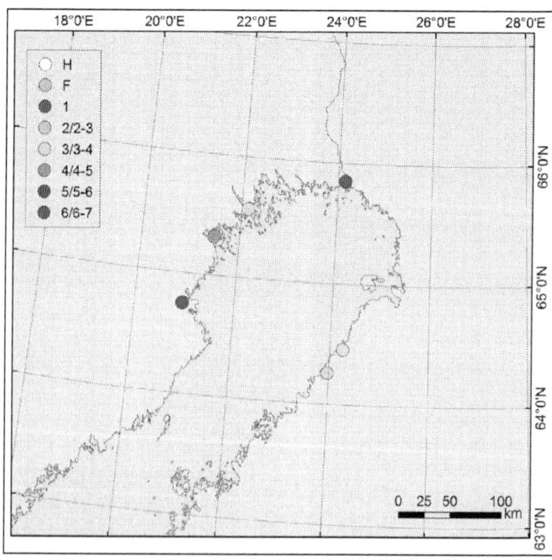

Fig. 3. A macroseismic map on the earthquake of 14 July 1765 at about 9 pm local time. Numbers are intensities given on the European Macroseismic Scale (Grünthal, 1998). At the time of the earthquake all area shown was Swedish territory.

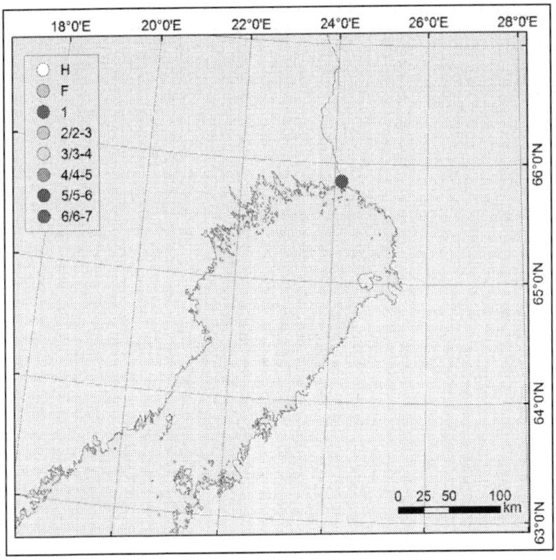

Fig. 4. A macroseismic map on the earthquake of 13 October 1780 at about 6:20 am local time. Filled circles denote intensities given on the European Macroseismic Scale (Grünthal, 1998). At the time of the earthquake all area shown was Swedish territory.

2.1.2 Earthquakes in the early 1900s

Swedish geologists and the Geographical Society of Finland were responsible for collecting felt reports of earthquakes in the study area in the first half of the 1900s. However, the activities were carried out parallel to each other and, as a rule, the outputs were not combined. There is usually little uncertainty about the timing of the observations in this century, so the chance of confusing different earthquakes with each other is rather low.

Sahlström (1911) described observations regarding the earthquake of 26 May 1907 (at about 11:33 am Swedish time) and draw the area of perceptibility. Later Renqvist (1930) reported that the event was felt in the archipelago and some locations east of the Gulf. Previously disregarded place names were found in the contemporary press in the present study. The new area of perceptibility is thus larger than that of Sahlström (1911). There are rather few classification criteria available at the lowest intensity values and thus the range of intensities is narrow (Fig. 5).

Sahlström (1911) named three communes where the earthquake of 31 December 1908 (at about 10:20 pm Swedish time) was felt west of the Gulf of Bothnia and outlined the respective area of perceptibility. Sederholm (1909, p. 65) reported that the event was felt on the eastern shore as well. More precise place names were uncovered in the contemporary press. They expanded the area of perceptibility westward; information about the eastern coast could not be augmented. The reports are not very detailed, so seismic intensity can be assessed only for a few places (Fig. 6). The reporting may have been overshadowed by the earthquake catastrophe in Messina, Sicily three days earlier, as news of the devastation there started to pour in at the beginning of the year 1909. The reportages were often accompanied by figures, which was not common for news coverage at that time.

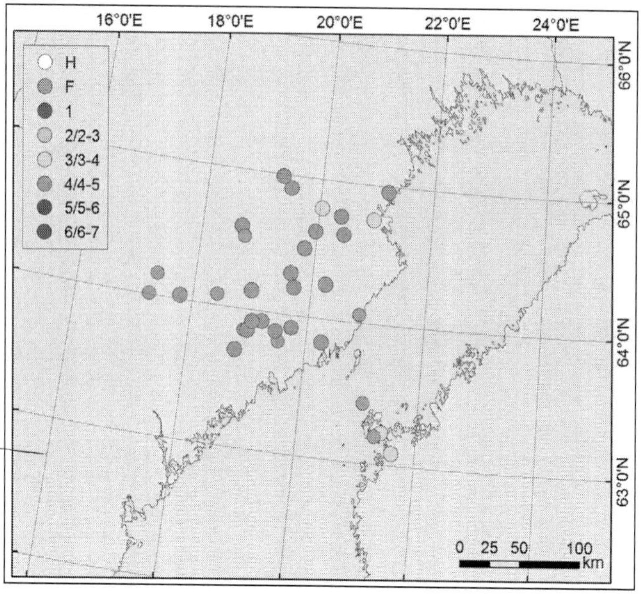

Fig. 5. A macroseismic map on the earthquake of 26 May 1907 at about 11:33 am Swedish time. Numbers are intensities on the European Macroseismic Scale (Grünthal, 1998). The letter H stands for heard, F for felt.

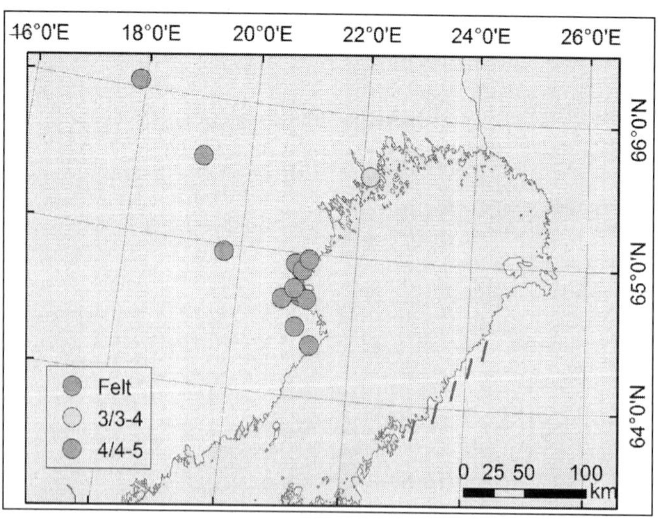

Fig. 6. A macroseismic map on the earthquake of 31 December 1908. Intensities are given on the European Macroseismic Scale (Grünthal, 1998). The earthquake was reportedly felt along the eastern coast (dashed area), but no specific locations can be named there.

The earthquake of 9 March 1909 was the largest event in the Gulf of Bothnia area in the 1900s (Fig. 7). The questionnaire data were presented and discussed by Sahlström (1911) for Sweden and by Rosberg (1912) for Finland. Both authors provided a map for the respective territory, but they were never combined. Båth (1956) gave the first parameters, including a magnitude estimate equal to 5, whereas Ahjos and Uski (1992) provided a macroseismic magnitude 4.6 for this event.

The contemporary press contained some letters from dwellers in the affected area, but most news were general descriptions of a given location. The newspaper reports were sometimes quite processed, providing the strength of earthquake effects relative to another location rather than the actual effects. The new map is thus based on the individual observations given on questionnaires and letters and the résumés prepared by unknown newspaper editors. However, the press provided helpful information in addition to the questionnaire data. The earthquake occurred at night (at about 1:20 am Swedish time), so information about the extent to which people were awakened was available. It is sometimes difficult to tell whether people were awakened by the accompanying earthquake sounds rather than actual

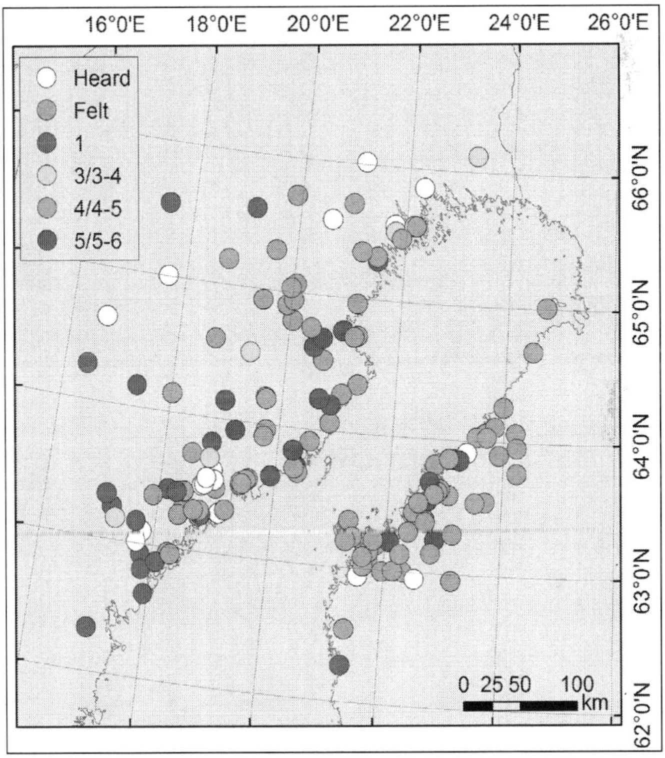

Fig. 7. Felt observations for the earthquake of 9 March 1909, the largest in the Gulf of Bothnia area in the 1900s. Filled circles show the villages and towns from where written documentary data are available. Numbers are seismic intensities on the European Macroseismic Scale (Grünthal, 1998).

ground shaking. Both shaking and sounds were often observed, whereas the heard-only observations tend to be located on the outskirts of the area of perceptibility (Fig. 7).
Estimation of epicenters for non-instrumental earthquakes relies on rather straightforward assumptions of symmetry – no strong dependency of ground motion on azimuth is assumed in the Gulf of Bothnia area until otherwise shown – and the proximity of the epicenter to the strongest effects. Symmetry assumptions may be hampered by the sea and incomplete area of perceptibility. It is commonly assumed that the strongest effects become included, in press reports in particular, as the possible damage is always a matter of concern. Since the density of population was low northward and inland of the Gulf in the early centuries, however, episodic strong ground motion could have been missed there. If reported, foreshocks and aftershocks may give insight into the location of the main shock.

2.2 Instrumental seismicity

Seven earthquakes of magnitude 3 or above occurred in the vicinity of the Gulf of Bothnia between 2001 and 2010 (Fig. 8). The largest had magnitude 3.5, the second largest magnitude

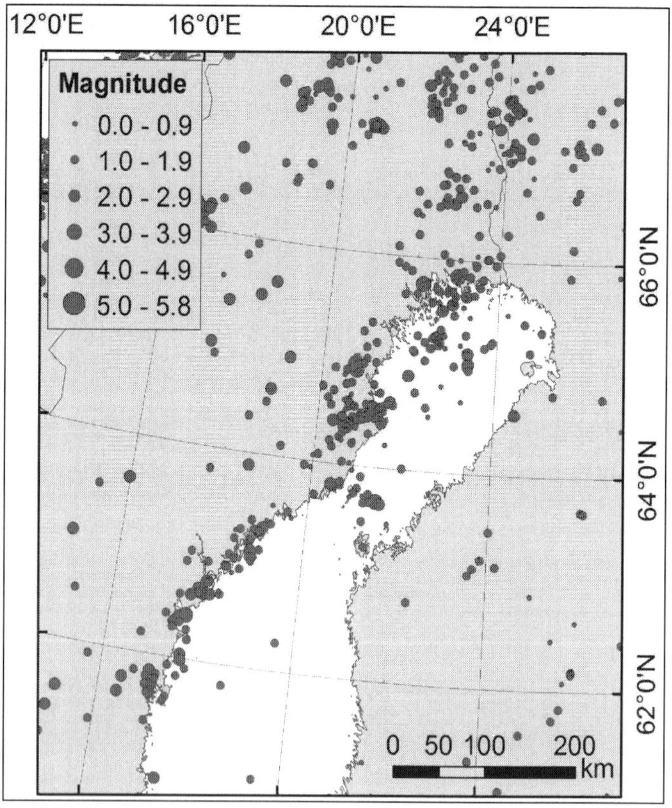

Fig. 8. Instrumental seismicity in the Gulf of Bothnia area between 2001 and 2010. Magnitude scale is M_L. Data: Institute of Seismology, University of Helsinki.

3.4. Two occurred at the bottom of the Gulf nearby Swedish Luleå, three close to Skellefteå, including the largest earthquake. Its coordinates were given as (21.32 E, 64.49 N). It was felt distinctly on the eastern coast including Vaasa and Kokkola (Fig. 1). One of the seven M>3 earthquakes occurred offshore in the narrowest part of the Gulf between Swedish Umeå and Finnish Vaasa, and one further southward on the western coast.

The estimated depths of M>3 events ranged between 6 and 19 km. The depth distribution of all earthquakes during this decade was wider, since the deepest micro-earthquakes were located at a depth of about 40 km. Many of the deepest quakes occurred along the seismicity trend that extends from the bottom of the Gulf northward.

Good-quality instrumental locations discern between onshore and offshore earthquakes. Moreover, the improved location accuracy means that the fuzzy clouds of micro-earthquakes sharpen up: the previously more or less continuous earthquake activity along the western coast of the Gulf of Bothnia south of latitude 64 N seems to be concentrated in smaller clusters. The most pronounced clusters can be found in the northern part of the Gulf. Very frequent micro-earthquake activity occurs in the vicinity of the town of Skellefteå, where it seems to be divided into two main clusters. The earthquake activity nearby Luleå is more diffuse with epicenters both inland and beneath the sea. Thus, the spatial distribution of earthquakes in the vicinity of the Gulf of Bothnia suggests that the probability of earthquake occurrence is not equal over the area.

2.3 Comparisons

The time span of the available earthquake catalogue, about three centuries, is too short for fundamental changes in the local stress field that constrains the occurring seismicity rates. Therefore similarities between the historical and instrumental data can be anticipated. Indeed, seismicity maps relying on observations stemming from the 1700s show how earthquakes occur along the coasts of the Gulf of Bothnia, a feature that accords with instrumental data. The historical catalogue includes several low-magnitude events (M<3) that may actually be relatively well known because of the proximity of epicenters to population centers. The main criterion for a near-by earthquake is the accompanying sound, which is heard in the vicinity of the epicenter and attributed to the seismic P wave (Tosi et al., 2000). However, the historical catalogue is incomplete at the lowest magnitudes and may also contain other types of events such as weather-related noise.

On the other hand, the longer non-instrumental catalogue may include rare events that did not occur during the brief instrumental era. Rare events are, for instance, earthquakes at unusual sites (e.g., sites not anticipated on the basis of more recent instrumental locations) and large earthquakes that occur far more seldom than small ones. The obvious difficulty is to tell whether an incompletely known historical area of perceptibility is evidence for a rare earthquake occurrence. A report describing felt effects may belong to a local or distant earthquake; a sparse report may not include any clue as to the type of ground motion (low or high frequency). The events that occurred outside the study area constitute one category of rare events. A distant offshore earthquake may be hard or impossible to recognize from the seismicity record (e.g., Musson, 2008). It is known from the historical seismicity record that earthquakes with epicenters away from the Gulf of Bothnia area have been felt around it. An example is the Lurøy earthquake of 31 August 1819, located on the coast of northern Norway (e.g., Muir Wood, 1988).

The narrowest part of the Gulf of Bothnia from Umeå to Vaasa spans about 80 km. The Gulf gets wider towards north; it is about 160 km from Piteå to Raahe (Fig.1). The most frequent micro-earthquakes are not felt over very long distances, so an earthquake felt both on the eastern and western coast of the Gulf is noteworthy. This can be used as a handy rule of thumb for detecting a rare event, rare meaning larger than usual. An earthquake occurring on the northernmost western coast has to be above M3, possibly closer to M3.5, in order to be felt on the eastern coast. It has to be borne in mind, however, that the historical felt observations may also be explained by an offshore event.

The earthquakes of 27 November 1757 (Fig. 2) and 13 October 1780 (Fig. 4) may be examples of repeated seismic activity at the bottom of the Gulf of Bothnia. This feature accords well with the instrumental data. An earthquake in the vicinity of Luleå would explain the distribution of felt observations shown in Fig. 2. However, other options can easily be constructed around a small number of felt observations. The population centers were small at that time, the density of population particularly low outside the coastal areas and river valleys, and the native Lapp people nomads, so occasional ground motion in the area may have easily been missed. Size estimates are obviously prone to error when the area of perceptibility is incomplete known. Since these earthquake observations were accompanied by sounds, however, they are assumed to be local events.

The earthquakes of 14 July 1765 (Fig. 3) and 31 December 1908 (Fig. 6) may be related to seismic activity in the vicinity of Skellefteå on the western coast. A more recent earthquake occurred there on 15 June 2010. Its magnitude was estimated at M_L3.5, and it was felt distinctly on the eastern coast as well. The earthquakes of 26 May 1907 (Fig. 5) and 9 March 1909 (Fig. 7) may be related to seismicity southward from the other examples, possibly closer to Umeå. These historical earthquakes seem to support the notion that seismicity in the area has preferred locations.

3. Seismo-tectonic setting

All earthquakes occur on faults, but within the continental interiors the networks of faults may be less well defined than on plate boundaries. A tectonic model for the Paleoproterozoic evolution of the Fennoscandian shield was presented by Lahtinen et al. (2005), based on petrological, geochronological, potential-field, deep seismic reflection and refraction, and geoelectric data. Major crustal-scale boundaries were inferred from lineaments on magnetic, electromagnetic, and Bouguer anomaly maps, where seismic reflection and refraction data were lacking (Fig. 9).

The Gulf of Bothnia is situated in the central part of the Fennoscandian shield, composed of Paleoproterozoic rocks. The central part is often referred to as the Svecofennian domain. Identified major boundaries in the area include the Baltic-Bothnian megashear that extends from the bottom of the Gulf towards north (Berthelsen and Marker, 1986) and the Piteå-Raahe shear zone crossing its northern part (place names in Fig.1). The Hassela shear zone crosses the southernmost Gulf. Paleoproterozoic units in Sweden include the Skellefte district in the vicinity of the town of Skellefteå, and the Bothnian basin.

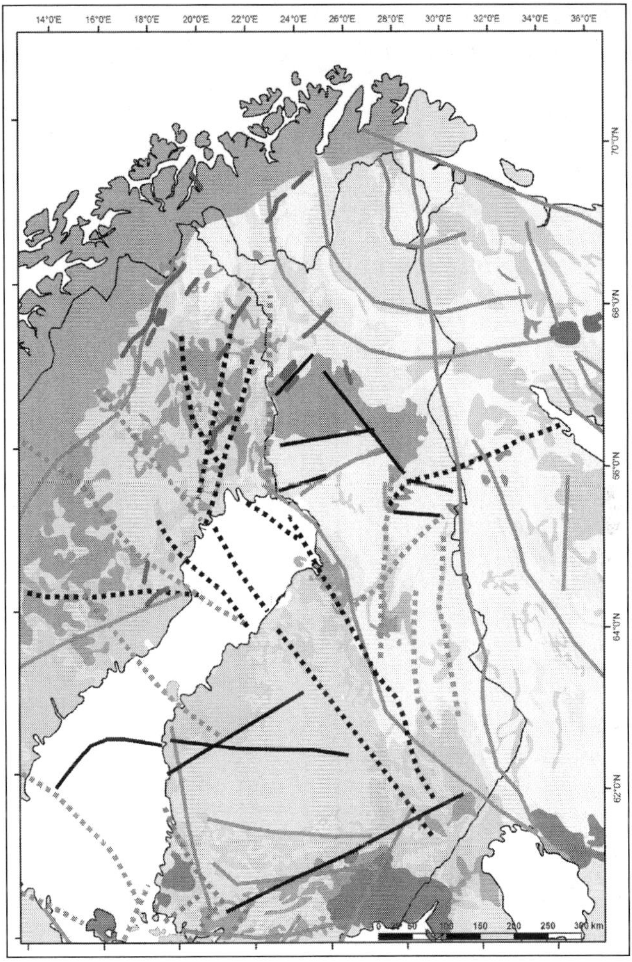

Fig. 9. Major aeromagnetic and Bouguer anomaly lineaments in the Fennoscandian shield. Continuous green lines denote magnetic lineaments, dotted green lines magnetic and Bouguer lineaments. Continuous black lines denote Bouguer lineaments, and dotted black lines Bouguer lineaments associated with shear zone at surface. Red lines denote mapped postglacial faults (from Lahtinen et al., 2005). The lineaments are shown on a simplified geological map. Igneous rocks and gneisses (denoted by light yellow) and supracrustal rocks (light green) are Archean rocks in the east. Supracrustal rocks of different ages (light and bright turquoise), the granulite belt (light blue), igneous rocks of different ages (light brown and pink), mafic intrusive rocks (spots of dark green), as well as granite and migmatite (reddish) are Paleoproterozoic rocks. The rapakivi granite association (purple) is Mesoproterozoic. The Caledonian orogenic belt (grey) and alkaline intrusions (bright purple) are Phanerozoic. Geology according to Koistinen et al. (2001).

4. Seismic hazard: how large magnitudes, where and when?

Seismic hazard assessment is confined to the location, time and maximum size of future earthquakes. Insufficient knowledge of ground motion is a secondary problem, if maximum magnitudes remain low. On many segments along plate boundaries, the location of the hazard is known with some probability, so the earthquake time constitutes the main concern. In contrast, in plate interiors even the locations may be poorly understood.

It is often assumed that the locations of small earthquakes are important indicators of the locations of future large earthquakes. Applying this scheme of things to the present study area means that the clusters of micro-earthquakes, discernible on the instrumental seismicity map (Fig. 8), should be looked at as sites where the hazard lies. As discussed in subchapter 2.3, there is some evidence of larger-than-usual earthquakes (at least M>3) that are located within the clusters, such as the earthquake of 15 June 2010 nearby Skellefteå. The historical epicenters cannot be pinpointed very accurately, but it is probable that the earthquakes of 14 July 1765 and 31 December 1908 are examples of repeat of M>3 events there. They have been given magnitudes of 3.9 and 3.7, respectively (Båth, 1956; Wahlström, 1990). These values were estimated using less information than presented in Figs. 3 and 6. However, the changes in the radius perceptibility affect the macroseismic magnitude slowly, because the dependency is logarithmic. The changes probably disappear within the magnitude uncertainty.

Determining the spatial distribution of the largest (M>4) earthquakes in the study area is hampered by uncertainties. When relying on the assumptions that the strongest earthquake effects, as well as foreshocks and aftershocks, occur in the vicinity of the epicenter, some of the largest historical earthquakes are located within or close to the clusters. For instance, Mäntyniemi (2008) proposed that the earthquake of 4 November 1898 may have occurred within the northern extension of the seismicity around the Gulf of Bothnia, which runs parallel to the Baltic-Bothnian megashear.

Båth (1956) gave the epicenter of the 9 March 1909 earthquake as (21.6 E, 64.0 N), with uncertainty in the range of 0.2-1.0 degrees (borders exclusive). This location is slightly offshore; Ahjos and Uski (1992) moved the epicenter eastward to (22.0 E, 64.0 N), which is clearly offshore. The previously unknown newspaper reports contain some remarks about small quakes preceding the main shock. If emphasis is placed on the foreshocks, the earthquake epicenter is shifted inland, to around longitude 19 E, while the latitude remains about the same. The distribution of felt effects shown in Fig. 7 could be explained by an inland epicenter: the westward portion of the area of perceptibility is larger than that eastward. There is no clear cluster of the strongest seismic intensities, which may indicate that the earthquake did not occur close to the ground surface. However, the modified epicenter is within one area of the largest seismic intensities. The modified epicenter seems to coincide with a crustal-scale boundary such as a magnetic lineament or a shear zone (Fig. 9), but the exact location remains uncertain. The possible connection to the enhanced seismicity around the postglacial fault north of the modified epicenter is unclear.

Magnitude 4 was not exceeded in the vicinity of the study area in seventy-four years until the Solberg earthquake of 29 September 1983. Kim et al. (1985) estimated its magnitude at

M_L=4.1, depth at 39 km, and located the epicenter near (17.5 E, 63.8N). This location is distant from the most pronounced micro-earthquake clusters; the Solberg earthquake did not occur where small quakes are frequent. Also instrumental locations are affected by uncertainties, but it is possible that the epicenter coincides with a mapped shear zone that commences (or ends) in the Skellefteå area.

Average recurrence time is somewhat conceptual in the case of M>4 earthquakes in the study area. There was a rather remarkable temporal cluster of these events in the late 1800s and early 1900s. It came to an end in 1909, and was followed only by the Solberg earthquake in 1983. The significance of this time variation of seismicity increases as more time passes and magnitudes remain low. Except for the 1909 earthquake, whose area of perceptibility totally covers that of the 1907 earthquake, the few available cases of M>4 occurred at different sites; the respective areas of perceptibility overlap scantily or partially. The seismic potential for M>4 earthquakes seems to be distributed widely over the area. It can be speculated whether earthquakes in this category will be repeated at the same sites in the future.

The seismicity in the study area follows the Gutenberg-Richter magnitude-frequency relationship with a large number of tiny earthquakes and a tiny number of "large" earthquakes. Earthquakes larger than shown by the short historical record could possibly occur. The discussion of the maximum magnitude in the area tends to be somewhat twofold. Great earthquakes (M>7 or above) have been associated to some of the mapped postglacial faults, especially in the north and north-west (Fig. 9). They have been dated to be less than 10 Ka old and have been explained by changes in the stress field due to deglaciation after the last glaciation. It is difficult to argue convincingly about the upper limit of magnitude in the current tectonic regime. However, it is known from areas of stable continental crust that earthquakes exceeding the largest observed magnitude have occurred. Such an example is known from east of Svalbard, northern Europe, where an M6.2 earthquake occurred on 21 February 2008.

5. Conclusion

The intraplate seismicity in the Gulf of Bothnia area, northern Europe exhibits enhanced micro-earthquake activity. The most pronounced clusters of micro-earthquakes are situated in places where several shear zones meet, for instance in the vicinity of the town of Skellefteå on the western coast. Earthquakes of M>3 have occurred within the clusters during the brief instrumental era, during the most recent decade as well, but earthquakes above magnitude 3.5 appear rare. Macroseismic maps of earthquakes felt on both the eastern and western shore of the Gulf are a practical means of detecting unusually large earthquakes. New maps making use of previously disregarded newspaper reports were prepared for a selection of earthquakes in the late 1700s and early 1900s. Some of the investigated earthquakes are interpreted to be repeat events within the clusters. For instance, the earthquakes of 14 July 1765 and 31 December 1908 are associated with the Skellefteå cluster. Their magnitude estimates are probably in the range 3.5-3.9. The earthquakes of 27 November 1757 and 13 October 1780 are regarded as local events at the bottom of the Gulf, where the pattern of instrumental seismicity is rather diffuse.

The few known earthquakes of magnitude M>4 stem mostly from the non-instrumental era. Some of the respective locations possibly coincide with the enhanced seismicity trends, but at least one M>4 earthquake occurred where small earthquakes are not that frequent. The location uncertainty precludes associating the historical epicenters with smaller structural features than crustal-scale boundaries, such as postglacial faults. It appears that the clusters of micro-earthquakes should not be looked at as the only sites where large earthquakes may occur in the area in the future.

6. Acknowledgment

The author thanks Kati Oinonen for plotting figures 1 and 6-8, Outi Valtonen for plotting figures 2-5, and Annakaisa Korja for permission to use figure 9.

7. References

Ahjos, T. & Uski, M. (1992). Earthquakes in Northern Europe in 1375-1989. *Tectonophysics*, Vol.207, No.1-2, (June 1992), pp. 1-23, ISSN 0040-1951

Båth, M. (1956). *An Earthquake Catalogue for Fennoscandia for the Years 1891-1950*. Sveriges Geologiska Undersökning, Ser.C, No.545, ISSN 1103-3371, Stockholm, Sweden.

Berthelsen, A. & Marker, M. (1986). 1.9-1.8 Ga old strike-slip megashears in the Baltic Shield, and their plate tectonic implications. *Tectonophysics*, Vol.128, No.3-4, (September 1986), pp. 163-181, ISSN 0040-1951

Bungum, H.; Havskov, J.; Hokland, B.Kr. & Newmark, R. (1986). Contemporary seismicity of northwest Europe. *Annales Geophysicae*, Ser. B, Vol.4, No.5, (October 1986), pp. 567-576, ISSN 0755-0685

England, P. & Jackson, J. (2011). Uncharted seismic risk. *Nature Geoscience*, Vol.4, No.6, (June 2011), pp. 348-349, ISSN 1752-0894

Fjeldskaar, W.; Lindholm, C.; Dehls, J.F. & Fjeldskaar, I. (2000). Postglacial Uplift, Neotectonics and Seismicity in Fennoscandia. *Quaternary Science Reviews*, Vol.19, No.14-15, (October 2000), pp. 1413-1422, ISSN 0277-3791

Gregersen, S.; Korhonen, H. & Husebye, E.S. (1991). Fennoscandian Dynamics: Present-day Earthquake Activity. *Tectonophysics*, Vol.189, No.1-4, (April 1991), pp. 333-344, ISSN 0040-1951

Grünthal, G. (Ed.) (1998). European Macroseismic Scale 1998, Cahiers de Centre Européen de Géodynamique et de Séismologie, Vol.15, ISBN N°2-87977-008-4, Luxembourg.

Kakkuri, J. (1997). Postglacial Deformation of the Fennoscandian Crust. *Geophysica*, Vol.33, No.1, (May 1997), pp. 99-109, ISSN 0367-4231, Available from http://www.geophysica.fi

Kim, W.-Y.; Kulhánek, O.; van Eck, T. & Wahlström, R. (1985). *The Solberg, Sweden, Earthquake of September 29, 1983*. Report No.1-85, Seismological Department, University of Uppsala, Sweden, ISSN 0280-3070

Koistinen, T.; Stephens, M.B.; Bogatchev, V.; Nordgulen, Ø.; Wennerström, M. & Korhonen, J. (2001). *Geological Map of the Fennoscandian Shield, 1:2000 000*. Geological Survey of Finland, Espoo; Geological Survey of Norway,

Trondheim; Geological Survey of Sweden, Uppsala; Ministry of Natural Resources of Russia, Moscow.

Lahtinen, R. ; Korja, A. & Nironen, M. (2005). Paleoproterozoic Tectonic Evolution, In: *Precembrian Geology of Finland – Key to the Evolution of the Fennoscandian Shield,* M. Lehtinen, P.A. Nurmi, O.T. Rämö (Eds.), pp. 481-532, Elsevier B.V., ISBN 978-0-444-51421-9, Amsterdam

Mäntyniemi, P. (2005). A Tale of Two Earthquakes in the Gulf of Bothnia, Northern Europe in the 1880s. *Geophysica,* Vol.41, No.1-2, (December 2005), pp. 73-91, ISSN 0367-4231, Available from http://www.geophysica.fi

Mäntyniemi, P. (2007). Town of Tornio in November 1898: a rare survey of earthquake damage in Finland. *Journal of Seismology,* Vol.11, No.2, (April 2007), pp. 177-185, ISSN 1383-4649

Mäntyniemi, P. (2008). Earthquake of 4 November 1898 in Northern Europe: New Insights. *Journal of Geophysical Research,* Vol.113, No.B11, doi:10.1029/2007JB005461, ISSN 0148-0227

Mäntyniemi, P.; Tatevossian, R.E. & Tatevossian, T.N. (2011). How to deal with sparse macroseismic data: Reflections on earthquake records and recollections in the Eastern Baltic Shield. *Annals of Geophysics,* Vol.54, No.3, doi:10.4401/ag-4786, ISSN 2037-416X

Muir Wood, R. (1988). The Scandinavian Earthquakes of 22 December 1759 and 31 August 1819. *Disasters,* Vol.12, No.3, (September 1988), pp. 223-236, ISSN 0361-3666

Musson, R.M.W. (2008). The Case for Large (M > 7) Earthquakes Felt in the UK in Historical Times, In: *Historical Seismology – Interdisciplinary Studies of Past and Recent Earthquakes,* J. Fréchet, M. Meghraoui, M. Stucchi (Eds.), pp. 187-207, Springer Science+Business Media B.V., ISBN 978-1-4020-8221-4

Renqvist, H. (1930). Finlands jordskalv (Earthquakes in Finland). *Fennia,* Vol.54, No.1, pp. 1-113, ISSN 1798-5617 (in Swedish)

Rosberg, J.E. (1912). Jordskalf i Finland 1904-1911 (Earthquakes in Finland 1904-1911). *Fennia,* Vol.32, No.5, pp. 1-25, ISSN 1798-5617 (in Swedish)

Sahlström, K.E. (1911). Jordskalf i Sverige 1907-1910 (Earthquakes in Sweden 1907-1910). Sveriges Geologiska Undersökning (Geological Survey of Sweden), Ser.C, No.238, ISSN 1103-3371, Stockholm, Sweden (in Swedish).

Sederholm, J.J. (1909). *Jordbäfningar och vulkaner* (Earthquakes and volcanoes), G.W. Eklunds Förlag, Helsingfors, Finland (in Swedish)

Sidenbladh, E. (1908). Sällsamma händelser i Sverige med Finland åren 1749-1801 och i Sverige åren 1821-1859 (Rare happenings in Sweden with Finland in the years 1749-1801 and in Sweden in the years 1821-1859). *Statistisk Tidskrift,* Supplement, pp. 110-126 (in Swedish)

Tosi, P.; De Rubeis, V.; Tertulliani, A. & Gasparini, C. (2000). Spatial Patterns of Earthquake Sounds and Seismic Source Geometry. *Geophysical Research Letters,* Vol.27, No.17, (September 2000), pp. 2749-2752, ISSN 0094-8276

Wahlström, R. (1990). A catalogue of earthquakes in Sweden in 1375-1890. *Geologiska Föreningens i Stockholm Förhandlingar,* Vol.112, No.3, (September 1990), pp. 215-225, ISSN 0016-786X

12

Probabilistic Method to Estimate Design Accelerograms in Seville and Granada Based on Uniform Seismic Hazard Response Spectra

José Luis de Justo[1*], Antonio Morales-Esteban[1*], Francisco Martínez-Álvarez[2] and J. M. Azañón[3]
[1]*Department of Continuum Mechanics, University of Seville*
[2]*Department of Computer Science, Pablo de Olavide University of Seville*
[3]*Department of Geodynamics, University of Granada*
Spain

1. Introduction

The response of a structure affected by an earthquake is the result of "filtering" the seismic signal through the structure. A dynamic analysis of a structure requires the previous definition of the accelerogram and the structure characteristics. A complete calculation implies working out the seismic response in all points of the structure; that is, calculating the seismic response in an infinite number of points and in an infinite number of instants. (Meirovitch, 1985) has demonstrated that, with an infinite number of points and instants, the problem has no numerical solution. To solve the numerical problem, models with a finite number of predeterminated points are defined.

The response of a structure subject to a seismic movement can be determined by two methods: either using the accelerograms recorded near the site, or using visco-elastic response spectra.

The first method can only be used in places where many accelerograms have been recorded, and needs a probabilistic calculation to ascertain the design accelerograms. This procedure can be used for linear and non-linear analyses. In both cases various records of a frequency similar to that expected at the location of the structure, may be used to obtain realistic calculation results. A structural analysis for all the accelerograms considered must be carried out in order to obtain a calculation envelope or carry out the probabilistic study. This procedure implies a significant work. This procedure has the difficulty of finding accelerograms at the location of the structure. In some regions, with a vast history of large earthquakes, such as Japan and California, a wide network of recording stations is available and provides many records for large earthquakes, for different type of soils and for a wide range of distances. In regions of minor seismicity, the network of recording stations is not so wide, or is not old enough, so that the number of records is insufficient. For the analysis of minor seismic activity regions, records from other regions are used, or artificial accelerograms are generated. Artificial accelerograms have the advantage that, from a minimum number of parameters, accelerograms can be obtained.

---

* Both Authors contributed equally

Once between 5 and 10, real or artificial, accelerograms have been obtained, they must be scaled to a level of severity. The most commonly used method consists of scaling the seismic peak acceleration up to a predetermined probabilistic level. However, the potential damage an earthquake can produce is not only a function of the peak acceleration so this method is not suitable for a linear calculation. There are many alternatives such as the Arias intensity and the spectral intensity of Housner, (Housner, 1975; Lin & Mahin, 1985) more related to the potential damage.

The use of visco-elastic response spectra is more adequate to obtain accelerograms in regions where the number of records is insufficient, as the response spectrum is the soil movement parameter better related to the structural response. This is the most commonly used method due to its simplicity and appropriate accuracy, as the response of structures, in the elastic linear range, can be obtained as the superposition of a few modes of vibration.

A probabilistic method for estimating the calculation accelerograms is presented in this paper. First, uniform seismic hazard response spectra in Seville and Granada are obtained. Based on them, the calculation accelerograms can be obtained.

2. Fundamentals

Seismic hazard can be defined as the probability of exceeding a parameter of the soil movement, produced by an earthquake, in a location and in a period of time. To unify criteria, UNESCO proposed the commonly accepted definition, given by (UNDRO, 1979). Hazard (*H*) is defined by a probability function of the characteristic parameter of the soil movement (*S*) at a location (*x*) according to:

$$H = P\left[S(x) \geq S_o; t\right] \quad (1)$$

P represents the probability of exceeding a threshold value (S_0) of the characteristic parameter of the soil movement during the time (*t*).

There are two methods for evaluating the seismic hazard: the deterministic method and the probabilistic method.

2.1 Deterministic method

The deterministic method assumes the hypothesis that the seismicity is stationary, considering that earthquakes in the future will be similar to those in the past and estimate the upper limit of the movement, expressed as the maximum value of the parameter. These earthquakes can be real earthquakes that in the past affected the location, or can be deduced from the seismic and tectonic characteristics of the area. The deterministic method can be divided into zoned or not-zoned, in function of how the seismicity distribution is considered. This method presents some advantages and disadvantages. Its main advantage is its easy application. It defines earthquakes that happened in the past and it is easy to suppose that similar earthquakes will happen in the future. However, the probability that those earthquakes will happen is, generally, unknown. The deterministic method estimates the largest earthquake that can affect the location, while the rest of earthquakes are not considered. The sources are characterized by the largest earthquake, and not by its recurrence law.

The calculation procedure is (Benito & Jiménez, 1999):

1. Definition of the influence area of the location, and identification of the seismic sources or faults within it.
2. Estimation of the largest earthquakes happened in the influence area, or at any of the source areas.
3. Estimation of the seismic parameter at the location, caused by the maximum potential earthquakes of every area or of the whole area.
4. Determination of the hazard at the location, taking the maximum value generated by the influence areas. Hazard is defined by the upper limit of the movement at the location.

2.2 Probabilistic method

The probabilistic methods sum up the contribution of all the possible earthquakes that can affect a location, and consider recurrence laws for them. As a result, the probability of exceeding every value of a parameter of the soil movement expected at the location, during a period of time, is estimated. The hazard is represented by probability curves. These methods are classified into parametric and non-parametric, according to the statistical distribution adopted.

2.2.1 Non-parametric methods

These methods analyze the hazard according to extreme value distribution functions. The most used were defined by (Gumbel, 1958). The method is based on the following steps:

1. Definition of the area of influence around the location.
2. Calculation of the values of the seismic parameter at the location, applying attenuation laws to the values of the parameter, that reflect the seismicity of the area during the period of time considered.
3. Adjustment to a distribution of extreme values of the random parameter, defined with the values of the estimated parameter, and estimation of the distribution coefficients.
4. Estimation of the probability of exceeding the extreme value, during the time considered, calculating, this way, the probability.

2.2.2 Parametric methods

The methodology was initially proposed by Cornell (1968). The method is based on the existence of different seismogenic areas. First, the influence area is divided into seismogenic areas and the seismicity of every area is adjusted according to a recurrence law. Later, the contribution of all the sources is summed up to obtain a probability function that represents the hazard at the location.

3. Methodology

3.1 Probabilistic method to estimate seismic hazard

Seismic hazard is presented by means of a hazard function (*H*) that indicates the characteristic parameter of the soil movement (*S*), according to the following formula:

$$H(S_0;t) = P(S \geq S_0;t) \tag{2}$$

Where $P(S \geq S_0;t)$ is the probability that the characteristic parameter of the soil movement will exceed a threshold value, S_0, at least once during the time, *t*.

The arrival of earthquakes to a location is assumed to follow a Poisson´s stationary process (Cornell, 1968; Veneziano et al., 1984). Under this hypothesis, the hazard function can be expressed as follows:

$$H(S_0;t) = 1 - e^{-\lambda(S_0)t} \tag{3}$$

where $\lambda(S_0)$ is the annual rate of times that the parameter (S_0) has been exceeded at the location.

To carry out a study of the seismic hazard, a database of earthquakes that can affect the location must be provided. If the value of the characteristic parameter of the soil movement, during every one of these earthquakes is known and the database is complete, even with the largest earthquakes that can affect the location, the annual rate of exceeding could be calculated according to the following formula:

$$\lambda(S_0) = \frac{1}{t_c}\sum_k \delta(S_k - S_0) \tag{4}$$

Where t_c is the duration of the database, S_k is the value of the characteristic parameter of the soil movement during the k earthquake of the database, and δ is Heaviside´s function:

$$\delta(S_k - S_0) = \begin{cases} 1 \rightarrow S_k - S_0 \geq 0 \\ 0 \rightarrow S_k - S_0 < 0 \end{cases} \tag{5}$$

The formula (4) provides a good estimation of $\lambda(S_0)$ only if all the earthquakes, that can affect the location, have been presented various times during the period of time the database covers. This implies that the database should have a very long duration, probably of thousands of years (Ebel & Kafka, 1999).

If an instrumental parameter is taken as characteristic of the soil movement, the annual rate of exceeding $\lambda(S_0)$ can't be obtained by means of (4) as there are no databases of enough duration.

In this case, the seismic sources and the attenuation laws of the soil movement, from the source to the location, must be analyzed. The steps to follow with this methodology were proposed by (Cornell, 1968):

1. Seismicity model: definition of the seismogenic areas that can affect the site. If the seismicity can be considered homogenous in the whole seismogenic area, a unique seismicity source of global influence can be defined.
2. Recurrence model. The recurrence model in every seismogenic area must be defined. If it is admitted that the seismicity is distributed randomly and it adjusts to the Gutenberg-Richter law with upper truncation, the parameters of the law (a and b) are characteristics of the model. Moreover, for every area maximum and minimum magnitudes are defined that establishes the validity limits of the model.
3. Attenuation law. Attenuation laws to obtain the selected parameter in function of the distance must be determined in order to evaluate the seismic hazard. The application of these laws over the seismicity of every area, represented by its recurrence law, allows obtaining the result over the site.
4. Probabilistic hazard equation. The estimation of the total hazard is obtained adding up the probabilities obtained by the result of all the areas that affect the site.

$$H = \sum_{i=1}^{n}\left(1 - e^{-t\lambda}\right) \tag{6}$$

Where λ is the annual rate of earthquakes, occurring in any area, that produce a parameter of the soil movement superior to the reference one at the studied site, n is the number of areas and t is the period of time, in years, considered.
The uniform seismic hazard response spectra are those that have the same probability of being exceeded in all the periods, obtained with the methodology proposed by (Cornell, 1968).

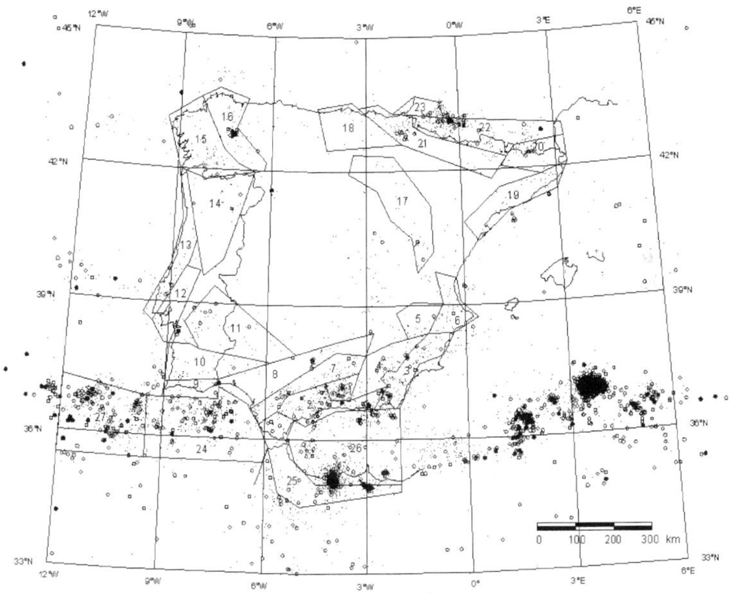

Fig. 1. Seismogenic areas of Spain and Portugal. Dots represent earthquakes of moment magnitude between 3.0 and 4.0, circles between 4.0 and 5.0, and solid circles of magnitude larger than 5.0.

3.1.1 Seismicity model

Seismicity is defined as the whole description of the seismic phenomena in its origin (Martín, 1984). Seismicity can be assimilated to a process of punctual events that result from the relaxation of stress that acts over an area. For its study, the spatial distribution of the earthquakes and its occurrence, according to time, must be known. The most appropriate way to study the irregularities of the temporal series of the earthquakes is through a statistical model.
The model used in this text is based on the seismogenic areas defined by (Martín, 1984). A seismogenic area is a source of earthquakes with homogenous seismic and tectonic characteristics. The process of earthquakes generation is spatially and temporary homogenous in every area. The twenty seven areas established for the Iberian Peninsula are based on tectonic, geological, seismic and gravimetrical data. The twenty seven areas are described in Table 1 and Figure 1.

3.1.2 Recurrence model

The seismicity in every seismogenic area is randomly distributed and it adjusts to the Gutenberg-Richter law. The Gutenberg-Richter law must be truncated, with upper and lower limits, in seismic hazard studies, to consider the magnitude, M_{max}, of the largest earthquake that can occur at the source, and to avoid considering earthquakes of magnitude less than M_{min} respectively. The probability density function of magnitude for Gutenberg-Richter law is:

$$f(M) = \beta \frac{e^{\beta(M-M_{min})}}{1-e^{\beta(M_{max}-M_{min})}} \quad (7)$$

The seismicity in every seismogenic area is defined by the following parameters:

1. The maximum and minimum magnitude.
2. The annual rate of earthquakes occurrence between M_{max} and M_{min}.
3. The *b*-value of the Gutenberg-Richter law.

Area	Description
1	Granada basin
2	Penibetic area
3	Area to the East of the Betic system
4	Quaternary Guadix-Baza basin
5	Area of moderate seismicity to the North of the Betic System
6	Area of moderate seismicity including the Valencia basin
7	Sub-betic area
8	Tertiary basin in the Guadalquivir depression
9	Algarve area
10	South-Portuguese unit
11	Ossa Morena tectonic unit
12	Lower Tagus Basin
13	West Portuguese fringe
14	North Portugal
15	West Galicia
16	East Galicia
17	Iberian mountain mass
18	West of the Pyrenees
19	Mountain range of the coast of Catalonia
20	Eastern Pyrenees
21	Southern Pyrenees
22	North Pyrenees
23	North-Eastern Pyrenees
24	Eastern part of Azores–Gibraltar fault
25	North Morocco and Gibraltar field
26	Alboran Sea
27	Western Azores–Gibraltar fault

Table 1. Seismogenic areas of Spain and Portugal

The maximum magnitude in every seismogenic area has been determined by (Martín, 1984) from seismic and tectonic considerations. The minimum magnitude in all areas is 5.0. Lower earthquakes are not considered dangerous. The *b*-value and the annual rate of earthquakes can be obtained from Table 2.

Area	*b*-value	Annual rate of earthquakes	Surface (km²)	Anual rate / surface (km²)
1	1.41	5.14	3835	1.34E-03
2	1.18	7.82	13979	5.59E-04
3	1.29	4.36	13251	3.29E-04
4	1.27	2.26	11957	1.89E-04
5	1.62	0.87	7066	1.24E-04
6	2.17	1.38	9735	1.42E-04
7	1.51	4.32	13954	3.10E-04
8	0.92	1.47	22228	6.63E-05
9	1.20	0.77	6371	1.21E-04
10	2.33	2.56	15717	1.63E-04
11	1.44	2.35	27694	8.47E-05
12	1.01	0.50	9803	5.08E-05
13	1.29	1.41	13029	1.09E-04
14	1.40	1.75	26049	6.71E-05
15	2.07	4.02	22597	1.78E-04
16	1.54	2.87	15475	1.85E-04
17	1.75	1.07	26993	3.96E-05
18	1.54	0.58	15738	3.70E-05
19	1.90	1.18	16032	7.35E-05
20	1.63	1.68	10622	1.58E-04
21	1.53	2.63	19946	1.32E-04
22	1.48	12.32	22383	5.51E-04
23	1.46	2.06	4301	4.78E-04
24	0.96	13.55	46329	2.92E-04
25	0.96	5.85	24600	2.38E-04
26	1.14	18.21	48669	3.74E-04
27	0.70	15.16	38955	3.89E-04

Table 2. Annual rate of earthquakes and *b*-value for the seismogenic areas of Spain and Portugal

3.1.3 Attenuation model

The characteristic parameter of the soil movement is the acceleration or relative velocity response spectrum for 0, 2, 5, 10 or 20 percent damping. The coefficients for the attenuation laws can be obtained from (Morales-Esteban, 2010).

3.1.4 Probabilistic hazard equation

It is admitted that the arrival at the site of earthquakes that exceed the reference value, log S_0, follows a Poisson stationary process, defined by Gutenberg-Richter law of constant λ_i:

$$\lambda_i = \upsilon_i \int_{M_{min}}^{M_{max}} P(\log S \geq \log S_0 / M,D) f(M) dM \tag{8}$$

The seismic rate of the punctual source is υ_i and *f(M)* is the probability density function of magnitude (equation 7).

If *N* punctual seismic sources hit simultaneously the site, the rate λ of arrivals at the location that exceed the reference value log S_0 is:

$$\lambda = \sum_{i=1}^{N} \lambda_i \tag{9}$$

The probability of exceeding the reference value $logS_0$ during a time *t* caused by the simultaneous action of *N* punctual seismic sources is:

$$P(\log S \geq \log S_0; t) = 1 - e^{-\lambda t} \tag{10}$$

Its return period can be obtained from:

$$T = \frac{1}{\lambda} = \frac{-t}{\ln(1 - P(\log S \geq \log S_0; t))} \tag{11}$$

Equation (10) can't be applied to the hazard calculation as the seismogenic areas have been modeled as areas and not as punctual seismic sources. To solve this problem, the seismogenic areas are divided into elements small enough to be assimilated to punctual seismic sources (Figure 2).

Every seismogenic area is divided into *N* square elements, small enough, so that the seismicity of every one of these squares is assumed to be concentrated in its center. Every seismogenic area is divided into *N* punctual seismic sources. Every one of them has an earthquake occurrence rate $\upsilon_i = \upsilon / N$, where υ is the seismic rate of the whole seismogenic area.

3.2 Resolution of the hazard probabilistic equation

A computer program that divides the seismogenic areas into punctual seismic sources that affect a site, that integrates numerically equation (8) and calculates the probability of exceeding with equation (10) has been developed.

With this method the probability of exceeding has been calculated for some values of the acceleration response spectra for an exposure time of 50 and 100 years for the cities of Seville as Granada. This way a hazard plot of seismic hazard is obtained and is shown in Figure 3. If this process is repeated for various periods of the structure and, for every hazard plot, the spectrum curve for the same probability of exceeding is obtained, and a uniform seismic hazard response spectrum can be obtained, point by point, as shown in Figure 4.

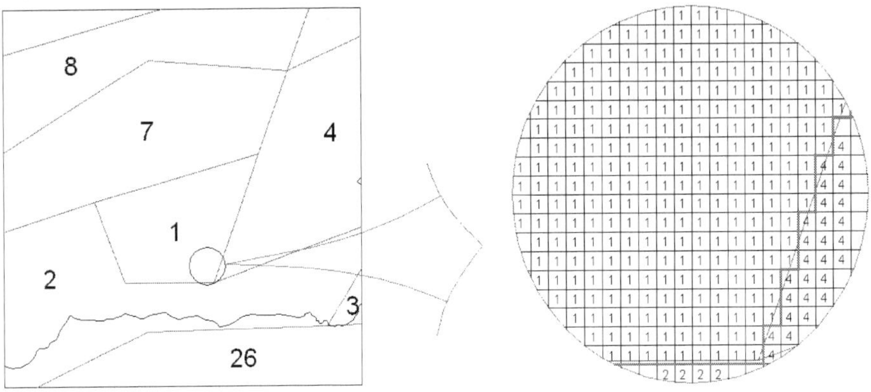

Fig. 2. Scheme of division of a seismogenic area into N punctual areas, through an orthogonal mesh.

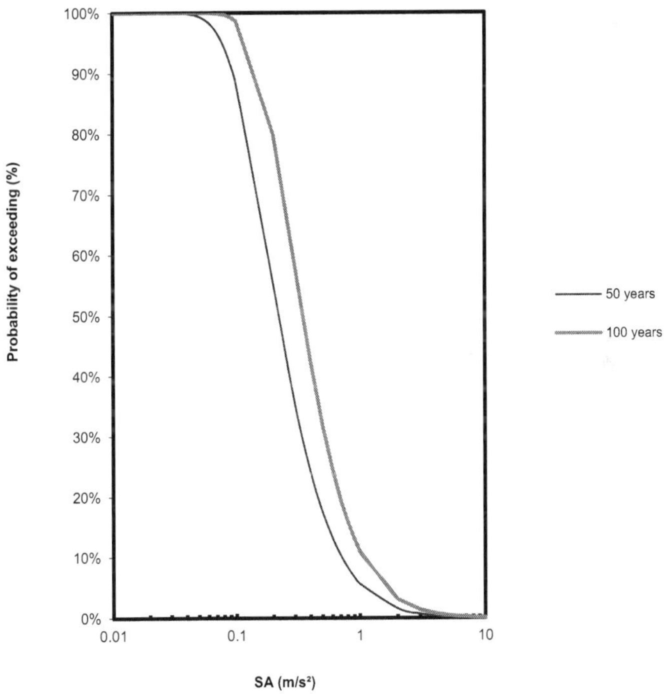

Fig. 3. Seismic hazard plot for Seville site; exposure times 50 and 100 years, respectively; natural period of structure 0.50 s; type of soil rock and relative damping 5%.

3.3 Application to Seville and Granada

Seismic hazard curves for the acceleration spectra in Seville and Granada have been obtained. Figures 5 and 6 show the curves for a period of 0.5 s, for a 5% relative damping and a time of exposure of 50 years, as a function of the soil type for Granada and Seville, respectively. Figure 7 compares the seismic hazard curves for Granada, as a function of the relative damping, for an acceleration response spectrum of period 0.20 s and for a time of exposure of 50 years. Figure 8 represents the hazard curves for Seville and Granada in hard soil for a period of 0.50 s, a relative damping of 5% and an exposure time of 50 years. Figure 9 compares the seismic hazard curves for Granada as a function of the time of exposure (50 and 100 years) in rock, for a period of 0.50 s and a relative damping of 5%.

Following the methodology described in this text, the uniform seismic hazard acceleration response spectra for the cities of Seville and Granada, for a relative damping of the 5%, a probability of exceeding of the 5% and a time of exposure of 50 years, which is equivalent to a return period of 975 years, have been obtained. In figures 10 and 11, the uniform seismic hazard response spectra for Granada and Seville, respectively, are compared as a function of the soil type. Figures 12, 13 and 14 compare the uniform seismic hazard response spectra in rock, hard soil and soft soil, respectively, for the two sites (Seville and Granada).

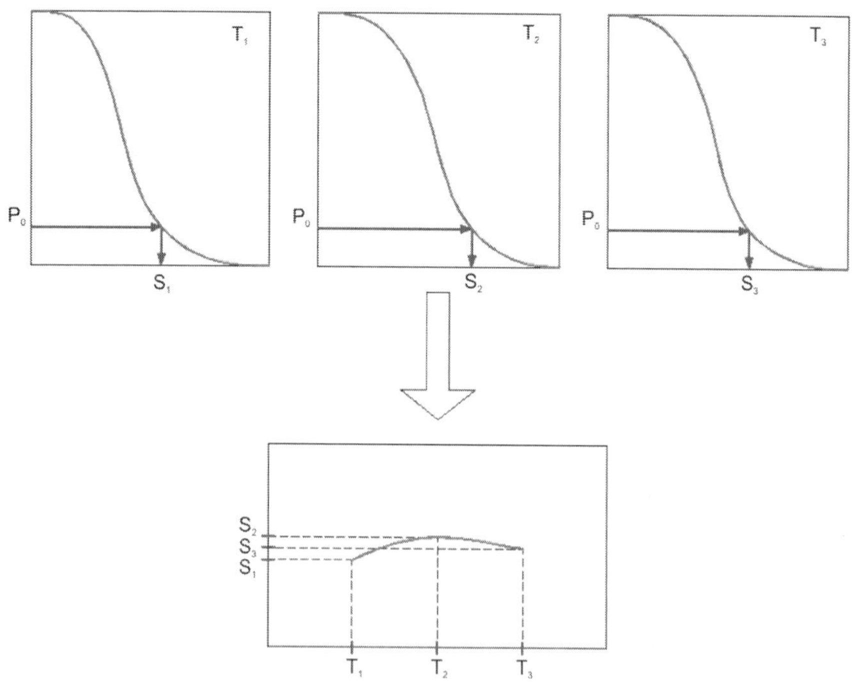

Fig. 4. Scheme of construction of a uniform seismic hazard response spectrum from the seismic hazard plots.

3.4 Assessment of design accelerograms

The procedure to estimate design accelerograms is:

1. The time of exposure of structure is established according to the hazard level.
2. The admitted probability of exceeding is established, normally a 5-10%, according to the hazard level.
3. The uniform seismic hazard acceleration response spectra for the site are calculated according to the type of soil and the required level of hazard (time of exposure and probability of exceeding).
4. In the database of accelerograms, registered in the same type of soil of the site, records are examined. The scale factor, *f*, between the logarithm of the calculated uniform seismic hazard spectrum and the logarithm of the response spectrum corresponding to the real spectrum, that minimizes the standard deviation, *s*, is calculated.

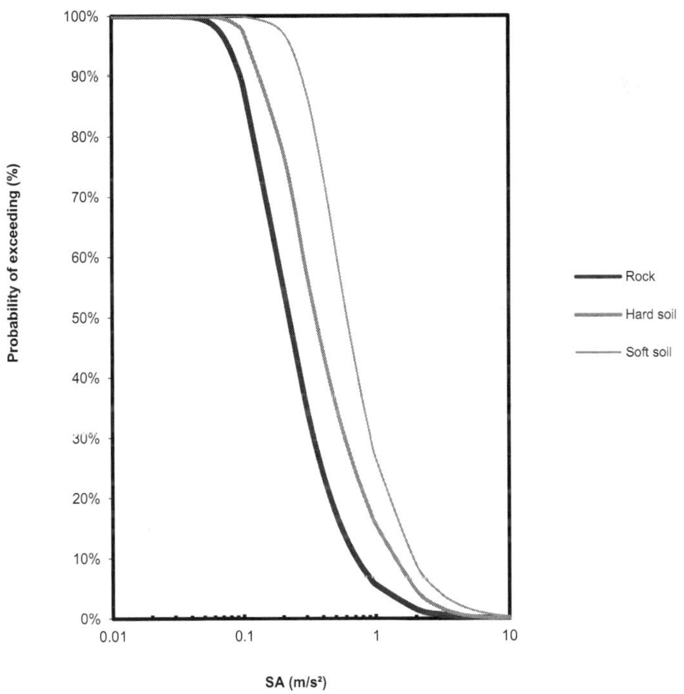

Fig. 5. Comparison of the seismic hazard plot of the acceleration response spectra of period 0.50 s for Granada for different soil types, for a 5% relative damping and a time of exposure of 50 years.

So, if S_R is the response spectrum correspondent to the real register, and S_C is the calculated response spectrum, the standard deviation is:

$$s = \sqrt{\frac{\sum\left[\log(f \cdot S_R) - \log(S_C)\right]^2}{25}} \tag{12}$$

The scale factor that minimizes the standard deviation is:

$$f = \frac{\Sigma \log S_C - \Sigma \log S_R}{25} \tag{13}$$

The sum is extended to the 25 periods for which the uniform seismic hazard response spectrum has been calculated.

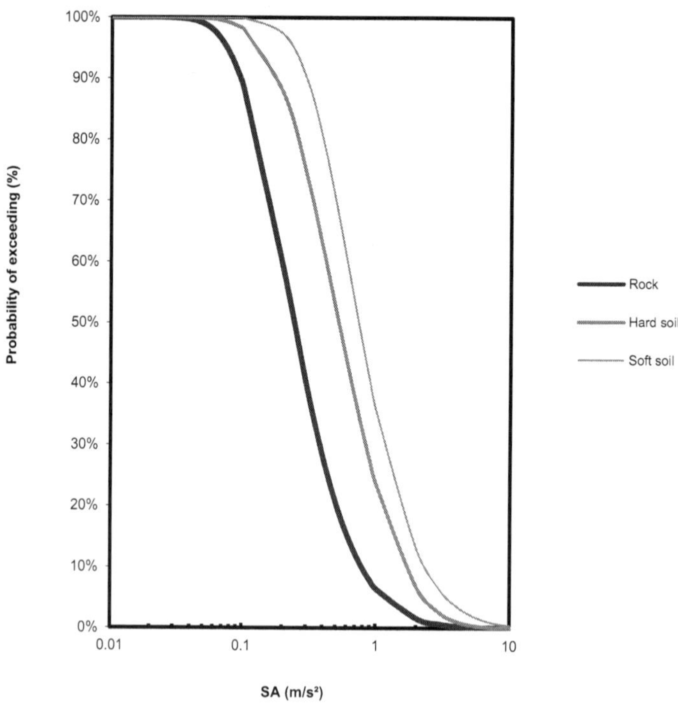

Fig. 6. Comparison of the seismic hazard plot of the acceleration response spectra of period 0.50 s for Seville for different soil types, for a 5% relative damping and a time of exposure of 50 years.

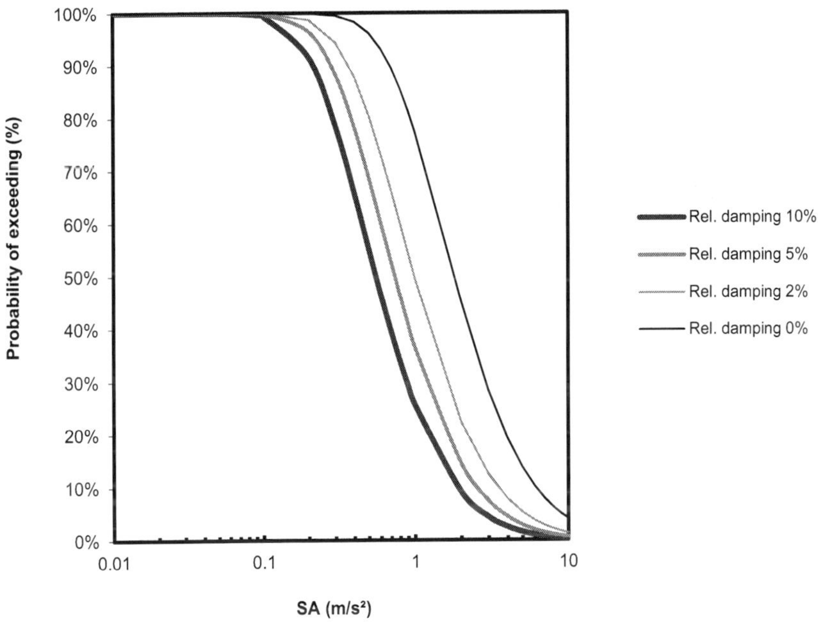

Fig. 7. Comparison for different damping ratios of the seismic hazard plot for the acceleration response spectrum. Period 0.20. s, in rock with a time of exposure of 50 years and for Granada.

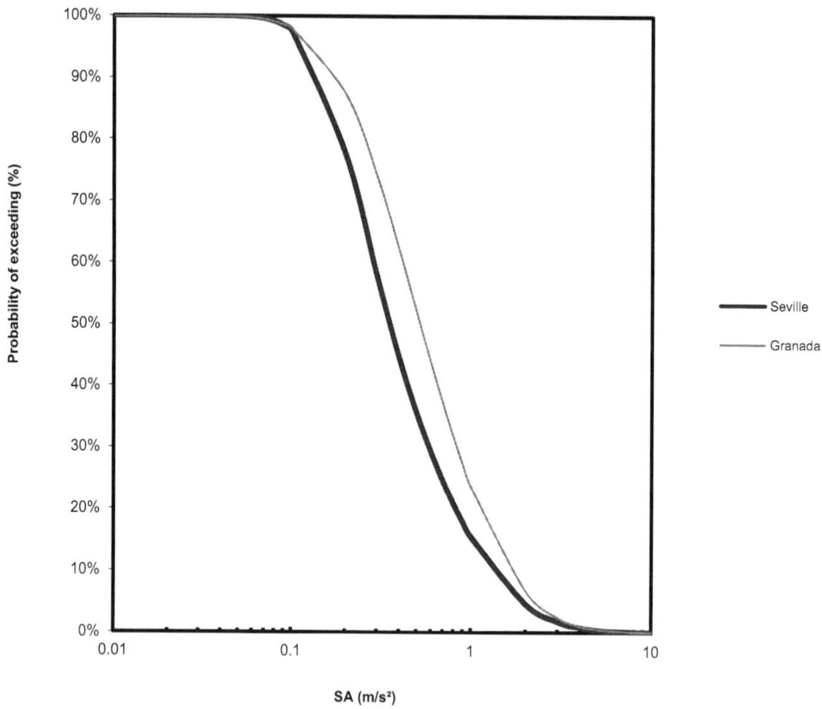

Fig. 8. Comparison of the seismic hazard plots of the acceleration response spectra between Seville and Granada. Period 0.50 s, hard soil, 5% relative dumping and a time of exposure of 50 years.

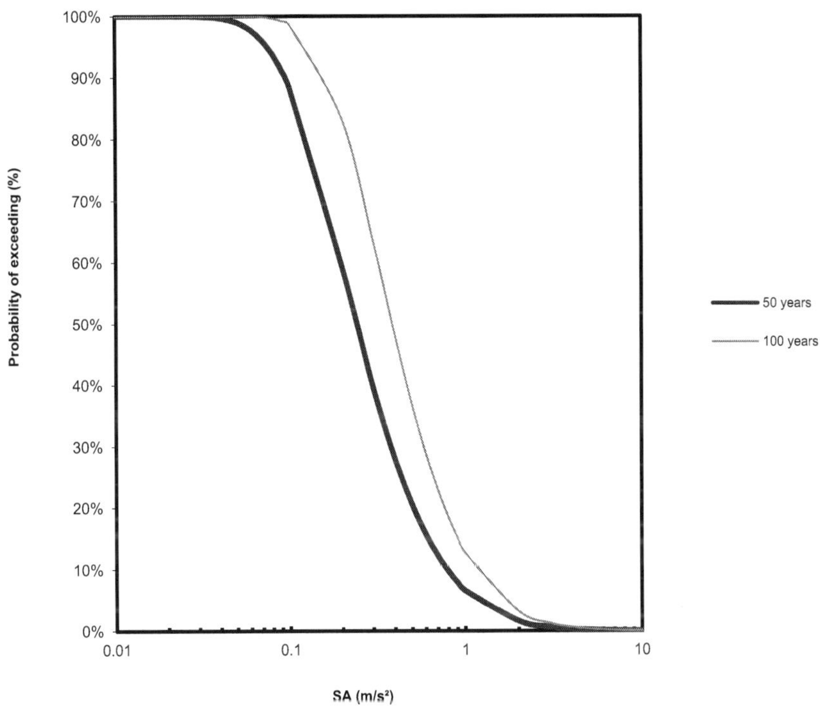

Fig. 9. Comparison for different exposure times of the seismic hazard plot of the acceleration response spectrum. Granada site for a 0.50 s period, in rock and a 5% relative dumping.

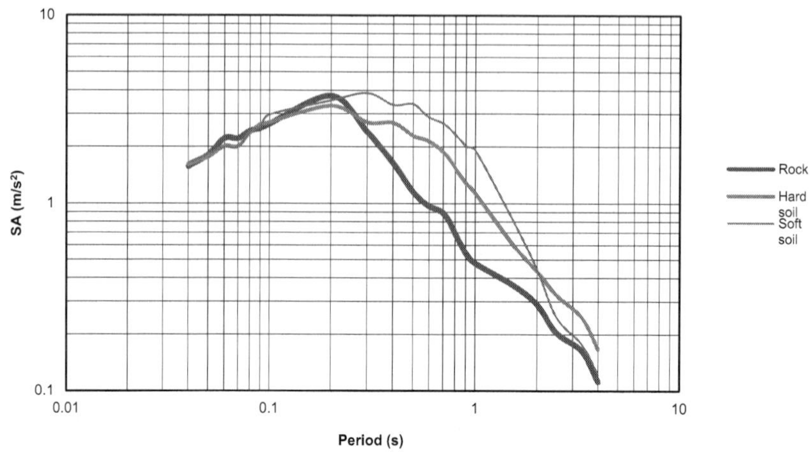

Fig. 10. Comparison of the uniform seismic hazard acceleration response spectra for different soil types. Granada site, probability of being exceeded 5%, relative damping 5% and time of exposure 50 years.

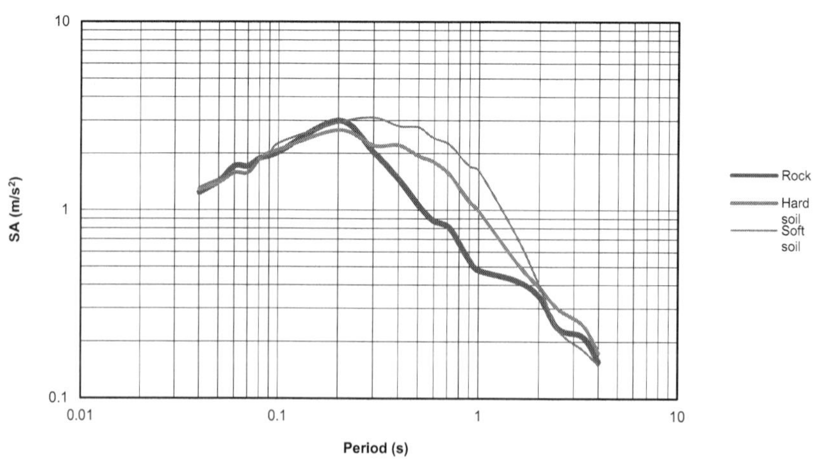

Fig. 11. Comparison of the uniform seismic hazard acceleration response spectra for different soil types. Seville site, probability of being exceeded 5%, relative damping 5% and time of exposure 50 years.

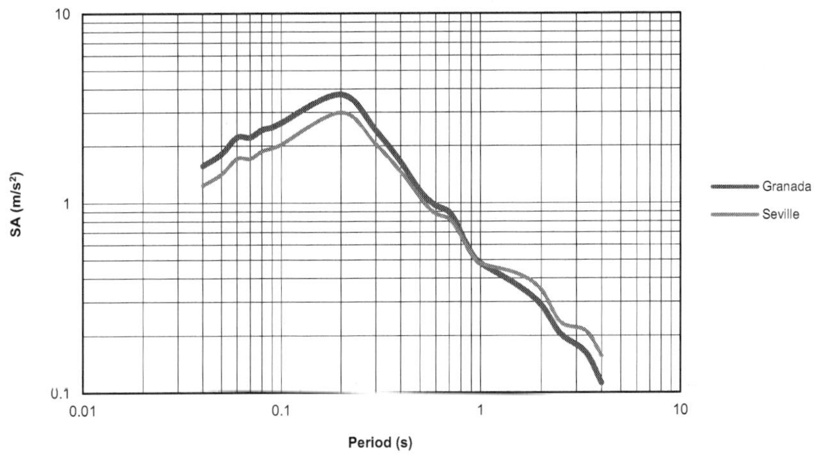

Fig. 12. Comparison between Seville and Granada of the uniform seismic hazard acceleration response spectrum in rock. Probability of being exceeded 5%, relative dumping 5% and time of exposure 50 years.

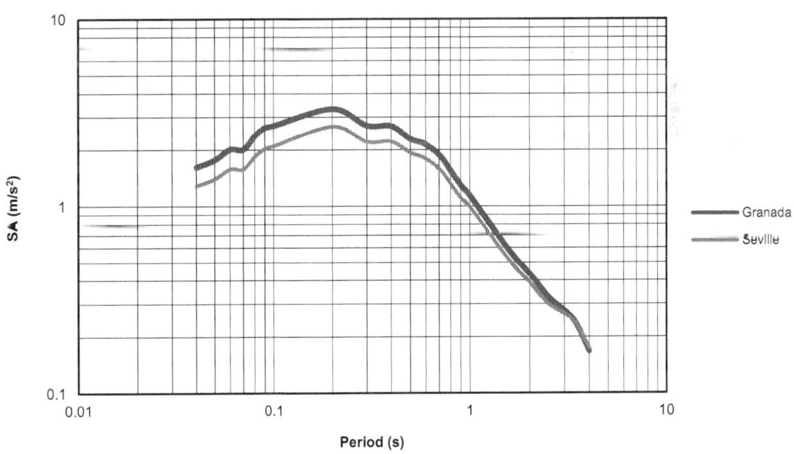

Fig. 13. Comparison between Seville and Granada of the uniform seismic hazard acceleration response spectrum in hard soil. Probability of being exceeded 5%, relative dumping 5% and time of exposure 50 years.

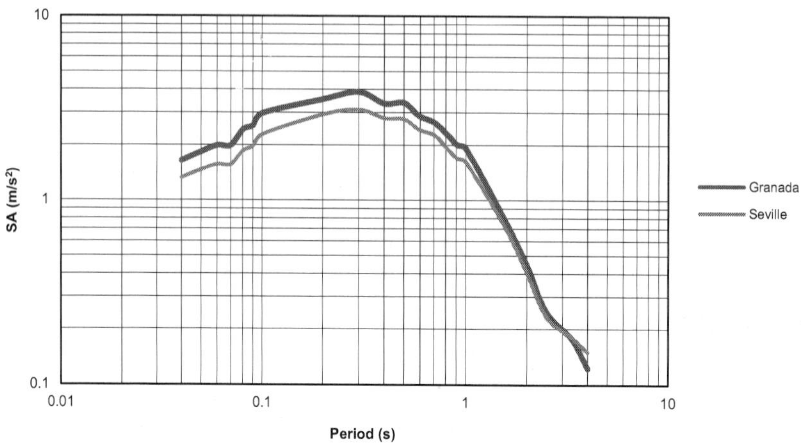

Fig. 14. Comparison between Seville and Granada of the uniform seismic hazard acceleration response spectrum in soft soil. Probability of being exceeded 5%, relative dumping 5% and time of exposure 50 years.

3.4.1 Estimation of calculation accelerograms for San Pedro Cliff at the Alhambra (Granada)

As an example of the methodology presented in this paper calculation accelerograms for San Pedro Cliff at the Alhambra of Granada are obtained. First, the acceleration response spectrum for San Pedro Cliff is calculated in rock (V_s≥750 m/s), for a probability of exceeding of the 5% and a time of exposure of 50 years, which is equivalent to a return period of 975 years. The accelerograms have been obtained from the European earthquake database that can be consulted from the internet: http://www.isesd.cv.ic.ac.uk/.

the Alhambra in Granada is one of the most important national monuments in Spain. This monument, a World Heritage site, is located on the top of a red hill that dominates a plain, the Granada basin, where most of the city is placed. One of the most important rivers of the region, River Darro, flows into the basin and is situated on the western part of the city. The Alhambra's walls are close to the escarpments generated by the incision of this river. Slope instability of the escarpments on this side of the Alhambra hill has been a critical problem since the construction of this palace. In this area, San Pedro Cliff (figure 15), a dihedral 65.5 m high, is the steepest escarpment of the Alhambra hill. This eroding cliff reaches to 23.8 m from the Alhambra palace wall. Retreat of this cliff has occurred through superficial slab falls mainly induced by the floods of the Darro River, the loosening produced by the extensional tectonic regime, erosion, seepage coming from the Alhambra palace and earthquakes.

Granada basin presents several sets of faults, most notably those E-W and NW-SE orientations. Conspicuous NW-SE faults are present in the eastern part of the basin, some of which limit the Granada basin. These faults are normal, mostly with a NW-SE orientation, and dipping towards the SW. These NW-SE faults cross-cut and displace previous E-W faults, defining the main subsiding areas of the central and eastern part of the basin.

Fig. 15. South view of San Pedro Cliff, showing to the right the fault line scarp. Above stand the Alhambra walls and, at the foot, River Darro and Albaicín houses

A site investigation was conducted by (Justo et al., 2008). The following layers appear from top to bottom in the geological profile:

1. Dense conglomerate. V_s=800 m/s (transverse wave velocity).
2. Very dense conglomerate. V_s=960 m/s.
3. Moderately dense conglomerate. V_s=800 m/s.
4. Very dense, gravelly and sandy conglomerate. V_s=1150 m/s.

4a. One meter thick clay layers, interspersed in layer 4. V_s=800 m/s.

Talus appears at the foot of the slope, composed of quartzose and phyllitic blocks, gravel and sand, with predominance of the sand fraction.

Figures 16 to 25 represent the uniform seismic hazard acceleration response spectra for San Pedro Cliff at the Alhambra of Granada and the response spectra of the real earthquakes with better adjustment. The scaled spectrum that minimizes the standard deviation has also been plotted. The records data that minimize the standard deviation are shown in Tables 3 and 4.

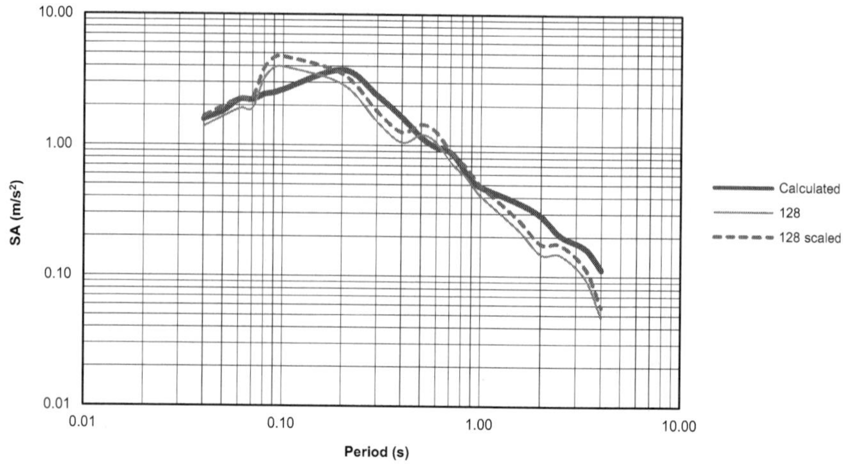

Fig. 16. Seismic hazard acceleration response spectra for San Pedro cliff at the Alhambra in Granada. Probability of being exceeded 5%, relative dumping 5%, rock, time of exposure 50 years. Comparison between the calculated uniform acceleration response spectrum, the spectrum corresponding to record 128 from the catalogue and the spectrum scaled to minimize the standard deviation.

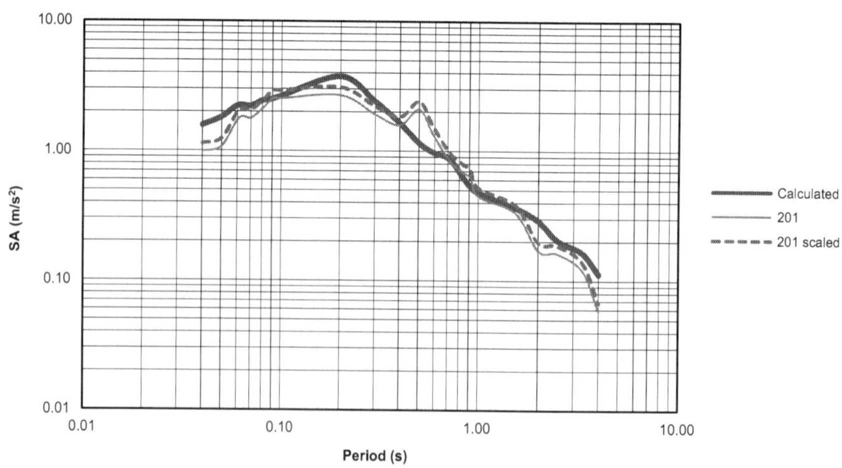

Fig. 17. Seismic hazard acceleration response spectra for San Pedro cliff at the Alhambra in Granada. Probability of being exceeded 5%, relative dumping 5%, rock, time of exposure 50 years. Comparison between the calculated uniform response spectrum, the spectrum corresponding to record 201 from the catalogue and the spectrum scaled to minimize the standard deviation.

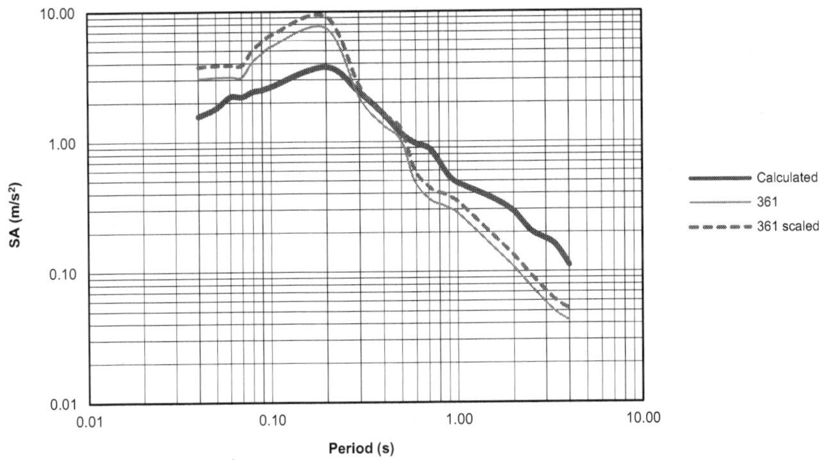

Fig. 18. Seismic acceleration response spectra for San Pedro cliff at the Alhambra (Granada). Probability of being exceeded 5%, relative dumping 5%, rock, time of exposure 50 years. Comparison between the calculated uniform response spectrum, the spectrum corresponding to record 361 and the spectrum scaled to minimize the standard deviation.

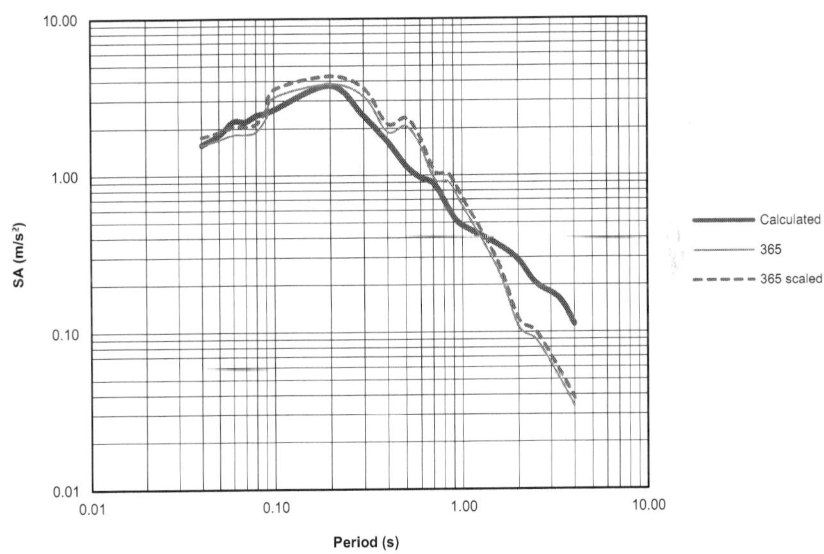

Fig. 19. Seismic acceleration response spectra for San Pedro cliff at the Alhambra (Granada). Probability of being exceeded 5%, relative dumping 5%, rock, time of exposure 50 years. Comparison between the calculated uniform response spectrum, the spectrum corresponding to record 365 and the spectrum scaled to minimize the standard deviation.

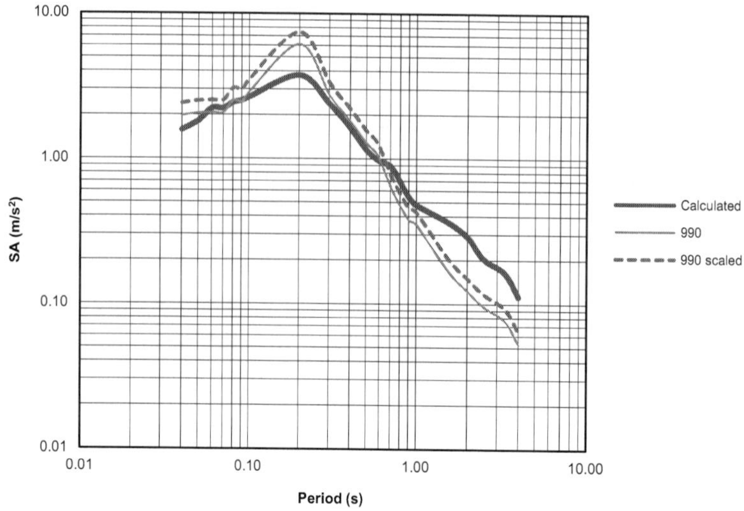

Fig. 20. Seismic acceleration response spectra for San Pedro cliff at the Alhambra (Granada). Probability of being exceeded 5%, relative dumping 5%, rock, time of exposure 50 years. Comparison between the calculated uniform response spectrum, the spectrum corresponding to record 990 and the spectrum scaled to minimize the standard deviation.

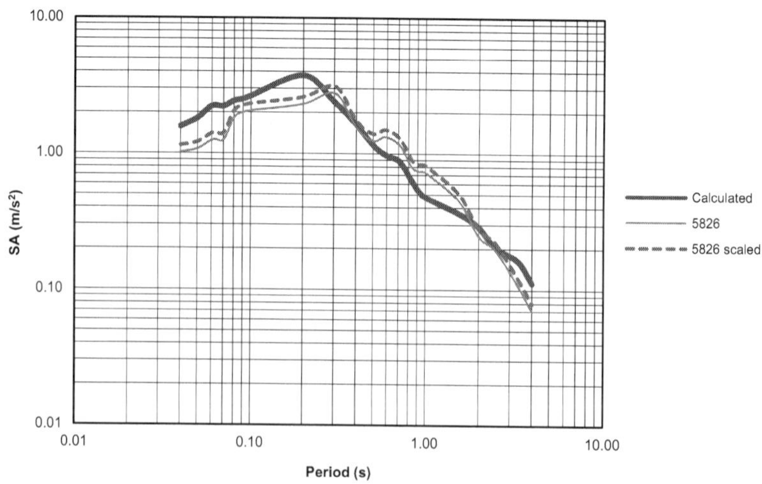

Fig. 21. Seismic acceleration response spectra for San Pedro cliff at the Alhambra (Granada). Probability of being exceeded 5%, relative dumping 5%, rock, time of exposure 50 years. Comparison between the calculated uniform response spectrum, the spectrum corresponding to record 5826 and this spectrum scaled to minimize the standard deviation.

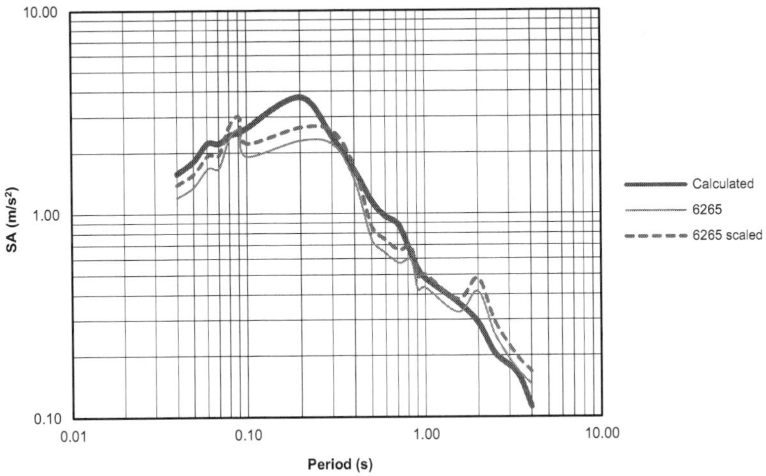

Fig. 22. Seismic acceleration response spectra for San Pedro cliff at the Alhambra (Granada). Probability of being exceeded 5%, relative dumping 5%, rock, time of exposure 50 years. Comparison between the uniform response spectrum, the spectrum corresponding to record 6265 and the spectrum scaled to minimize the standard deviation.

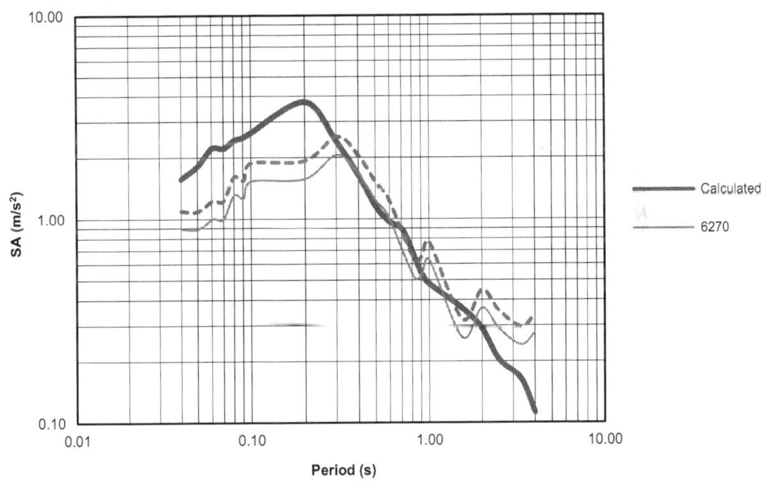

Fig. 23. Seismic acceleration response spectra for San Pedro cliff at the Alhambra (Granada). Probability of being exceeded 5%, relative dumping 5%, rock, time of exposure 50 years. Comparison between the calculated uniform response spectrum, the spectrum corresponding to record 6270 and the spectrum scaled to minimize the standard deviation.

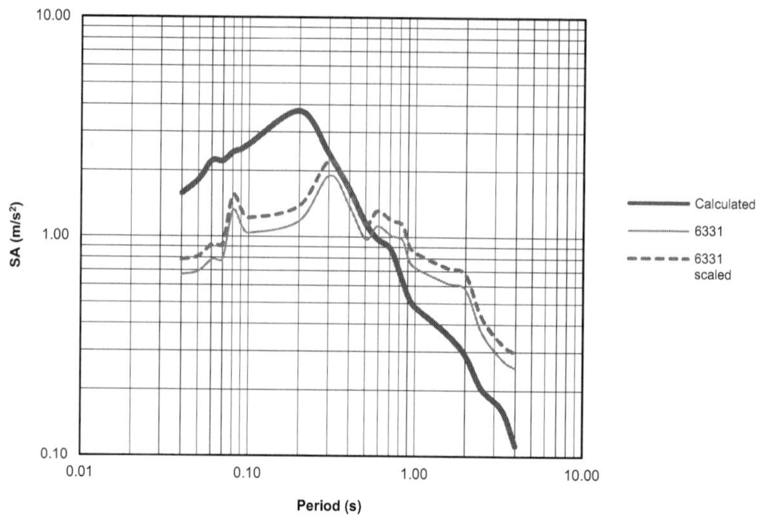

Fig. 24. Seismic acceleration response spectra for San Pedro cliff at the Alhambra (Granada). Probability of being exceeded 5%, relative dumping 5%, rock, time of exposure 50 years. Comparison between the calculated uniform response spectrum, the spectrum corresponding to record 6331 and the spectrum scaled to minimize the standard deviation.

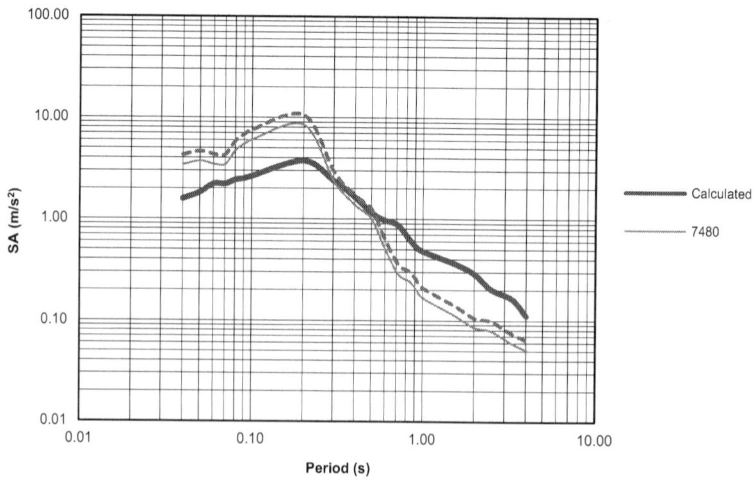

Fig. 25. Seismic acceleration response spectra for San Pedro cliff at the Alhambra (Granada). Probability of being exceeded 5%, relative dumping 5%, rock, time of exposure 50 years. Comparison between the calculated uniform response spectrum, the spectrum corresponding to record 7480 and the spectrum scaled to minimize the standard deviation.

Earthaquake record	128
Seismic data	
Earthquake:	Friuli (Northern Italy)
Date:	9/15/1976
Magnitude:	6.0 M_w
Record data	
Station:	Robic (Slovenia)
Type of soil:	Rock
Fault distance:	19
f	1.2
s	0.016
$SA_{max}(m/s^2)$	3.96

Earthaquake record	201
Seismic data	
Earthquake:	Montenegro (Adriatic Sea)
Date:	4/15/1979
Magnitude:	6.9 M_w
Record data	
Station:	Dubrovnik-Pomorska School (Croatia)
Type of soil:	Rock
Fault distance:	61
f	1.15
s	0.087
$SA_{max}(m/s^2)$	2.68

Earthaquake record	361
Seismic data	
Earthquake:	Umbria (Center of Italy)
Date:	4/19/1984
Magnitude:	5.6 M_w
Record data	
Station:	Nocera Umbra (Italy)
Type of soil:	Rock

Fault distance:	19
f	1.23
s	0.028
$SA_{max}(m/s^2)$	7.49

Earthaquake record	365
Seismic data	
Earthquake:	Lazio Abruzzo (Southern Italy)
Date:	5/7/1984
Magnitude:	5.9 M_w
Record data	
Station:	Atina (Italy)
Type of soil:	Rock
Fault distance:	11
f	1.12
s	0.13
$SA_{max}(m/s^2)$	3.83

Earthaquake record	990
Seismic data	
Earthquake:	Lazio Abruzzo, aftershock (Southern Italy)
Date:	5/11/1984
Magnitude:	5.5 M_w
Record data	
Station:	Atina-Pretura Terrazza (Italy)
Type of soil:	Rock
Fault distance:	13
f	1.22
s	0.011
$SA_{max}(m/s^2)$	6.11

Table 3. Information about the records whose typical deviation is minor in relation to the uniform seismic hazard acceleration response spectra for San Pedro Cliff at the Alhambra in Granada for a 5% probability of being exceeded, an exposure time of 50 years over rock and a relative damping of the 5%

Earthaquake record	5826
Seismic data	
Earthquake:	Strofades (Jonic Sea)
Date:	11/18/1997
Magnitude:	6.6 M_w
Record data	
Station:	Kyparrisia-Agriculture Bank (Greece)
Type of soil:	Rock
Fault distance:	65
f	1.13
s	0.129
$SA_{max}(m/s^2)$	2.76

Earthaquake record	6265
Seismic data	
Earthquake:	Southern Iceland
Date:	6/17/2000
Magnitude:	6.5 M_w
Record data	
Station:	Burfell Hydroelectric Station (Iceland)
Type of soil:	Rock
Fault distance:	25
f	1.16
s	0.076
$SA_{max}(m/s^2)$	2.58

Earthaquake record	6270
Seismic data	
Earthquake:	Suthern Iceland
Date:	6/17/2000
Magnitude:	6.5 M_w
Record data	
Station:	Ljosafoss Hydroelectric Station (Iceland)
Type of soil:	Rock

Fault distance:	32
f	1.23
s	0.026
$SA_{max}(m/s^2)$	2.06

Earthaquake record	6331
Seismic data	
Earthquake:	Southern Iceland, aftershock
Date:	6/21/2000
Magnitude:	6.4 M_w
Record data	
Station:	Flagbjarholt (Iceland)
Type of soil:	Rock
Fault distance:	22
f	1.17
s	0.057
$SA_{max}(m/s^2)$	1.90

Earthaquake record	7480
Seismic data	
Earthquake:	St. Die (France)
Date:	2/22/2003
Magnitude:	4.7 M_w
Record data	
Station:	Bremgarten (Germany)
Type of soil:	Rock
Fault distance:	-
f	1.25
s	0.058
$SA_{max}(m/s^2)$	8.45

Table 4. Information about the records whose typical deviation is minor in relation to the uniform seismic hazard acceleration response spectra for San Pedro Cliff at the Alhambra in Granada for a 5% probability of being exceeded, an exposure time of 50 years over rock and a relative damping of the 5%

4. Conclusions

The proposed method to obtain design accelerograms provides real accelerograms, registered in the same type of soil of the location and compatible with the uniform seismic hazard response spectra calculated at the site.

It can be observed from the comparison of the uniform seismic hazard acceleration response spectra for Seville and Granada, as a function of the type of soil (Figures 10 and 11), that for periods minor to 0.20 s there is almost no difference between the different type of soils. The difference has a maximum for the intermediate periods (0.40 to 1.00 s) and disappears for the periods over 2.00 s. The maximum value for the acceleration response spectrum is obtained for periods between 0.20-0.30 s.

From the comparison between Seville and Granada for different soil types (fig. 12 to 14) the following can be concluded: for longer periods the acceleration response spectra converge. The maximum value for rock and hard soil is obtained for a period of 0.20 s. For soft soil and very soft soil, the maximum of the acceleration response spectra is for 0.30 s.

The proposed method to estimate calculation accelerograms has been used for San Pedro Cliff at the Alhambra in Granada for a return period of 975 years. The results are presented in Figures 16 to 25. The type of soil at the site is rock. The site investigation has shown a transverse wave velocity for the cliff of over 800 m/s. Tables 3 to 4 provide information about the earthquake records that better fit the uniform seismic hazard spectrum. It can be observed the good adjustment obtained as the standard deviation is very low, with a minimum of 0.011 for the register 990.

5. Acknowledgements

The financial support given by the Spanish Ministry of Science and Technology, projects BIA-2004-01302 and BIA2010-20377 is acknowledged. The authors also want to acknowledge Mr. Antonio Jesús Martín for providing the Spanish database of earthquakes.

6. References

Benito, B. & Jiménez, M. E. (1999). Peligrosidad sísmica, *Física de la Tierra*, 11, pp. 1-28.

Cornell, C. A. (1968). Engineering seismic risk analysis, *Bulletin of the Seismological Society of America*, Vol.58, pp. 1583-1606.

Ebel, J. E. & Kafka, A. L. (1999). A Monte Carlo approach to seismic hazard analysis, *Bulletin of the Seismological Society of America*, Vol.89, pp. 854-866

Gumbel, E. J. (1958). *Statistics of extremes*, Columbia university press, New York, 1958.

Housner, G. W. (1975). Spectrum intensities of strong motion earthquakes, *Proceedings of the symposium on earthquake and blast effects on structures*, EERI.

Justo, J. L.; Azañón, J.M.; Azor, A.; Saura, J.; Durand, P.; Villalobos, M.; Morales-Esteban, A. & Justo E. (2008), Neotectonics and slope stabilization at the Alhambra, Granada, Spain, *Engineering Geology*, Vol.100, pp. 101-119.

Lin, J. & Mahin, S.A. (1985). Effect of inelastic behavior on the analysis and design of earthquake resistant structures, *Earthquake engineering*.

Martín, A. J. (1984). Riesgo sísmico en la Península Ibérica, *Thesis*, Universidad Politécnica de Madrid.

Meirovitch, L. (1985). *Introduction to dynamics and control*, John Wiley & Sons, Inc., New York.

Morales-Esteban, A. (2010). Peligrosidad sísmica. Leyes de atenuación y análisis de las series temporales de los terremotos. Aplicación al Tajo de San Pedro de la Alhambra de Granada, *Thesis*, Universidad de Sevilla, Sevilla.

UNDRO, 1979. Natural disasters and vulnerability analysis, *Report expert group meeting*, Geneva.

Veneziano, D.; Cornell, C. A. & O´Hara, T. (1984). *Historical methods of seismic hazard analysis*, Electric power research institute, Palo Alto, California, Research Report NP-3438.

13

Development of On-Site Earthquake Early Warning System for Taiwan

Chu-Chieh J. Lin[1], Pei-Yang Lin[2], Tao-Ming Chang[2], Tzu-Kun Lin[2], Yuan-Tao Weng[2], Kuo-Chen Chang[2] and Keh-Chyuan Tsai[3]

[1]*SC Solutions, Inc.*
[2]*National Center for Research on Earthquake Engineering*
[3]*National Taiwan University*
[1]*USA*
[2,3]*Taiwan*

1. Introduction

Taiwan is located between Euro-Asian and Philippines tectonic plates on the Pacific Earthquake Rim; therefore, Taiwan has suffered from the threatening of moderate earthquakes for a long time. The earthquake usually caused tremendous damages to human beings and these irreversible damages include loss of human lives, public and private properties, as well as huge adverse economic impacts. It is very difficult to avoid the damages caused by earthquake due to its widely destructive power. However, if people can receive the warning for the coming of the earthquake even by only a few seconds, the damages can be reduced due to possible appropriate reaction. The earthquake early warning system (EEWS) makes it possible to issue warning alarm before the arrival of S-wave (severe shaking) and then to provide sufficient time for quick response to prevent or reduce casualty and damages.

The idea of EEWS was originated in the U.S. (Cooper, 1868) based on the principle that transmission of the electronic signal is faster than the earthquake wave, and the typical research project goes ahead mainly in the California. Up to now, there are three types Earthquake Early Warning System (EEWS). The first type is based on the earthquake locating of local seismometer network systems, the second type is based on an on-site warning of a single seismometer, and the third type is a mixed combination of the first two types. The first type EEWS is a traditional seismological method which locate earthquake, determine magnitude using local seismometer network readings then estimate strong ground motion for other sites. In 1985, the very beginning of Personal Computer (PC) era, Heaton proposed a seismic computerized alert network model which will provide short-term warning (tens of seconds) for large epicentral distance region while a major earthquake happen. In Japan, Prof. Hakuno showed an idea of the earthquake early warning at an earlier stage. Also, JR's UrEDAS (Nakamura, 1988) is famous for their practical system. However, most seismic networks in the world cannot reach such goal. During the 1994 Northridge, 1995 Kobe earthquakes, seismic center took 30 minutes to hours to locate earthquakes. In 1999 921 Chi-Chi Taiwan earthquake, the critical information was determined within 102 seconds. Since then, this type EEWS become mature and applicable. In 2007, Japan announced to public the first EEWS system in the world that can commercially provide earthquake warning information before the large S-wave amplitude arrives. The Real-time Earthquake Information by JMA is based on the source information as a point source and therefore the accuracy of the predicted ground motion is limited especially for a large scale earthquake. However, the limitation of first type EEWS produces a large blind zone where no warning will be received before S-wave arrives. Therefore the second type EEWS (on-site) is designed for such region. The third type EEWS is a hybrid use of regional and on-site warning methods. Although there is no real practical example, this is a reasonable research direction because the limitations of the first two types EEWS is somehow complementary. The regional EEWS is accurate but slow, the on-site EEWS is fast but less accurate. In this chapter, the development of the on-site EEWS for Taiwan is introduced. As part of the total solution of seismic hazard mitigation, an on-site earthquake early warning system (EEWS) has been developed for Taiwan. It provides time-related information including the magnitude of the earthquake, the expected arrival time of strong shaking, the seismic intensity and the peak ground acceleration (PGA) of the shaking, the dominant frequency of the earthquake and the estimation of structural response.

The development of the on-site EEWS is divided into 2 stages. The 1st stage provides a basic prediction of the earthquake, and in the 2nd stage the response of the structure is estimated. In the 1st stage, the P-wave predicated PGA method and neural networks were used to model the nonlinearities caused by the interaction of different types of earthquake ground motion and the variations in the geological media of the propagation path, and learning techniques were developed for the analysis of the earthquake seismic signal. The earthquake characteristics (PGA, amplitude, arriving time and dominate frequency… etc.) were then predicted at stage I. In the 2nd stage, two different approaches are used to satisfy the different demands for the rapid estimation of structural responses. Both modules can estimate the structural response rapidly using the output of the first stage. This rapid-estimation of structural response modulus can do the estimation online in a very short time. The user can get more information about what will happened in the coming earthquake. For different type of usage, two different modules are developed. The general modulus, which only uses the common data of the structure (height, structure type, floor, address …etc), is proposed to provide a low-cost, general-application and rapid estimation of the structural responses. For the user who needs more accurate and detail estimation of structural response, such as the hi-tech facilities, hi-raised building, power plant …etc. The customized modulus (scenario-based response predictor) provides a more accurate and detailed structural response estimation. Moreover, it can connected to the automatic control system, do the adequate decision under different levels of structural responses. With this customized modulus, the economic loss will be dramatically reduced. In order to build the two rapid estimation of the structural responses modulus, a wide range of real structural response data are needed. Only with these data, both modulus can be generated and verified. In this study, the Tai-Power building is used as the target structure; the refined FEM model is build by using FEM analysis software PISA3D. The recorded data of the structural responses from the CWB are used to refine the FEM model. After that, more than 50 on-site data and 200 free-field data are used as inputs in the FEM analysis. All the simulated structural responses from the FEM are collected into the database. This database not only can be used in the development rapid estimation of the structural responses modulus, but also can be used in the structural health monitoring studies. By utilizing the large database collected by the Central Weather Bureau (CWB), rapid and precise results can be obtained after any major earthquake. In addition, the customized modulus can also include the actuation system, which can automatically respond and reduce the economic losses due to the earthquake.

The neural network models are applied during this development of the on-site EEWS in both stages. These neural networks are used to analyse the first-arrival of the earthquake signals in as early as 3 seconds after the first ground motion is registered by the sensors at a rate of 50 samples per second. Then, the on-site EEWS instantaneously provides a profile of information consisting of the estimates of the hazard parameters at the 1st stage and the structure response at the 2nd stage. The system is trained using the seismogram data from 2371 earthquakes recorded in Taiwan. By producing accurate and informative warnings, the system has shown the potential to significantly minimize the hazards caused by the catastrophic earthquake ground motions.

2. On-site earthquake early warning system for Taiwan

The EEWS were expected to work efficiently at the sites with certain distance from the epicenter of the earthquake and also from the observation point where the earthquake motion (P-wave) is firstly observed. However, the EEWS was classified into 'on-site warning' and 'regional warning' by Kanamori (2003). Since 2007, Japan Meteorological Agency (JMA) began the general operation of the real-time earthquake information, which is composed of earthquake occurring time and hypocenter information (the magnitude and the earthquake location) and is expected to provide warning and to substantially reduce the human and physical damage for earthquakes. The EEWS by JMA could be called as 'national warning' due to the use of JMA-NIED system (nationwide earthquake observation system). One of the technical limits is applicability to the near-source earthquakes. The 'regional' EEWS was proposed and developed first by utilizing the difference of the velocity for the beginning P-wave and destructive S-wave of the earthquake as well as the epicenter locating technology. If the earthquake epicenter can be located within several seconds, more response time for the sites with certain distance from the epicenter of the earthquake can be obtained. Although the velocity of the waves depends on density and elasticity of the medium penetrated, the typical speed for the P-wave is around 5 km/sec and the speed for the S-wave is around 3 km/sec. If the area is far from the epicenter, say 100 kilometer, then we will have at most 15 seconds reaction time before the S-wave arrives if the sensor at the observation station was able to pick up the earthquake signal (P-wave) right away and locate the epicenter within 18 seconds, as shown in Eq.1.

$$100\ (\text{km}) \div 3\ (\text{km/sec}) - 18\ (\text{sec}) = 15\ (\text{sec}) \tag{1}$$

However, if the location is less than 50 kilometer from the epicenter of the earthquake, then the 'regional' (traditional) EEWS is almost useless since reaction time is less than one second. In the other words, there are so-called "blind zone", where within the 50 kilometers radius from the epicenter, existed for the 'regional' EEWS. Therefore, the "on-site" EEWS has become increasingly important for areas located in the blind zone of the "regional"

EEWS. In addition, the effort to integrate the regional warning with on-site warning to become a more robust EEWS is noticed.

The EEWS developed in Taiwan by Central Weather Bureau (CWB) is similar to the one by JMA with so-called 'regional warning' or 'national warning'. While working with the sensor array from Taiwan Strong Motion Instrumentation Program (TSMIP), the EEWS was under testing by cooperative research institutes since 2006, but the warning is only useful for the area located outside the 50 km radius from the earthquake epicenter. Therefore the 'on-site warning' is needed for a near-source earthquake and the regional system has been developed comprising 6 observation points on a circle with radius of 30km around a nuclear power plant in Lithuania to save a lead time of 4-8 seconds. These systems do not use source information but a threshold value for issuing an alarm. In addition, the effort to integrate the regional warning with on-site warning to become a more robust EEWS is noticed.

The on-site EEWS (EEWS), as part of the total solution of seismic hazard mitigation, was under development to provide a series of time related parameters such as the magnitude of the earthquake, the time until strong shaking begins, and the seismic intensity of the shaking (peak ground acceleration). Interaction of different types of earthquake ground motion and variations in the elastic property of geological media throughout the propagation path result in a highly nonlinear function. We use P-wave predicated PGA method and neural networks to model these nonlinearities and develop learning techniques for the analysis of earthquake seismic signal. This warning system is designed to analyze the first-arrival from the earthquake signals in as little as 3 seconds after first ground motion is

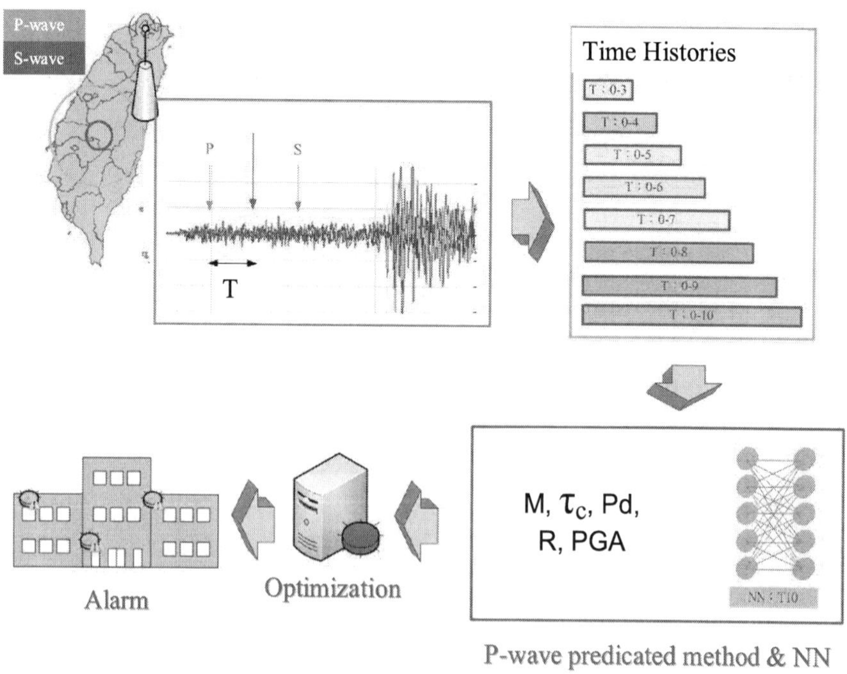

Fig. 1. The framework of the 1st stage for the on-site EEWS

felt at the sensors at a rate of 50 samples per second. Then the EEWS instantaneously provide a profile consists of the estimates of hazard parameters, such as magnitude, dominate frequency, arrival time of S-wave, and maximum seismic intensity (peak ground acceleration, PGA). The system is trained using seismogram data from more than 1000 earthquakes recorded in Taiwan. The proposed EEWS can be integrated with distributed networks for site-specific applications. By producing accurate and informative warnings, the system has the potential to significantly minimize the hazards of catastrophe ground motion. (Figure 1)

3. P-wave predicated PGA method

The major concept of the second type EEWS (on-site) is that the beginning of ground motion recorded by local seismometer can be used to predict ensuing ground motion at the same site. This is based on the physics of earthquake rupture process. Big earthquakes will have larger rupture areas and will produce longer "period" than small earthquakes if the fault rupture velocities are similar for different size earthquakes. Kanamori review the theory (hereafter τc method). The definition is

$$\tau c = \frac{1}{\sqrt{\langle f^2 \rangle}} = \frac{2\pi}{\sqrt{r}}, \quad where \quad r = \frac{\int_0^{\tau_0} \dot{u}^2(t)dt}{\int_0^{\tau_0} u^2(t)dt} = 4\pi^2 \langle f^2 \rangle \tag{2}$$

Usually τ0 is taken as 3 seconds. The ratio of first 3 second velocity ground motion with respect to displacement ground motion is related to the "period" (τc) of the initial portion of an earthquake rupture process. The log(τc) verse Mw magnitude follows a linear trend which is useful for estimating earthquake magnitude for using only one single seismometer. Wu and Kanamori applied this method to Taiwan strong motion data and provided some useful regression results.

$$M=3.088\log(\tau_c)+5.300 \tag{3}$$

$$\log(Pd)=-3.801+0.722M-1.444\log(R) \tag{4}$$

Here Pd is the maximum absolute amplitude of first 3 seconds P-wave displacement waveform, R is the epicentral distance. The attenuation relationship used in this study is provided by Dr. Jean (NCREE, Taiwan).

$$PGA=0.00284\exp(1.73306M)[R+0.09994\exp(0.77185M)]^{(-2.06392)} \tag{5}$$

The practical computation flow is described as follows.

1. Get the serial data from acceleration seismometer.
2. Compute the data stream is triggered by an earthquake or not.
3. If triggered, accumulate 3 seconds vertical component acceleration data.
4. Remove the mean and trend from data.
5. Integrate data into velocity and displacement.
6. Compute τc from integrated velocity and displacement data, then estimate earthquake magnitude.
7. Compute the maximum absolute amplitude of displacement data.
8. Estimate the epicentral distance using magnitude and amplitude information.
9. Compute the P-wave S-wave travel time difference.

10. Compute PGA from attenuation relationship by giving magnitude and distance.
11. Issue earthquake warning based on the estimated PGA

This algorithm has been used to test Taiwan strong motion acceleration data (95000 free field records). The success rate is approximate 60%. It also has been tested using building array data. The results show that the algorithm works well for seismometer installed in the basement or roof of buildings. This research result proofed that in the future application, the seismometer can be installed in any floor in a building and the prediction algorithm will not fail. Figures 2-5 show testing results using 1999 921 Chi-Chi Taiwan earthquake data recorded in the roof and basement of Taipower building which is a 26-floor with 3-floor underground structure.

Fig. 2. Photo of Taipower building.

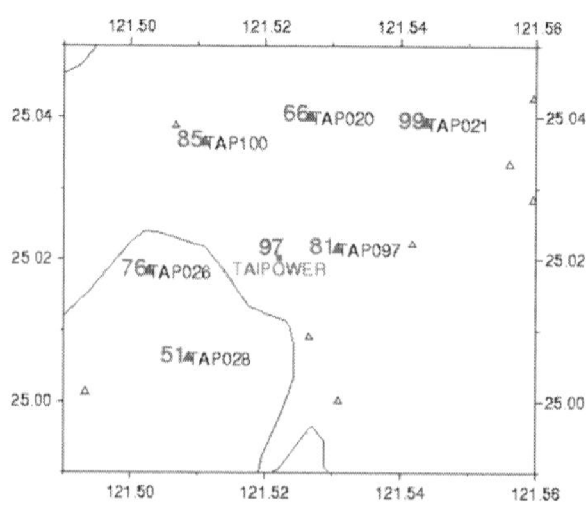

Fig. 3. The estimated ground PGA (97 gal) using Taipower building basement recorded data is reasonable comparing with the neighborhood recorded ground PGA (red number). The estimated ground PGA using roof data is 61 gal which is not shown on map.

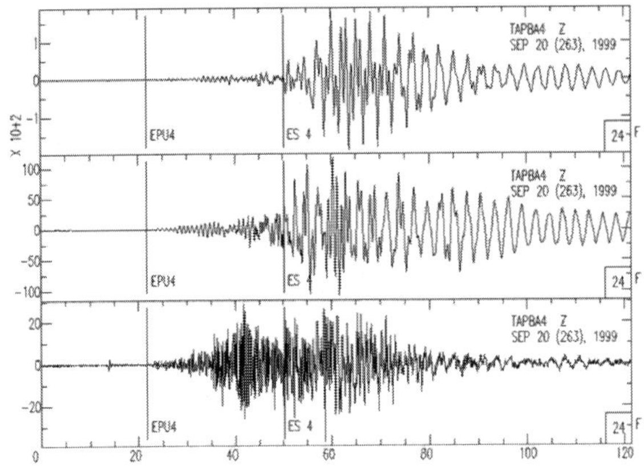

Fig. 4. The recorded 1999 921 earthquake data from the roof of Taipower building.

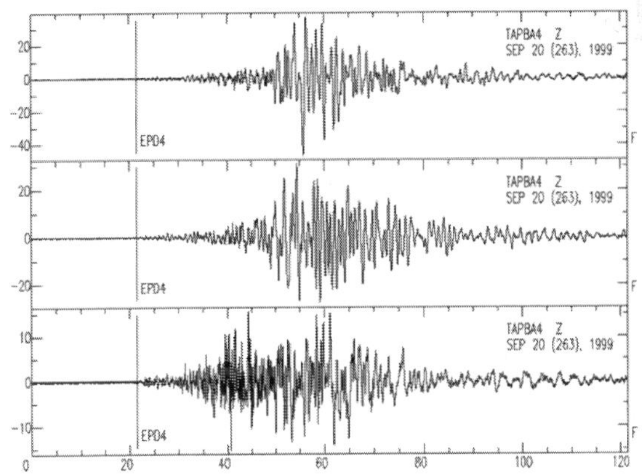

Fig. 5. The recorded 1999 921 earthquake data from the basement of Taipower building.

4. Neural network

Neural networks, that possess a massively parallel structure, are well known as a biologically inspired soft computing tool. Their learning capabilities, which differ them from other mathematically formulated methods, are provided by the unique structure of neural networks and allow the development of neural network based methods for certain mathematically intractable problems. Neural networks are formed by many interconnecting artificial neurons. Signals propagate along the connections and the strength of the transmitted signal depends on the numerical weights that are assigned to the connections. Each neuron receives signals from the incoming connections, calculates the weighted sum of the incoming signals, computes its activation function, and then sends signals along its outgoing connections. Therefore, the knowledge learned by a neural network is stored in its connection weights. To solve difficult engineering problems, it is necessary to design a task-specific neural network. Therefore, the neural networks program developed by Lin using Fortran were used in this study. A combination of the Quick-Prop algorithm and the local adaptive learning rate algorithm were applied to the multiple-layer feed-forward (MLFF) neural networks to speed up the convergence rate of the networks. In addition, a mechanism to avoid over-training the neural networks for certain patterns, the developed algorithm was designed to monitor and equalize the influence of each pattern in the training case on the connection weights during each epoch. The average root-mean-square output error of the networks became lower while maintaining the generalization ability of the neural networks when using this adaptive process (Lin, 1999). The Neural Network has been applied to the ground motion prediction and generation since 1997. The previous researches shows that the neural network makes it possible to provide more accurate, reliable and immediate earthquake information for society by combining the national EEWS and to be applied to the advanced engineering application as well as planning of hazard mitigation (Kuyuk and Motosaka, 2009). The on-site EEWS developed for Taiwan has used initial part of P-waveform measured in-situ and neural networks for forward forecasting of ground motion parameters (Magnitude, PGA, estimated arrival time for strong motion) before S-wave arrival. The estimated ground motion information can be used as warning alarm for earthquake hazard reduction. The validity and applicability of this method have been verified by using the CWB observation data sets of 2505 earthquakes occurred in Taiwan area.

A state-of-the-art neural networks based methodology is presented for forward forecasting of ground motion parameters before S-wave arrival using initial part of P-waveform measured on-site. The estimated ground motion information can be used as warning alarm for earthquake damage reduction. The neural networks program developed by Lin using FORTRAN was used in this study. A combination of the quick-prop algorithm and the local adaptive learning rate algorithm were applied to the multiple layer feed-forward back-propagation neural networks to speed up the learning of the networks (Lin, 1999). The supercomputer is also used to train the neural networks. The validity and applicability of the method have been verified using the CWB observation data sets of 1012 earthquakes occurred in Taiwan.

Furthermore, a new concept of grouping neural networks called Expert Group Neural Network (EGNN) is also used in this study. The EGNN behaved like a group of experts, who grew up from different backgrounds with individual expertise, and were able to provide the appropriate comment when working together as a committee (Lin et al., 2006). The optimal solution among the comments will be chosen while solving this kind of problem. Eight feed forward back-propagation neural networks trained by different inputs constituted the EGNN as a committee of experts to provide the time related information from earthquake accelerograms. The architecture of each neural network among EGNN is set to be different. It consisted of one input layer with 450 to 1500 neurons, two hidden layers and one output layer with 11 neurons (as shown in Fig.6). Each of the EGNN was used to analyze the relationship between the initial few seconds of the earthquake accelerogram and the earthquake waveform information of that specific earthquake.

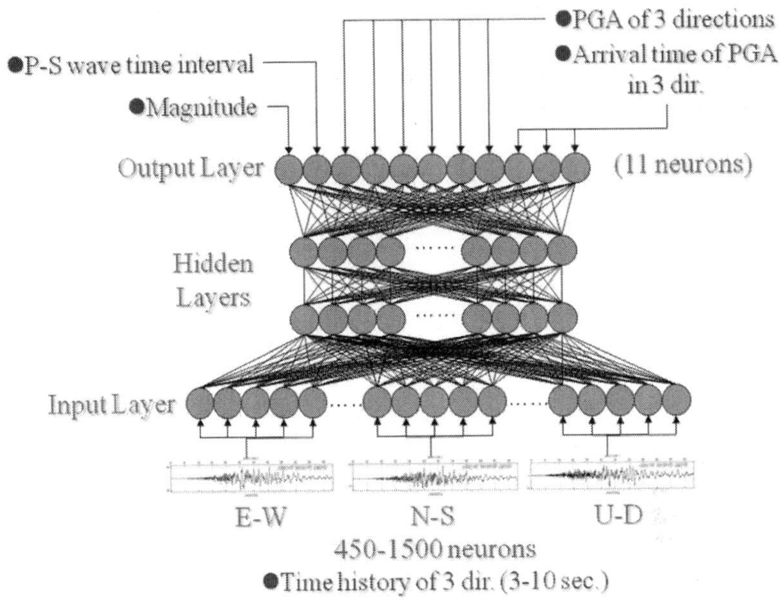

Fig. 6. The architecture of neural networks $NN_{T0\text{-}3\sim10}$

The neural networks were used in both stages of the on-site EEWS development. At the first stage development of the on-site EEWS while the parametric analysis study was conducted, the P-wave predicated PGA method (Wu and Kanamori 2005) and the neural networks were used to provide a basic prediction of the earthquake information such as the earthquake magnitude, seismic intensity, peak ground acceleration, and arrival time of S-wave for free field. In addition, the neural networks were able to predict the arrival time of PGA. The proposed method has been verified its validity and applicability. Furthermore, the neural networks were used at the 2nd stage development of the on-site EEWS to predict the seismic intensity, peak acceleration, arrival time of S-wave, and arrival time of peak acceleration for the roof of the specific building (structure response). In the first stage, the EGNN were trained with the data from first 3 seconds to 10 seconds of the earthquake accelerograms separately,. The earthquake magnitude, PGA (seismic intensity), arrival time of S-wave and arrival time of PGA were predicted using the waveform data from in-situ sensors. In this case, when the real-time information measured from the in-situ sensors is verified as earthquake using 1 second of time history after the arrival of P-wave at the site, then the initial 3 seconds of the earthquake accelerogram (P-waveform) was used as the input for the neural network (NN:T-3) to estimate the magnitude of the earthquake, the PGA (seismic intensity) in three directions, and the arrival time of the S-wave as well as those of PGA. At the same time, the sensors are recording and the initial 4, 5, .., and up to 10 seconds of P-waveform were used as the input for the neural networks (NN:T-4, NN:T-5, .., NN:T-10) to estimate the parameters for the on-site EEWS consequently. The best prediction was then chosen from these 8 results through certain optimization algorithm or time factor. The emergency response actions can be activated right after receiving the warning due to the seismic intensity predicted for the site as well as the remaining time before the strong S-wave or PGA occurred.

The training and testing (validation) data were prepared using the earthquake accelerograms recorded through TSMIP in Taiwan from 1992 through 2006, the magnitude of these earthquake ranged from 4.0 to 8.0 on the Richter Scale. There are a total of 50149 recorded accelerograms from 2505 earthquakes. Among them, the training data were randomly chosen using 40539 earthquake records (80% of the total) from 2371 recorded earthquakes while the test data were prepared using the remaining 9610 earthquake records (20% of the total) from 1012 recorded earthquakes

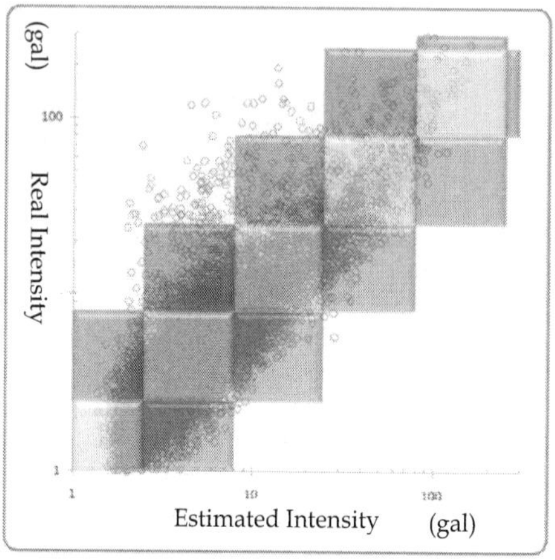

Fig. 7. Comparison of the real and est. seismic intensity (NN_{T0-3})

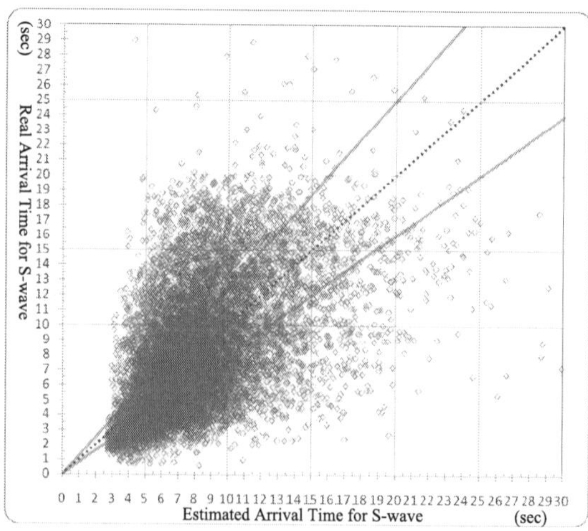

Fig. 8. Comparison of the real and est. arrival time for S-wave (NN_{T0-3})

Figs. 7-8 compare the results of the prediction of seismic intensity as well as the arrival time of the S-wave for the model NN:T-3. The accuracy of the seismic intensity prediction (green area) is around 60% while the accuracy for the ± one degree seismic intensity (red, green and purple areas) is around 95% (Fig.3 R^2=0.6977). Figure 4 shows that R^2=0.28 as the accuracy for the estimated arrival time of the S-wave. If a tolerance of ±20% is considered feasible for warning people on the arrival time of S-Wave, then the accuracy is around 70%.
Figs. 9-10 compare the results of the prediction of seismic intensity as well as the arrival time of the S-wave for the model NN:T-10. The accuracy of the seismic intensity prediction (green area) is around 68% while the accuracy for the ± one degree seismic intensity (red, green and purple areas) is around 98% (Fig.3 R^2=0.7714). Figure 6 shows that R^2=0.48 as the accuracy for the estimated arrival time of the S-wave. However, the accuracy rises up to around 80% within ±20% tolerance for the purpose of warning people.

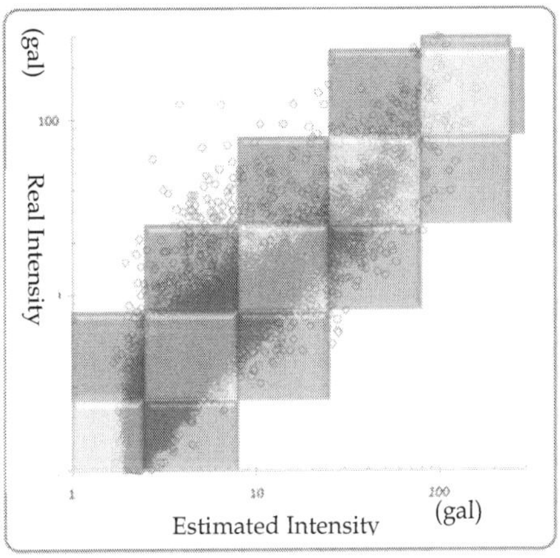

Fig. 9. Comparison of the real and est. seismic intensity (NN_{T0-10})

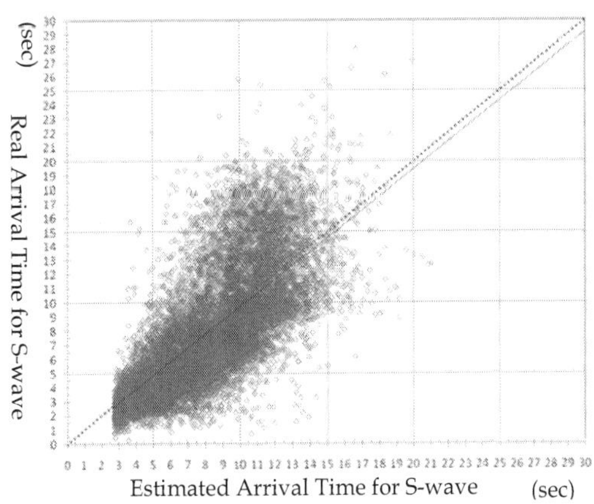

Fig. 10. Comparison of the real and est. arrival time for S-wave (NN_{T0-10})

It has been found that the accuracy of the predicted peak ground motion is drastically improved compared the results of NN:T-10 to NN:T-3 since more earthquake information (7 seconds more of earthquake accelerogram) is provided for the neural network (NN:T-10), as shown in Tables 1 and 2. However, the disadvantage for model NN:T-10 is 7 seconds loss of time. Therefore, the result from NN:T-3 is recommended for the purpose of hazard mitigation and emergency response as long as the precision requirement is tolerable for certain applications. Besides, the results of the neural networks can be improved if more earthquake records can be obtained and more training can be done.

EQ	Real	T-3	T-4	T-5	T-6	T-7	T-8	T-9	T-10
A	4.75	10.00	6.77	7.41	9.36	5.84	5.79	9.24	4.59
B	10.33	9.53	9.12	5.79	14.59	8.57	8.15	5.06	11.14
C	9.29	5.85	6.44	9.93	5.98	7.84	5.78	5.29	8.98
D	5.50	7.61	8.64	2.95	6.72	6.05	7.56	8.40	5.57
E	6.79	8.85	11.07	11.36	9.98	10.35	10.38	10.08	6.58
F	11.43	8.64	14.75	10.32	12.91	7.28	8.49	6.26	11.15
G	5.07	3.29	3.51	4.31	4.37	5.13	4.13	4.68	5.47
H	9.94	7.23	9.04	7.24	4.62	5.32	3.86	3.98	9.00
I	8.99	9.19	9.40	6.52	11.09	4.94	7.71	9.49	8.43
J	9.69	10.92	11.60	9.99	13.73	9.10	7.38	11.34	11.38
K	8.18	8.19	5.16	3.74	4.38	7.02	4.60	8.33	7.42

Table 1. Comparison of the arrival time for S-wave from the neural networks with several recorded earthquakes. (Unit: second)

EQ	Real	T-3	T-4	T-5	T-6	T-7	T-8	T-9	T-10
A	2	2	3	2	2	2	3	3	2
B	3	3	3	3	3	2	3	3	3
C	2	2	2	3	3	2	2	2	2
D	4	4	4	4	3	3	4	4	4
E	2	3	3	3	2	2	2	3	2
F	3	3	3	3	4	3	3	2	4
G	3	3	4	3	3	3	3	2	3
H	5	4	4	4	4	4	4	3	5
I	4	4	4	3	4	4	4	4	4
J	3	3	3	3	3	3	3	2	3
K	3	3	3	3	3	3	3	3	3

Table 2. Comparison of the seismic intensity from the neural networks with several recorded earthquakes. (Unit: degree)

5. Rapid estimation of the structural responses / general modulus

In this section, the non-linear un-damped vibrations of an shear beam model subjected to a harmonic motion along its base are investigated. This model is applied to simulate the seismic responses of a building when peak ground acceleration (PGA) and earthquake significant frequency are predicted in part I. It is assumed that the linear part of the shear modulus of the building is uniform along the building height. Transforming the Ramberg-Osgood model to a suitable for the relationship of the shear stress versus shear strain, a non-linear partial differential equation is obtained as the governing equation. In the method presented here, the multi-story building is modelled as an equivalent continuum with non-uniform stiffness consisting of a combination of a shear cantilever beam deforming in shear configurations. The base of this building model makes as harmonic horizontal motion Asin(wt). Since the height is normalized by the total building height and nonlinear deformations occur along the transverse horizontal direction. As a result, the governing equation can be written as

$$\rho\frac{\partial \upsilon_x(z,t)}{\partial t}=\frac{\partial \tau_{xz}(z,t)}{\partial z},\ \forall z\in(0,H),\ t\in(-\infty,+\infty) \tag{6}$$

where ρ is the mass density, $\upsilon x(z,t)$ is the lateral velocity, z is the co-ordinate measured from the top and t is the time. Herein, the Ramberg-Osgood model is widely used as it can be transform to a suitable form for analytical solution. Rearranging this transformed model and considering the relationship between the shear stresses and strains, the non-linear stress-strain relation can be written as

$$\gamma_{xz}(z,t)=\frac{\tau_Y}{G}\{\frac{\tau_{xz}(z,t)}{\tau_Y}+\alpha[\frac{\tau_{xz}(z,t)}{\tau_Y}]^3\},\ \forall z\in(0,H),\ t\in(-\infty,+\infty) \tag{7}$$

where τxz it the shear stress, τY it the yielding shear stress, G is the linear shear modulus and α is the post-yielding stiffness ratio. Then the equation of motion for the shear building model can be written as

$$\frac{\partial \gamma_{xz}(z,t)}{\partial t}=\frac{\partial \upsilon_{xz}(z,t)}{\partial z},\ \forall z\in(0,H),\ t\in(-\infty,+\infty) \tag{8}$$

The shear building is subjected to a harmonic motion Asin(wt) at the base and traction free at the top. Thus, the boundary conditions can be written as

$$\tau_{xz}(0,t)=0,\ t\in(-\infty,+\infty) \tag{9}$$

$$\upsilon_x(H,t)=\frac{\partial}{\partial t}\left[A\cdot\sin(\omega t)\right],\ t\in(-\infty,+\infty) \tag{10}$$

In this study, the response of the shear building is assumed to be harmonic with the period 2π/ ω, as the forcing term. Thus the periodicity condition is written as

$$\tau_{xz}(z,t)=\tau_{xz}(z,t+2\pi/\omega),\ \forall z\in(0,H),\ t\in(-\infty,+\infty) \tag{11}$$

$$\upsilon_x(z,t)=\upsilon_x(z,t+2\pi/\omega),\ \forall z\in(0,H),\ t\in(-\infty,+\infty) \tag{12}$$

It is convenient to rewrite the above equations in terms of dimensionless quantities, which are defined as follows:

$$c=\sqrt{\frac{G}{\rho}},\bar{z}=\frac{z}{H},\bar{t}=\frac{ct}{H},\gamma_Y=\frac{\tau_Y}{G},\bar{A}=\frac{A}{\gamma_Y H},\Omega=\frac{\omega H}{c} \tag{13}$$

where $\bar{z}$ and $\bar{t}$ are the dimensionless co-ordinate and time. $\bar{A}$ is the dimensionless displacement.
Then the system total energy E can be expressed as follows:

$$E=\int_0^1\left(\frac{1}{2}\left[\frac{\partial\hat{\phi}(z,\hat{t})}{\partial z}\right]^2+\frac{\Omega^2}{2}\left[\frac{\partial\hat{\phi}(z,\hat{t})}{\partial\hat{t}}\right]^2+\frac{3\alpha\Omega^4}{4}\left[\frac{\partial\hat{\phi}(z,\hat{t})}{\partial\hat{t}}\right]^4\right)_{t=0}dz \tag{14}$$

Due to the non-linear governing equations described above is impossible to find an exact solution. For this reason, an approximate solution is sought by using the perturbation method. Among a few variants of the perturbation method, the Lindsted-Poincaré technique seems to be a suitable one. For the use of this technique, the potential energy function $\hat{\phi}$, the system total energy E and normalized excitation frequency Ω are expanded into perturbation series in terms of ε as follows:

$$\hat{\phi}=\phi_0+\varepsilon\phi_1+\varepsilon^2\phi_2+\varepsilon^3\phi_3+..... \tag{15}$$

$$E=E_0+\varepsilon E_1+\varepsilon^2E_2+\varepsilon^3E_3+..... \tag{16}$$

$$\Omega^2=\Omega_0^2+\varepsilon\Omega_1^2+\varepsilon^2\Omega_2^2+\varepsilon^3\Omega_3^2+..... \tag{17}$$

And the system total energy can be expressed as follow:

$$\begin{aligned}E&=\int_0^1\left(\frac{1}{2}\left[\frac{\partial\hat{\phi}(z,\hat{t})}{\partial z}\right]^2+\frac{\Omega^2}{2}\left[\frac{\partial\hat{\phi}(z,\hat{t})}{\partial\hat{t}}\right]^2\right)_{t=0}dz=\frac{\Omega_0^2A^2}{2\Omega_0^2\cos^2(\Omega_0)}\int_0^1\left[\cos(\Omega_0 z)\right]^2dz\\&=\frac{\Omega_0^2A^2}{4\cos^2(\Omega_0)}\left[1+\frac{\sin(2\Omega_0)}{2\Omega_0}\right]\end{aligned} \tag{18}$$

5.1 An example

Assume that a example RC building is located in the site which seismic zone factor Z is 0.33g. The total building height H is 30 meter. The corresponding empirical period T_0 = 0.07H3/4=0.8923 sec. According the seismic force requirements of the Taiwan 1989 seismic provision, the corresponding spectral acceleration $Sa(T_0)$ is equal to 0.33(9.8)1.2/ $\left(T_0^{2/3}\right)$ =4.1715 m/sec2. the spectral displacement SD=Sa(T_0/2π)2=0.0851 meter. If assume the ductility factor R of this building is 3.0, and the post-yielding stiffness ratio is 0.05 along the lateral direction, then the system yielding displacement Δy= SD/R=0.0284 m. Then the perturbation factor ε can be calculated as follow:

$$\varepsilon = 0.01 \times \frac{(1-\alpha)(R-1)}{\left[1+\alpha(R-1)\right]^3} = 0.0143 \tag{19}$$

If the "P-Wave Predicted PGA Method" predicted the PGA value and the dominant frequencyω are 59.1gal and3.44 rad/sec, respectively, the peak base displacement amplitude

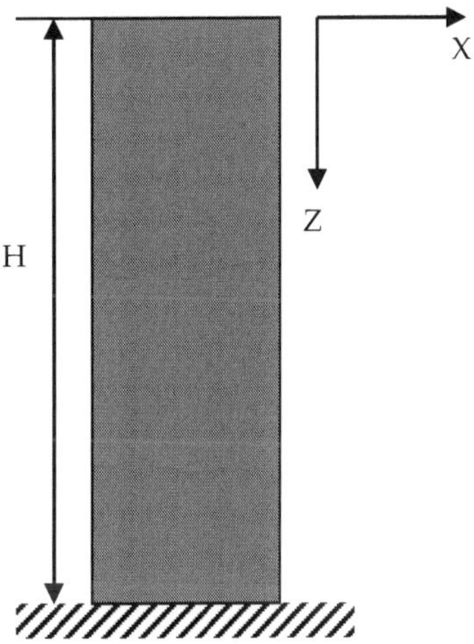

Fig. 11. Simplified shear building model

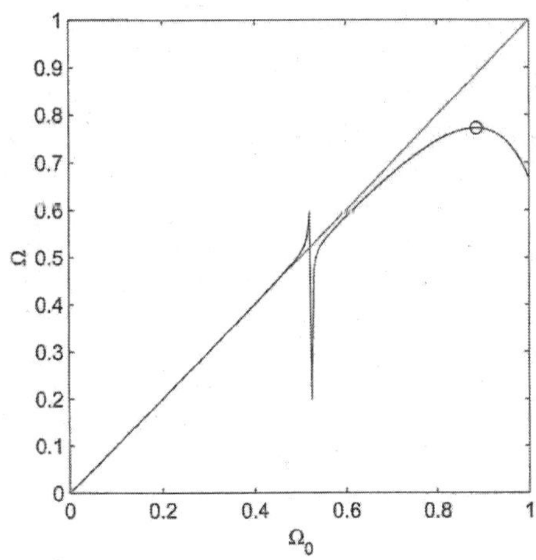

Fig. 12. The relationship between the Ω and Ω0

A is 0.05m, Substituting z=1 and A=0.05m, then the relationship between the dimensional excitation frequency Ω and the linear Ω0 can be shown as Figure2. In Figure 12, when the Ω0 is equal to the maximum value 0.885, the corresponding Ω value is 0.7715. Then the maximum story drift distribution can be shown as Figure 13. Figure 14 shows the relative displacement between the base and top of this example building. (no figure 11)

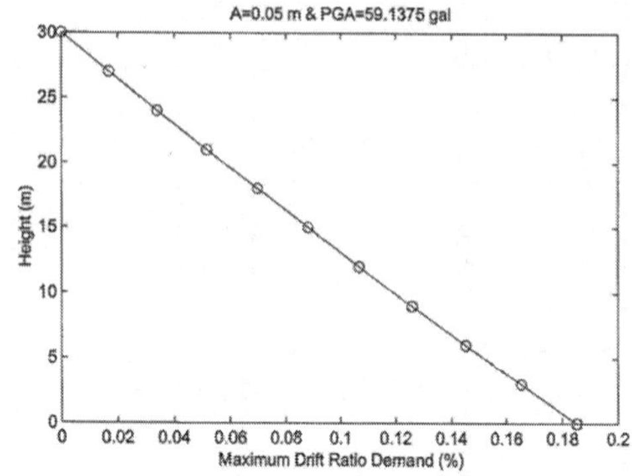

Fig. 13. The maximum drift ratio of the example building

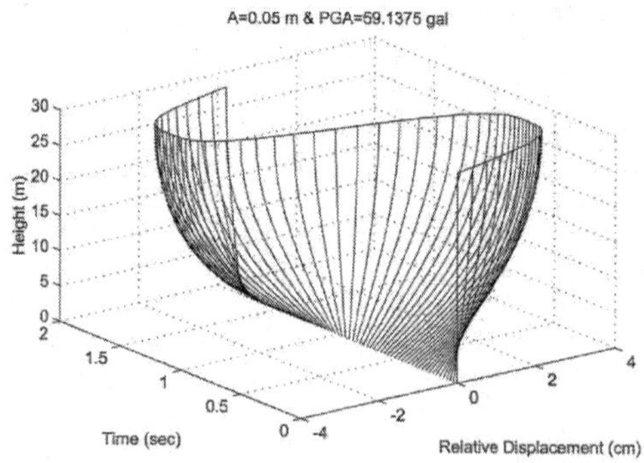

Fig. 14. The relative displacement between the base and top

6. Rapid estimation of the structural responses / customized modulus

The main target of this research is to rapidly estimate the amplification factor of structural response with the basic earthquake characteristics supported from other sub-projects. Ideally, the developed system should be able to predict the structural response accurately within 0.1 sec calculation time to protect the life and property of the whole island.

The Tai-power Building is used as the objective building in the research. By using a fine-tuned finite element model of the Tai-power Building with 60 earthquake time histories recorded in the Taipei basin, all the parameters including the PGA, earthquake magnitude, and distance between the epicenter and the local site are collected as the database for the scenario-based repression analysis.

By considering the covariance's of the amplification factors with PGA, earthquake magnitude, and epicenter, obvious relationship can be found between epicenter distance and the roof amplification factor. As the result, the epicenter distance is chosen as the governing factor to establish modules.

6.1 Regression method

The Quadratic Response Surface Model (QRSM) is used as the regression model in this scenario-based technique and the general form can be expressed as

$$y(x) = a_0 + \sum_{i=0}^{N} a_i x_i + \sum_{i<j}^{N} a_{ij} x_i x_j + \sum_{i=0}^{N} a_{ii} x_i^2 + ... \quad (20)$$

By choosing this model, a polynomial model with multi-variables can be easily established than the conventional nonlinear analysis process.

6.2 Analysis result

The regression results of the amplification factor on the roof floor in both X and Y directions are expressed from Figures 15 to 18.

According to the analysis result, the average error percentage of the amplification factor in the X direction is approximately 21.71% when distance between the epicenter and the site is less than 120 km, and the average error percentage in the Y direction is approximately 21.8%. Moreover, when the distance is larger than 120 km, the average error percentages reach 20% in the X direction and only 13.92% in the Y direction. In short, although some fluctuations still exist in few cases, the established regression models can generally estimate the amplification factor within several seconds after the earthquakes happen.

By using the scenario-based technique, the amplification factor of structural response can be successfully predicted by the regression models using QRSM. The theoretical analysis has demonstrated that the prediction error can be kept in an acceptable range while the result is obtained within few seconds after the major earthquakes.

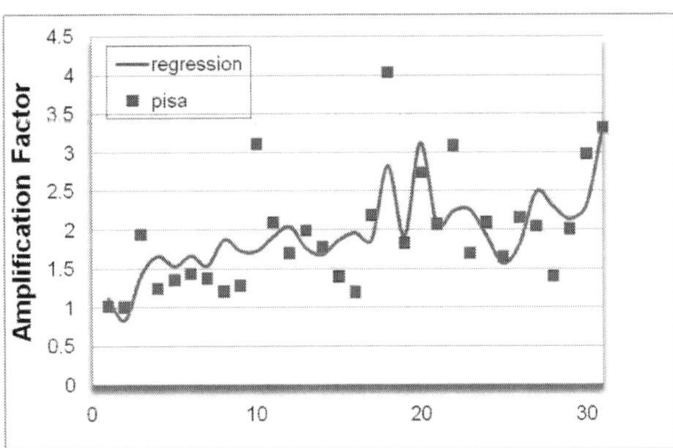

Fig. 15. Distance<120 km (X direction)

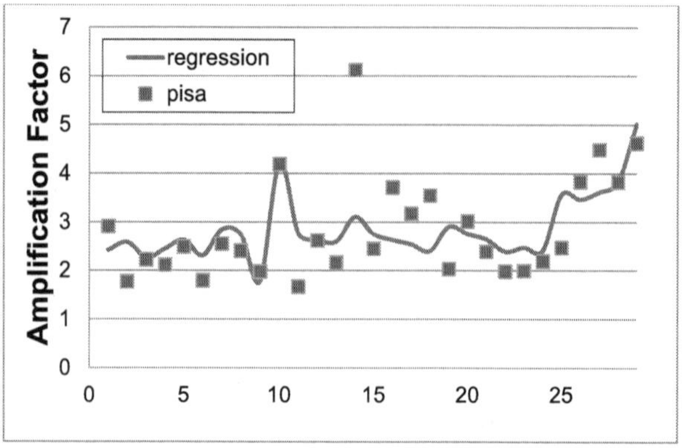

Fig. 16. Distance>120 km (X direction)

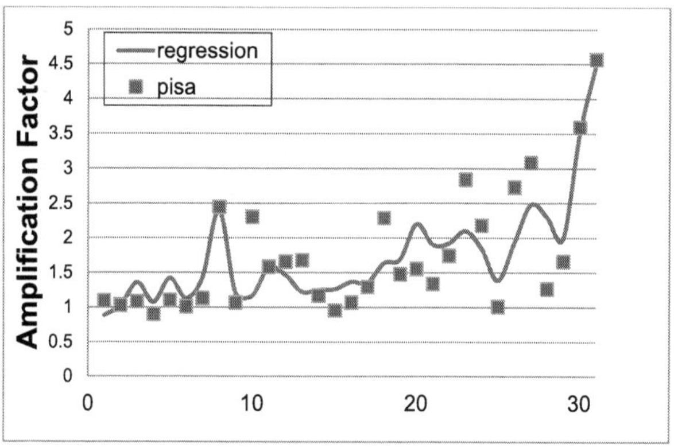

Fig. 17. Distance<120 km (Y direction)

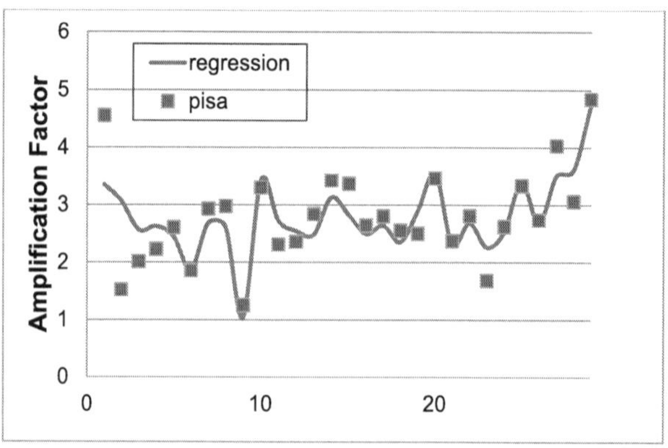

Fig. 18. Distance>120 km (Y direction)

To improve the performance of the system, the dominant frequency of the earthquake and the time difference between the P-wave and the S-wave will be considered as major factors in the regression models. Meanwhile, nonlinear model method will also be used to develop an alternative regression model. By combing these two models, it is believed that a reliable structural response prediction system can be expected in the near future.

7. Database for rapid estimated structural response

Sstructures suffered from earthquakes is concerned more and more for engineers. The earthquakes are always occurred immediately; however, how to estimate the structural response accurately and fast before the earthquake arrived is a great challenge. To achieve this end, a database contained seismic characteristics and dynamic analysis of FEM model is presented for fast estimated response of structure and the Tai-Power Building is opted as a full-scale FEM model which is constructed by using FEM analysis software PISA3D developed by NCREE.

7.1 Modification of PISA3D model

In order to estimate accurate response of structure, verification of the response between PISA3D model and reality structure should be considered. The famous Chi-Chi earthquake was chosen to compare the response of original PISA3D model and time history records of Tai-power Building. It can be seen in Figure 19 that the simulated acceleration response and

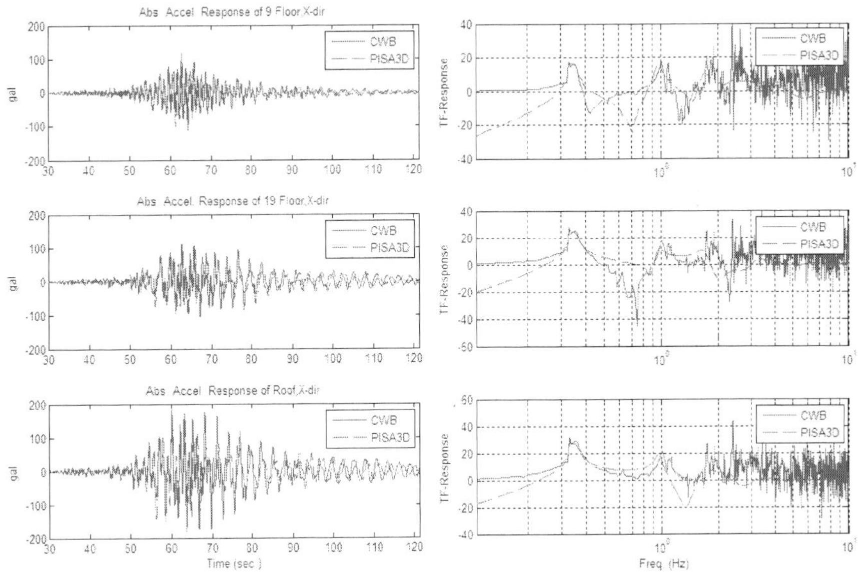

Fig. 19. Time history and transfer function of CWB records and PISA3D model in strong axis. (Excitation: Chi-Chi earthquake)

transfer function in the strong axis can be slightly regarded as consistent with the records. But as shown in Figure 20, it says that the structural response of PISA3D model in the weak axis was not estimated well as the reality structure.

From the time history and transfer function of weak axis, it figures out the structural stiffness (natural frequency) of FEM model is not sufficient as the reality structure. Therefore, modification of PISA3D model should be considered. It has several methods to modify the PISA3D model. In this case, the effect of rigid end zone and shear wall will be applied to the model to increase the stiffness of the model appropriately.

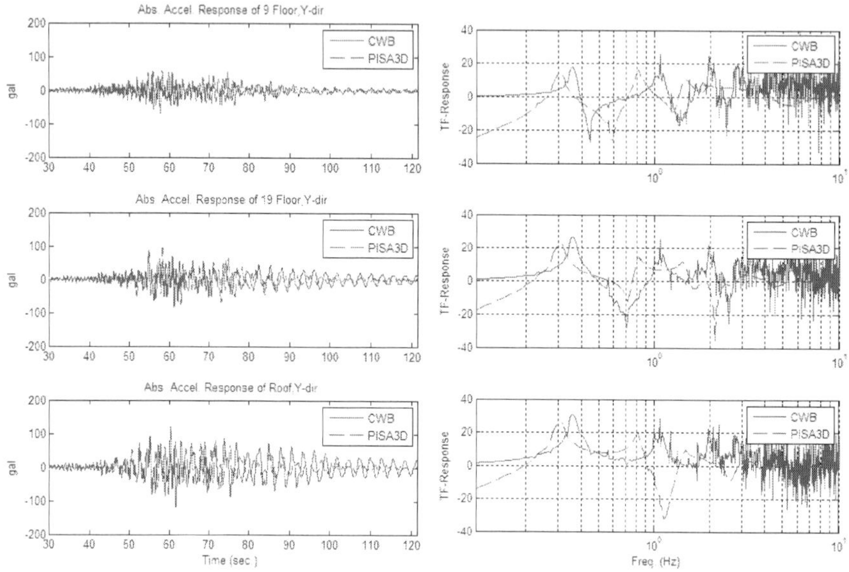

Fig. 20. Time history and transfer function of CWB records and PISA3D model in weak axis. (Excitation: Chi-Chi earthquake)

Consideration of rigid end zone and shear wall (in weak axis):

The effect of rigid end zone and shear wall shows better estimated structural response both in strong axis and weak axis. It can be shown in the Figure 21.

Besides, it's worth to mention that the dynamic characteristics of reality structure show different natural frequencies when structure suffered from different PGA of excitation. This phenomenon can be simply separated about PGA 25 gal. Two different PISA3D models, therefore, were proposed to simulate this phenomenon appropriately.

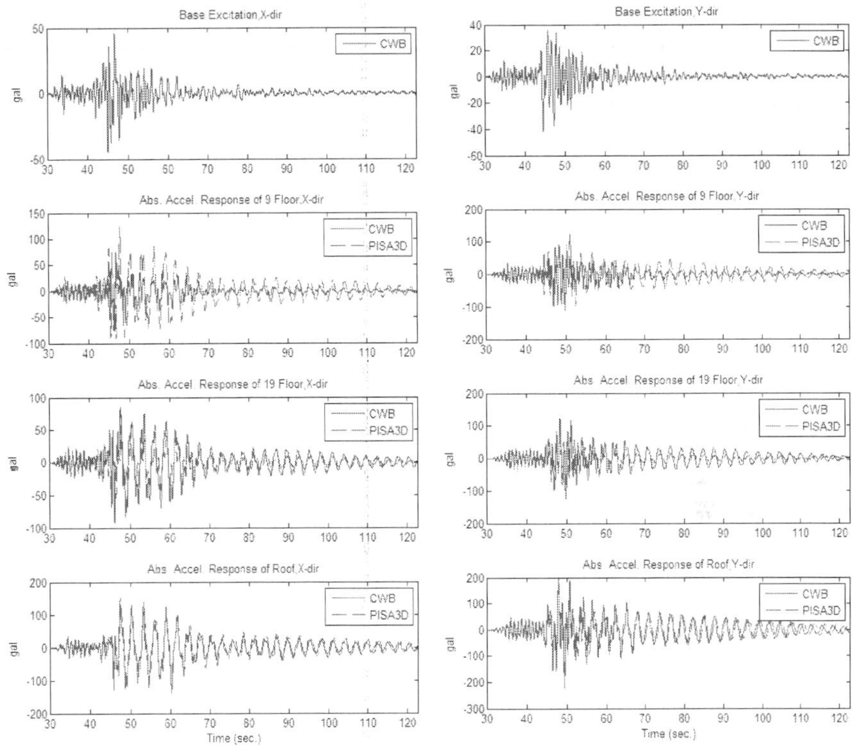

Fig. 21. Time History of CWB records and PISA3D model in strong and weak axis. (Excitation: 331 earthquake)

7.2 Construction of database

One of main purposes to construct the database is to understand structural characteristics for fast estimated response of structure suffered from different earthquakes; the other is to provide for other research such as SHM, and system identification, etc.

The contents of seismic characteristics are shown as following:

1. Coordinates of earthquake epicenter,
2. Coordinates of sensor stations,
3. Distance between epicenter and site,
4. Earthquake magnitude,
5. Site intensity,
6. Main frequency of earthquake
7. Time difference between P-wave and S-wave arrived.

And the dynamic analysis of finite element model includes:

1. Time history of structural response at specific floor
2. Time history of ground motion,
3. Peak floor acceleration
4. And PGA.

For fast estimated response of structure, the number of database should be large than a certain amount of data. However, the on-site records can not be provided as large as a database. Therefore, In Figure 22, it shows the concept to construct the database and to simulate the different site condition in Taiwan. It uses about 50 records of strong earthquake motion of Tai-power Building to verify and modify the PISA3D model, and uses more than thousands free field ground motion to excite PISA3D model to simulate the different site condition.

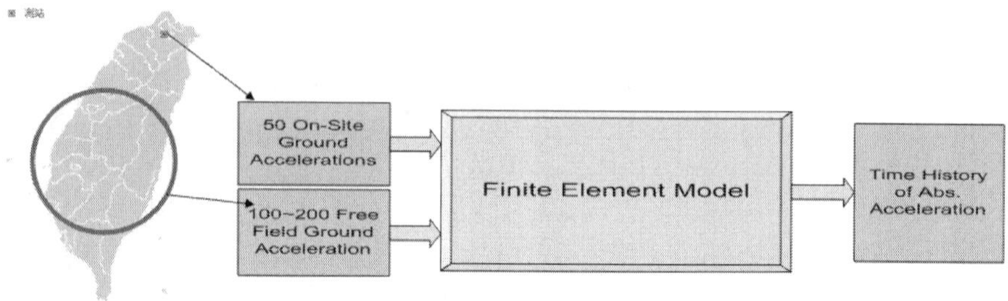

Fig. 22. Simulation of site condition in Taiwan.

7.3 Structural response prediction

This study use the recorded earthquake free field time histories collected from 1992-2006. Each earthquake record was marked P-wave and S-wave by automatically P-wave trigger program and verified manually. Then these earthquake accelerograms were consolidated into the earthquake free-field accelerogram database. Among the 59 observed buildings within the Taiwan Strong Motion Instrumentation Program (TSMIP) by CWB, the Tai-Power building is chosen since its significant features. There are totally 73 earthquake records from Tai-Power building. There are 26 time histories recorded from each sensor installed for each earthquake record. It is the tallest building when it is constructed. There are totally 26 tri-axial strong motion sensors installed in the building. Since the data recorded is not enough for training a neural network, the time history structural analysis software PISA-3D is used to build a numerical model for Tai-Power building. The 73 recorded earthquake time histories from Tai-Power building were used to modify and calibrate the numerical model of Tai-Power building. Then the chosen earthquake records from the earthquake free-field accelerogram database were used as input to run the PISA-3D and the response time histories were obtained to form the earthquake response accelerogram database. The structural response is assumed to be linear (elastic) behavior.

Usually the time history analysis for large-scale structure is time consuming because of its large linear and nonlinear analysis will affect the efficiency of calculation. Therefore, the structural analysis software PISA-3D, developed by National Center for Research on Earthquake Engineering (NCREE), is used for calculating the structural response of Tai-Power building under earthquakes due to its computational advantages for Large-scale structure and therefore the computing time required for analysis is reduced.

To increase the records of structural response for the neural network training, all the earthquake records in the earthquake free-field accelerogram database were considered and the distribution of the peak ground acceleration (PGA) were considered. The records with

PGA range from 5-500 gal were chosen to run the PISA-3D for structural response. There are totally 10,097 records were used for further analysis (Chang, K.C. et al., 2010). Figure 23 shows the flowchart of numerical building model analysis using PISA3D. All the results from PISA-3D were collected into the building response database for further research.

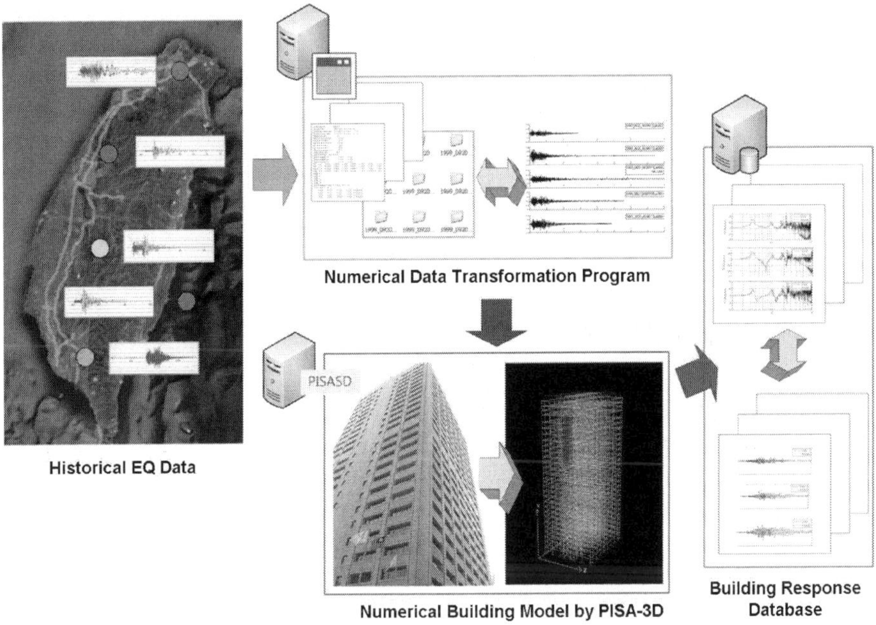

Fig. 23. Flowchart of numerical building model analysis using PISA3D

The architecture of neural networks used in this study is set to be different. Two neural networks, NN_{T0-3} and NN_{T0-10}, were used for on-site EEWS with the Fast Fourier Transform (FFT) of the initial 3 seconds and 10 seconds sensed earthquake waveform as input respectively. Both models consisted of one input layer with 129 (NN_{T0-3}) or 257 (NN_{T0-10}) neurons, two hidden layers and one output layer with 6 neurons (as shown in Figure 24). Each of the neural networks was used to analyze the relationship between the initial three or ten seconds of the sensed earthquake accelerogram and structural response of the Tai-Power building for that specific earthquake.

In this method, the numerical model of Tai-Power building is built using PISA-3D. There are 73 earthquake records between 1994 and 2006 recorded from the sensors installed on the Tai-Power building were used to modify and calibrate the numerical model for structural response. The simulated responses for the roof of the Tai-Power building were within 10% of error. Then the 10097 earthquake records from the database of CWB were chosen as input to run the time history analysis using PISA-3D. The structural response of these 10097 earthquake records were then integrated into building response database.

The neural networks were used to learn (analyze) the relationship between free field ground motion and the structural response on the roof of the Tai-Power building. The 10097 earthquake records were divided into training group and testing group randomly. 8082 earthquake records (80% of the total) were used to train the neural networks while 2015 earthquake records (20% of the total) were used to test and validate the trained neural networks. The Fast Fourier Transform (FFT) of the digitized signal from the first 3 seconds of the earthquake time history after p-wave were used as input to the model $NN_{T0\text{-}3}$ while the FFT from first 10 seconds of the earthquake time history were used as input to the model $NN_{T0\text{-}10}$. The structural responses for the roof of the Tai-Power building from the building response database were used as output for the neural networks.

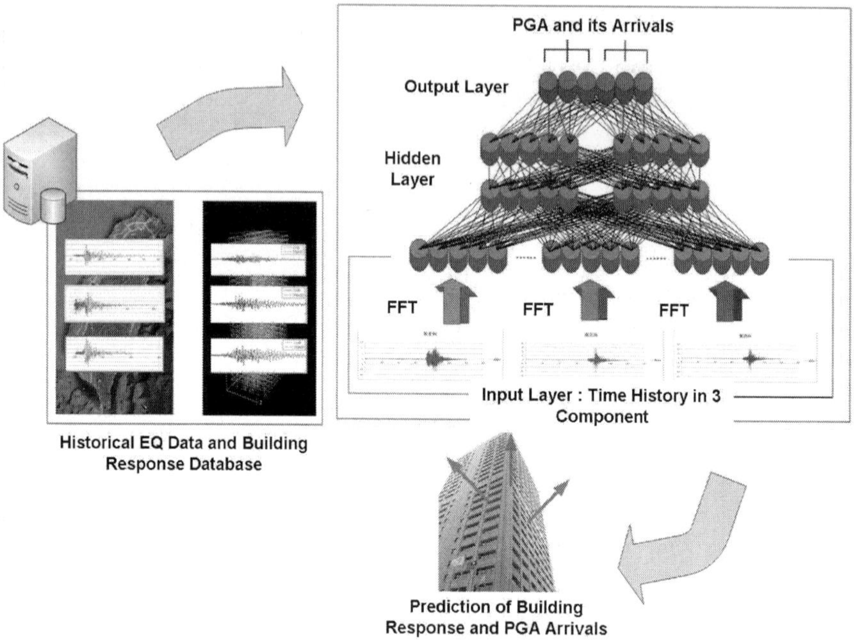

Fig. 24. Predicting structural response using neural networks

The comparison of the real and estimated seismic intensity for $NN_{T0\text{-}3}$ and $NN_{T0\text{-}10}$ were shown in figures 25 and 26. The results of NN estimated PGA versus the real PGA from 2015 novel testing cases is plotted in the figures. The green area means that the intensity of the NN estimated PGA is the same as the intensity of the real PGA. The red area means that the intensity of the NN estimated PGA is one grade less than the intensity of the real PGA. The purple area means that the intensity of the NN estimated PGA is one grade larger than the intensity of the real PGA. The accuracy of the intensity estimation is shown in table 3.
Both the results from $NN_{T0\text{-}3}$ and $NN_{T0\text{-}10}$ are acceptable and the regression analysis R^2 is 0.638 for $NN_{T0\text{-}3}$ and 0.787 for $NN_{T0\text{-}10}$. It also shows better convergence in figure 4. Which shows more input information the neural network model, more accuracy result can be obtained from the neural networks. Which means $NN_{T0\text{-}10}$ is doing better than $NN_{T0\text{-}3}$. If the acceptable range for the intensity prediction is set to be ±1 grade, the accuracy for $NN_{T0\text{-}3}$ will be 91.5% (1844/2015) and 93.7% (1888/2015) for $NN_{T0\text{-}10}$.

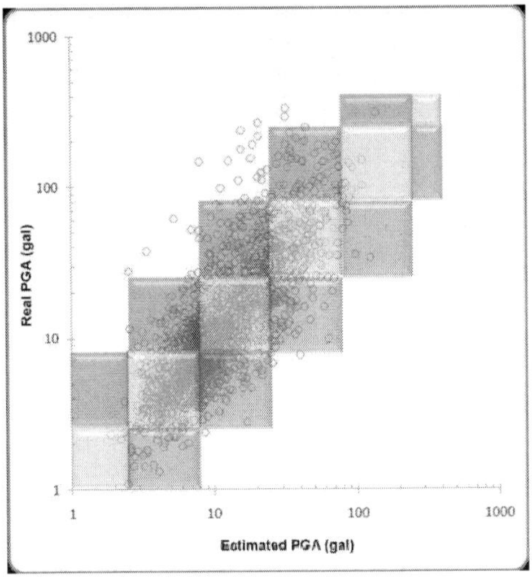

Fig. 25. Comparison of the real and est. seismic intensity (NN_{T0-3})

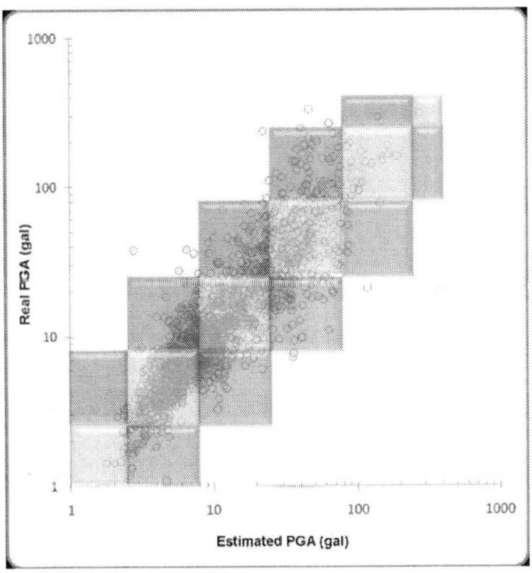

Fig. 26. Comparison of the real and est. seismic intensity (NN_{T0-10})

NN model \ Error (grade)	-2	-1	0	1	2
NN_{T0-3}	162	**781**	**885**	**178**	2
NN_{T0-10}	120	**846**	**930**	**112**	3

Table 3. Accuracy of the intensity estimation

The comparison of the real and estimated arrival time for PGA measured on the roof in two directions (horizontal and vertical) from both NN models ($NN_{T0\text{-}3}$ and $NN_{T0\text{-}10}$) were shown in figures 27-30. The results have shown convergence and the regression analysis R^2 is 0.67 for $NN_{T0\text{-}3}$ and 0.70 for $NN_{T0\text{-}10}$ in East-West direction. As for North-South direction, the regression analysis R^2 is 0.644 for $NN_{T0\text{-}3}$ and 0.662 for $NN_{T0\text{-}10}$. As for Up-Down (vertical) direction (figures 7 and 8), the regression analysis R^2 is 0.676 for $NN_{T0\text{-}3}$ and 0.7 for $NN_{T0\text{-}10}$. These results show that the performance of $NN_{T0\text{-}10}$ is better than $NN_{T0\text{-}3}$. The conclusion can be made that more input information (longer earthquake time history) to the neural network; more accuracy of the prediction can be increased.

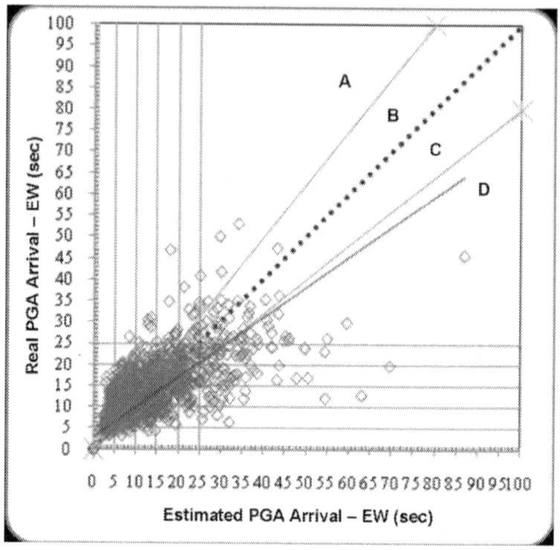

Fig. 27. Comparison of the real and est. arrival time for PGA-EW ($NN_{T0\text{-}3}$)

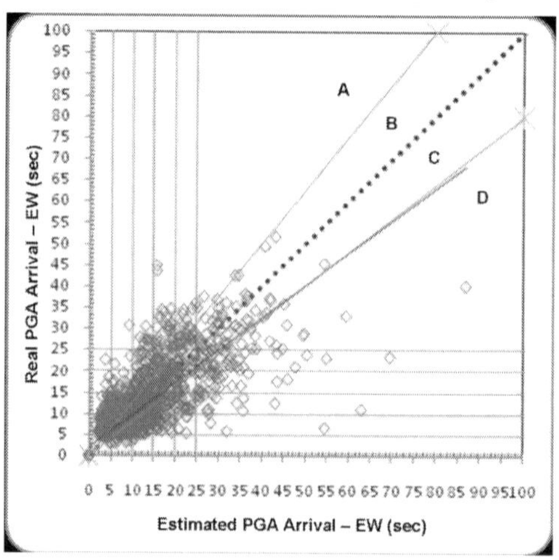

Fig. 28. Comparison of the real and est. arrival time for PGA-EW ($NN_{T0\text{-}10}$)

In figures 27-30, the blue lines indicated the relative error of plus or minus 20% of the real values. The comparison of the real and estimated arrival time for PGA is shown in Table 4. The B area means the estimation is slightly small than the real value within 20% of error. The C area means the estimation is slightly larger than the real value within 20% of error. If the allowable error range is set to be plus or minus 20% of the real values (B and C areas indicated in figures), then the average accuracy for NN_{T0-3} is 28.3% and 31.6% for NN_{T0-10}. However, in the sense of early warning, the A area should be considered acceptable since the estimation is less than the real value, i.e. the warning is still effective to the people. Therefore, the allowable error range can be set to be A, B, and C areas indicated in figures, the average accuracy for NN_{T0-3} is 60.2% and 66.9% for NN_{T0-10}.

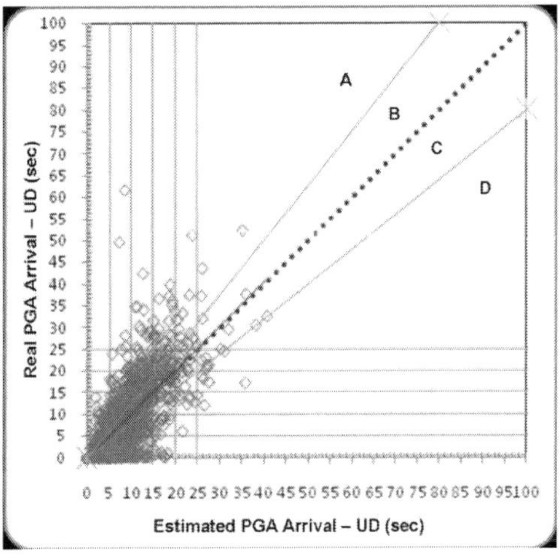

Fig. 29. Comparison of the real and est. arrival time for PGA-UD (NN_{T0-3})

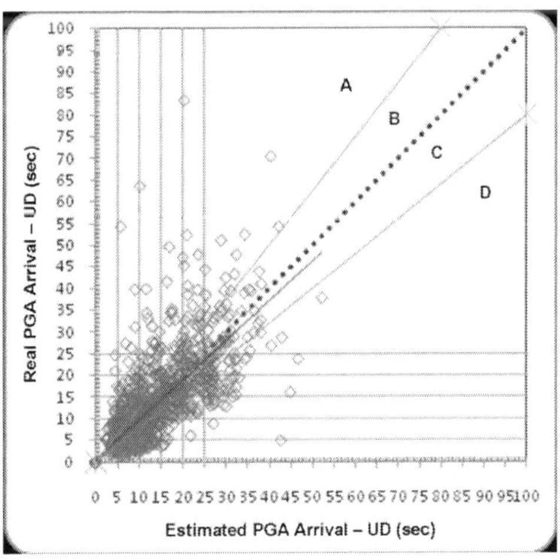

Fig. 30. Comparison of the real and est. arrival time for PGA-UD (NN_{T0-10})

	NN	A	B	C	D	Total
EW	T0-3	406	289	301	1019	
		20.15%	14.34%	14.94%	50.57%	
	T0-10	494	305	338	878	
		24.52%	15.14%	16.77%	43.57%	
NS	T0-3	1109	316	231	359	
		55.06%	15.68%	11.47%	17.83%	2015 100%
	T0-10	1182	350	219	264	
		58.66%	17.37%	10.87%	13.10%	
UD	T0-3	412	263	311	1029	
		20.45%	13.05%	15.43%	51.07%	
	T0-10	460	340	356	859	
		22.83%	16.87%	17.67%	42.63%	
A: Est.< Real*80%, B: Real*80% < Est.< Real, C: Real < Est.< Real*120%, D: Real*120%< Est.						

Table 4. Comparison of the real and estimated arrival time for PGA

8. Conclusion

Preparedness is crucial when a severe earthquake occurs since most obstacles and dangers can be determined beforehand. In this chapter, the authors presented the development of the on-site EEWS on Taiwan using several different methodology including neural networks. The time issue is the key countermeasure during a large earthquake, thus a good optimization algorithm to determine immediately when and what information must be provided. Therefore, the challenge of using only 1 second of earthquake acceleration time history signal to predict earthquake information is under development. The accuracy and reliability of earthquake information is of the utmost importance and is of immense benefit in the mitigation of earthquake hazards. Furthermore, the estimation of the structural response before the arrival of the S-wave has been studied in the 2nd stage of the on-site EEWS development. The authors also presented two rapid-estimation of structural responses modulus. The general modulus, which assumed the structure as a shear beam model, can be widely applied to different type of structures. The general modulus is low-cost, widely application field and rapid estimation. In the other hand, the customized modulus can provide the more accurate and detail estimation of the structural responses. Moreover, it can connect the actuation system to dramatically reduce the economic loss due to earthquake hazard. During the development of the two rapid-estimation of structural responses modulus, the FEM analysis of the Tai-Power building is made. The real dynamic structural responses from the CWB are used to refine and verify the FEM in PISA3D. Both the on-site and free field data are used as inputs and feed into the FEM. All the simulated structural responses are collected into the database. This database can be provided to the development of the rapid-estimation of structural responses modulus. Finally, it will be released to public for the researches of the structural health monitoring to develop and verify their algorithms.

In the future, the verification of the reliability of the communication lines as well as the system is needed to ensure reliable operation of the EEWS. Therefore, the EEWS is able to consequently bring huge benefits on the earthquake hazard mitigation. With further research on the use of the observed earthquake records, and enhancing the accuracy and immediate response of the real-time ground motion prediction, the possibility of the on-site EEWS is on the horizon.

9. Acknowledgments

We are grateful to the National Center for High-performance Computing (NCHC) for computer time and facilities used for training of the neural networks in this study.

10. References

Baer, M., and U. Kradolfer (1987). An automatic phase picker for local and teleseismic events. *Bull. Seism. Soc. Am.* 77, 1437-1445.

Chang, K.C., P.Y. Lin, C-C.J. Lin, T.K. Lin, Y.B. Lin, J.L. Lin, Y.T. Weng, T.M. Chang, S.K. Huang, Z.P. Shen, and Y.C. Lin, 2010, *Database for Fast Estimated Response of Structure,* NCREE-10-007, National Center for Research on Earthquake Engineering.

Cooper, J. D. (1868). *San Francisco Daily Evening Bulletin,* November 3.

Heaton, T. H. (1985). A model for a seismic computerized alert network, *Science* 228, 987-990.

Horiuchi, S., H. Negishi, K. Abe, A. Kamimura, and Y. Fujinawa (2005). An Automatic Processing System for Broadcasting Earthquake Alarm, *Bull. Seism. Soc. Am.* 95, 708-718.

Hsiao, N.G., 2006, *The Application of Real-Time Strong-Motion Observations on the Earthquake Early Warning in Taiwan,* PhD Thesis, Graduate Institute of Geophysics, National Central University

Kanamori, H., 2003, "On-site Earthquake Early Warning," *2nd Symposium on the real-time earthquake information transmission system,* pp.14-16.

Kanamori, H. (2005). Real-time seismology and earthquake damage mitigation, *Annu. Rev. Earth Planet. Sci.* 33, 195-214.

Kuyuk H.S., and M. Motosaka (2008). Spectral Forecasting of Earthquake Ground Motion using Regional and National EEWS for Advanced Engineering Application against Approaching Miyagi-ken Oki Earthquakes, *14th World Conference on Earthquake Engineering,* Beijing, China. S05-03-013

Kuyuk, H.S., and M. Motosaka (2009). Forward spectral forecasting of ground motion with the information of earthquake of earthquake early warning systems for structural control, *J. Japan Ass. Earthquake Engrg.* 9(3).

Lin, C-C.J. (1999). A Neural Network Based Methodology for Generating Spectrum Compatible Earthquake Accelerograms, *Ph.D. thesis, Department of Civil Engineering,* University of Illinois at Urbana-Champaign, Urbana, IL

Lin, C.-C.J., and Z.P. Shen, 2010, "Application of Neural Networks on Recent Development of the Earthquake Early Warning System for Taiwan", *Proceeding, the 9th US*

National and 10th Canadian Conference on Earthquake Engineering, (2010EQConf), Toronto, Canada.

Nakamura, Y. (1988). On the Urgent Earthquake Detection and Alarm System (UrEDAS), *Proc. 9th World Conference on Earthquake Engineering*, 673-678.

Nakamura, Y. (2005). The system development for the preparation against the unpredictable earthquake disaster, *Future Material*, 71-74. (in Japanese)

Tsai, K.C., P. Y. Lin, C-C.J. Lin, T. K. Lin, Y. B. Lin, J. L. Lin, Y. T. Weng, T. M. Chang, Z. P. Shen, and C. S. Chung, 2009, *Preliminary Study of Added-Value Information and Analysis for Earthquake Early Warning System*, NCREE-09-013, National Center for Research on Earthquake Engineering.

Wieland, M., M. Griesser, and C. Kuendig (2000). Seismic Early Warning System for a Nuclear Power Plant, *12th World Conference on Earthquake Engineering.*

Wu, Y.-M., and H. Kanamori (2005). Experiment on an onsite early warning method for the Taiwan early warning system, *Bull. Seism. Soc. Am.* 95, 347-353.

Yoshioka, K., 2006, "Outline of the Real-time Earthquake Disaster Prevention System," *Kenchiku Bosai*, pp. 22-27. (in Japanese)

Wu, Y.M. and H. Kanamori, 2005, "Experiment on an onsite early warning method for the Taiwan early warning system" *Bull. Seism. Soc. Am.*, 95, pp.347-353.

Wu, Y.M., H. Kanamori, R. Allen, and E. Hauksson, 2007, "Determination of earthquake early warning parameters, c τ and d P , for southern California," *Geophys. J. Int.*, 170, pp.711-717.

Part 4

Earthquake Geology

14

Extracting Earthquake Induced Coherent Soil Mass Movements

Kazuo Konagai, Zaheer Abbas Kazmi and Yu Zhao
University of Tokyo
Japan

1. Introduction

In the earthquake engineering practice, fragility curves are useful tools for predicting the extent of probable damage. They show the probability of damage to a particular-style structure as a function of strong motion parameters, and they allow estimation of a level of damage probability for a known ground motion index. The problem, however, is that strong ground motion records are often unavailable in seriously devastated areas, thus frustrating all attempts to rationally deduce the whole picture of the devastation at a great cost of many lives and properties. Other than that, we should recognize that not only intense shakes but also ground deformations can be equally or often more responsible for the devastation. Aftermaths of an earthquake are often more devastating than its immediate effect, especially in mountainous terrains. Large strains built up in soils and rocks along a dislocated fault can trigger post-earthquake disasters such as landslides and debris flows, which can last long causing serious problems for rehabilitations and land conservations. Therefore one of what required of us is to deduce as much hidden signs as possible from observable change of landforms.

Earthquake-induced landform changes have a wide range of ground movement. Among them, tectonic deformations under the action of deep-seated forces may hint the presence of a zone of deformed rock along the exposed and/or hidden fault, namely the zones which became more susceptible to landslides than they had been in the past. Coherent mass movements are generally less catastrophic than chaotic mass movements. However, they can surely cause long-lasting problems for rehabilitations. Moreover even a chaotic mass movement can be preceded by a slow and coherent mass movement at its early stage.

Recent development of remote sensing technologies has enabled us to detect precise landform changes in a scientific manner. However, the methods allow us to detect displacements only in the Eulerian description, in which the description of motion is made in terms of the spatial coordinates which does not follow the motion of soil particles. Discussions of earthquake-inflicted geotechnical issues require more direct description of soil particle movements because soils are typically history-dependent materials. The first half of this chapter presents a method to extract Lagrangian components of displacements from available set of elevation data. The method is then applied to an active folding zone affected by the Mid-Niigata Prefecture Earthquake of October 23rd 2004. The surface Lagrangian displacements and the cause-and-effect relationships for the reported damages are discussed in detail.

2. Method to extract lagrangian soil displacements

Nowadays, most common methods to monitor landform changes include GPS control measurement, level measurement, long baseline interface measurement etc. But these measurements are limited to certain areas and only monitor discrete points. Due to the remoteness of disaster struck areas, usually high mountains, and due to its spatial extension, remote sensing is a cost efficient and more practical tool for monitoring terrain dynamics.

Recent development of remote sensing technologies such as Laser Imaging Detection and Ranging Technology (LIDAR), Interferometric Synthetic Aperture Radar (InSAR) and Differential-InSAR has enabled the acquisition of images of landforms and the changes in elevation with high precision. However, the methods allow us to detect displacements only in the Eulerian description. Rather than the Eulerian displacements, we need to extract Lagrangian displacements of soils, whose behaviours are typically history-dependent.

Konagai et al. (2009) proposed a method to estimate Lagrangian components of tectonic displacement considering rigid-body-translation of three consecutive soil patches, their centres arranged in a triangular pattern, with an underlying assumption that tectonic displacement varies gently in space. The method has been improved by the second author incorporating vertical rotation of the soil patch and filtering out landslides and manmade changes.

Assuming that a small patch of ground, having a node $i1$ mapped upon it, with inclinations of θ_{x1} and θ_{y1} in x and y directions, respectively, moves to a new position, changing the inclinations to θ_{x2} and θ_{y2} (Figure 1). Describing the method in two dimensional setting, the concept can be made simpler to understand (Figure 2). Referring to Figure 1 and 2, Eulerian change in elevation $(\Delta_{z,i1})$ is expressed in terms of the Lagrangian vector $\{\Delta_{x,k} \quad \Delta_{y,k} \quad \Delta_{z,k}\}$ of the movement of a particular soil particle, k, on this patch as;

$$\begin{bmatrix} t_{x2,i1} & t_{y2,i1} & 1 \end{bmatrix} \begin{Bmatrix} \Delta_{x,k} \\ \Delta_{y,k} \\ \Delta_{z,k} \end{Bmatrix} = \Delta_{z,i1} \tag{1}$$

Where $t_{x2,i1} = -\tan\theta_{x2}$ and $t_{y2,i1} = -\tan\theta_{y2}$, i.e. direction tangents of the soil patch in its new position.

Arranging three soil patches, $i1$, $i2$ and $i3$, immediately next to each other in a triangular fashion (Figure 3), and using the displacement of its center $\{\Delta_{x,k} \quad \Delta_{y,k} \quad \Delta_{z,k}\}^T$ as the representative Lagrangian displacement vector of the triangle, the following set of solvable simultaneous equations is obtained.

$$\begin{Bmatrix} \Delta_{z,i1} \\ \Delta_{z,i2} \\ \Delta_{z,i3} \end{Bmatrix} = \begin{bmatrix} t_{x2,i1} & t_{y2,i1} & 1 \\ t_{x2,i2} & t_{y2,i2} & 1 \\ t_{x2,i3} & t_{y2,i3} & 1 \end{bmatrix} \begin{Bmatrix} \Delta_{x,k} \\ \Delta_{y,k} \\ \Delta_{z,k} \end{Bmatrix} = \mathrm{T} \begin{Bmatrix} \Delta_{x,k} \\ \Delta_{y,k} \\ \Delta_{z,k} \end{Bmatrix} \tag{2}$$

Solving the above set of simultaneous equations (2) for all triangles within the target zone, Lagrangian components of displacement vectors $\{\Delta_{x,k} \quad \Delta_{y,k} \quad \Delta_{z,k}\}^T$ can be obtained for the entire target zone.

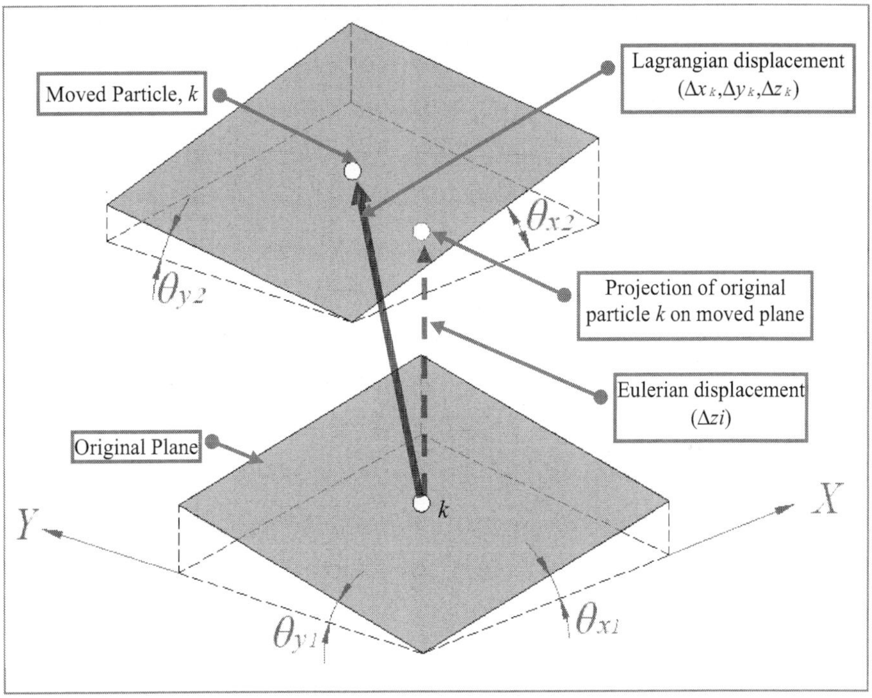

Fig. 1. Description of the method to convert Eulerian change in elevation to Lagrangian displacements

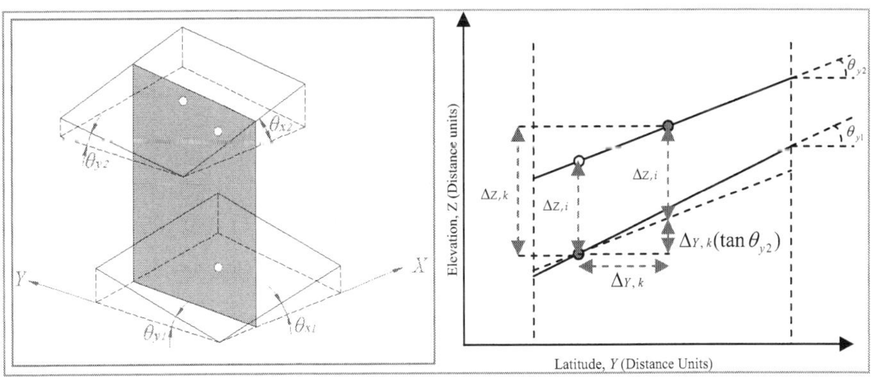

Fig. 2. Description of the method in two dimensional settings to simplify the concept. $\Delta_{z,i}$ is the Eulerian change in elevation while $\{\Delta_{x,k}\ \Delta_{y,k}\ \Delta_{z,k}\}$ are the Lagrangian displacement components.

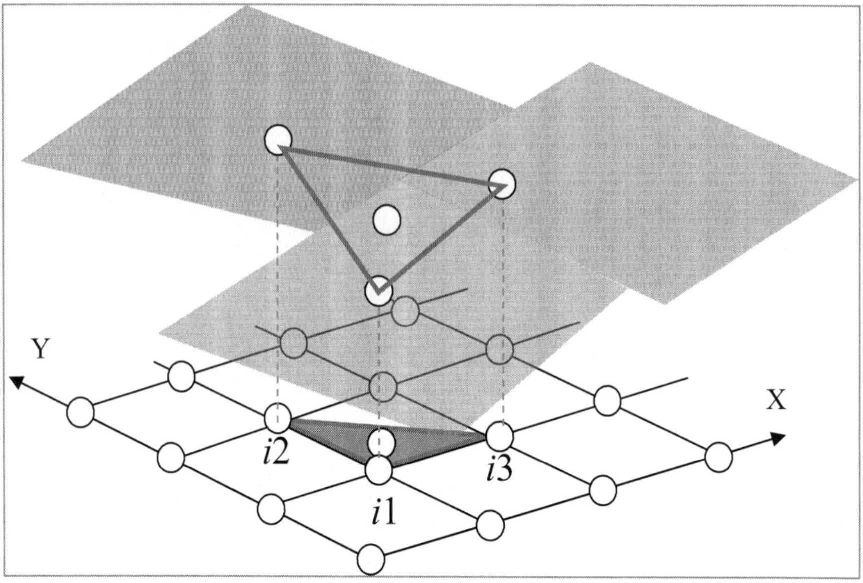

Fig. 3. Three consecutive soil patches arranged in a triangular pattern to solve a set of simultaneous equations for Lagrangian components of displacement.

The obtained Lagrangian vectors show soil particle motions on the ground surface, which can be greatly affected by landslides and manmade changes. To separate tectonic deformations from the entire Lagrangian displacements, landslides and large scale manmade changes are filtered out by limiting the Lagrangian elevation change below a threshold value. This threshold value is set such that the centre of moved plane remains within the projected boundary of the original plan. The following filtering criterion is used;

$$\Delta_{z,k} > \text{Threshold for Lagrangian vertical displacement}$$

$$= \left\{ t_{x2,i} \quad t_{y2,i} \quad 1 \right\} \left\{ \Delta_{x_{\text{lim}}} \quad \Delta_{y_{\text{lim}}} \quad \Delta_{z_{\text{lim}}} \right\}^T \tag{3}$$

Where, $\Delta_{x_{\text{lim}}}$ and $\Delta_{y_{\text{lim}}}$ are half the planner dimensions of the soil patch in X and Y direction, respectively. $\Delta_{z_{\text{lim}}}$ is set to a constant value; e.g. while applying to the Mid-Niigata Prefecture Earthquake, this can be determined as 1.36 meters from the maximum vertical displacement of the main shock in this earthquake (Hikima K. and Koketsu K., 2005).

However, the obtained Lagrangian displacement components often show a remarkable scatter. Manmade changes during the time between two DEM's and presence of some non-surface objects on digital surface models might be most prominent among the possible causes. Therefore, the moving average method was used for overall features of displacements. Assuming that the scattered values follow the Gaussian distribution within a square window, the most frequent value (mode) is interpreted to be the real vector of the soil displacement for this area. Sweeping the entire zone with this square window, one can obtain the whole picture of the deformation.

The window size is desirable to be larger than the largest hidden landslide in the target area for the discussion of tectonic deformations to minimize the effect of the hidden coherent landslides. For discussing smaller scale soil movements such as hidden coherent landslides, the window size is to be minimized, and yet to be substantially larger than abovementioned manmade changes.

2.1 Ill condition

To obtain a reliable solution, the condition of coefficient matrix T in equation (2) is important. A system of equations is considered to be well-conditioned if a small change in the coefficient matrix or a small change in the right hand side results in a small change in the solution vector. This small change in the solution vector is namely important digits or tolerance of accuracy. The matrix condition can be determined as;

$$Cond(T) = a_{tol} / \varepsilon_{mach} \tag{4}$$

Where machine epsilon, ε_{mach}, gives an upper bound on the relative error due to rounding in floating point arithmetic.

If MATLAB is used for calculation, the machine epsilon of default data type (double precision) is obtained as $\varepsilon_{mach} = 2.2201 \times 10^{-16}$ with "eps" command. Setting the tolerance of calculation accuracy to any desired level, the condition of matrix T can be obtained by equation (4). All the sets of equations not fulfilling this criterion need to be filtered out to get a reliable solution. For the example case described in the following section, a_{tol} is set at 0.1.

3. Extracting coherent mass movements

3.1 Active folding zones as target

Taking a look at history, it is found that earthquakes in active folding areas have distinctive features and can trigger long lasting geotechnical issues. In an active folding zone, the action of deep-seated forces has been shortening sedimentary rock layers causing folded geomorphic surfaces to appear and develop. Looking at a fold surface in profile (Figure 4) upslope and downslope flanks of the fold join together at anticlines and synclines, respectively. Since the up-folded rocks along anticlines have been expanded and cracked over centuries, anticlines frequently have their crests deeply eroded, with a number of debris deposits rimming the eroded hollows. Large-scale landslides are found even on gentle mountain sides dipping towards synclines because their toes are often deeply eroded by rivers. The active folding regions can be thus one of the most landslide-prone zones.

The May 8th, 1847 Zenkoji Earthquake (M=7.4) jolted the active folding mountainous terrain west of Nagano, central Japan. Devastations were serious along the entire 50km stretch of the Nagano western basin-edge fault that appeared along the skirts of the active folding mountains (Figure 5). The earthquake caused about 44,000 landslides to occur on the hanging wall side of the fault. In 1884, 37 years after the earthquake, a big crack appeared near the southern summit of the Chausu-yama twin peaks and an 800m long soil mass started moving. After the heavy rains of 1930, the entire soil mass began to creep down the slope exhibiting thick, wet and sticky features, and the maximum speed of 93 m/year was reached in 1932-1934. The slope was finally stabilized in the 1970s with a tremendous amount of drainage works that were started in 1965 (Kato H. and Akabane H., 1986).

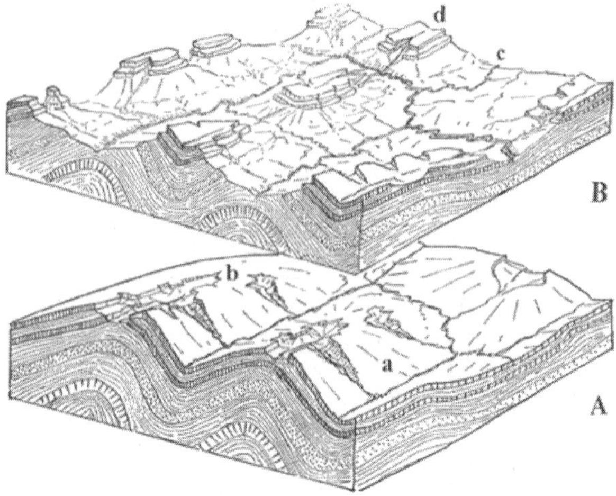

Fig. 4. Erosion of geological fold (Original figure from de Martonne, 1927). Erosion develops from A to B in such a way that syncline valley (a) and anticline ridge (b) in A to become anticline valley (d) and syncline ridge (c) in B, respectively.

A M7.1 earthquake jolted a low-rising mountainous terrain of Senboku, Akita, Japan on March 15th, 1914. Though the intensities registered at major cities were not surprisingly large, the reported deaths of 84 among the total 94 were concentrated locally within the 10km x 10km Senboku area. Landslides in Senboku area that appeared in the report of Imperial Earthquake Investigation Committee, No. 82 (1915) and those from the Mid-Niiata Prefecture Earthquake of 2004, which will be explained later, are exactly alike in that just surface laminar sedimentary rocks were detached and slipped down the planar layer boundaries as shown in Figure 6.

The magnitude 6.8 Mid-Niigata Prefecture Earthquake of October 23rd, 2004 jolted one of the areas where active folding geological structure was most clearly recognized (K.H. Yoong and A. Okada, 2005). The hypocenter of the main shock was located at 37.29°N, 138.87°E at focal depth of 13km. The maximum acceleration of 1500 cm/s^2 was recorded at Ojiya K-net station which is about 10km west of the epicenter. The main shock was followed by a large number of aftershocks with four being over M6. The focal mechanisms of these strong shocks, estimated by Hi-net and F-net (Honda et al., 2005), were reverse fault type that is concordant with pre-existing fold axis. The epicentres of the aftershocks were distributed along the NNE and SSW direction within a length of about 30km (Honda et al., 2005). This earthquake reportedly triggered and/or reactivated thousands of landslides. The economic loss due to these landslides was initially estimated at 8 billion US dollars, making this one of the costliest landslide events in history (Kieffer et al., 2006) and with all similar examples compiled in active folding zones, the quick stabilization of slopes in the Mid-Niigata mountainous terrain was considered to be a pressing need.

In case of Mid-Niigata Prefecture Earthquake, not only landslides but also surface tectonic displacements have caused some problems for rehabilitating the affected areas. As will be discussed hereafter, the tectonic movements have caused the middle part of both the Shinano and Uono rives to be raised upward by about 0.5 to 1.0 meters. Probably due to this

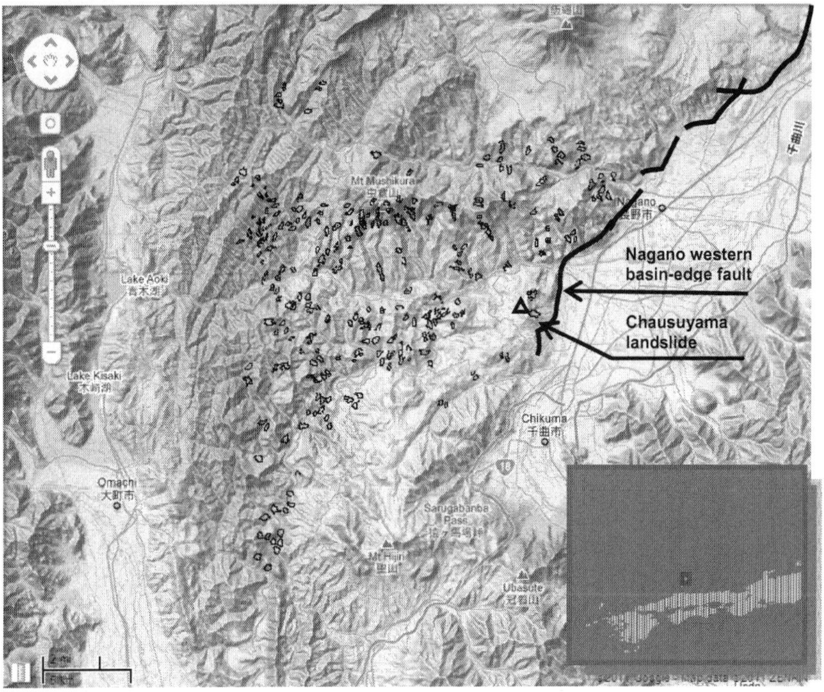

Fig. 5. Landslides caused by the Zenjkoji Earthquake of 1847: The Nagano western basin edge fault is considered to have been responsible for the earthquake. Earthquake-induced landslides distribution is taken from Fig. 1-12 (by Akabane), the Report of Zenkoji Earthquake, Central Disaster Prevention Council (2008). A crack near the southern summit of the Chausu-yama twin peaks (Elevation 730m, Location: 37.294694°N, 138.875492°E, triangle mark) began to open wide, which was an early sign of a long-lasting landslide.

Fig. 6. Similarities between Landslides in the March 15th, 1914 Senboku Earthquake and the October 23rd, 2004 Mid-Niigata Prefecture Earthquake a) Nunomata Landslide in Senboku Earthquake (39.473419,140.311219) b) Uragara landslide in Mid-Niigata Prefecture Earthquake (39.473419,140.311219)

tectonic deformation, the upper stream reach of the Uono River was flooded in the heavy rain of June, 2005, about eight months after the earthquake. Therefore, the first and essential step for research to proceed and before any rational and scientific discussions on remedial measures to take place, should be to separate soil deformations caused by the tectonic movement of the active folding zone from the overall soil deformations observed on the ground surface. As mentioned above, Interferometric Synthetic Aperture Radar (InSAR) is one of the most advanced technologies to study landform changes (USGS, 2005). However for the Mid-Niigata Prefecture region, thick vegetation and thousands of landslides have made fringe patterns too complicated for extracting pure elevation changes from the available C-band (5.405 GHz) InSAR interferogram from RADSAT, a Canada's commercial SAR satellite (Ozawa et al., 2005). Therefore, digital elevation models (DEMs hereafter) for the mountainous terrain before and after the earthquake were first obtained as raster graphic images with pixels arranged in a 2m x 2m square grid; each pixel has information of its elevation. The DEM for the pre-earthquake time of (1) 1975-1976 was obtained using stereoscopy, and DEMs for the post-earthquake times of (2) Oct. 24, 2004, (3) Oct. 28, 2004, (4) May 2005, (5) May 2006, and (6) May 2007 (JSCE active folding project 2008) were obtained using Laser Imaging Detection and Ranging technology (LIDAR). Changes in elevation in Eulerian description for DEM were first converted to Lagrangian displacements, and then the moving average method was adopted to obtain the whole picture of landform change.

3.2 Application of the method

3.2.1 Surface tectonic displacement

Target zone is an 11 × 7 km active folding area of Yamakoshi mountainous terrain (Figure 7). There were only 13 triangulation points within the target zone and 11 out of those 13 points were not considered to have been affected by landslides. However, the points are so sparsely distributed that a thorough discussion can't be made upon a clear image of displacement. Therefore, precise digital elevation models (2m resolution) before and one day after the Mid-Niigata Prefecture Earthquake (Oct. 24) were used for the analysis (Figure 8).

The size of the soil patch (Figure 3) is to be determined from the ground features and the objective of the analysis. As is shown in Figure 9, there were a number of manmade changes of landform over the 28 years interval between DEMs prepared in 1976 and 2004. These changes include creation of ponds for Koi-fish farm business, road widening etc. Reflecting these changes, an 8m × 8m soil patch, least square regressed from the points in the square domain surrounding the center point, is used.

First of all large changes in elevation were filtered out by applying the filtering criteria described in section 2. Distribution of the filtered points gives good correlation with the landslide map (Figure 10) (Oyagi et al., 2008). At the lower left corner of the target zone, there is Shinano river and much flatter topography (Figure 8 and 10). Large bunch of filtered points in this area is possibly due to ill conditioned tangential matrices in the flat land.

Figure 11 shows horizontal components of surface tectonic displacement extracted from the DEMs, sweeping the whole target zone with a 1400m smoothening window. Shinano River Office, Hokuriku Regional Bureau of the Ministry of Land, Infrastructure and Transport (MLIT), has been measuring exact locations of bench marks along both the Shinano and Uono rivers on regular basis. Lateral components of the bench marks' displacements due to the earthquake were also plotted on Figure 11. There is a NNE-SSW trending 1 to 2 km wide belt of large eastward movement to the west of and along the Kajigane syncline. This belt of

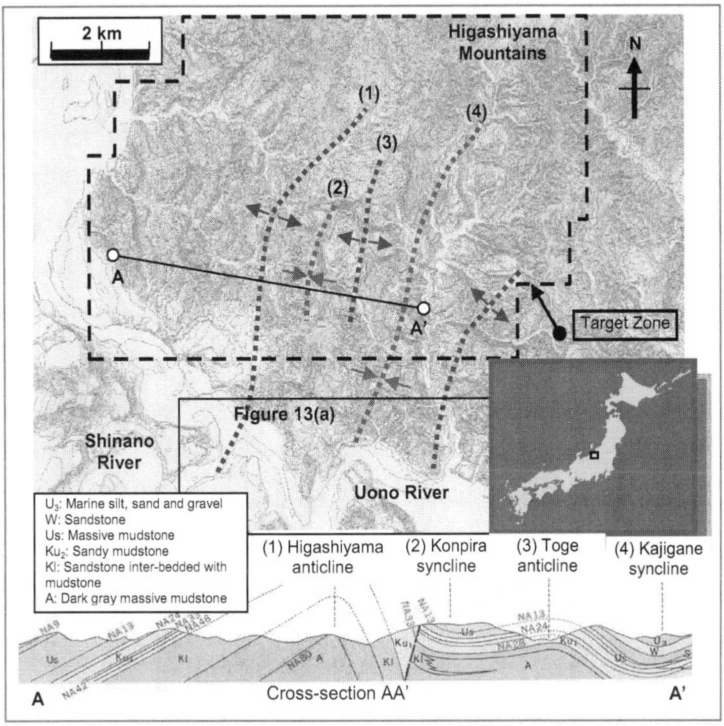

Fig. 7. Target zone on Zone VIII of the Japanese National Grid System

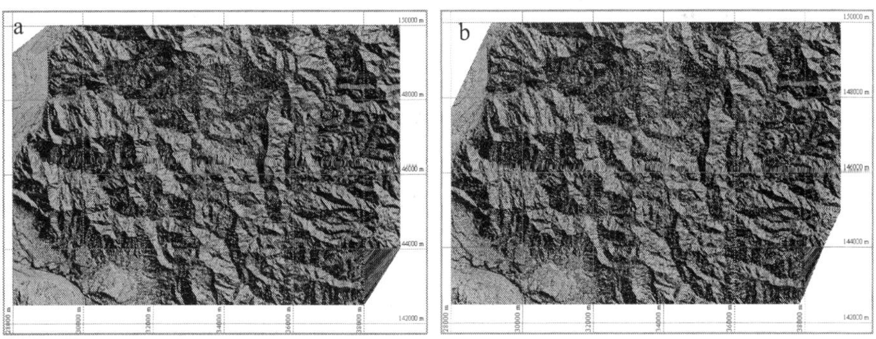

Fig. 8. Topography of target zone prepared from a) 1976 DEM b) 2004 DEM (Zone VIII of the Japanese National Grid System)

lateral displacements seems to have appeared on the hanging wall side of the line of intersection between the ground surface and the straight extension of the hidden deep-dipping fault rupture plane for the major event, whose geometry was estimated by Hikima and Koketsu (2005). The second cluster of large lateral displacement vectors that has appeared 4 to 5 kilometers west of the Kajigane Syncline, is near the projection on the ground surface of the hidden fault rupture plane for the first largest aftershock of M6.3, which took place at 18:03 JST, about 7 minutes after the main event (Hikima K. and Koketsu K., 2005). While comparing the lateral components of tectonic displacements with landslide map of Figure 10(a), it is found that large clusters of tectonic displacements are concordant with thick clusters of landslides.

Fig. 9. Improvement of farmland in Yamakoshi: Location of the lower left corner of each photo is 37.337928°N, 138.878678°E. The left and right photos were taken respectively in 1976 (Geospatial Information Authority of Japan) and on Oct. 24th, 2004 (JSCE Active folding Project, 2008). Some of the terraced paddy fields were converted to Koi fish ponds.

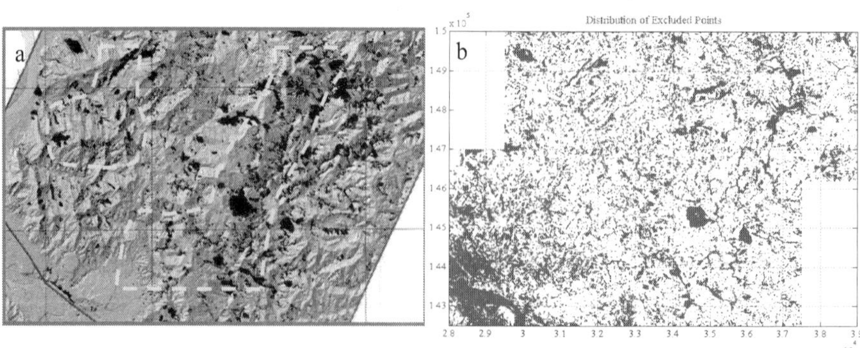

Fig. 10. a) Landslide map of the target zone and b) distribution of the filtered points for landslides and manmade changes (Broken yellow lines show two thick clusters of landslides which will be compared to large lateral displacements shown in Figure 11).

Figure 12 shows vertical components of estimated tectonic displacement. It is notable that there are two areas in the target zone, which have been pushed up by 0.5 to 1.0 meters. The most remarkable hump spreads wide across the southwestern part of the target zone where the Uono River joins the Shinano River. The Uono River, after flowing straight west through a flat wide spread valley of Horinouchi, meets the sedimentary silty sand rock ridge (Figure 13(a)). The river then abruptly changes its direction, from SE-NW to NE-SW, along this rock ridge, making a sharp down-folded bend. Then it forces its way through the narrow and lowest points among the mountains making a sharp up-folded bend. On the geological map of this area (Yanagisawa et al., 1986), the approximately 2 km-long stretch of the Uono River between these two bends continues straight to both the Kajigane and the Kodaka synclines at its north and south ends, respectively, suggesting that this 2 km-long stretch of the river is a part of the large Kajigane syncline.

As was explained in 3.1, areas along the upper reach of this part of the Uono River were flooded due to heavy rainfall of June 27th-28th, 2005, about 8 months after the earthquake (Figure 13). Assuming that the same amount of water in the 2005 rain flowed down the Uono River as existed before the earthquake (ignoring the landform changes caused by the Mid-Niigata Prefecture Earthquake), possible water depths at all bench marks along the 57.5km-long flooded zone (from BM No. 15 at 37.26642°N, 138.862209°E, to BM No. 72.5 at

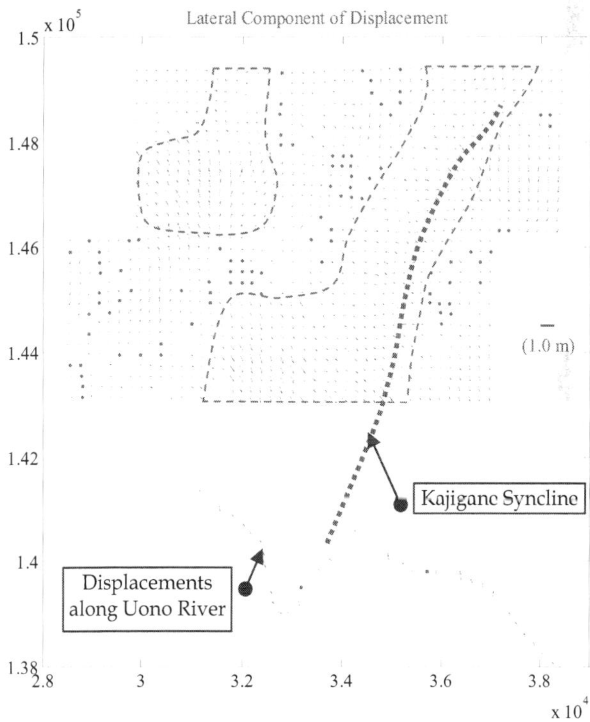

Fig. 11. Lateral components of surface tectonic displacement of the target zone on Zone VIII of the Japanese National Grid System. Broken green lines show two clusters of large lateral displacement which are consistent with thick clusters of landslides shown in Fig. 10(a).

37.25923°N, 138.899975°E) were estimated by using the Manning empirical equation (open circles in Figure 14 (Dooge J.C.I., 1992)). For this estimation, precise dimensions for the river cross-sections and inclinations at all benchmarks before and after the earthquake were provided by the Shinano River Office, Hokuriku Regional Bureau of the Ministry of Land, Infrastructure and Transport (MLIT).

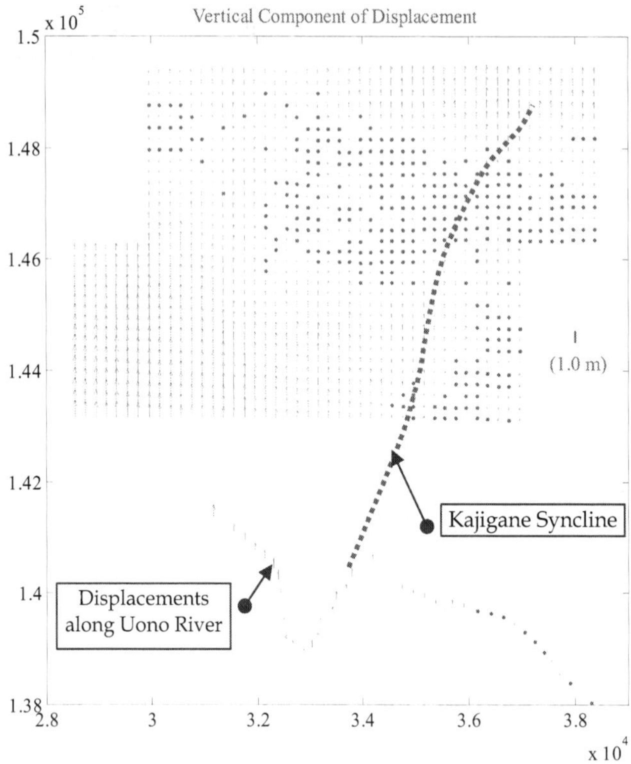

Fig. 12. Vertical components of surface tectonic displacement of the target zone on Zone VIII of the Japanese National Grid System.

Solid circles in Figure 14 show the actual water levels at all bench marks reached in the 2005 flood, while open circles show virtual water levels calculated for the Uono River as it existed before the earthquake. At almost all points, the virtual water levels (open circles) are lower than those (solid circles) reached in the 2005 real flood. Actual water levels were higher than the high water levels (HWL) at bench marks # 37.5, # 52.5 and # 62.5, while virtual water levels at these points do not reach the high water levels. This figure thus suggests that there was a cause-and-effect relation between the earthquake-induced tectonic deformation and the flooding of June, 2005.

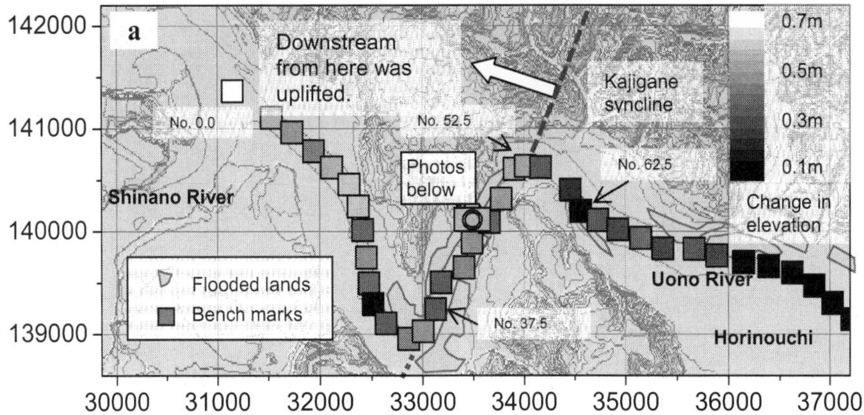

Fig. 13. Farmlands flooded in the heavy rain of June, 2005, about eight months after the earthquake. (a) Locations of flooded farm lands (Hokuriku Regional Agricultural Administration Office, Ministry of Agriculture, Forestry and Fisheries, and Uonuma City), and change in elevations at benchmarks along Uono River on Zone VIII of the Japanese National Grid System (Shinano River Office, Hokuriku Regional Bureau of MLIT) (b) Flooded area near Benchmark No. 42.5km (Photo by Kotajima, S., 28th June, 2005) and (c) a photo at a later date from the same location (37.259828°N, 138.876801°E) as above (Photo by Konagai, K., 19th September, 2010).

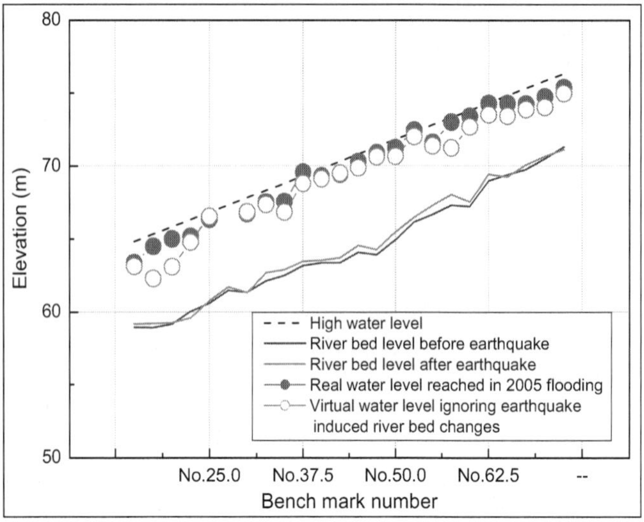

Fig. 14. Water level reached during the flood and the virtual water level estimated from manning's formula.

3.2.2 Shallow soil displacement

A strong earthquake usually disturbs the soil/rock mass, especially near the ground surface. Although there is demand of immediate rehabilitation and reconstruction of the damaged facilities, it is very important to know whether the soil mass around that facility is stable enough to start the reconstruction work. Otherwise, we will have to face the same problems repeatedly. The Mid-Niigata Prefecture Earthquake triggered and/or reactivated a large number of landslides and many hamlets have been rendered uninhabitable (Kieffer et al., 2006). The authorities responsible for the reconstruction were especially careful about the reactivation of hidden landslide during the snow melt seasons. Reconstruction/ rehabilitation of Kizawa tunnel (Konagai et al., 2009) is selected as an example case to discuss the shallow soil displacement.

A 300m long ring shaped Kizawa tunnel skims the NW-SE trending branch of Futagoyama Mountain ridge (Figure 16). During the earthquake, the tunnel experienced significant damage. The damage was mainly cracking (Figure 15). The cracks formed two parallel pairs, through the east and west walls, near the north entrance of the tunnel while its crown was shifted sideways by about a half meter (Konagai et al., 2009). Kizawa locality lies on the exposed slip surface of an old landslide whose escarpment is marked by blue broken line in Figure 16 (National Research Institute of Earth Science and Disaster Prevention, NIED, 2008).

Core samples collected from the boreholes near the cracked portion and hydro-geological properties of the soil/rock suggested presence of a thin shear plane of mud stone which was crushed significantly (Konagai et al., 2009). This shear plane, dipping about 10° south, was found parallel with the intact bed stratum.

Northern 40 to 80 meter length of the tunnel was embedded below this shear plane while the remaining southern part lies in the upper disturbed soil mass. The tunnel was thus obliquely bisected by this shear plane. Measurement of the change in alignment of road center line showed that the length of tunnel above the shear plane shifted 0.5 to 1 meter east to southeast wards (Nagaoka Regional Development Bureau, Niigata Prefecture Government, 2008). This suggests that the upper soil mass slipped coherently down this shear plane, gripping southern part of the tunnel which lies in that upper disturbed soil mass.

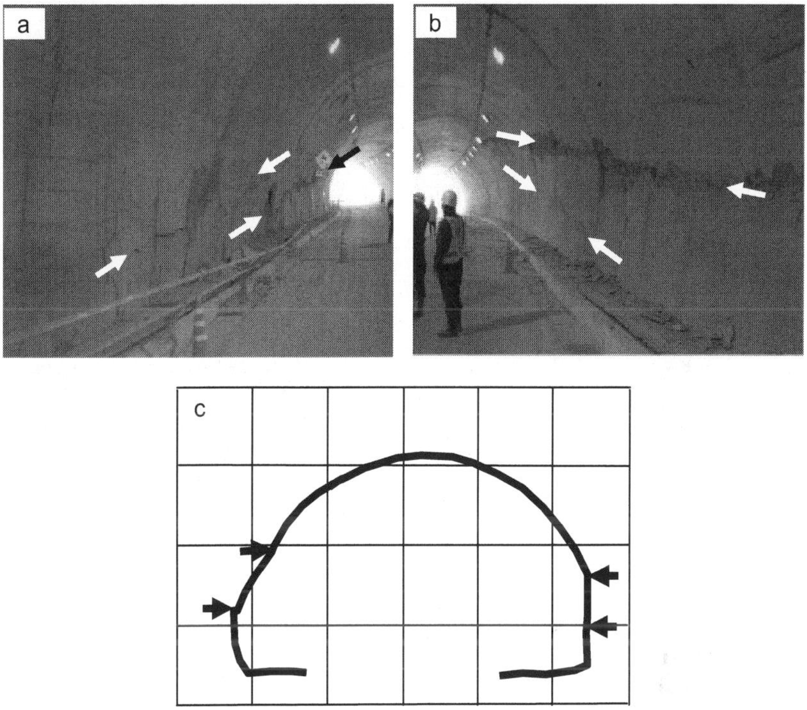

Fig. 15. Diagonal cracks on both walls of Kizawa Tunnel (a) the west wall of the tunnel (b) the east wall of the tunnel c) cross section at 57m from the north entrance

In addition to the Kizawa tunnel, two drainage wells (Well A and B in Figure 16) were found dislocated. Well A was dislocated by 0.15m at a depth of 34m below ground level. The shaft above this dislocation point was inclined indicating 0.5m southeast wards movement of the ground. Although the direction of dislocation could not be investigated for such a narrow well, well B was also dislocated at a depth of 20m below ground level.
These two dislocated wells and the diagonal cracks that appeared in Kizawa Tunnel make up a large triangle. This triangle infers a large hidden shear plane extending far beyond the south mouth of Kizawa Tunnel with the azimuth of its strike and dip angle being about 100° and 6 °, respectively. This plane is almost parallel with the exposed planar slip surface of an old landslide in Kizawa locality, and its azimuth and dip angle are not much different from those for the cutting plane inferred from the diagonal cracks that appeared in Kizawa Tunnel. However, the problem was to know the extent of this shear plane and to ensure the stability of the upper disturbed soil mass against further slippage during the snow melt season.

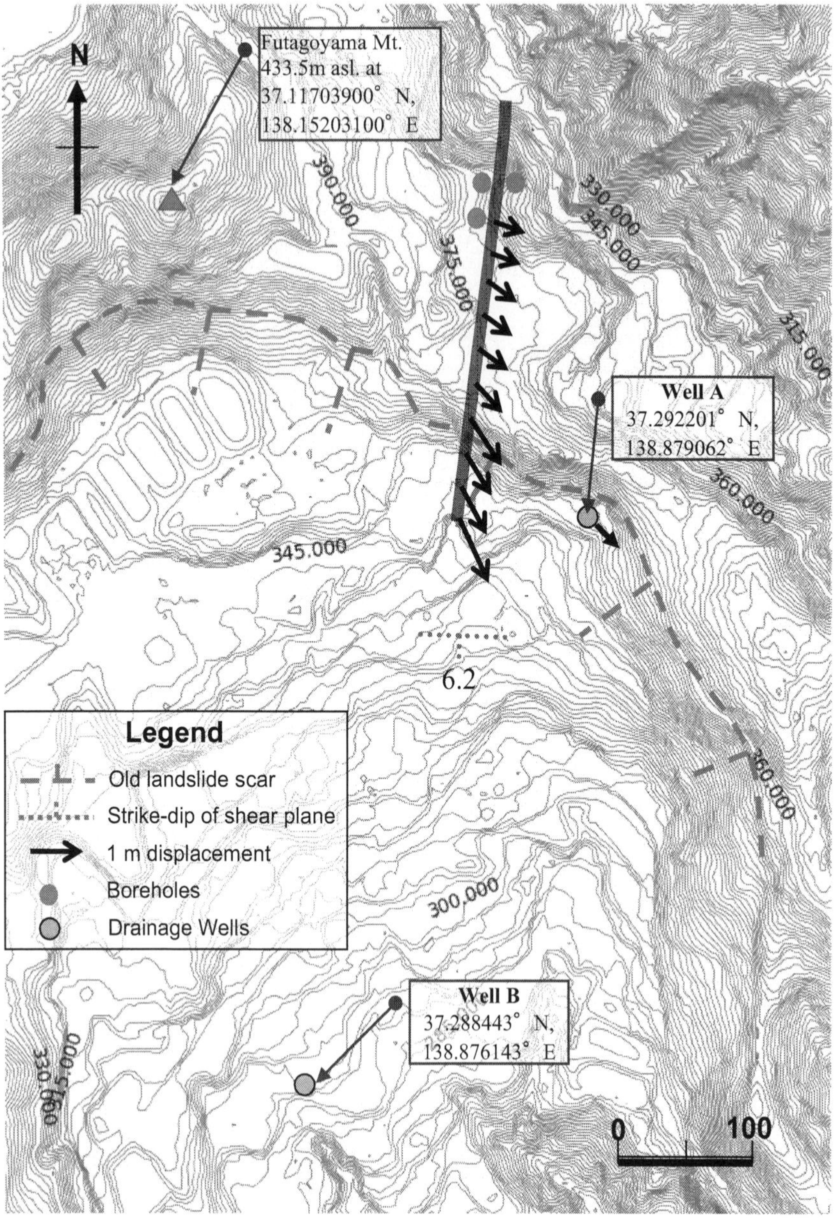

Fig. 16. Topography of the Kizawa hamlet showing inferred shear plane in the interior of soil mass.

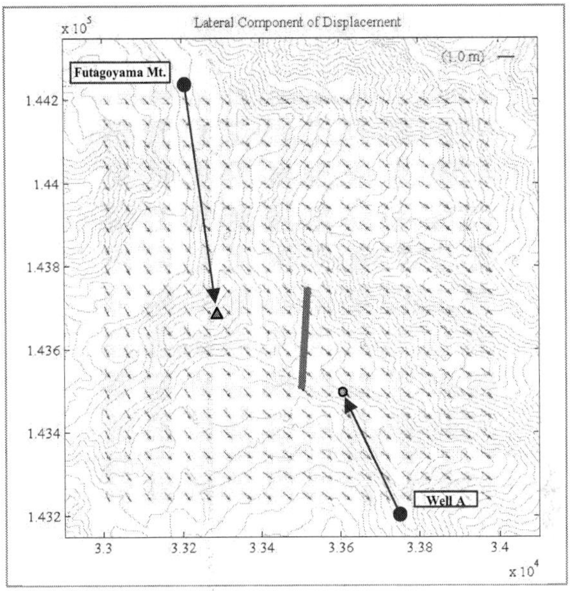

Fig. 17. Lateral component of Surface tectonic displacement of the Kizawa hamlet on Japanese National Grid System, Zone VIII

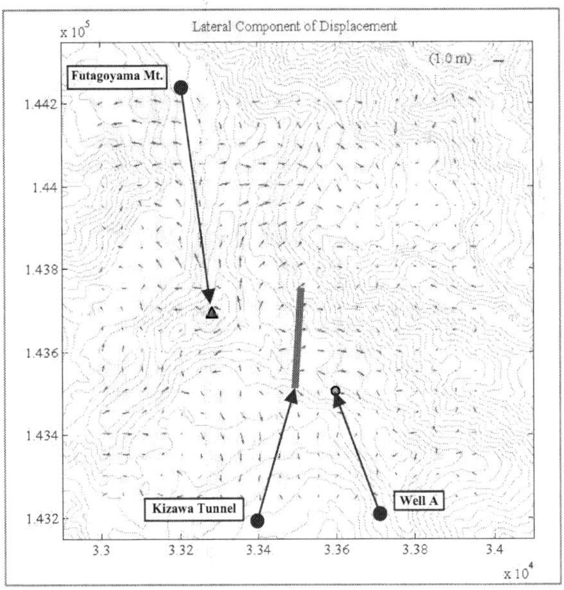

Fig. 18. Lateral component of shallow soil displacement of the Kizawa hamlet on Japanese National Grid System, Zone VIII

Focusing only to Kizawa hamlet and finding out the Lagrangian components of the estimated tectonic displacements, it is found that the whole area has moved laterally in southeast direction (Figure 17). Scaling down the smoothing window to 200m x 200m, an image of shallow soil displacement is obtained. Subtracting the surface tectonic displacements, obtained through the 1400m square smoothening window, form those of soil displacements obtained through the 200m square smoothening window, gives clearer image of shallow soil displacements only (Figure 18). It suggests that soil masses on all sides of Futagoyama Mountain have moved down the slopes by 0.5 to 1m. The displacement vectors are well consistent with the direction of dislocation of drainage well and Kizawa tunnel.

4. Conclusions

Aftermaths of an earthquake are often more devastating than its immediate effect, especially in the mountainous terrain where most of the earthquakes occur. Large strains built up in soils and rocks along a dislocated fault can also trigger post-earthquake disasters such as landslides and debris flows, which can last long causing serious problems for rehabilitations and land conservations. It is, therefore, important to monitor the landform changes to understand the possibility of any hazard. The Mid-Niigata Prefecture Earthquake provided us a good opportunity to study terrain dynamics. A method is proposed and updated to extract Lagrangian components of soil displacement from a set of DEM's before and after the earthquake.

For establishing well conditioned set of equations and smoothness of the result, landslides and large scale manmade changes were filtered out. While comparing with the landslide map of the target zone, it can be deduced that the filtering criteria traces out landslides and large scale manmade changes successfully. The Lagrangian components of tectonic surface displacement were then determined using 1400m square smoothening window. Clusters of large lateral displacement vectors are concordant with the thick clusters of landslides. Vertical components show that the southwestern corner of the target zone is lifted up by 0.5 to 1m. This is the confluence point of Shinano and Uono Rivers and the benchmark measurements along Uono River show that river bed is elevated by almost the same magnitude. Thus the vertical components of surface tectonic displacements in the target zone also correlate with the field measurements/observations. Flooding of the upper reach of Uono River during heavy rains of 2005 strengthens the evidence of raised river bed.

Shallow soil displacements can be obtained by subtracting the displacement vectors obtained from a 1400m square window from that of a 200m square window. The shallow soil displacements were obtained for Kizawa locality. In this Kizawa locality, Kizawa tunnel was damaged by two pairs of parallel cracks. Furthermore, two drainage wells were also found dislocated in this locality. This all suggest the presence of a hidden shear plane. The shallow soil displacement vectors obtained for this locality are well consistent with the slip direction of the shear plane.

5. References

De Martonne, E. 1927. Regions of interior basin drainage. Geographical Review 17:397-414.

Dooge, J.C.I. in (1992): Yen BC, editor. Channel wall resistance: *Centennial of Manning's formula. Littleton, Colorado, Water Resources Publications.*

Honda, R., Aoi, S., Sekiguchi, H., Morikawa, N., Kunugi, T., and Fujiwara, H. (2005) Ground motion and rupture process of the 2004 Mid Niigata Prefecture Earthquake obtained from strong motion data of K-net and KiK-net. *Earth Planets Space*, 57, 527–532.

JSCE Active folding Project (2008) Earthquake Damage in Active-Folding Areas: Creation of a comprehensive data archive and suggestions for its application to remedial measures for civil-infrastructure systems (2005-2007) Research and Development Program for Resolving Critical Issues, Special Coordination Funds for Promoting Science and Technology, (Core research organization: JSCE, Leader: K. Konagai).

Kato, H. and Akabane, H. (1986) Geology of Nagano District, with geological sheet map at 1:50,000. Geological Survey of Japan. 120.

Kazuhito Hikima and Kazuki Koketsu. (2005). Rupture processes of the 2004 Chuetsu (mid-Niigata prefecture) earthquake, Japan: A series of events in a complex fault system. *Geophysical Research Letters*. VOL. 32, L18303, 5.

Kieffer, D.S., Jibson, R., Rathje, E.M. and Kelson, K. (2006) Landslides triggered by the 2004 Niigata Ken Chuetsu, Japan, Earthquake. *Earthquake Spectra*. 22, S47-S73.

Kim Haeng Yoong1 and Atsumasa Okada (2005): Surface deformations associated with the October 2004 Mid-Niigata earthquake: Description and discussion. *Earth Planets Space*, 57, 1093–1102

Konagai, K., Fujita, T., Ikeda, T. and Takatsu, S. (2009) Tectonic deformation buildup in folded mountain terrains in the October 23, 2004, Mid-Niigata earthquake. *Soil Dynamics and Earthquake Engineering*. 29(2), 261-267.

Konagai, K, Takatsu, S. Kanai, T., Fujita, T., Ikeda, T. and Johansson, J. (2009). Kizawa tunnel cracked on 23 October 2004 Mid-Niigata earthquake: An example of earthquake-induced damage to tunnels in active-folding zones. *Soil Dynamics and Earthquake Engineering*. 29(2), 394-403.

Nagaoka Regional Development Bureau, Niigata Prefectural Government, and Honma Corporation (2007) Construction Report of Kizawa Tunnel of the Ojiya-Kawaguchi-Yamato Line, D 16(1903 00 00 03) [in Japanese].

National Research Institute for Earth Science and Disaster Prevention (NIED) and Japan Science and Technology Agency (2008) Landslide Map Database. http://lsweb1.ess.bosai.go.jp/jisuberi/jisuberi_mini_Eng/jisuberi_top. htmlS.

National Institute of Earth Science and Disaster Prevention. (2004) Mid Niigata Earthquake. http://www.hrr.mlit.go.jp/saigai/H161023/1023_top.htmlS

Oyagi, N., Uchiyama, S. and Inokuchi, T. (2008) Map of Landslides Caused by the 2004 Niigata-ken Chuetsu (Mid Niigata) Earthquake (MJMA=6.8). *Technical Note of the National Research Institute for Earth Science and Disaster Prevention*, 317.

Ozawa, T., Nishimura, S., Wada, Y. and Okura, H. (2005) Coseismic deformation of the Mid Niigata prefecture earthquake in 2004 detected by RADARSAT/InSAR. *Earthquake Planets Space*. 57, 423–428.

Report of Zenkoji Earthquake, Central Disaster Prevention Council (2008), http://www.bousai.go.jp/jishin/chubou/kyoukun/rep/1847-zenkoJISHIN/

United States Geological Survey (2005) "What is InSAR?" Earthquake Hazard Program. http://quake.usgs.gov/research/deformation/modeling/InSAR/S, 2005.

Yanagisawa, Y., Kobayashi, I., Takeuchi, K., Takeishi, M., Chihara, K. and Kato, H. (1986) Geology of the Ojiya District. Quadrangle Series Scale 1:50,000, Niigata (7), 50. *Geological Survey of Japan*.